International Conference on Advanced Materials and Chemical Engineering (AMCE2021)

Journal of Physics: Conference Series Volume 2194

Online
3-5 December 2021

ISBN: 978-1-7138-5161-5
ISSN: 1742-6588

Printed from e-media with permission by:

Curran Associates, Inc.
57 Morehouse Lane
Red Hook, NY 12571

Some format issues inherent in the e-media version may also appear in this print version.

Institute of Physics
Dirac House, Temple Back
Bristol BS1 6BE UK

Phone: 44 1 17 929 7481
Fax: 44 1 17 920 0979

techtracking@iop.org

Additional copies of this publication are available from:

Curran Associates, Inc.
57 Morehouse Lane
Red Hook, NY 12571 USA
Phone: 845-758-0400
Fax: 845-758-2633
Email: curran@proceedings.com
Web: www.proceedings.com

TABLE OF CONTENTS

Preface

The 2021 International Conference on Advanced Materials and Chemical Engineering (AMCE 2021) was held on December 03-05, 2021 as a virtual conference due to the growing concerns over the coronavirus outbreak (COVID-19), and in order to protect the well-being of our attendees, partners, and staff as our number one priority.

AMCE 2021 is to bring together innovative academics and industrial experts in the field of Advanced Materials and Chemical Engineering to a common forum. The primary goal of the conference is to promote research and developmental activities in Advanced Materials and Chemical Engineering and another goal is to promote scientific information interchange between researchers, developers, engineers, students, and practitioners working all around the world. The conference will be held every year to make it an ideal platform for people to share views and experiences in Mechanical Engineering, Intelligent Manufacturing, and related areas.

This scientific event brings together more than 120 national and international researchers in advanced materials and chemical engineering. During the conference, the conference model was divided into three sessions, including oral presentations, keynote speeches, and online Q&A discussion. In the first part, some scholars, whose submissions were selected as the excellent papers, were given about 5-10 minutes to perform their oral presentations one by one. Then in the second part, keynote speakers were each allocated 30-45 minutes to hold their speeches.

In the second part, we invited four professors as our keynote speakers. Prof. Jiajun Wang, Harbin Institute of Technology. And then we had Prof. Huan Wang, from Northeast Petroleum University. Prof. Jiawei Yuan, General Research Institute for Nonferrous Metals. Lastly, we had Prof. Qingle Zeng, from Chengdu University of Technology. Their insightful speeches had triggered heated discussion in the third session of the conference. Every participant praised this conference for disseminating useful and insightful knowledge.

The proceedings are a compilation of the accepted papers and represent an interesting outcome of the conference. Topics include but are not limited to the following areas: New Materials, Material Physical Chemistry, and more related topics. All the papers have been through rigorous review and process to meet the requirements of International publication standard.

We would like to acknowledge all of those who supported AMCE 2021. The help and contribution of each individual and institution was instrumental in the success of the conference. In particular, we would like to thank the organizing committee for its valuable inputs in shaping the conference program and reviewing the submitted papers.

The Committee of AMCE 2021

Journal of Physics: Conference Series **2194** (2022) 011001 doi:10.1088/1742-6596/2194/1/011001

Committee

Committee Chair
David BASSIR, Guangdong Technion - Israel Institute of Technology, France

Organizing Committee
Prof. Liang Li, Anhui University, China
Prof. Tao Yu, Tongji University, China
Prof. Wenbo Liu, Sichuan University, China
Prof. Xuegong Yu, Zhejiang University, China
Prof. Sheng-Rui Jian, I-Shou University, China
Prof. Chunxiang Cui, Hebei University of Technology, China
Prof. Ir. Dr. Ts. Abdul Talib bin Din, Universiti Teknikal Malaysia Melaka, Malaysia
Prof. Dongfang Yang, College of Chemistry and Environmental Science, Guizhou Minzu University, China
Dr. Hao Peng, Yangtze Normal University, China

Technical Committee
Prof. Xuegong Yu, Zhejiang University, China
Prof. Sheng-Rui Jian, I-Shou University, China
Prof. Xiaotao Hao, Shandong University, China
Prof. Guangyu Ren,Wuhan University, China
Prof. Qiang Zhao, Nanjing University of Posts and Telecommunications, China
Prof. Haifeng Ling, Nanjing University of Posts and Telecommunications, China
Prof. Yuwu Zhong, Institute of Chemistry, Chinese Academy of Sciences, China
Dr. Hao Peng, Yangtze Normal University, China

Peer Review Statement

All papers published in this volume have been reviewed through processes administered by the Editors. Reviews were conducted by expert referees to the professional and scientific standards expected of a proceedings journal published by IOP Publishing.

- **Type of peer review:** Single Anonymous
- **Conference submission management system:** Morressier
- **Number of submissions received:** 86
- **Number of submissions sent for review:** 57
- **Number of submissions accepted:** 57
- **Acceptance Rate (Submissions Accepted / Submissions Received × 100):** 66.3
- **Average number of reviews per paper:** 2
- **Total number of reviewers involved:** 8
- **Contact person for queries:**
  Name: Xuexia Ye
  Email: xx.ye@keoaeic.org
  Affiliation: AEIC Academic Exchange Information Centre

An analysis of new materials and their effects on improving fuel efficiency

Anran Li[1], Yanchao Qiao[2*], Shenghong Fu[3], Yichen Gu[4]
[1]Baradene College of the Sacred Heart, Auckland 1050, New Zealand.

[2]Weihai IVY Foreign Language School, Weihai, Shandong 264201, China.

[3]Malvern college, Qingdao, Shandong 266106, China.

[4]DonBosco Technical Institute, California 91770, USA.

*Corresponding author Email: qiaoyanchao0631@163.com.

Abstract: This paper introduces the use of new materials in the replacement of traditional materials in automotive vehicles. Firstly, the properties such as tensile strength, elongation at break, Young's modulus, density were discussed and analyzed. Also, the benefits of using new materials are discussed by comparing the physical properties. Then the advantages of AZ91D magnesium alloy, T1000G carbon fiber were compared with traditional automotive materials and the improvements that were achieved where the new materials were proven to have increased strength and decreased weight. Furthermore, the concept of automotive lightweight is applied to demonstrate the relationship between new materials and fuel economy. Finally, the internal and external structures of the materials are analyzed such as the anisotropy of carbon fiber, and it was found that the benefits outweigh the drawbacks like the high production cost of carbon fiber, then suggestions are given for future investigations.

1. Introduction

Vehicle fuel economy has been a hot topic of discussion in recent years. With the increase in fuel prices due to inflation, driving has become less affordable for many households [1]. Climate changes and greenhouse gases is also a major issue that our society faces today. Transport accounts for over 24% of global CO_2 emission, where a typical passenger vehicle produces about 4.6 metric tons of carbon dioxide a year [2]. This means improving fuel economy will also bring benefits for our environment.

New materials have been researched to reduce the mass of the vehicle and maintain costs in achieving fuel economy. Lightweight materials were shown to have great effect in reducing fuel consumption as it directly correlates to energy consumption. However, safety must also be ensured in a vehicle, which means materials must be selected based on their physical and chemical properties such as strength, ductility, hardness, reactivity, and melting points to derive a suitable candidate [3]. Other attributes that must be considered with new materials are their durability (warranty), damageability, repair, and crashworthiness [4].

Efforts have been made to analyze the use of different materials for components of the body. Masaaki Saito reviewed various physical properties of aluminum, which shows that using aluminum for the vehicle body will reduce its weight by 50% [5]. While Eneyw Gardie has focused on carbon fiber, in which 80.4% percent of the weight of wheel application was reduced [6]. P.K. Mallick shows the advantages of magnesium alloy, such as high special mechanical properties, high castability [7]. B.R.

Powell, P.E. Krajewski, A.A. Luo's article introduces magnesium, the lightest automotive metal structure [8], the appropriate amount of magnesium in the automotive metal.

Although studies have been made in analyzing a specific material to ensure the stability and safety of the car while also improving fuel efficiency by reducing the body weight, this paper has put together 2 commonly discussed new materials to compare their use in vehicle production. This paper starts off introducing the properties of magnesium alloy (AZ91D) and Carbon Fiber (T1000G) under experimental conditions. Then a detailed analysis of various kinds of carbon fibers was also made to derive the most suitable candidate for automotive manufacturing. Furthermore, the data was compared with those from the traditional materials to derive an optimal material for each of the bumper, chassis, and door panels of a vehicle. Limitations were discussed on current production of these new materials and possible future steps were offered to ensure the accessibility in the automotive industry.

2. Discussion

2.1 Properties of new materials

AZ91D is a magnesium alloy that is newly introduced into the automobile industry due to its excellent mechanical properties, corrosion resistance, and castability. The corrosion resistance comes from the composition of the alloy, in which over 90% is magnesium as mentioned in *Table 1* and the use of iron and nickel is limited to under 0.005% limiting oxidation. AZ91D has a relatively high tensile strength of 230MPa, with a Young's Modulus of 44.8GPa. This makes the alloy strong and ductile to ensure safety and castability. It has a low density of 1.81 g/cm^3 means that the alloy is lightweight and would effectively reduce the weight of the whole vehicle if AZ91D is widely used. Magnesium is currently being used in gearbox, front end and IP beams, steering column, airbag housings as well as in steering wheels, seat frames, and fuel tank covers. Steel components in vehicles can be replaced by a single cast piece of magnesium adding to the strength of the material and allowing for housings to be cast into place. This castability also requires less tooling and gauges which lowers manufacturing consumption.

Table 1 The composition of AZ91D (MAGNESIUM ALLOY) [9].

AZ91D	Percentage (%)
Aluminum	8.30-8.90
Magnesium	90
Iron	0.005
Nickel	0.002
Zinc	0.35-0.10
Silicon	0.10max

Carbon fiber is a new type of fiber material with high strength and modulus. It contains more than 95% of carbon, consists of flake graphite microcrystalline organic fibers stacked along the axial direction of the fiber after carbonization and graphitization treatment of microcrystalline stone ink material. The biggest advantage of carbon fiber is its lightweight. Carbon fiber has a strength 20 times greater than that of iron, and it is the only material whose strength does not decrease at high temperatures up to 2000℃. Its impact absorption capacity is 4-5 times than that of the ordinary metals and has the desirable characteristics of fatigue resistance and corrosion resistance. Carbon fiber has a low coefficient of thermal expansion, allowing it to drop from temperatures of several thousand degrees Celsius to room temperature without bursting. Carbon fiber composites have a large thermal expansion coefficient, so when they are under the same heat as other materials, they are less likely to deform, making it a useful material in many high-precision structural parts [10]. In *fig 1*, there is a comparison of different kinds of materials where carbon fiber has shown to have a higher tensile strength and lower density. T1000G carbon fiber (T1000G) has the highest tension strength, along with other remarkable performances in elasticity and density as shown in *table 2*, which makes it a good substitute and improving the quality of the frame also guarantees the vehicle safety. Therefore, this paper will be focusing on T1000G as it is a better choice for vehicle production.

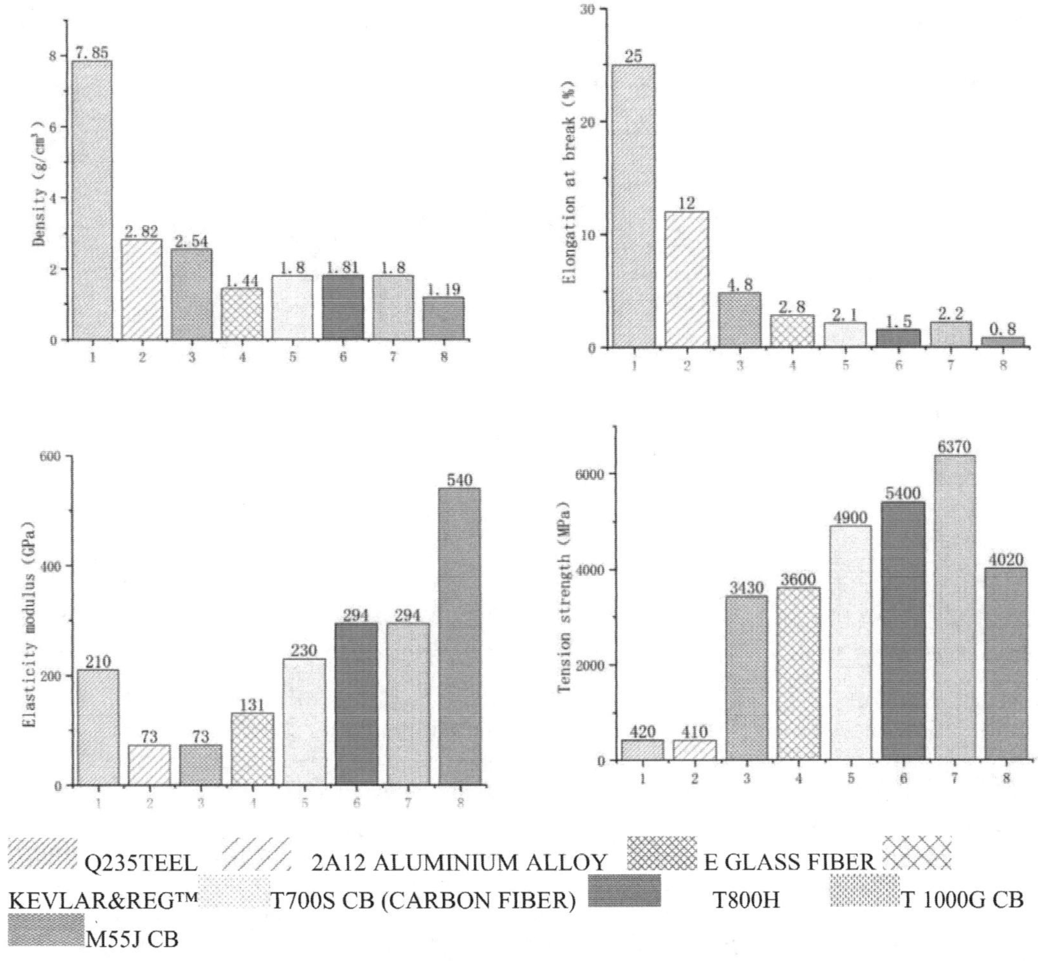

Fig 1 the density, elongation at break, Elasticity modulus and Tension strength of different materials.

2.2 A comparison of traditional with new materials

The traditional material that was used for the vehicle's chassis was aluminum alloy but compared to the new materials, it has a higher density of 2.67 g/cm^3 compared to AZ91D and T1000G of around 1.8 g/cm^3, this means that by replacing the traditional material, the chassis would become lighter. Carbon fiber is over 27 times stronger than the current aluminum alloy while AZ91D is sitting at around the same tensile strength of 230MPa, although AZ91D isn't significantly stronger it shows that it does have enough strength to replace the current material while significantly reduce the weight to achieve better fuel economy. The Young's modulus shows the resistance to damage and deformation, after comparing the data, the aluminum alloy has only a modulus o 27.2GPa whereas AZ91D is 44.8GPa and T1000G is sitting at 294GPa, making the new materials far safer options to use.

TPO was commonly used in vehicle bumpers today, but our new materials are better replacements because although their densities are very similar varying between 1.80-1.82 g/cm^3, but the purpose of a bumper is to protect the vehicle and the passengers in a crash. Therefore, the significant increase in tensile strength and young's modulus of AZ91D and T1000G improves the safety and quality of the vehicle and reduces the likeliness that the bumper is damaged in small accidents, which reduces replacement costs and waste production.

The door panels on the vehicle are mainly built with Q235 steel, it has a high tensile strength of 420MPa and young's modulus of 210GPa, which is far stronger than the AZ91D, but not as much as T1000G. However, it does have a high density of 7.85 which is over 4 times as dense as the new materials, adding a significant amount of weight to the vehicle considering its size. In this case, the use of T1000G would decrease the weight of the vehicle significantly, hence increasing fuel efficiency, but research must be done to reduce the costs of producing the carbon fiber on a large scale, or else it would be impractical.

Table 2 the data of traditional materials of chassis, bumper, doors and new materials

Materials	Aluminum alloy	Thermoplastic Polyolefins (TPO)	Q235 Steel	AZ91D Magnesium alloy	T1000G Carbon fiber	Gen3AHSS
Components	Chassis	Bumper	Doors	/	/	/
Density ($g/cm^{3)}$)	2.67	1.82	7.85	1.81	1.80	7.83
Tensile strength (MPa)	234	57	420	230	6370	980
YOUNG'S modulus (GPa)	27.2	2.10	210	44.8	294	200

2.3 Effects of using new materials

Weight reduction has a significant effect on improving vehicle fuel economy. A 10% reduction in vehicle weight will lead to an approximately 8% to 10% increase in power performance at 100km, and a 2-7m reduction in braking distance which can bring many advantages such as the reduction in fuel consumption and so on.

Lightweight materials are necessary for propelling the fuel economy of modern cars while ensuring their safety and performance. Less energy will be used for the acceleration of a lighter car than a heavier one, thus, lightweight materials are more promising for both automakers and auto buyers. Magnesium and carbon fibers have better energy absorption than the traditional metal materials such as steel, thus providing more protection upon impact. Beyond that, lightweight vehicles enhance performance, especially steering. They are usually more maneuverable to park in limited space. Assembling vehicles with the components made of light-weight materials normally can reduce the weight of any vehicles from10% to 60%. Life cycle assessment (LCA) is the most common methodology for assessing how the products or services influence the environment. Plenty of the examples of comparative LCAs which related to automotive lightweight context has been exist already [11] and they provide two main approaches: Proportions-based and FRV-based.

The proportions-based approach is founded on the assumption of a linear dependence between the quota of FC saving with respect to total vehicle FC and mass saving with respect to total vehicle mass through a proportionality constant, as shown by the equation below:

$$\frac{\Delta FC}{\Delta M} = \frac{FC_{ref_{veh}}}{M_{ref_{veh}}} * c \quad (1)$$

Where ΔFC is Fuel Consumption reduction achieved through vehicle mass reduction [l/100 km].ΔM is vehicle Mass reduction [kg]. FC_{ref_veh} is Fuel Consumption of reference vehicle [l/100 km].M_{ref_veh} is Mass of reference vehicle [kg]. c is proportionality constant [12].

In this approach the proportionality constant fixed which has a value 0.6, as suggested by Lynne Ridge. And adopting the same value of c means the ratio still unaltered ignoring vehicle technical features of each specific application, thus, it may lead to a high degree of uncertainty.

In the FRV-based approach FC reduction achievable through light weighting is determined by the following relation:

$$\Delta FC = \Delta M * FRV * 0.01 \quad (2)$$

FRV is Fuel Reduction Value coefficient [l/100 km*100 kg] [12]

Unlike proportions-based approach, which is founded on a rigid proportionality, FRV-based approach is more accurate estimation. With these two algorithms the fuel economy can be calculated whether it's in large quantities or in specific cases, so the data can be clearly compared to get the highest fuel economy.

Magnesium alloys have a high strength-to-weight ratio. 30-70% weight is reduced if replaced with the traditional material, specifically, they are 36% lighter per unit volume than aluminum and 78% lighter than iron and 75% lighter than steel withstanding 10 times more strength than aluminum, titanium, or steel. This will likewise cause a 24-60% increase in fuel efficiency.

If the traditional steel were replaced with magnesium or carbon fiber composite materials, the weight of the car could be reduced by approximately 300 kilograms. According to Francesco Del Pero, Massimo Delogu, Marco Pierini [12], the number of FRV is between 0.02and 1, use the averaging formula, FRV is 0.707, so ΔFC would be 2.12.

2.4 Analyze

The increase in the utilization of magnesium in automobile production will bring positive effects to the environment. Magnesium is the eighth-most abundant element of around 2.5% of the Earth's crust, which represents an unlimited supply that is sustainable. Magnesium is recyclable, therefore, instituting a recycling system would extend supplies, reduce waste, and save energy. This is especially important when around 12 million cars are destroyed around the world each year, producing over 15 million tons of metal wastes. In addition, magnesium can also reduce vibration and overall noise through the tuning of parts, so it also benefits the environment in reducing noise levels. However, the strength, hardness, and modulus of elasticity of magnesium-based materials decrease with increasing temperature. Therefore, if the vehicle is constantly under high temperatures, it will speed up the aging and reduce its life span.

Carbon fiber can be distinguished from the isotropic properties of steel and aluminum alloys because it can be customized due to orientation. For example, an object made of steel can withstand the same amount of pressure from any direction. When a force is applied perpendicular to the fibers, the material can withstand a great deal of force, which means the geometry of the fiber can be engineered for the best application for some vehicle components like a bumper beam. While carbon fiber is light and strong, it also has a weakness: it can be pierced by sharp objects because carbon fiber material is essentially a cloth woven with thread. Its hardness and toughness are based on its large area of stress, when confronted with sharp objects making it is easy to puncture. The other weakness is that carbon fiber is not only expensive but also impossible to repair. If the original material is damaged, the entire carbon fiber will have to be replaced. So even if there's a small accident, it costs the vehicle owner a great deal to fix, for this reason, currently only 0.007% of carbon composites are used in vehicle as shown in *fig 2.*

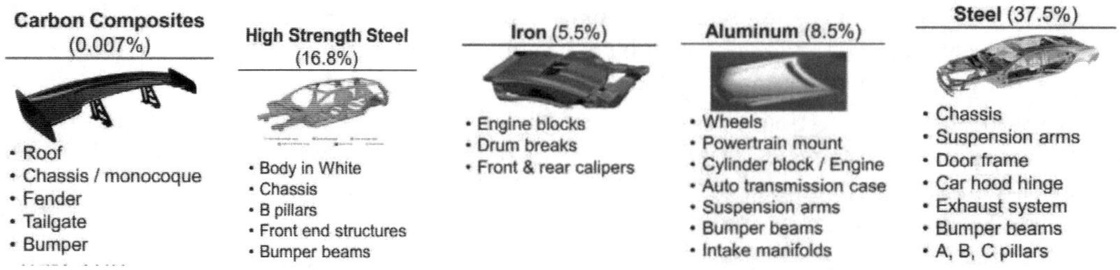

Fig 2 Composition of automobile parts

Another reason why carbon fiber components are not widely used is their high cost of production, they are about 20 times more expensive to produce than steel as shown in *fig 3*, making it impractical to produce at an industrial level. However, these issues can be addressed when further research and developments are made so that carbon fiber can be standardized to lower the costs of production.

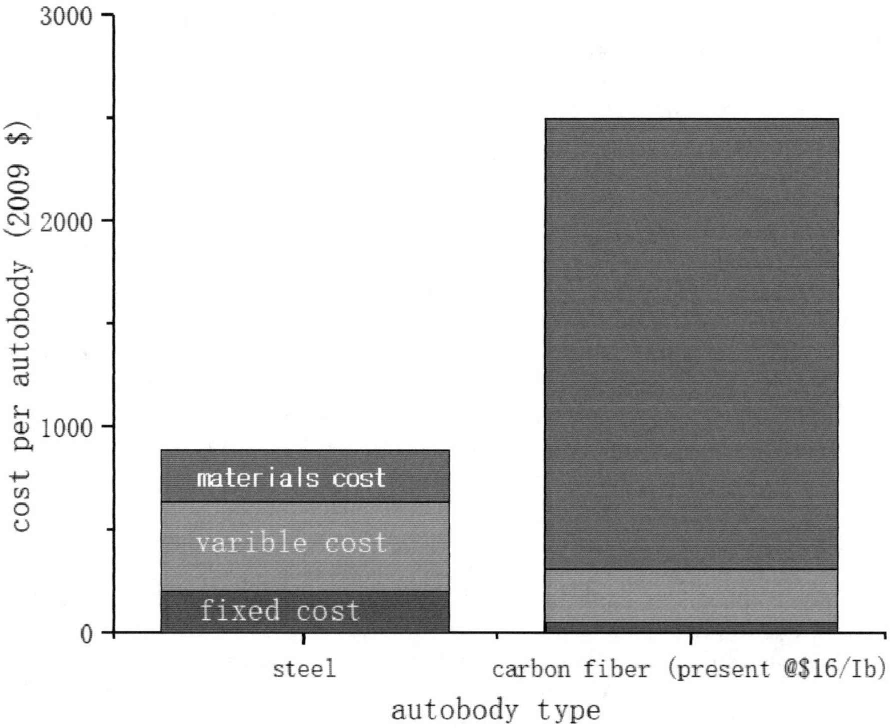

Fig 3 Cost breakdown of carbon fiber body (right) versus steel body (left) for 2009, from Rocky 2011 Mountain Institute's Industry Report.

3. Conclusion

After close analysis of the two new materials: AZ91D and T1000G, it was shown that they can be used as excellent substitutes for traditional automotive materials. They both have the desirable high strength, low density, and good ductility that will work together to reduce the weight of the vehicle by over 30-70% and enhance its performance. This will achieve the goal of increasing fuel efficiency by approximately 25-60% while ensuring safety. In turn, this improvement can promote the protection of the environment by saving energy and reducing emissions, which is beneficial for addressing the constant environmental issues that society is facing today. Overall, through the comparison of data for the traditional and new materials, it was found that the replacement of traditional materials in automobile production with new lightweight materials will bring positive social and environmental impacts, however, future studies are encouraged for solutions in reducing the cost of production.

REFERENCES

[1]. Petrol prices at eight-year high amid fuel crises. https://www.bbc.com/news/business-58718148, BBC, September 2021

[2]. Hannah Ritchie. Cars, planes, trains: where do CO2 emissions from transport come from? https://ourworldindata.org/co2-emissions-from-transport, October 2020

[3]. Taub AI, Krajewski PE, Luo AA, Owens JN. The evolution of technology for materials processing over the last 50 years: the automotive example. JOM 59(2):48–57, 2007

[4]. G. S. Cole and A. M. Sherman. Lightweight Materials for Automotive Applications ELSEVIER Ford Motor Company, Dearborn, Ml 48121-2053,1995

[5]. Masaaki Saito, Shuuichiro Iwatsuki, Kunihiro Yasunaga, Kimitaka Andoh. Development of aluminum body for the most fuel-efficient vehicle. JSAE Review, Vol. 21, Issue 4, Pages 511-516, 2000 ISSN 0389-4304

[6]. Eneyw Gardie, Velmurugan Paramasivam, Habtamu Dubale, Ewnetu Tefera Chekol, Senthil Kumaran Selvaraj, Numerical analysis of reinforced carbon fiber composite material for lightweight automotive wheel application. Materials Today: Proceedings, Vol. 46, Part 17, Pages 7369-7374, 2021, ISSN 2214-7853.

[7]. P.K. Mallick. Chapter 10 - Designing lightweight vehicle body. Editor(s): P.K. Mallick, In Woodhead Publishing in Materials, Materials, Design and Manufacturing for Lightweight Vehicles (Second Edition), Woodhead Publishing, Pages 405-432, 2021, ISBN 9780128187128

[8]. B.R. Powell, P.E. Krajewski, A.A. Luo. Chapter 4 - Magnesium alloys for lightweight powertrains and automotive structures. Editor(s): P.K. Mallick, In Woodhead Publishing in Materials, Materials, Design and Manufacturing for Lightweight Vehicles (Second Edition), Woodhead Publishing, Pages 125-186, 2021, ISBN 9780128187128.

[9]. Automotive Uses of Magnesium Alloys: Part one. https://www.totalmateria.com/page.aspx?ID=CheckArticle&site=ktn&NM=246, July 2010

[10].Francisco G. Emmerich, "Young's modulus, thermal conductivity, electrical resistivity and coefficient of thermal expansion of mesophase pitch-based carbon fibers," Vol. 79, Pages 274-293, 2014, ISSN 0008-6223]6.[Do-Hyoung Kim, Hyun-Gyung Kim, Hak-Sung Kim, "Design optimization and manufacture of hybrid glass/carbon fiber reinforced composite bumper beam for automobile vehicle, Composite Structures," Vol. 131, Pages 742-752, 2015, ISSN 0263-8223

[11].Mayyas, A., Qattawi, D., Omar, M., Shan, D. Design for sustainability in automotive industry: a comprehensive overview Renew. Sustain. Energy Rev.16 (4), 1845-1862, 2012

[12].Francesco Del Pero,Massimo Delogu,Marco Pierini. The effect of lightweighting in automotive LCA perspective: Estimation of mass-induced fuel consumption reduction for gasoline turbocharged vehicles. J. Clean. Pros., 154, 566-77, 2017

Amino-Modified Hollow Mesoporous Silica Microspheres Loaded with Cresol Red-Boric Acid for Potent Reversible Thermochromic Performance

Tingting Yu[1a*], Xu Wang[1b], Liang Cheng[1c*], Yuehang Hu[1d], Haoya Wang[2e], Hao Wang[1f]

[1] Research and Development of Centre of China Tobacco Yunnan Industrial Co., Ltd., Kunming 650231, China

[2] Yunnan Industrial Tobacco HI-TECH Material Co., Ltd, Kunming 650106, China

[a*]353461052@qq.com, [b]sageplaton@163.com, [c*]1161556017@qq.com, [d]364756539@qq.com, [e]175491817@qq.com, [f]wanghao9999@126.com

Abstract— In this study, two reversible thermochromic composites were obtained, hollow mesoporous silica (HMS) and amino group-modified hollow mesoporous silica (HMS-NH_2) microspheres loaded with cresol red-boric acid (CR-BA), respectively. Owing to the rigid structure of hollow interior and mesoporous shell layers, CR-BA was easily encapsulated into a relatively high loading capacity micro-carrier. The composites analysed were characterized by Fourier transform infrared spectroscopy transmission electron microscopy, scanning and transmission electron microscopy and dynamic light scattering. By amino functionalization, a more obvious thermal colour-change was observed in HMS-NH_2 loaded with CR-BA. In addition, both HMS and HMS-NH_2 loaded with CR-BA illustrated a sever colour reversibility, which were 5min and 3min respectively.

1. Introduction

Thermochromic materials has been received significant attention in recent years due to their potential applications, the related research is ranging from the original simple metal oxides, double salt, complex development to the present all kinds of organic matters, such as liquid crystal, polymer and biological macromolecules[1-3]. The different colour-changing mechanisms can be attributed to their different properties and thermodynamic characteristics. Otherwise mainly low temperature inorganic thermochromic materials are always reversible, such as metal, metal halide[4], transition metal complexes[5], metallic oxide[6] and double salts[7]. However, as most of them or their composites are heavy metals, the high biosafety risks lead restrictions in practical applications. Compared with inorganic substances, the simple mechanism and easy preparation lead to the organic thermochromic dyes well developed. They included polymer[8], Schiff bases[9], spiropyran and spirooxazine[10], lactone derivatives [11] and liquid crystals[12]. Mainly organic substance's thermochromic mechanisms were achieved by chemical structural change, included the opening and closing of ring structures or conformation transition of chemical bonds during heating, and after cooling and the original structure is restored. In addition, the colour-changing process involves the gain and loss of electrons or a change in pH value. However, as organic thermochromic materials in general with benzene ring are easy to produce new harmful gases and substances when heating by high temperature heating and burning, alternative environmentally friendly thermochromic systems are still the target of further research.

By using an acid-base indicator cresol red as colour former and boric acid as chromogenic agent, a reversible thermochromic dye system with low-toxicity, good reversibility, low colour change temperature has been developed[13]. To enhance the stability and biosafety, such thermochromic dye system can combine with base materials, solvents and additives with less polarity[14].

Hollow-structured mesoporous silica (HMS), by integrated mesoporous shell layer structure with hollow interiors structure, was observed the high loading capacity amount[15].To enhance the adsorption performance of mesoporous silica, certain organic group can functionalize the surface of hollow-structured mesoporous silica[16]. Hence, the amino-group modification on mesoporous silica can provide access to enhance the thermochromic performances of hollow-structured mesoporous silica (HMS) that combined with dye system.

In this study, we selected cresol red-boric acid (CR-BA) to load with the base material hollow-structured mesoporous silica (HMS) and 3-aminopropyltriethoxysilane (APTES) modified hollow-structured mesoporous silica (HMS-NH_2). The two groups of the reversible thermochromic composites were termed as HMS/CR-BA and HMS-NH_2/CR-BA, respectively. Transmission electron microscopy (TEM) was used to observe the composite microspheres and thermogravimetric analysis (TGA) was used to measure the loading amount of CR-BA composites, dynamic light scattering (DLS) was used to investigated the increase loading capacity. After the APTES on the surface of the HMS microspheres to introduce amino groups (-NH_2), the load capacity of CR-BA can be enhanced. Furthermore, CR-BA endowed the composites with excellent high rapid thermochromic reversibility. Because of the low-temperature thermochromic reversibility and low toxicity, the composites obtained in this work have certain application possibility in indicator, sensor, package coating, decoration and anti-counterfeiting packaging industry.

2. Materials and Method

2.1. Materials

Analytical reagent styrene (St) and azobisisobutyronitrile (AIBN) provided by Shanghai Chemical Reagents Co., Ltd., Shanghai, China, were purified by vacuum distillation and by recrystallization from ethanol, respectively. Analytical reagent polyvinylpyrrolidone (PVP), ammonium hydroxide ($NH_3 \cdot H_2O$), hexadecy trimethyl ammonium bromide (CTAB), tetraethyl orthosilicate (TEOS), ethanol, cresol red (CR) and boric acid (BA) were directly purchased from Shanghai Chemical Reagents Co., Ltd., Shanghai, China. 3-aminopropyltriethoxysilane (APTES) was purchased from Aladdin Reagents (Shanghai) Co., Ltd., Shanghai, China. Deionized water was used in all experiments.

2.2. Preparation of polystyrene (PS) microspheres

PVP (1.0 g) was dissolved in a solution of ethanol (100 mL) and water (10 mL) under mechanical stirring. Then St (10 g) and azobisisobutyronitrile (AIBN) (0.158 g) were added into the solution under N_2 protection. The mixture were heated at 70°C for 24 hand cool down to the room temperature, then the PS microspheres were synthesized. The PS microspheres were collected by centrifugation and washed with ethanol twice and dried in a vacuum oven at 50°C for 24 h.

2.3. Preparation of PS@SiO_2 microspheres

CTAB (2.7 g) was dissolved in a solution of water (45 mL) and ethanol (60 mL). Ethanol (30 mL) was added into an aqueous dispersed solution of PS microspheres (10 wt%)（50 ml） and slowly dropped into the above mixture under mechanical stirring. Then $NH_3 \cdot H_2O$ (3 mL) and TEOS (60 mL) were added into the reaction system. After reaction at room temperature for 48 h, the PS@SiO_2 microspheres were collected by centrifugation and washed with ethanol twice, and the microspheres dried in a vacuum oven at 50°C for 24 h.

2.4. Preparation of HMS

The PS@SiO_2 microspheres were heated inside a muffle furnace from the room temperature to 550°C (heating rate of 5°C/min) and keep for 5 h then cooled naturally. All the processes were performed in air. The final product was noted as HMS.

2.5. Amino group modification on HMS

Anhydrous toluene (50 mL) and HMS (1.0 g) were placed in a flask and the mixture was sonicated for 30 min, followed by the addition of APTES (2 mL). The mixture was stirred at 120°C for 24 h with nitrogen gas protection. The resultant microspheres were collected by centrifugation 600rpm and washed with toluene and ethanol three times each, and dried under vacuum. The final product was HMS-NH_2.

2.6. Preparation of HMS-NH_2/CR-BA and HMS/CR-BA composites

CR and BA were added into a solution of ethanol (1.0 mL) and water (5.0 mL), followed by heating to 80°C. Then microspheres (HMS and HMS-NH_2 microspheres) were added rapidly into the above dye system. The mixture was sonicated for 5 min and kept at 80°C for 5 min, followed by centrifugation while the system was still hot. Finally, the collected products were dried in the air naturally. The detailed composition of the HMS-NH_2/CR-BA and HMS/CR-BA composites prepared is listed in Table 1. All the composites were heated to 70°C and cooled naturally. Colour changes were recorded by using a digital camera.

Table 1 Recipes of HMS-NH_2/CR-BA and HMS/CR-BA composites

HMS-NH_2/CR-BA composites				HMS/CR-BA composites			
Label	HMS-NH_2(mg)	CR(mg)	BA(mg)	Label	HMS(mg)	CR(mg)	BA(mg)
A	30	1	100	E	30	1	100
B	50	1	100	F	50	1	100
C	70	1	100	G	70	1	100
D	90	1	100	H	90	1	100

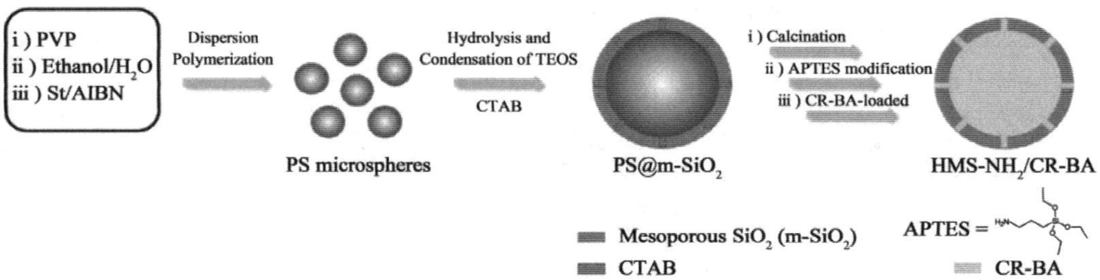

Scheme 1. Illustration of the preparation processes of HMS-NH_2/CR-BA

The preparation process of HMS-NH_2/CR-BA was illustrated in Scheme 1. In the first step, PS microspheres were synthesized by PVP under the solution of ethanol/water and St/AIBN. In the second step, the mesoporous structure PS@m-SiO_2 microspheres was prepared by hydrolysis and TEOS with CTAB loaded on the surfuce. In the third step, APTES modified PS@m-SiO_2 and CR-BA loaded into the microspheres (HMS-NH_2/CR-BA).

2.7. Characterization

Transmission electron microscopy Morphological (TEM) was used for observation of the mesoporous silica microspheres (Hirachi H-7650, Tokyo, Japan, 100 KV). Thermogravimetric analyzer (TGA) was used for Q5000IR Thermogravimetric analysis (Washington, US). Zeta-potential values were determined by a Malvern Zetasizer Nano ZS-2S90 dynamic light scattering (DLS, CORDOUAN, Paris, France) system at 25°C. Fourier transform infrared spectra (FTIR) were conducted by using a Bruker Tensor 27 FTIR spectrometer (Scotia, NY, US). Digital camera (Nikon, NY, US) was used for record colour changes.

3. Test Results and Discussions

3.1. Characterization of HMS-NH_2/CR-BR and HMS/CR-BR composites

According to the Transmission electron microscopy Morphological (TEM) observation, the PS microspheres were synthesized with an average diameter of 1.07 μm (Fig. 1(a)). After the hydrolysis of TEOS, uniform microspheres with a porous shell and a mean diameter of 1.31 μm are obtained, as observed by TEM (Fig. 1(b)). Subsequently, the HMS were produced by calcining at 550°C for 5 h the corresponding TEM image is shown in Fig, 1(c), which clearly shows that a hollow structure with a well-defined mesoporous shell was obtained. The average diameter of the microspheres was about 1.26 μm and the average thickness of the mesoporous shell was estimated to be 190 nm. The mesoporous were homogeneously arranged, generating efficient channels for CR-BA diffusion and loading.

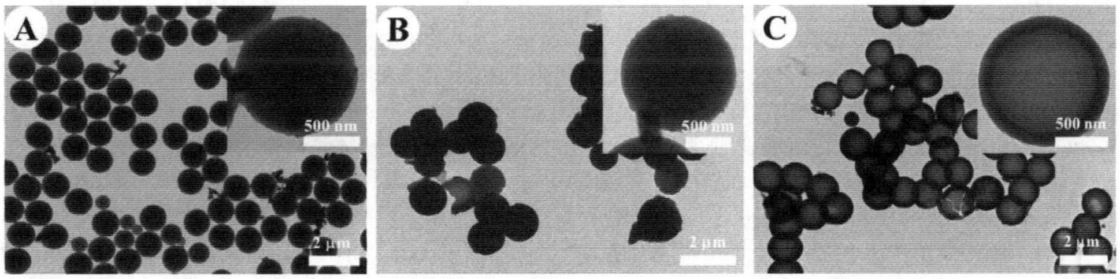

(a) PS microspheres (b) PS@m-SiO_2 (c) hollow mesoporous SiO_2 (HMS)

Fig. 1. TEM images of mesoporous silica microspheres

Figure 2(a) presents the FT-IR spectra of Cresol red, boric acid, HMS, HMS-NH_2, HMS/CR-BA and HMS-NH_2/CR-BA. All of the samples were recorded at room temperature from 400 to 4000 cm^{-1} in transmittance mode. After modification by APTES, the HMS microspheres were functionalized to gain amino group (-NH_2), the curve of HMS-NH_2 microspheres appeared a new peak at ~950 cm^{-1} from the HMS group. It should be attributed to the N-H deformation vibration of the primary amine After the HMS-NH_2 microspheres were loaded with CR-BA, it was observed that the extending vibration peak of O-H at 3226 cm^{-1} from the HMS/CR-BA and HMS-NH_2/CR-BA composites, attributed to hydroxyl group of boric acid. It can be indicated the CR-BA was loaded with the HMS-NH_2 microspheres.

TGA analysis was conducted to determine the relative ratio of the different components in the composites, as shown in figure 2(b). The results reveal that the weight losses of HMS, HMS-NH_2, HMS/CR-BA and HMS-NH_2/CR-BA were 2%, 15%, 29% and 39%, respectively. The TGA curves of HMS and HMS-NH_2 showed that when the temperature is above 325°C, an obvious weight loss occurred, due to the decomposition of APTES modified on the surface of HMS microspheres. As for HMS/CR-BA, the mass loss above 100°C represents the decomposition of CR-BA. Thus, the weight loss between 100 and 325°C on the TGA curves of HMS-NH_2/CR-BA belonged to the decomposition of CR-BA, while the weight loss above 325°C should be attributed to the decomposition of APTES.

As a result, the APTES modified HMS microspheres increased the load capacity of CR-BA. The laoding capacity of the HMS-NH_2 and HMS microspheres were 0.51 and 0.38 g/g, respectively.

It was observed that the zeta potential became positive after the modification of amino groups (Fig. 2(c)), suggesting that the APTES was successfully loaded into the HMS microspheres. Otherwise the HMS microspheres and CR-BA are negatively charged in deionized water. It was contributed by the electrostatic repulsion hinders the adsorption of the HMS microspheres to CR-BA. After the introduction of amino groups, the zeta potential of the HMS-NH_2 microspheres became positive, and the loading capacity to CR-BA increased.

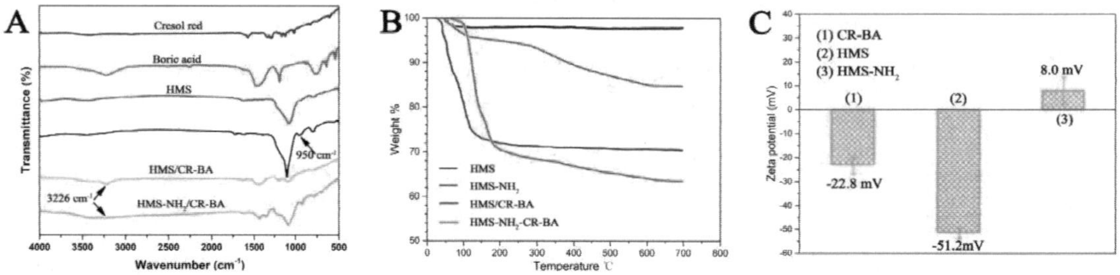

(a) FT-IR spectra (b) TGA curves (c) Zeta potential values

Fig. 2 Characterization of mesoporous silica microspheres and composites

3.2. Thermochromic performances of HMS/CR-BA and HMS-NH_2/CR-BA composites

Owing to the APTES modification, the original colour of HMS-NH_2/CR-BA composites presented as green-yellow, whereas HMS/CR-BA composites was light pink and at the room temperature 20 °C in this study, shown in Fig 3. 1-3 g composites were taken from the eight prepared groups of HMS-NH_2/CR-BA and HMS/CR-BA composites (Table 1), and placed on plates, then heated the plates at the rate of 5°C/min. The colour changes were recorded by digital camera at different temperatures. It can be seen from figure 4(a), when heated to 50 °C, the HMS-NH_2/CR-BA appeared an obvious colour change, and when the temperature increased to 70 °C, the green-yellow powder turned to rose-purple. Furthermore, by increasing the mass ratio of HMS-NH_2, significant colour change occurred at lower temperatures. For the composites of HMS /CR-BA (Fig.4 (c)), the light pink showed an obvious colour darken at the temperature of 40 °C, and when heated to 70 °C, the composited changed into rose-purple. The mass ratio of HMS also affected the colour change temperature and time as shown in Fig. 4(c) and (d). When cooling the composites, the HMS-NH_2/CR-BA composites required only 3 min to recover their initial colour. As a contrast, the HMS/CR-BA required approximate 5 min to recover their initial colour as shown in Fig. 4(d), and the colour stayed light pink after the temperature maintained at room temperature. The prominent change in colour and rapid recovery colour features for the HMS-NH_2/CR-BA composites endows an excellent loading capacity for dye pigment and rapid colour change ability.

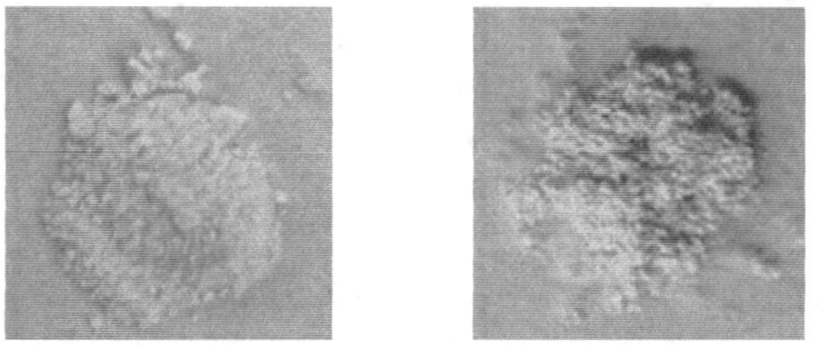

(a) HMS/CR-BA powder (b) HMS-NH_2/CR-BA powder

Fig. 3 The original appearance of HMS/CR-BA and HMS-NH_2/CR-BA composites

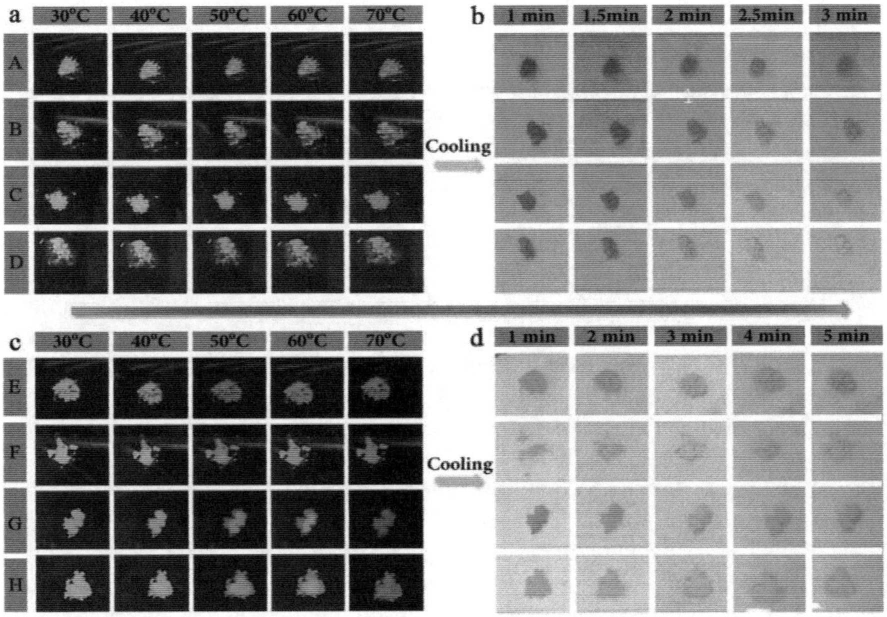

Fig. 4 Heating Colour changes of $HMS-NH_2/CR-BA$ (a) and HMS/CR-BA composites (c) Cooling Colour changes of $HMS-NH_2/CR-BA$ (b) and HMS/CR-BA composites (d)

4. Conclusion

In this work, the low toxic reversible thermochromic composites that showed high temperature sensitivity and rapid reversible performance were built in simple process. FT-IR spectra indicated the amino group and hydroxyl group of boric acid were loaded into hollow-structured mesoporous silica. According to TGA analysis, with the APTES modification, the obtained $HMS-NH_2$ microspheres possessed an excellent CR-BA loading capacity of 0.51g/g which was higher than HMS. The DLS determined the zeta-potential values and showed that APTES was successfully modified the HMS microspheres. From the thermochromic performances observation, both $HMS-NH_2/CR-BA$ and HMS/CR-BA had pleased colour change and good sensibility to the temperature change. Furthermore, the amino-group modification endowed the $HMS-NH_2/CR-BA$ composites with a more prominent colour change of colour in response to elevated temperatures (from green-yellow to rose-purple to green-yellow) and a rapid reversibility of colour once it cools down (only 3 minutes).

Compared with other thermochromic materials[3,8,11], the colour change of $HMS-NH_2/CR-BA$ and HMS/CR-BA composites can appear at relative low temperature (50 to 70 °C), which provides a thermal-safety way in indicator and sensor applications. In this study, the thermochromic performances study offset the thermochromic reverse time research limitation in former research[9,12]. The thermochromic materials synthesized in this work can be applied in expected applications owing to the characteristics of high sensitive, low toxicity and rapid reversibility.

Acknowledgments

This work was financially supported by the Science and technology Project of China Tobacco Yunnan Industry Cooperation (2018 No.2018CL01).

References

[1] Pitchaimani, J., Karthikeyan, S., Lakshminarasimhan, N., Anthony, S. P., Moon, D., & Madhu, V. (2019). Reversible Thermochromism of Nickel (II) Complexes and Single-Crystal-to-Single-Crystal Transformation. ACS omega, 4(9), 13756-13761.
[2] Choi, J. W., Cho, N., Woo, H. C., Oh, B. M., Almutlaq, J., Bakr, O. M., ... & Kim, J. H. (2019).

Investigation of high contrast and reversible luminescence thermochromism of the quantum confined Cs_4PbBr_6 perovskite solid. Nanoscale, 11(12), 5754-5759.

[3] Chiang, H. K., & Chu, L. K. (2014). Solvent isotope effect on the dark adaptation of bacteriorhodopsin in purple membrane: Viewpoints of kinetics and thermodynamics. The Journal of Physical Chemistry B, 118(10), 2662-2669.

[4] Pareja-Rivera, C., & Solis-Ibarra, D. (2021). Reversible and Irreversible Thermochromism in Copper-Based Halide Perovskites. Advanced Optical Materials, 2100633.

[5] Liu, H., Yuan, L., Wu, X., Hou, X., Tang, M., Hou, C. & Feng, S. (2020). Reversible thermochromic property of Cr, Mn, Fe, Co-doped $Ca_{14}Zn_6Ga_{10}O_{35}$. Journal of Materials Chemistry C, 8(28), 9615-9624.

[6] Nguyen, D. K., Bach, Q. V., Kim, B., Lee, H., Kang, C., & Kim, I. T. (2019). Synthesis of Cr-doped Al_2O_3 by Pechini sol-gel method and its application for reversible thermochromic sensors. Materials Chemistry and Physics, 223, 708-714.

[7] Mehta, S., Kushwaha, A., Kisannagar, R. R., & Gupta, D. (2020). Fabrication of a reversible thermochromism based temperature sensor using an organic–inorganic composite system. Rsc Advances, 10(36), 21270-21276.

[8] Jurewicz, I., King, A. A., Shanker, R., Large, M. J., Smith, R. J., Maspero, R., ... & Dalton, A. B. (2020). Mechanochromic and thermochromic sensors based on graphene infused polymer opals. Advanced functional materials, 30(31), 2002473.

[9] Zbačnik, M., Pičuljan, K., Parlov-Vuković, J., Novak, P., & Roodt, A. (2017). Four thermochromic o-hydroxy Schiff bases of α-aminodiphenylmethane: solution and solid state study. Crystals, 7(1), 25.

[10] Seiler, V. K., Tumanov, N., Robeyns, K., Wouters, J., Champagne, B., & Leyssens, T. (2017). A structural analysis of spiropyran and spirooxazine compounds and their polymorphs. Crystals, 7(3), 84.

[11] Kim, M. J., Muthukumar, P., Angupillai, S., & Son, Y. A. (2018). Synthesis of Novel Morpholine Appended Crystal violet lactone Derivatives and an Investigation of Their Thermochromic Behavior. Textile Coloration and Finishing, 30(1), 1-8.

[12] Mu, B., Zhao, Y., Li, X., Quan, X., & Tian, W. (2020). Enhanced conductivity and thermochromic luminescence in hydrogen bond-stabilized columnar liquid crystals. ACS applied materials & interfaces, 12(8), 9637-9645.

[13] Hao, Z. F., Yang, Y., & Yu, J. (2004). Study of Reversible Thermochromic Paint of Cresol Red-Boric Acid. FINE CHEMICALS-DALIAN-, 21(4), 241-244

[14] Lo, C. Y., Zhao, Y., Ma, Y., Wu, S., Alsaid, Y., Peet, M. M., ... & He, X. (2021). Bioinspired Sensors and Actuators Based on Stimuli-Responsive Hydrogels for Underwater Soft Robotics. In Bioinspired Sensing, Actuation, and Control in Underwater Soft Robotic Systems (pp. 99-115). Springer, Cham.

[15] Yang, Z., Fan, W., Zou, J., Tang, W., Li, L., He, L., ... & Chen, X. (2019). Precision cancer theranostic platform by in situ polymerization in perylene diimide-hybridized hollow mesoporous organosilica nanoparticles. Journal of the American Chemical Society, 141(37), 14687-14698.

[16] Khan, M. A., Wallace, W. T., Islam, S. Z., Nagpure, S., Strzalka, J., Littleton, J. M., ... & Knutson, B. L. (2017). Adsorption and recovery of polyphenolic flavonoids using TiO_2-functionalized mesoporous silica nanoparticles. ACS applied materials & interfaces, 9(37), 32114-32125.

Study on the PI/SPEEK nanofiber composite proton exchange membrane for fuel cells

Peng Wei [1,2], Dong Huang [1,2], Xi Li [1,2], Bensheng Zhu [3], Chuanbo Cong [1,2], Xiaoyu Meng [1,2a*], Qiong Zhou[1,2b*]

[1]Department of Materials Science and Engineering, New Energy and Material college, China University of Petroleum-Beijing, Beijing 102249, People's Republic of China

[2]Beijing Key Laboratory of Failure, Corrosion, and Protection of Oil/Gas Facilities, China University of Petroleum-Beijing, Beijing 102249, People's Republic of China

[3]China Helicopter Research and Development Institute, Aviation Industry Corporation of China, Limited, Jingdezhen Jiangxi 333001, People's Republic of China

[a*] xymeng800418@sohu.com, [b*] zhouqiong_cn@163.com

Abstract—Sulfonated poly ether ether ketone (SPEEK) can be selected as the candidate for proton exchange membrane (PEM) used in the fuel cell for its cheap price, great proton conductivity and stability. To further improve its performance, more SO_3H were induced into SPEEK but would result in excessive swelling and losing its stability. Polyimide (PI) nanofibers have excellent mechanical properties and can be a supporter to improve stability. In this work, the PI nanofiber was put in the middle of the SPEEK to create the PI/SPEEK nanofiber composite membranes (NCMs). Its morphology of PI/SPEEK NCMs observed by SEM showed that PI nanofiber was almost in the middle of these NCMs. The proton conductivity and swelling ratio of 3%PI/SPEEK were 271.7 $mS \cdot cm^{-1}$ and 20.0% at 60°C and 100% RH, which is 29.6% higher and 90.4% lower than that of pristine SPEEK. It was intact after the single-cell test in the fuel cell and its power density was 199.1 $mW \cdot cm^{-2}$, but for pure SPEEK there was some broken area on it. The reason was that PI/SPEEK can maintain the durability of the channels for the proton to transport and it can be attributed to an interaction at their interface of PI nanofiber and SPEEK. The 3%PI/SPEEK NCM will have a broad application prospect in PEM fuel cells.

1. Introduction

Proton exchange membrane fuel cells (PEMFCs) have fascinated considerable interests due to their great efficiency of energy conversion and environmental friend [1, 2]. The proton exchange membrane (PEM) was served as electrolytes for the protons to transport and prevent fuel permeability [2]. Sulfonated poly (ether ether ketone) (SPEEK) is an ideal PEM for its excellent strength and high proton conductivity. But sulfonic acid directly linked to the main chain makes it hard to form a continuous proton transport channel [3]. A common method to raise its proton conductivity was to introduce more sulfonic acid groups to the main chain but it will lead to excessive swelling [4]. Based on the above mentioned, it is important for high-performance PEM in PEMFC to equilibrate proton conductivity and stability [1].

A facile way to solve the above problem is to introduce inorganic polymer to form organic-inorganic nanocomposite PEM [5]. SiO_2 [6-9], TiO_2 [10-12], ZrO_2 [13], g-C_3N_4 [14-16] were incorporated into PEM to improve their performance. But it caused the accumulate of fillers, which degrades the performance. Under the magnetic field [17, 18] and electric field [19], the fillers can form aligned functional nanocomposite PEM with highly conductive well-aligned proton channels. If conductive proton channels can be formed in PEM, the proton conductivity will be improved. Based on this assumption, we found electrospinning can form long-range proton transport channels at micro and nanoscales [20]. Due to its versatility and easily scale-up, it has been extensively used in nanostructured PEM [21, 22].

Electrospinning nanofiber can be divided into two categories, one is easy to dissolve into chemical solvent and the other is chemical resistant [23]. For the dissolving one, polymers such as chitosan [24] or Nafion [25] used water as the solvent. And for the chemical-resistant one, the structure of the nanofiber must be stable and steady in the chemical solvent, such as Polyimides (PI) [26-29] and polybenzimidazoles (PBI) [30]. PI as an aromatic imide monomer are well known because of its higher mechanical strength, high chemical and thermal stability [28]. The ultrafine PI nanofibers prepared by electrostatic spinning have a large specific surface area and great durability, thus improving the performance of composite membranes [26]. On this basis, some researchers placed ultrafine PI nanofibers in the middle of SPEEK [31] or SPAES [32] for the efficient microbial electrolysis cell.

In our work, pyromellitic dianhydride (PMDA) and 4,4-oxydianiline (ODA) in the N, N-dimethylformamide (DMF) were used to form poly (amic acid) (PAA) solutions. Then the PAA solutions were electrospun into PAA ultrafine nanofiber when the DMF was evaporated. PI nanofibers were prepared by thermal imidization [26-29, 33]. Then the PI was placed into the center of the SPEEK, so the PI/SPEEK composite membrane was fabricated. For the great chemical stability of PI nanofibers, the PI/SPEEK membrane has high performance. Compared with pure SPEEK, PI/SPEEK had higher stability at high temperature with PI as the framework of the nanofiber composite membrane, and their proton channels will not collapse. And PI didn't occupy the -SO_3H in the PI/SPEEK nanofiber composite membrane (NCM), its proton conductivity won't decrease. Its proton conductivity was higher than that of pure SPEEK under high temperatures.

2. Materials and methods

2.1 Materials

Poly (ether ether ketone) (PEEK) powder was supplied by Victrex. PMDA and ODA were provided by Aldrich. Beijing Chemical Reagent Factory provided dimethyl sulfoxide (DMSO), DMF and H_2SO_4 (95-98 wt%) with no further purification.

2.2 Fabrication of x-PI/SPEEK NCMs

SPEEK and PI nanofiber was fabricated and can be seen in our previous work [34, 35]. SPEEK was dissolved in DMSO solution to acquire the clean solutions. PI (x=1%, 2%, 3% and 4%) nanofiber was placed in the middle of the SPEEK solution as displayed in Scheme 1, and the x-PI/SPEEK membrane was constructed after removing the DMSO solution at 80°C for 48 hours.

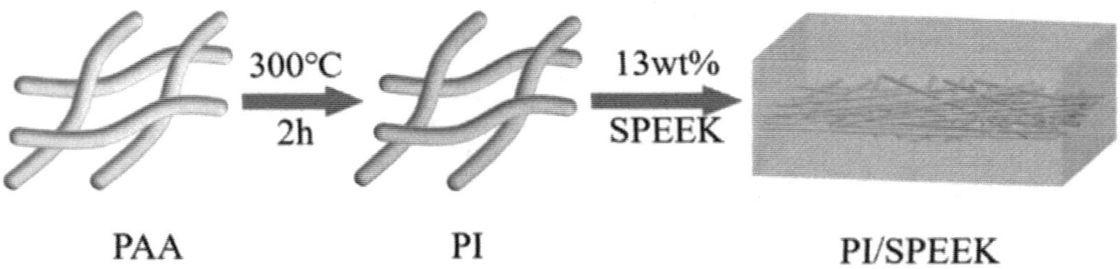

Scheme 1 The preparation method of PI/SPEEK NCMs

2.3 Characterization
The surface of PI nanofiber and the cross-sectional morphology of all NCMs were inspected by scanning electron microscope (SEM, SU8010). Its chemical structure was determined through Fourier Transform Infrared Spectroscopy (FTIR, Bruker Tensor II/Hyperion). The thermal stability of all membranes was analyzed by a thermogravimetric analyzer (TGA, TA Q5000). Mechanical stability was characterized through the machine (SHIMADZU, AGS-X). The test conditions and processing methods of IEC values, proton conductivity, E_a, swelling ratio (SR), water uptake (WU) and fuel cell performance are all the same compared with our previous work [34, 35].

3. Results and discussion

3.1 Cross-sectional morphologies of x-PI/SPEEK NCMs
The x-PI/SPEEK NCMs were prepared by the casting method mentioned above. To further explore the location of nanofiber in these NCMs, the cross-sectional morphology was observed after fracture in the liquid nitrogen through SEM as displayed in Fig. 1. These nanofibers were basically in the middle position of the nanofiber composite membrane. It was flat and there were no defects for the cross-sectional morphology of SPEEK. But for x-PI/SPEEK membranes, the PI nanofibers were pulled out from the SPEEK matrix due to their difference in mechanical performance. Most PI nanofibers even lay in the surface of the cross-sectional surface, leaving some holes in the SPEEK matrix. This phenomenon suggests that PI nanofibers did have excellent mechanical properties compared with the SPEEK matrix.

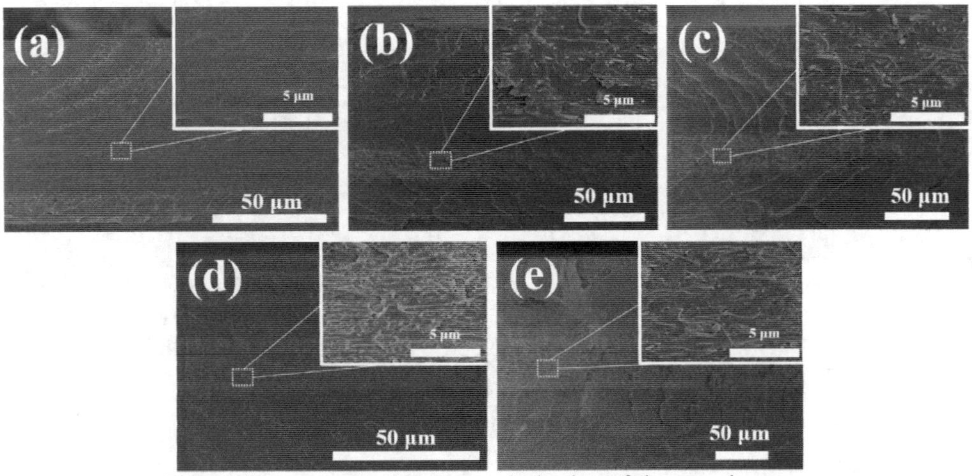

Fig. 1 The cross-sectional morphologies of the membranes

3.2 Thermal performance of x-PI/SPEEK NCMs
The thermal performances of the x-PI/SPEEK membranes were detected through TGA and can be seen in Fig. 2. There was only one disassemble step for the PI nanofiber, which occurred at 450°C and above. But for all the NCMs, it can be divided into three steps [36, 37]. The DMF evaporated below 200°C results in degradation at the first step. The fracture of SO_3H in SPEEK started to decompose at 200~350°C in the second step. The main chain of SPEEK and PI was decomposed at 450°C and above results in the third disassemble step. All membranes can be used at the service condition of PEM because there was no change in thermal stability for these NCMs after the PI nanofibers were incorporated in the SPEEK matrix.

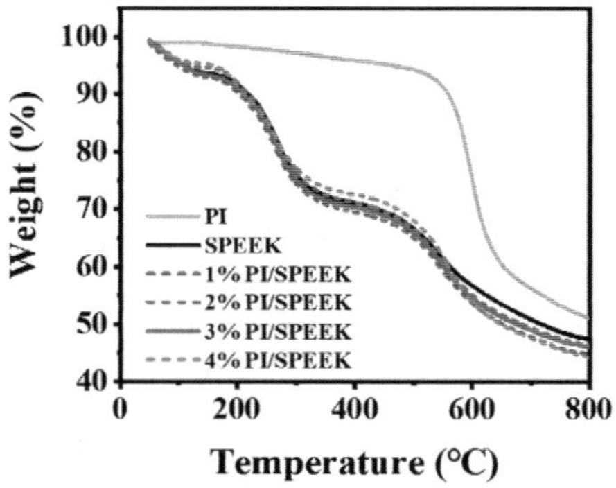

Fig. 2 TGA curve of PI nanofibers the membranes

3.3 Mechanical Properties of x-PI/SPEEK NCMs

Mechanical properties of x-PI/SPEEK NCMs at different conditions have been experimented and the outcome was depicted in Fig. 3 and table 1. Yield strength (YS) and elongation at break (EB) of SPEEK were 45.9 MPa and 35.1% at dry state. As the increase of PI nanofiber incorporation content, YS and EB increased at the outset then declined. The YS and EB of 2%PI/SPEEK NCM were 53.7 MPa and 85.3%, and it was 17% and 143% higher than that of original SPEEK, respectively. PI could improve the stability of composite membranes after incorporating them into the SPEEK matrix at the dry state.

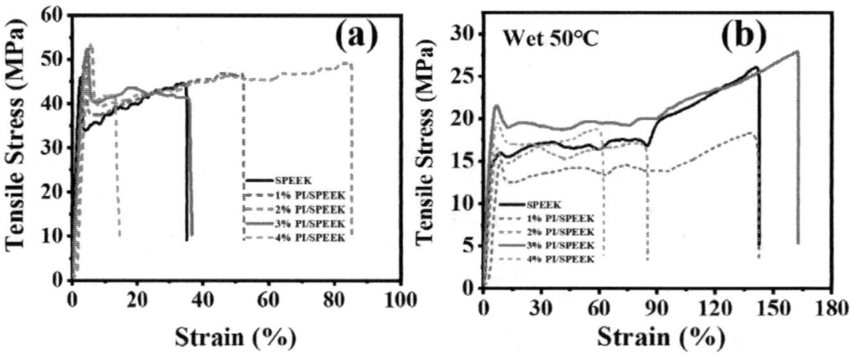

Fig. 3 Mechanical properties of dry (a) and wet (b) condition of the membranes

Table 1 Mechanical properties of x-PI/SPEEK NCMs

Specimen	YS (MPa)		EB (%)	
	Dry	Wet (50°C)	Dry	Wet (50°C)
SPEEK	45.9	16.0	35.1	142.6
1%PI/SPEEK	50.4	15.1	52.5	142.6
2%PI/SPEEK	53.7	16.6	85.3	85.3
3%PI/SPEEK	52.3	21.5	36.6	162.6
4%PI/SPEEK	48.4	19.6	14.4	62.4

PEM was used in hydrate conditions and high temperatures. So all membranes in 50°C at hydrate conditions were tested. It has the same tendency as tested in dry conditions but different results. The YS of pristine SPEEK was 16 MPa, which is 65% lower than that of SPEEK at dry state. Absorbing excessive water can result in losing its usage in the PEM. However, the YS and EB of 3%PI/SPEEK were 21.5 MPa and 162.6%, which were 34% and 14% higher than that of pure SPEEK. PI nanofiber soaked in SPEEK solution could be selected as a potential candidate material for PEM.

3.4 WU and SR of x-PI/SPEEK NCMs

All sulfonic acids are on the side chain of SPEEK[38] can serve as transport sites for the protons to transport by Grotthuss mechanism, and the water as vehicles for protons to transport by vehicle mechanism in PEM [39, 40]. IEC values mean the quantity of SO_3H in the SPEEK and more sulfonic acids mean high proton transportability but it would cause a high SR of NCM and low mechanical performance [41]. Therefore, it is important for PEM to have a proper SR and WU. IEC values of all NCMs are displayed in Table 2.

Table 2 IEC values of the NCMs

Content of PI nanofibers (wt%)	0	1	2	3	4
IEC (mmol/g)	1.654	1.589	1.588	1.554	1.524

The SR and WU of x-PI/SPEEK NCMs were detected and displayed in Fig. 4. For pure SPEEK, it has a higher sulfonated degree and its SR and WU reached 209% and 485% at 60°C and 100% RH. As the increase of PI nanofibers in the SPEEK, their SR decreased from 50% to 20% at 60°C and 100% RH and 3%PI/SPEEK have the lowest SR. The decreased tendency of SR and WU was nearly consistent. PI nanofibers acted as the framework in PEM had great stability would descend their change in SR and WU. 3%PI/SPEEK had the highest comprehensive performance and its SR was 20% and WU was 135%, which would satisfy the demand of PEM in the fuel cell.

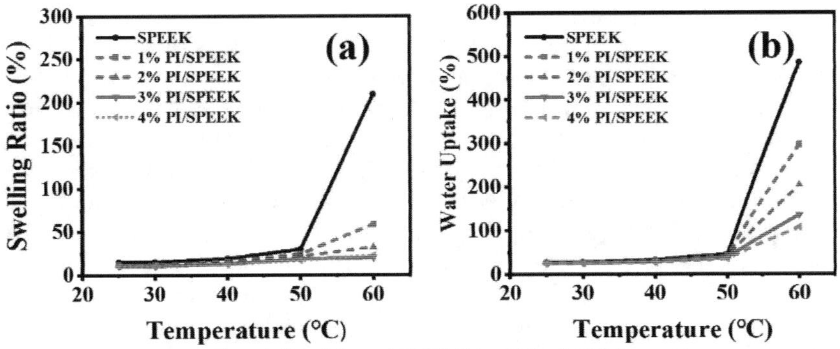

Fig. 4 SR (a) and WU (b) of the NCMs

3.5 Proton conductivity of x-PI/SPEEK NCMs

It is crucial for the x-PI/SPEEK membranes to have higher proton conductivity because high proton conductivity means high power density. The proton conductivity of x-PI/SPEEK NCMs was tested range from 25°C to 60°C at 100% RH. All the NCMs were soaked at 100% RH for 24 hours for equilibration. Membranes were fixed on the homemade mold to test their proton conductivities through CHI 660E with the same procedure in 100% RH, as shown in Fig. 5a. The addition of PI nanofibers into the SPEEK would decrease the proton conductivity of the NCMs before 50°C. When the temperature reached 60°C, the proton conductivities of all NCMs were higher than that of the original SPEEK, which was 209.7 $mS \cdot cm^{-1}$ at 60°C and 100% RH. The proton conductivities of 1%PI/SPEEK, 2%PI/SPEEK, 3%PI/SPEEK and 4%PI/SPEEK were 226.4, 288.0, 271.7 and 252.8 $mS \cdot cm^{-1}$ at 60°C and 100% RH respectively, and it was better than that of original SPEEK at the same condition. The proton conductivity was confirmed by the content of sulfonic acids and the stability of proton transport channels, their synergistic effect determines the proton conductivity of the NCMs. Moreover, it can be attributed to an interaction at their interface of PI nanofiber and SPEEK could be proved by the FTIR part in the next chapter.

E_a was applied to express the minimum energy for protons to transport and could be calculated as shown in Fig. 5b. The E_a of pristine SPEEK was 24.7 kJ/mol, and it is almost near to the value [42]. The E_a of 1%PI/SPEEK, 2%PI/SPEEK, 3%PI/SPEEK and 4%PI/SPEEK were 24.0, 23.8, 23.7 and 21.6 kJ/mol, respectively. The ability of proton transport is increased with the increase of PI nanofiber. Indeed, it can be attributed to an interaction at their interface of PI nanofiber and SPEEK, and it can decrease the value of E_a.

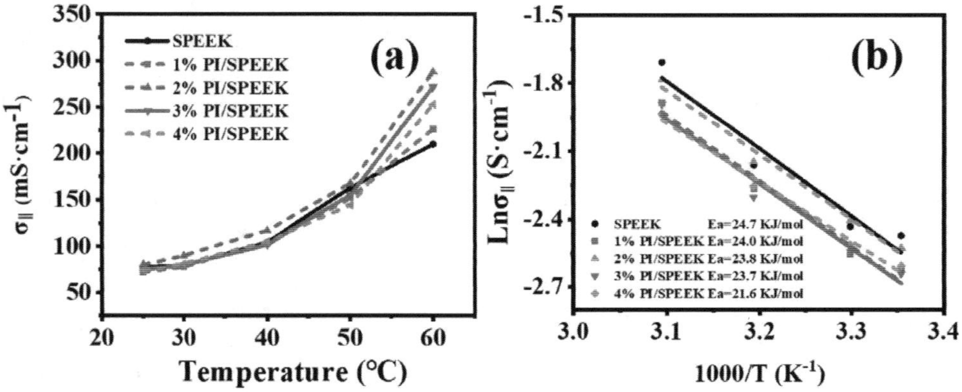

Fig. 5 The curve of proton conductivity (a) and E_a (b) of x-PI/SPEEK composite membrane changed with temperature

3.6 Fuel cell test of x-PI/SPEEK NCMs

The fuel cell test was conducted at 60°C and 95% RH to test their performance. 3%PI/SPEEK was chosen to test single-cell performance for its high proton conductivity and stability. SPEEK was also tested for comparison, as displayed in Fig. 6a. The maximum current density of SPEEK and 3%PI/SPEEK is 338.9 and 332.5 mA·cm^{-2} when the power density reaches maximum. But the maximum power density of SPEEK is 185.2 mW·cm^{-2} and for 3%PI/SPEEK, it's 199.1 mW·cm^{-2}. The addition of PI nanofibers into the SPEEK indeed increased fuel cell property according to the result of proton conductivity. This can be attributed to an interaction at their interface of PI nanofiber and SPEEK, which is favorable for the proton to transport in PEM. And PI nanofibers indeed can improve the stability of PI/SPEEK NCMs as displayed in Fig. 6, which are the photos of MEA (membrane electrode assembly) after the test. The outcome displayed that MEA made of SPEEK can't stand in the test environment and there were some broken areas on it which are circled in Fig. 6b. But for MEA that is made of 3%PI/SPEEK as shown in Fig. 6c, it was intact after the test in the fuel cell. In a word, 3%PI/SPEEK has a better performance with the balance of the fuel cell performance and stability.

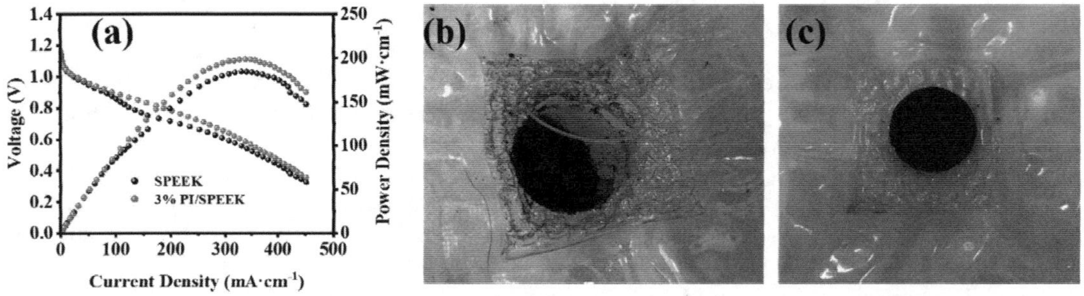

Fig. 6 Fuel cell test (a) of SPEEK and 3% PI/SPEEK in fuel cell and photos of SPEEK (b) and 3%PI/SPEEK (c) after single-cell test in fuel cell

3.7 FTIR spectra of SPEEK and x-PI/SPEEK NCMs

The vibration of C=C in benzene ring on SPEEK can be found at 1500 cm^{-1}, which is stable during fabrication of membranes, was used for all band areas to normalize and the FTIR spectra were displayed in Fig. 7. The axial stretching and bending vibration of C-N-C were found at 1376 and 724 cm^{-1}, respectively, which can be assigned to PI nanofibers and can be found in 4%PI/SPEEK. But there are no characteristic peaks in other composite membranes this is because the addition of 4% PI nanofibers to the SPEEK was excessive for the PI/SPEEK composite membrane. This phenomenon can also explain why 3%PI/SPEEK has the best comprehensive performance. The intensity of symmetric vibration of O=S=O, stretching vibration of S=O and S-O is increased with the addition of PI nanofibers. This can be attributed to an interaction at their interface of PI nanofiber and SPEEK [43].

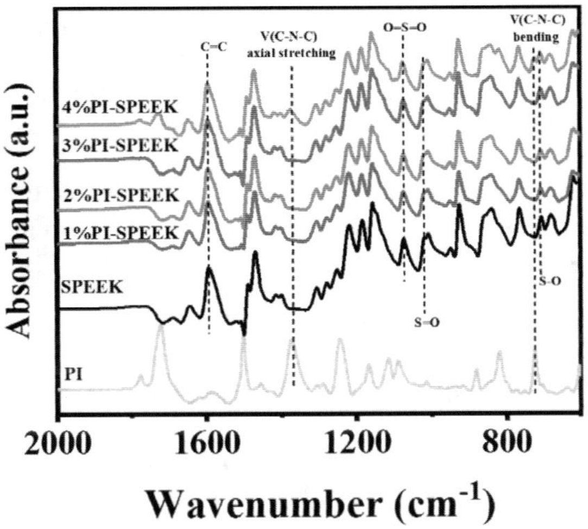

Fig. 7 The FTIR spectra of PI nanofibers and all NCMs

4. Conclusions

The PI/SPEEK NCMs fabricated by immersing PI nanofibers to the SPEEK solutions were used in the PEM for the fuel cell. PI nanofiber was almost at the center of the PI/SPEEK NCM. The incorporation of PI nanofibers indeed has higher performance at high temperatures. PI/SPEEK NCM was intact during the fuel cell test. The PI nanofibers had excellent mechanical performance and would keep the stability of the channels for the protons to transport. Also, this can be attributed to an interaction at their interface of PI nanofiber and SPEEK. As a result, the PI/SPEEK NCM was an essential candidate used in PEM to balance proton conductivity and stability. And the PI/SPEEK NCM can be used in high-temperature PEM, filtration and separation, and so on.

Acknowledgment

Thanks to the National Key Research and Development Plan (Grant No: 2016YFC0303705) to support this work.

References

[1] D.W. Shin, M.D. Guiver, Y.M. Lee, Hydrocarbon-based polymer electrolyte membranes: Importance of morphology on ion transport and membrane stability, Chem. Rev., 117 (2017) 4759-4805.

[2] G.L. Liu, W.C. Tsen, S.C. Jang, F.Q. Hu, F. Zhong, H. Liu, G.J. Wang, S. Wen, G.W. Zheng, C.L. Gong, Mechanically robust and highly methanol-resistant sulfonated poly(ether ether ketone)/poly(vinylidene fluoride) nanofiber composite membranes for direct methanol fuel cells, J. Membr. Sci., 591 (2019) 117321.

[3] X. Wu, X. Wang, G. He, J. Benziger, Differences in water sorption and proton conductivity between Nafion and SPEEK, Journal of Polymer Science Part B-Polymer Physics, 49 (2011) 1437-1445.
[4] M. Ahmed, I. Dincer, A review on methanol crossover in direct methanol fuel cells: challenges and achievements, Int J Energ Res, 35 (2011) 1213-1228.
[5] B.P. Tripathi, V.K. Shahi, Organic-inorganic nanocomposite polymer electrolyte membranes for fuel cell applications, Prog. Polym. Sci., 36 (2011) 945-979.
[6] G. He, J. Zhao, C. Chang, M. Xu, S. Wang, S. Jiang, Z. Li, X. He, X. Wu, Z. Jiang, Molecular engineering of organic-inorganic interface towards high-performance polyelectrolyte membrane via amphiphilic block copolymer, J. Membr. Sci., 563 (2018) 1-9.
[7] T. Kamjornsupamitr, T. Sangthumchai, S. Youngme, S. Martwiset, Proton conducting composite membranes from crosslinked poly(vinyl alcohol) and poly(styrene sulfonic acid)-functionalized silica nanoparticles, Int. J. Hydrog. Energy, 43 (2018) 11190-11201.
[8] T. Ko, K. Kim, S.-K. Kim, J.-C. Lee, Organic/inorganic composite membranes comprising of sulfonated Poly(arylene ether sulfone) and core-shell silica particles having acidic and basic polymer shells, Polymer, 71 (2015) 70-81.
[9] J. Zeng, P.K. Shen, S. Lu, Y. Xiang, L. Li, R. De Marco, S.P. Jiang, Correlation between proton conductivity, thermal stability and structural symmetries in novel HPW-meso-silica nanocomposite membranes and their performance in direct methanol fuel cells, J. Membr. Sci., 397 (2012) 92-101.
[10] C. Dong, Z. Hao, Q. Wang, B. Zhu, C. Cong, X. Meng, Q. Zhou, Facile synthesis of metal oxide nanofibers and construction of continuous proton-conducting pathways in SPEEK composite membranes, Int. J. Hydrog. Energy, 42 (2017) 25388-25400.
[11] K. Ketpang, S. Shanmugam, C. Suwanboon, N. Chanunpanich, D. Lee, Efficient water management of composite membranes operated in polymer electrolyte membrane fuel cells under low relative humidity, J. Membr. Sci., 493 (2015) 285-298.
[12] B.R. Matos, E.I. Santiago, J.F.Q. Rey, A.S. Ferlauto, E. Traversa, M. Linardi, F.C. Fonseca, Nafion-based composite electrolytes for proton exchange membrane fuel cells operating above 120 degrees C with titania nanoparticles and nanotubes as fillers, J. Power Sources, 196 (2011) 1061-1068.
[13] K. Ketpang, B. Son, D. Lee, S. Shanmugam, Porous zirconium oxide nanotube modified Nafion composite membrane for polymer electrolyte membrane fuel cells operated under dry conditions, J. Membr. Sci., 488 (2015) 154-165.
[14] C. Dong, Q. Wang, C. Cong, X. Meng, Q. Zhou, Influence of alkaline 2D carbon nitride nanosheets as fillers for anchoring HPW and improving conductivity of SPEEK nanocomposite membranes, Int. J. Hydrog. Energy, 42 (2017) 10317-10328.
[15] M. Gang, G. He, Z. Li, K. Cao, Z. Li, Y. Yin, H. Wu, Z. Jiang, Graphitic carbon nitride nanosheets/sulfonated poly(ether ether ketone) nanocomposite membrane for direct methanol fuel cell application, J. Membr. Sci., 507 (2016) 1-11.
[16] Y. Shi, S. Jiang, K. Zhou, C. Bao, B. Yu, X. Qian, B. Wang, N. Hong, P. Wen, Z. Gui, Y. Hu, R.K.K. Yuen, Influence of g-C3N4 Nanosheets on Thermal Stability and Mechanical Properties of Biopolymer Electrolyte Nanocomposite Films: A Novel Investigation, ACS Applied Materials & Interfaces, 6 (2014) 429-437.
[17] X. Liu, Y. Li, J.D. Xue, W.K. Zhu, J.F. Zhang, Y. Yin, Y.Z. Qin, K. Jiao, Q. Du, B.W. Cheng, X.P. Zhuang, J.X. Li, M.D. Guiver, Magnetic field alignment of stable proton-conducting channels in an electrolyte membrane, Nature Communications, 10 (2019) 842.
[18] X. Liu, J.F. Zhang, C.Y. Zheng, J.D. Xue, T. Huang, Y. Yin, Y.Z. Qin, K. Jiao, Q. Du, M.D. Guiver, Oriented proton-conductive nano-sponge-facilitated polymer electrolyte membranes, Energy Environ. Sci., 13 (2020) 297-309.

[19] S. Gahlot, V. Kulshrestha, Dramatic Improvement in Water Retention and Proton Conductivity in Electrically Aligned Functionalized CNT/SPEEK Nanohybrid PEM, Acs Applied Materials & Interfaces, 7 (2015) 264-272.
[20] R. Sood, S. Cavaliere, D.J. Jones, J. Roziere, Electrospun nanofibre composite polymer electrolyte fuel cell and electrolysis membranes, Nano Energy, 26 (2016) 729-745.
[21] P. Kallem, N. Yanar, H. Choi, Nanofiber-Based Proton Exchange Membranes: Development of Aligned Electrospun Nanofibers for Polymer Electrolyte Fuel Cell Applications, Acs Sustainable Chemistry & Engineering, 7 (2019) 1808-1825.
[22] S. Cavaliere, S. Subianto, I. Savych, D.J. Jones, J. Roziere, Electrospinning: designed architectures for energy conversion and storage devices, Energy & Environmental Science, 4 (2011) 4761-4785.
[23] S.-H. Yun, J.-J. Woo, S.-J. Seo, L. Wu, D. Wu, T. Xu, S.-H. Moon, Sulfonated poly(2,6-dimethyl-1,4-phenylene oxide) (SPPO) electrolyte membranes reinforced by electrospun nanofiber porous substrates for fuel cells, Journal of Membrane Science, 367 (2011) 296-305.
[24] J. Wang, Y. He, L. Zhao, Y. Li, S. Cao, B. Zhang, H. Zhang, Enhanced proton conductivities of nanofibrous composite membranes enabled by acid-base pairs under hydrated and anhydrous conditions, Journal of Membrane Science, 482 (2015) 1-12.
[25] J. Wang, P. Li, Y. Zhang, Y. Liu, W. Wu, J. Liu, Porous Nafion nanofiber composite membrane with vertical pathways for efficient through-plane proton conduction, Journal of Membrane Science, 585 (2019) 157-165.
[26] E. Han, D. Wu, S. Qi, G. Tian, H. Niu, G. Shang, X. Yan, X. Yang, Incorporation of Silver Nanoparticles into the Bulk of the Electrospun Ultrafine Polyimide Nanofibers via a Direct Ion Exchange Self-Metallization Process, Acs Applied Materials & Interfaces, 4 (2012) 2583-2590.
[27] L. Kong, Y. Wang, H. Yu, B. Liu, S. Qi, D. Wu, W.-H. Zhong, G. Tian, J. Wang, In Situ Armoring: A Robust, High-Wettability, and Fire-Resistant Hybrid Separator for Advanced and Safe Batteries, Acs Applied Materials & Interfaces, 11 (2019) 2978-2988.
[28] G. Gong, K. Gao, J. Wu, N. Sun, C. Zhou, Y. Zhao, L. Jiang, A highly durable silica/polyimide superhydrophobic nanocomposite film with excellent thermal stability and abrasion-resistant performance, J. Mater. Chem. A, 3 (2015) 713-718.
[29] C. Kim, Y.O. Choi, W.J. Lee, K.S. Yang, Supercapacitor performances of activated carbon fiber webs prepared by electrospinning of PMDA-ODA poly(amic acid) solutions, Electrochim. Acta, 50 (2004) 883-887.
[30] M. Tanaka, Y. Takeda, T. Wakiya, Y. Wakamoto, K. Harigaya, T. Ito, T. Tarao, H. Kawakami, Acid-doped polymer nanofiber framework: Three-dimensional proton conductive network for high-performance fuel cells, J. Power Sources, 342 (2017) 125-134.
[31] K.-J. Chae, K.-Y. Kim, M.-J. Choi, E. Yang, I.S. Kim, X. Ren, M. Lee, Sulfonated polyether ether ketone (SPEEK)-based composite proton exchange membrane reinforced with nanofibers for microbial electrolysis cells, Chem. Eng. J., 254 (2014) 393-398.
[32] S.-G. Park, K.-J. Chae, M. Lee, A sulfonated poly(arylene ether sulfone)/polyimide nanofiber composite proton exchange membrane for microbial electrolysis cell application under the coexistence of diverse competitive cations and protons, J. Membr. Sci., 540 (2017) 165-173.
[33] Q. Ding, Y.-E. Miao, T. Liu, Morphology and Photocatalytic Property of Hierarchical Polyimide/ZnO Fibers Prepared via a Direct Ion-exchange Process, Acs Applied Materials & Interfaces, 5 (2013) 5617-5622.
[34] P. Wei, Y. Sui, X. Li, Q. Liu, B. Zhu, C. Cong, X. Meng, Q. Zhou, Sandwich-structure PI/SPEEK/PI proton exchange membrane developed for achieving the high durability on excellent proton conductivity and stability, J. Membr. Sci., (2021) 120116.
[35] B.S. Zhu, Y. Sui, P. Wei, J.H. Wen, H. Cao, C.B. Cong, X.Y. Meng, Q. Zhou, NH_2-UiO-66 coated fibers to balance the excellent proton conduction efficiency and significant dimensional stability of proton exchange membrane, J. Membr. Sci., 628 (2021) 119214.

Journal of Physics: Conference Series **2194** (2022) 012003 doi:10.1088/1742-6596/2194/1/012003

[36] G. Rambabu, S.D. Bhat, Sulfonated fullerene in SPEEK matrix and its impact on the membrane electrolyte properties in direct methanol fuel cells, Electrochim. Acta, 176 (2015) 657-669.
[37] H. Gao, C. Dong, Q. Wang, H. Zhu, X. Meng, C. Cong, Q. Zhou, Improving the proton conductivity of proton exchange membranes via incorporation of HPW-functionalized mesoporous silica nanospheres into SPEEK, Int. J. Hydrogen Energy, 43 (2018) 21940-21948.
[38] M.J. Parnian, S. Rowshanzamir, F. Gashoul, Comprehensive investigation of physicochemical and electrochemical properties of sulfonated poly (ether ether ketone) membranes with different degrees of sulfonation for proton exchange membrane fuel cell applications, Energy, 125 (2017) 614-628.
[39] X. Hou, Z. Liu, Y. Wei, Q. Zhao, J. Dong, B. Liu, Z. Sun, T. Shi, M. Zhang, W. Hu, Proton conducting nanocomposite membranes of nanocellulose reinforced poly(arylene ether ketone)s containing sulfonic/carboxylic groups, Solid State Ionics, 311 (2017) 31-40.
[40] J. Wang, X. Yue, Z. Zhang, Z. Yang, Y. Li, H. Zhang, X. Yang, H. Wu, Z. Jiang, Enhancement of proton conduction at low humidity by incorporating imidazole microcapsules into polymer electrolyte membranes, Adv. Funct. Mater., 22 (2012) 4539-4546.
[41] Z. Zhang, J. Ren, J. Xu, Z. Wang, W. He, S. Wang, X. Yang, X. Du, L. Meng, P. Zhao, Adjust the arrangement of imidazole on the metal-organic framework to obtain hybrid proton exchange membrane with long-term stable high proton conductivity, Journal of Membrane Science, 607 (2020) 118194.
[42] S. He, S. Liu, W. Dai, S. Zhai, J. Lin, Nanocomposite Proton Exchange Membranes Incorporating Phosphotungstic Acid Anchored on Imidazole-Functionalized Halloysite Nanotubes, Journal of the Electrochemical Society, 165 (2018) F951-F958.
[43] F.Y. Hao, J.H. Wang, S.L. Qi, G.F. Tian, D.Z. Wu, Structures and properties of polyimide with different pre-imidization degrees, Chin. J. Polym. Sci., 38 (2020) 840-846.

Determination of 11 kinds of ultraviolet absorbents in plastic food contact materials by high performance liquid chromatography

Yue Qiu*, Qing Zhang, Qian Yang, Chanlan Tan, Lu Zhang

Chongqing Academy of Metrology and Quality Inspection, Chongqing, China.

*Corresponding author's e-mail: qiuyuecqu@foxmail.com

Abstract: A method for the determination of 11 kinds of ultraviolet (UV) absorbents in food contact materials was established based on high performance liquid chromatography. The target compounds were separated on a Agilent SB C18 (250 mm×4.6 mm×5 μm) column with gradient elution using acetonitrile and water as mobile phase, which analyzed by diode array detector and quantified by external standard method. The results showed that the linear range of the 11 UV absorbents was between 0.2 and 50.0 mg/L ($R^2 \geq 0.9985$), the limits of detection (LODs) were in the range of 0.01-0.05 mg/L, and the limits of quantification (LOQs) were 0.05-0.20 mg/L. The recoveries (R) of blank samples at 0.2, 10 and 50 mg/kg were from 80.2% to 97.5%, and the relative standard deviations (RSD) were ranged from 0.37% to 4.15%. The method had low detection limit, high stability and good repeatability, which could be used for daily detection of UV absorbents in food contact materials.

1. Introduction

UV absorbers can absorb ultraviolet light in the range of 280-400 nm and release light energy in the form of heat energy to reduce the photo degradation and aging of plastic products and extend the service life, so they are widely used in the manufacturing process of food contact materials [1,2]. Studies have shown that UV absorbents have certain toxicity, which can cause dermatitis and skin allergy. Moreover, UV absorbents have bio-accumulation, which will affect the length and quality of organisms and result in imbalance of thyroid system and fatty acid metabolism imbalance [3,4].

Currently, GB 9685-2016 standard [5] stipulates the safe usage and migration amount of some UV absorbents, but there is no corresponding standard method to detect UV absorbents, which will not guarantee the quality of food contact materials and seriously threaten food safety and human health. Therefore, it is urgent to establish a method for the detection of UV absorbent in food contact materials.

In this paper, a method for the determination of 11 common UV absorbents in food contact materials was established based on high performance liquid chromatography. The method has low detection limit, high stability and good repeatability, which can be used for daily detection of UV absorbents in food contact materials.

2. Materials and Methods

2.1 Materials and reagents. 11 UV absorbents were all supplied from Dr. Ehrenstorfer (Augsburg, Germany). HPLC grade acetonitrile was purchased from Merck (Darmstadt, Germany). The water used in the experiment was ultra-pure water.

2.2 Instruments and equipment. Agilent 1260 HPLC with DAD detector; MS105DU electronic analysis balance (Mettler Toledo International TradeCo., LTD.); N-evap-112 Nitrogen Blower (Organomation, USA); KH3200DE Ultrasonic cleaner (Kunshan Hechuang Ultrasonic Instrument Co., LTD.)

2.3 Preparation of standard solution. Each standard substance of 10 mg was accurately weighed (accurate to 0.1 mg) and placed in a 10 mL volumetric flask, and then dissolved with mixed organic solvent (methanol: acetone: n-hexane =2:1:1) to prepared into UV absorbent single standard reserve liquid of 1000 mg/L. An appropriate amount of 11 kinds of UV absorbent single standard reserve solution was removed and diluted with mixed solvent into standard working solutions with concentrations of 0.2 mg/L, 1.0 mg/L, 10.0 mg/L, 25.0 mg/L and 50.0 mg/L.

2.4 Sample pretreatment. The sample was cut into small pieces with an area no larger than 5 mm×5 mm. The 1.0 g sample was accurately weighed into a 25 mL colorimetric tube, and mixed solvent of 10 mL was added for ultrasonic extraction for 40 min. The sample was cooled to room temperature, and the supernatant was filtered through 0.45 μm organic system filter membrane for UPLC analysis.

2.5 Chromatographic condition. The separation was performed on an Agilent SB C18 column (250 mm ×4.6 mm, 5 μm) with water as mobile phase A and acetonitrile as mobile phase B. Column temperature was 30°C, sample volume was 10 μL, the detection wavelength was 300 nm, and the flow rate was 1.0 mL/min. The gradient elution procedure was shown in Table 1.

Table 1 Determination of elution gradient of UV absorbent by liquid chromatography

Time (min)	A (%)	B (%)
0	15	85
3.0	15	85
12.0	75	25
17.0	100	0
30	100	0
31	15	85
35	15	85

3. Results and Discussions

3.1 Optimization of chromatographic conditions

The effects of chromatographic column, mobile phase, column temperature, elution gradient and wavelength on the separation of UV absorbent were investigated. The results showed that Agilent SB C18 column had high response strength, good chromatographic peak shape and complete baseline separation. Through the comparison of acetonitrile-water, methanol-water and ethyl acetate-water, it was found that acetonitrile-water obtained better separation effect. In addition, the experiment also optimized the column temperature and other influencing factors, finally Agilent SB C18 column in acetonitrile-water system was selected for the separation of UV absorbent. The best column temperature was set at 30°C, and absorption wavelength was 300 nm with flow rate of 1.0 mL/min. Under the optimal conditions, 11 UV absorbents could achieve good baseline separation within 30 min, as shown in Figure 1.

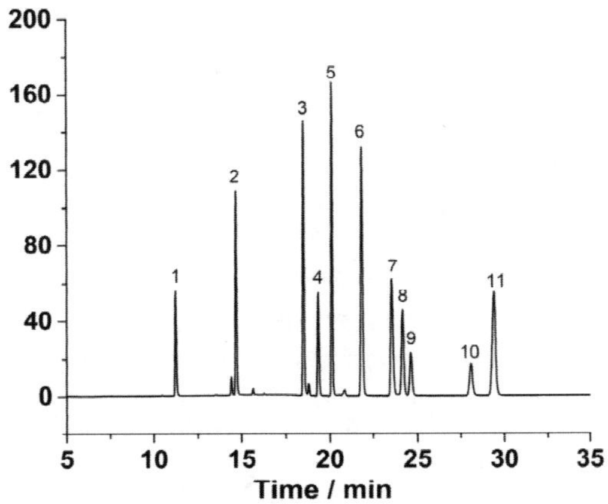

Fig. 1 Chromatograms of standard solutions with 11 kinds of UV absorbents

3.2 Standard curve and detection limit

The mixed standard solution series were determined under the optimal chromatographic conditions, and the mass concentration of each compound was linearly regression based on the peak area. The results showed that the 11 target compounds showed good linear relationship in the range of 0.2-50 mg/L, and the linear correlation coefficients R^2 were all greater than 0.9985. The method limit of detection (LOD) and method limit of quantification (LOQ) of 11 UV absorbents were calculated using 3 times signal-to-noise ratio (S/N≥3) and 10 times signal-to-noise ratio (S/N≥10), and the results were shown in Table 2. This indicates that the method can be used for the qualitative and quantitative analysis of 11 kinds of UV absorbents in plastic food contact materials.

Table 2 The linear regression equation, R^2, LOD and LOQ of UV absorbent

No.	Compound	linear regression equation	R^2	LOD (mg/L)	LOQ (mg/L)
1	UV-2	Y=5.2804X-2.5916	0.9994	0.02	0.10
2	UV-24	Y=10.8880X-5.9158	0.9993	0.01	0.05
3	UV-71	Y=15.0832X-6.9006	0.9994	0.01	0.05
4	TBS	Y=6.1525X-3.5478	0.9988	0.02	0.10
5	UV-3030	Y=15.8745X-8.4015	0.9993	0.01	0.05
6	UV-1579	Y=19.9449X-9.3209	0.9995	0.01	0.05
7	UV-329	Y=11.1812X-6.7698	0.9991	0.02	0.10
8	UV-234	Y=8.3885X-5.3021	0.9990	0.02	0.10
9	UV-326	Y=4.3749X-2.8150	0.9989	0.05	0.20
10	UV-327	Y=4.2273X-2.0482	0.9993	0.05	0.20
11	UV-1577	Y=14.8150X-7.0527	0.9994	0.02	0.10

3.3 Precision and recovery

The standard solution with different concentrations was added to the negative sample, and the recovery and precision were tested. The standard addition levels were 0.2, 10.0 and 50.0 mg/L, and each concentration level was tested 6 times in parallel. The experimental results (Table 3) showed that the recoveries (R) of 11 UV absorbents at high, medium and low spiked levels were 80.2%-97.5%, and the relative standard deviations (RSD) were 0.37%-4.15%, which could meet the actual detection requirements.

Table 3 Recoveries (R) and relative standard deviations (RSD) (n=6) for 11 UV absorbents

No.	Compound	Spiked levels (mg/kg)	R (%)	RSD (%)
		0.2	83.6	1.45
1	UV-2	10.0	88.6	0.80
		50.0	95.9	0.50
		0.2	83.3	1.53
2	UV-24	10.0	88.3	1.05
		50.0	95.6	0.63
		0.2	82.5	4.15
3	UV-71	10.0	89.9	1.95
		50.0	96.9	0.80
		0.2	80.2	3.18
4	TBS	10.0	81.6	1.61
		50.0	90.4	0.43
		0.2	82.9	2.51
5	UV-3030	10.0	87.9	0.86
		50.0	95.2	0.44
		0.2	80.4	1.43
6	UV-1579	10.0	90.4	1.00
		50.0	96.6	0.59
		0.2	83.2	1.25
7	UV-329	10.0	87.2	1.63
		50.0	96.0	0.51
		0.2	81.5	1.46
8	UV-234	10.0	88.4	2.23
		50.0	97.5	0.60
		0.2	80.6	1.00
9	UV-326	10.0	88.1	2.13
		50.0	97.0	0.65
		0.2	81.7	1.69
10	UV-327	10.0	89.3	0.93
		50.0	96.4	0.63
		0.2	81.3	0.37
11	UV-1577	10.0	89.1	0.81
		50.0	96.4	0.57

3.4 Actual sample testing

The method was applied to the determination of 10 kinds of plastic packaging materials, and the results showed that three kinds of them detected a certain amount of UV absorbent residues. UV-326 residue was found in Sample 1, UV-329 and TBS were detected in sample 2, and UV-24 was detected in Sample 3. The content was ranged from 0.28 to 2.14 mg/kg.

4. Conclusions

A HPLC method was established for the determination of 11 kinds of UV absorbents in plastic food contact materials through the investigation and optimization of chromatographic conditions. The method could effectively separate 11 kinds of UV absorbents within 30 min with the limits of detection (LODs) of 0.01-0.05 mg/L and quantification (LOQs) of 0.05-0.20 mg/L. The method had the advantages of simple operation, high sensitivity, high separation and high accuracy, which could provide technical reference for the analysis and detection of UV absorbent in plastic food contact materials in the future.

Acknowledgments
This research was funded by Scientific Research Project of Chongqing Market Supervision Administration [CQSJKJ2019005].

References
[1] Nakata, H.; Murata, S.; Filatreau, J. Occurrence and Concentrations of Benzotriazole UV Stabilizers in Marine Organisms and Sediments from the Ariake Sea[J], Japan. Environ. Sci. Technol. 2009, 43, 6920-6926.
[2] Vidal, L.; Chisvert, A.; Canals, A.; Salvador, A. Ionic LiquidBased Single-Drop Microextraction Followed by Liquid Chromatography-Ultraviolet Spectrophotometry Detection to Determine Typical UV Filters in Surface Water Samples[J], Talanta. 2010, 81, 549-555.
[3] Zhang, D.; Liu, C.; Yang, Y. Determination of UV Absorbers and Light Stabilizers in Food Packing Bags by Magnetic Solid Phase Extraction Followed by High Performance Liquid Chromatography[J], Chromatographia 2016, 79, 45-52.
[4] Qiu Y.; Ruan Y.; Tan C. L.; Li G. R.; Long M.; Xia Z. N. Rapid analysis of 14 ultraviolet absorbents in plastic food contact materials by supercritical fluid chromatography on Sub-2-micron particles[J], J. Liq. Chromatogra. R. T. 2020, 11-12(43), 1080-1087.
[5] GB 9685-2008 Hygienic standards for uses of additives in food containers and packaging materials [S].

Journal of Physics: Conference Series **2194** (2022) 012005 doi:10.1088/1742-6596/2194/1/012005

Selective Separation of Gases by Metal Organic Framework Materials (Mofs)

Jiayi Hu [1, †], **Yining Wang** [2, †], **Leyi Wu** [3,*, †]

[1] IB Department, High School Affiliated to Nanjing Normal University, Nanjing 210003, China

[2] Zhenhai High School of Ningbo, Ningbo 330200, China

[3] Hangzhou Foreign Languages School, Hangzhou 310000, China,
[*]guanghua.ren@gecacademy.cn

[†] These authors contributed equally.

Abstract —Gas separation is used to separate one or more gases from a mixture, which is crucial in the chemical industry. Compared with conventional porous materials, the distinctive physicochemical traits of MOFs play a dominant role in gas storage and separation. This review focuses on MOF materials in selective gas separation and current challenges. According to research, the usage of porous materials has achieved effective separation and greatly reduced energy consumption in nitrogen/oxygen (N2/O2) separation, acetylene/ethylene (C2H2/ C2H4) separation, carbon capture and toxic gas separation. MOF's unique structural features lead to the promising status for gas storage and separation. However, MOF materials still have some problems and challenges, e.g., the lack of practice for scientists to easily find a high-performance MOFs, high cost of MOF materials and shaping process. In the future, researchers around the world are looking to synthesize more and more artificial MOFs. Overall, these results offer a guideline for improvement of MOF materials.

1. Introduction

Gas separation is a widely used technique for separating one or more gases from a mixture. It is one of the key steps of the chemical manufacturing that is closely associated with all facets of human society. Separation or purification of crucial products in gas mixtures accounts for approximately 15% of global energy consumption [1]. At present, people have a strong interest in the development and implementation of technologies to reduce carbon dioxide emissions, thereby preventing the most serious consequences of global warming, which is an important application of gas separation [2]. Some of these separation processes are carried out on a large scale. Compared with some traditional porous materials (e.g., activated carbon and zeolite), the special structural characteristics of MOFs make them of great significance in gas storage and separation to meet the increasing demand for energy-saving, high-efficiency and environmentally-friendly separation processes [3], which has great potential in future application and development.

In adsorptive separation, gas separation is accomplished by the difference in the adsorption capacity of adsorbents for different substances in the mixture. Gas adsorptive separation of porous materials are normally realized through one or several mechanisms, including size exclusion, equilibrium separation, kinetic separation etc. [4]. Then, the strong adsorption group is desorbed, i.e., the adsorbent can be reused. Thus, a good adsorbent for gas adsorptive separation ought to have good adsorption capacity and selectivity, but also good regeneration capacity. In order to meet these requirements, the adsorbent should have higher surface areas and larger pore size to allow the adsorbed molecules to approach the inner surface.

For the review of MOF materials in the field of selective gas adsorption, the review published in 2009 by Li et al. [5] discussed the development of early MOF materials that are mainly in laboratory. A review published in 2018 by Wang et al. [4] listed in detail the practical applications of a large number of MOF materials in various gas separations. In order to provide a clarified MOF analysis based on actual application scenarios, this review will start from the perspective of practical applications and analyze the advantages of MOFs in several currently mainstream gas separation applications.

This review will present the recent advances of MOFs in selective gas separation, including the application of MOFs in Nitrogen/Oxygen (N_2/O_2) separation, acetylene/ethylene (C_2H_2/ C_2H_4) separation, carbon capture and toxic gas separation, including the development and mechanism of each gas separation application, and reviewing the characteristics of specific MOFs. This review will also pay attention to the problems and challenges that still exist to give a more complete and systematic review on selective gas separation.

2. Application of MOF in gas selective adsorption and separation

2.1 N_2/O_2 separation

O_2 now plays a key role in many areas, e.g., industrial-scale O_2 used in oxygen fuel combustion technology. High purity oxygen is mostly obtained through low temperature air separation process, which is expensive and high energy consumption, the use of porous materials as a substitute can greatly reduce energy consumption, but Pressure Swing Adsorption (PSA) technology to obtain the purity of oxygen will be limited, MOF material has recently been used to improve O_2 adsorption selectivity and separation from the air O_2 purity [6]. MOF adsorbents have highly polarized or high-electric field gradient surfaces that distinguish gas molecules by polarization, dipole instantaneous and quad polar moments. Open metal sites (OMS) (seen from Fig.1) have the most significant effect on the adsorption properties of major gases, even non-polar gases such as Hydrogen (H_2) and Methane (CH_4) [7].

Using discrete Fourier Transformation (DFT) simulations, Nenoff and colleagues studied two MOFs types containing OMSs, where O_2 binding is significantly stronger than late transition metals, while N2 binding is moderate in the transition metal series [8]. Hallapally and colleagues used composite Fc@MIL-101 to separate O_2 and N_2 and reported strategies to improve O_2 adsorption and O_2/air separation selectivity at room temperature. The results show that in the gas mixture experiment of 21/79 O_2/ N_2, N_2 broke through immediately, while O_2 broke through after 40 minutes. Besides, O_2 absorbs very well, almost four times as much as composite materials [9].

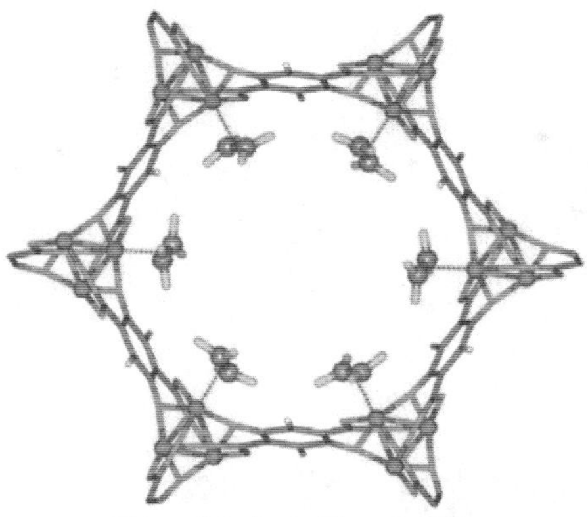

Fig.1 An open metal site for a single site (Fe_2(dobdc)) and a double point for each metal (UTSA-74-Zn) [10].

Perhaps the most noteworthy OMS series MOFs are the M_2 (dobdc) (MOF-74, M -Mg, Mn, Fe, Co, Ni, Zn.) [10]. These materials have 1D hexagonal channels, size about 11 Å -wide[11], demonstrating strong affinity [11]. In olefin/paraffin separation, OMS in MOF-74 is usually the primary binding point for the priority recognition of olefins with p-orbit, as confirmed by crystallography studies [12].

2.2 C_2H_2/C_2H_4 separation
Eliminating acetylene (C_2H_2) from ethylene (C_2H_4) is essential in industry, especially in oil refining and petrochemical related industries [13]. C_2H_4 must be separated from alkynes like C_2H_2 before being converted into polymers, because C_2H_2 impurities deactivate the catalysts and terminate the polymerization of C_2H_4 [13]. Therefore, separation for the mixture of C_2H_2 and C_2H_4 is necessary. Compared to the state-of-the-art in line with the partial hydrogenation of alkynes over supported Pd-catalysts[14], MOFs materials in C_2H_2/C_2H_4 have relatively high selectivity and lower price. However, traditional molecular sieving or porous materials are unable to separate them.

The important breakthrough for alkyne/olefin separations by MOFs occurred in 2008 [15], and with few years' development, new MOF materials such as SIFSIX MOFs, [$Cu_2(pzdc)_2(pyz)$], and NOTT-300 are synthesis to take advantage in the application of alkyne/olefin separations including C_2H_2/C_2H_4 separation. So far, the separation mechanisms for MOFs are in the light of sieving effect [15]. For dual functionalities, the pore opening of a functionalized cage of many MOFs in C_2H_2/C_2H_4 separation is reduced to improve the sieving effect and functionalize the cavity for high gas uptake, especially for those featuring fluosilicate moieties.

SIFSIX MOFs (as illustrated in Fig.2) include SIFSIX-3-Zn (3=pyrazine), SIFSIX-1-Cu (1=4,40-bipyridine), SIFSIX-2-Cu (2= 4,40-dipyridylacetylene) and so on, exhibiting pore sizes of about 3.84, 8.0, and 13.0Å respectively[15]. SIFSIX-14-Cu-i is one of best-performing MOFs for it achieved nearly molecular sieving for C_2H_2/C_2H_4. It can be synthesized from replacing 4,40-dipyridylacetylene by slightly shorter linker 4,40-azopyridine in SIFSIX-2-Cu-i. The micro pore size (≈3.3-4.0Å) of SIFSIX-14-Cu-i might completely block C_2H_4 molecules. Meanwhile, the functional fluosilicate sites create strong affinity toward C_2H_2 molecules [16]. SIFSIX-2-Cu-i has a relatively compact pore aperture size of 3.4Å, and nearly complete exclusion of C_2H_4 can be achieved with selectivity of 6,000 for the 1/99 C_2H_2/C_2H_4 mixture at 298 K [16]. SIFSIX MOFs contains nucleophilic nature sites which serve as hydrogen bonding acceptors for gases with acidic hydrogen atoms like C_2H_4 molecules to interact with Fluorine atoms in the MOFs [15]. The strong C_2H_2 binding through the Hydrogen bond can open pores under environmental conditions, indicating that the absorption of C_2H_2/C_2H_4 is significantly different [16]. SIFSIX-2-Cu-i is a kind of small-pore-variant SIFSIX MOFs that owns the dual functionality for both optimal pore size and strong binding sites in the application like C_2H_2/C_2H_4 separation, which can reduce the C_2H_2 concentration to a level of below 2 ppm [15]. It also shows its advantage on low-pressure uptake.

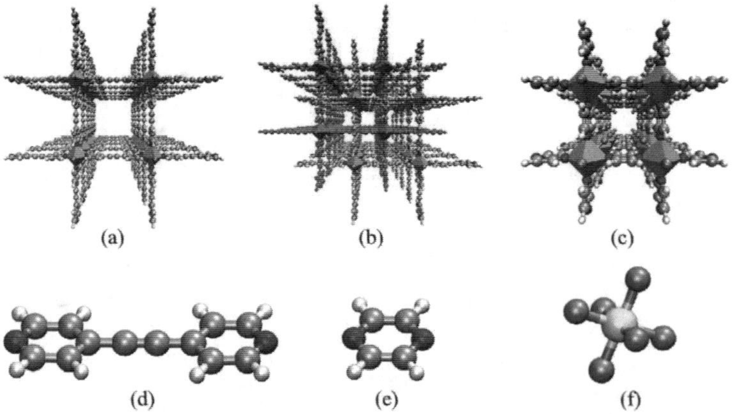

Fig.2 Structure of the MOFs from the SIFSIX series. (a) SIFSIX-2-Cu (b) SIFSIX-2-Cu-I (c) SIFSIX-3-Cu. Organic ligand. (d) 4,4'-dipyridylacetylene (e) pyrazine [5].

In addition to SIFSIX MOF, many other types of MOF materials are also applied in C_2H_2/C_2H_4 separation, and they also have unique advantages. For example, [Cu2(pzdc)2(pyz)] with threshold pressure of 1.1 kPa for C_2H_2 adsorption at 270 K has potential to be used in C_2H_2/C_2H_4 separation. It's a kind of pillared-layer MOF with the use of gate-opening effects [15]. NOTT-300 reported by Schröder and co-workers [17]. Its high uptake capacity and selectivity for unsaturated hydrocarbons is unique and useful, which result in the relatively large difference between the uptakes of C_2H_2 and C_2H_4. In general, the use of interpenetration and the varied change in length of ligands promote the molecular sieving effect of MOFs, and the featured functional group and the exploration of dual functionality makes the separation of C_2H_2 and C_2H_4 become possible and reliable.

2.3 Carbon capture

In the 21st century, the problem of global temperature rise caused by burning fuels containing carbon is becoming more and more serious, the demand for carbon capture is also growing. Introducing heteroatoms into the main chain of MOF or as part of covalent bond binding function, especially heteroatoms with high polarity and nucleophilic properties in some cases, have shown strong interaction with CO_2 molecules. Coordinatively unsaturated metal sites, as exemplified by the secondary building unit in Mg-MOF-74 makes its high CO_2 uptakes value is pretty notable at a pretty low pressure and temperature under a conditions without water (5.28 mmol/g at 40°C and 0.15 bar) [1, 18], which stem primarily from the high density of open metal sites (OMSs). In the meantime, due to the radius of Mg^{2+} which is smaller than transition metal, the adsorption enthalpy could be greater than others.

In 2012, Long's group constructed functionalizing Mg_2 (dobpdc) (H4dobpdc = 4,4'-dihy droxy-3,3'-biphenyldicarboxylic acid) with N, N' -dimethylethyle nediamine (mmen) via the coordination bonds formed between the OMSs (as illustrated in Fig. 3a). on Mg_2 (dobpdc) and amine groups [19]. The foregoing examples all concern alkylamine modification on Mg_2 (dobpdc), an expanded variant of Mg-MOF-74, because the sizes of the alkylamines molecules necessitate an isostructure with larger channels to avoid steric hindrance or intermolecular hydrogen bond formation [18]. For each (mmen) molecule, one end is coordinated to an unsaturated Mg center, while the other end is dangling in the channel to capture CO_2 in a chemisorptive manner (Fig. 3b). The high density of amine groups in mmen-Mg_2 (dobpd) results in its good selectivity for CO_2 even in the presence of moisture [20]. Compared with regular MOF sorbents, the adjustable phase-change step in the CO_2 isotherm of mmen-M_2 (dobpdc) help it successfully circumvent the disadvantage of large temperature swings in regeneration (Fig. 3d), making it extremely favorable for industrial CO_2 capture [18].

MOF can be used either as an adsorbent or as a membrane material. In the case of adsorbents, the main effort is to improve the affinity of carbon dioxide or O_2. This enhances the selectivity of oxygen fuel capture after combustion. Some CO_2 selective MOFs modified by alkylamines exhibit high CO_2 working power and are well recoverable under the desired capture conditions. Additionally, their performance is not affected by the presence of moisture, suggesting that they may be used in real industrial cases. With long-term stability and drug resistance in flue gas, as well as reduced regenerative energy and significant synthesis costs, these MOF adsorbents are likely to play a critical role in coffers capture.

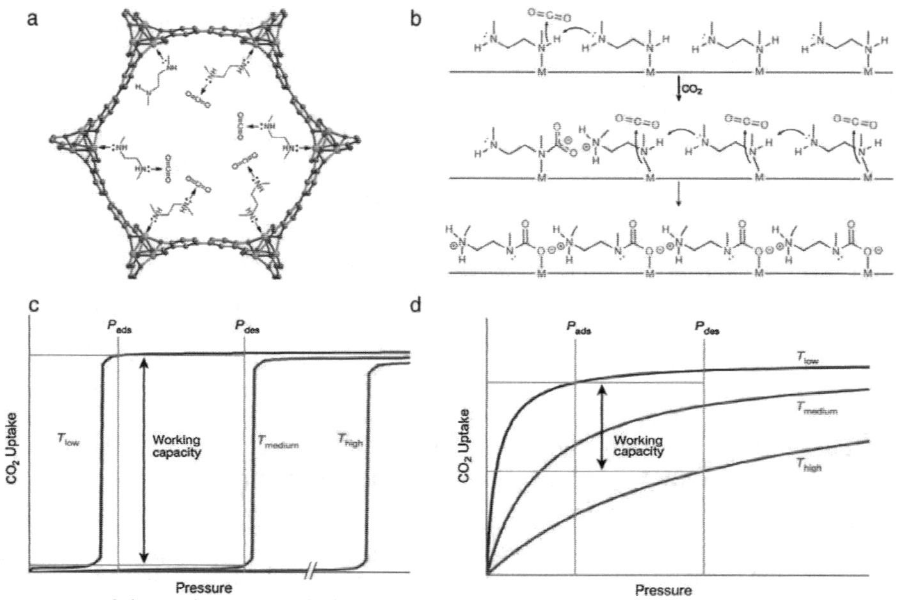

Fig.3 Structure of the amine-appended CO_2 adsorbent mmen-Mg_2(dobpdc).. (b) Illustration of the cooperative insertion mechanism for CO_2 adsorption in mmen-M_2(dobpdc). Comparison of the idealized CO_2 adsorption isotherms of (c) a phase-change adsorbent and (d) a typical microporous adsorbent at different temperatures.

2.4 Toxic gas separation

The development in storage and separation of toxic gases (e.g., carbon monoxide (CO), ammonia (NH_3) and sulfur dioxide (SO_2)) that are closely related to health, environment conservation and industrial production, has been aroused [21].

2.4.1 CO purification

CO has large impact on the production of ammonia and fuel cells by poisoning the catalysts. Moreover, the removal of CO from H_2 in syngas is considerably significant. For the increasing demand of a more power-saving porous materials for CO purification, MOF adsorbents can be a choice at ambient temperature [11].

2.4.2 NH_3 removal

NH_3 is highly toxic, and it's crucial to remove and achieve adsorptive separation of NH_3 from air [22]. As a reducing agent, NH_3 is promising to be better adsorbed by the functionalization of pore surface of a MOF by introducing functional groups with redox ability. Creating increased density of open metal sites in MOF materials while maintaining the structural stability can be achieved by using postsynthetic synthesis methods. A series of effective MOFs are produced for efficient NH_3 removal, including HKUST-1, Zn(INA)$_2$ and a series of UiO-66 analogs [4].

2.4.3 SO_2 removal

Residual SO_2 in flue gas can react with amines and thus the efficiency of the CO_2 scrubbing process will decrease. Complete removal of traces of SO_2 from flue gas is therefore of fundamental importance. Reversible physisorption can be theoretically efficient for the removal. Theoretical analysis listed two possible adsorption configurations of adsorbed SO_2 molecules. Moreover, interactions between SO_2 molecules and the hydroxyl groups or aromatic rings on the adsorbents can be capitalized to rationalize the observed selectivity [23]. Representing MOFs include NOTT-202a, MFM-300(In) ad SIFSIX MOFs [4].

3. Problems and challenges for using MOFs in gas separation

Although MOFs hold great promise for gas separation applications, there still have some challenges and difficult tasks to deal with. Firstly, it is not practical for scientists to easily find a high-performance MOFs material for gas separation. Although people only want to detect the separation of a specific and single gas, it takes several weeks to conduct experimental tests and completely unlock the performance of a given material [18]. This testing process is not high-performing, and researches may be unable to select the appropriate and best MOFs materials immediately.

The second challenge is the high cost of MOFs materials (e.g., Mg-MOF-74), which is quite expensive. The shaping process is also a challenge for the industry. In addition, MOF shaping processes is pretty essential and important for the development of new MOF products.

Scientists spend a lot efforts on air separation, mainly focusing on the O_2/N_2 separation. Because it is important for the carbon capture in process of oxyfuel combustion [18]. On the other hand, this separation is one of the most difficult separations. Because oxyfuel combustion is the ignition of fossil fuels in almost pure oxygen condition. After removing water from the gas that are generated, the generated flue gas is almost CO_2 entirely [24], which makes the CO_2 capture step much easier . In this method, the separation goal is to produce pure O_2 from air. Nevertheless, the typical low-temperature distillation requires a large amount of energy input. If it is to be widely implemented to capture carbon dioxide from oxygen fuel combustion, this solution is not feasible [24]. Therefore, society needs a strategy that consumes less energy but can still effectively separate O_2/N_2. On the other hand, there is only one computational study on predicting separation performance of MOF membranes in which membrane selectivity of 44 MOFs were found to be in the range of 1.5–11 [25]. As a result, even though the number of MOFs has been rapidly increasing and we still do not know separation performance limits of pure MOF membranes and MOF/polymer MMMs they have [26].

4. Conclusion

In summary, MOFs materials are strongly associated with the whole of human society with diverse fields. In this paper, we discuss the usage of MOFs in selective gas separation, including the practical applications of the materials in N_2/O_2 separation, C_2H_2/C_2H_4 separation, carbon capture, and toxic gas treatment, as well as the current problems and challenges of MOFs. In the field of carbon capture, the performance of MOFs materials is not influenced by water, indicating that they can have excellent performance in practical industry. However, inefficient testing, expensive costs and unknown separation performance limits make MOFs still have a lot of room for development in the future. In the future, researchers should focus on solving the above-mentioned shortcomings and develop more efficient and economical MOFs materials for large-scale industrial use. This paper puts forward some suggestions for the wide use of MOFs materials and the further development trend of MOFs materials. These results pave a path for the improvement of MOF materials.

References

[1] H. Li, K. Wang, Y. Sun, C.T. Lollar, J. Li, H.-C. Zhou, Recent advances in gas storage and separation using metal–organic frameworks, Materials Today 21(2) (2018) 108-121.

[2] T.M. McDonald, W.R. Lee, J.A. Mason, B.M. Wiers, C.S. Hong, J.R. Long, Capture of Carbon Dioxide from Air and Flue Gas in the Alkylamine-Appended Metal–Organic Framework mmen-Mg (dobpdc), Journal of the American Chemical Society (2012) Medium: X; Size: p. 7056-7065.

[3] J.R. Li, R.J. Kuppler, H.C. Zhou, Selective gas adsorption and separation in metal-organic frameworks, Chem Soc Rev 38(5) (2009) 1477-504.

[4] B. Wang, L.-H. Xie, X. Wang, X.-M. Liu, J. Li, J.-R. Li, Applications of metal–organic frameworks for green energy and environment: New advances in adsorptive gas separation, storage and removal, Green Energy & Environment 3(3) (2018) 191-228.

[5] W.G. Guimaraes, Jr., G.F. de Lima, Investigating greenhouse gas adsorption in MOFs SIFSIX-2-Cu, SIFSIX-2-Cu-i, and SIFSIX-3-Cu through computational studies, J Mol Model 26(7) (2020) 188.
[6] D.F. Sava Gallis, K.W. Chapman, M.A. Rodriguez, J.A. Greathouse, M.V. Parkes, T.M. Nenoff, Selective O_2 Sorption at Ambient Temperatures via Node Distortions in Sc-MIL-100, Chemistry of Materials (2016) acs.chemmater.6b00249.
[7] M.V. Parkes, D. Gallis, J.A. Greathouse, T.M. Nenoff, Effect of Metal in $M_3(btc)_2$ and M_2 (dobdc) MOFs for O_2 /N_2 Separations: A Combined Density Functional Theory and Experimental Study, The Journal of Physical Chemistry C 119(12) (2015) 6556-6567.
[8] Enhanced O_2 Selectivity versus N_2 by Partial Metal Substitution in Cu-BTC, Chemistry of Materials (2015).
[9] Redox-Active Metal–Organic Composites for Highly Selective Oxygen Separation Applications, Advanced Materials 28(18) (2016).
[10] E.D. Bloch, W.L. Queen, R. Krishna, J.M. Zadrozny, C.M. Brown, J.R. Long, Hydrocarbon Separations in a Metal-Organic Framework with Open Iron(II) Coordination Sites, Science 335(6076) (2012) 1606.
[11] Reversible CO binding enables tunable CO/H_2 and CO/N_2 separations in metal-organic frameworks with exposed divalent metal cations, Journal of the American Chemical Society 136(30) (2014) 10752-61.
[12] J.R. Li, J. Yu, W. Lu, L.B. Sun, H.C. Zhou, Porous materials with pre-designed single-molecule traps for CO_2 selective adsorption, Nature Communications 4 (2013) 1538.
[13] L. Jiang, P. Wang, M. Li, P. Zhang, J. Li, J. Liu, Y. Ma, H. Ren, G. Zhu, Construction of a Stable Crystalline Polyimide Porous Organic Framework for C_2H_2/C_2H_4 and CO_2/N_2 Separation, Chemistry – A European Journal 25(38) (2019) 9045-9051.
[14] Y. Chai, X. Han, W. Li, S. Liu, S. Yao, C. Wang, W. Shi, I. da-Silva, P. Manuel, Y. Cheng, L.D. Daemen, A.J. Ramirez-Cuesta, C.C. Tang, L. Jiang, S. Yang, N. Guan, L. Li, Control of zeolite pore interior for chemoselective alkyne/olefin separations, Science 368(6494) (2020) 1002-1006.
[15] R.-B. Lin, S. Xiang, W. Zhou, B. Chen, Microporous Metal-Organic Framework Materials for Gas Separation, Chem 6(2) (2020) 337-363.
[16] B. Li, X. Cui, D. O'Nolan, H.-M. Wen, M. Jiang, R. Krishna, H. Wu, R.-B. Lin, Y.-S. Chen, D. Yuan, H. Xing, W. Zhou, Q. Ren, G. Qian, M.J. Zaworotko, B. Chen, An Ideal Molecular Sieve for Acetylene Removal from Ethylene with Record Selectivity and Productivity, Advanced Materials 29(47) (2017) 1704210.
[17] S. Yang, A.J. Ramirez-Cuesta, R. Newby, V. Garcia-Sakai, P. Manuel, S.K. Callear, S.I. Campbell, C.C. Tang, M. Schröder, Supramolecular binding and separation of hydrocarbons within a functionalized porous metal–organic framework, Nature Chemistry (2014) Medium: ED; Size: p. 121-129.
[18] H. Li, K. Wang, Y. Sun, C.T. Lollar, J. Li, H.C. Zhou, Recent advances in gas storage and separation using metal–organic frameworks, Materials Today (2017) S1369702117302407.
[19] T.M. Mcdonald, W.R. Lee, J.A. Mason, B.M. Wiers, S.H. Chang, J.R. Long, Capture of carbon dioxide from air and flue gas in the alkylamine-appended metal-organic framework mmen-Mg_2(dobpdc), Journal of the American Chemical Society 134(16) (2012) 7056.
[20] T.M. Mcdonald, J.A. Mason, X. Kong, E.D. Bloch, J.R. Long, Cooperative insertion of CO_2 in diamine-appended metal-organic frameworks, Nature 519(7543) (2015) 303-8.
[21] Barea, E., Montoro, C., Navarro, J., R., Toxic gas removal - metal-organic frameworks for the capture and degradation of toxic gases and vapours, CHEMICAL SOCIETY REVIEWS (2014).
[22] A.J. Rieth, A.M. Wright, M. Dincă, Kinetic stability of metal–organic frameworks for corrosive and coordinating gas capture, Nature Reviews Materials 4(11) (2019) 708-725.

[23] M. Savage, Y. Cheng, T.L. Easun, J.E. Eyley, S.P. Argent, M.R. Warren, W. Lewis, C. Murray, C.C. Tang, M.D. Frogley, G. Cinque, J. Sun, S. Rudić, R.T. Murden, M.J. Benham, A.N. Fitch, A.J. Blake, A.J. Ramirez-Cuesta, S. Yang, M. Schröder, Selective Adsorption of Sulfur Dioxide in a Robust Metal–Organic Framework Material, Advanced Materials 28(39) (2016) 8705-8711.
[24] K. Sumida, D.L. Rogow, J.A. Mason, T.M. Mcdonald, E. Bloch, Z.R. Herm, T.H. Bae, J.R. Long, Carbon Dioxide Capture in Metalorganic Frameworks (2012).
[25] E. Gulcay, I. Erucar, Bio-compatible MOFs for Storage and Separation of O_2: A Molecular Simulation Study, Industrial & Engineering Chemistry Research (2019).
[26] H. Daglar, I. Erucar, S. Keskin, Exploring the performance limits of MOF/polymer MMMs for O_2/N_2 separation using computational screening, Journal of Membrane Science (2020) 118555.

Progress and trend of PbTe based Thermoelectric Materials

Yeshiyuan Zhou[1a*]

[1]Physics Department, Southern University of Science and Technology, Shenzhen, Guangdong Province, China, 518055,

[a*]11810627@mail.sustech.edu.cn

Abstract—Energy shortage and environmental pollution have become severe problems that can not be ignored in every country. As a new energy conversion material, thermoelectric material can realize thermoelectric conversion with the advantages of safety, energy saving and environmental protection. This paper summarizes the progress of some PbTe based thermoelectric materials in recent ten years, including the influence of doping elements such as Na, S, Sr and I on their thermoelectric properties, and introduces a method to balance their mechanical and thermoelectric properties. It can be seen that PbTe based thermoelectric materials have good development and broad application prospects in the future.

1. Introduction

Thermoelectric material is a functional material. It transfers heat and electric energy to each other. The Seebeck effect (revealed in 1821) and the Peltier effect (revealed in 1834) pave the way for the application of thermoelectric energy converters and thermoelectric refrigeration. Therefore, the material is very promising in market applications. Today, environmental pollution and energy crises are also threatening the survival of mankind at all times, which also shows that the research on this new type of thermoelectric material has very important practical significance.

The purpose of this paper is to summarize the previous methods to improve the properties of lead telluride based thermoelectric materials and the mechanism of performance improvement. This study promotes the understanding of the mechanism of thermoelectric materials and contributes to the further exploration of the mechanism of improving the properties of thermoelectric materials.

2. Thermoelectric effect

It shows three main principles of thermoelectric effect: Seebeck effect, Peltier effect and Thompson effect[1-6]. In practical application, "thermoelectric merit" (ZT) which is dimensionless is used to measure the thermoelectric properties of thermoelectric materials. The higher the ZT value, the higher the energy conversion efficiency of thermoelectric devices. The expression of ZT is as follows:

$$ZT = \frac{\sigma S^2}{\kappa} T = \frac{PF}{\kappa} T \qquad (1)$$

Where: σ Is conductivity; S is Seebeck coefficient; κ Is the thermal conductivity; T is the temperature; PF is a power factor used to characterize the electrical properties of thermoelectric materials.

The three parameters, including Seebeck coefficient, conductivity and thermal conductivity, affecting the thermoelectric properties of materials are not independent. They all depend on the electronic structure of materials as well as the transport and scattering of carriers. Therefore, scientists tend to look for materials with a high Seebeck coefficient, and while improving their conductivity, they expect to reduce their thermal conductivity to obtain a higher ZT value. Generally, researchers will dope thermoelectric materials with excellent properties. That helps to advance the carrier concentration and carrier mobility of the materials, improve their phonon scattering ability to some degrees, and cut down the thermal conductivity of these materials.

As a traditional thermoelectric material in the middle temperature range, PbTe based thermoelectric material has isotropic electrical and thermal properties, and the band gap is only 0.32ev, making it easy to show p-type or n-type conductivity after doping. It has been a research hotspot in thermoelectric materials since the middle of the 20th century. In the last two decades, both n-type and p-type PbTe based thermoelectric materials have made good development, as shown in Figure 1. Several doping improvements for lead telluride based thermoelectric materials will be introduced below.

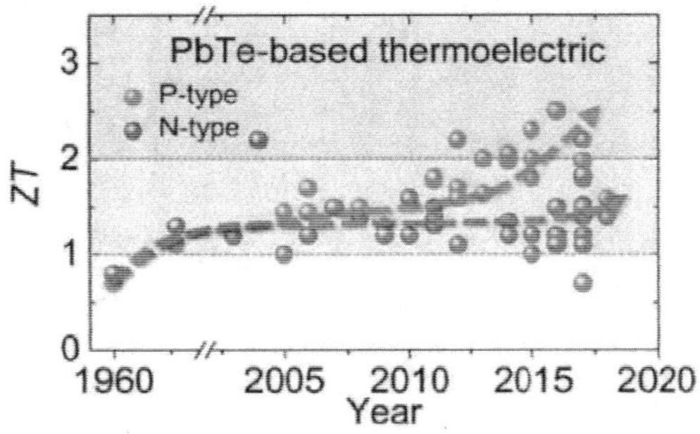

Figure 1 the developments of n-type and p-type PbTe based thermoelectric materials[7]

3. Several Doping of lead telluride based thermoelectric materials

3.1. Na-doped PbTe–PbS Thermoelectric Materials

Steven's work[8] confirms that by adding PbS and Na to PbTe, there are two effects: the change of the electronic structure and the decrease of the lattice thermal conductivity, which together contribute to an excellent quality factor. They build a successful thermoelectric material system that combined nanostructure to reduce lattice thermal conductivity and density-of-states(DOS) engineering to enhance electronic transport properties. The highest ZT value of 2% Na doped PbTe and PbTe PBS 12% is 1.8(shown in Figure 2).

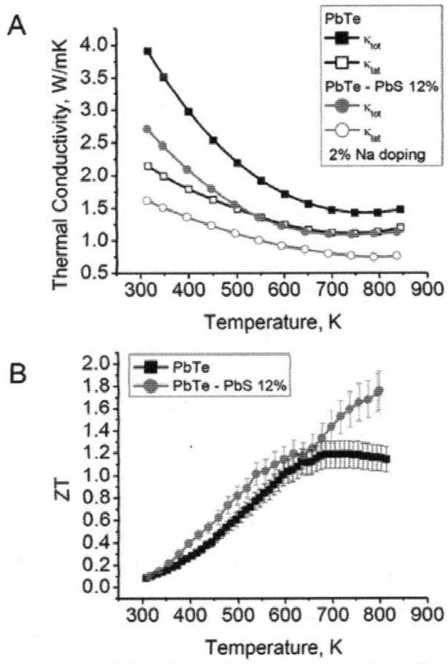

Figure 2 High-temperature (A) total and lattice thermal conductivity and (B) ZT of 2% Na-doped PbTe and PbTePbS 12%.

3.2. Na-Doped PbTe–SrTe Thermoelectric Material

In He's work[9], some conduct chalcogenide semiconductors, such as PbTe, PbSe, and PbS coped with various concentrations of Na lead to Na-rich precipitates, and boost the power factor. Later, the highest ZT of Na doped PbTe–SrTe system reached 2.2 in several researches[10-12](Figure 3 shows Biswas's work[10]). Recently, a research[13] studied the effect of annealing on the thermoelectric properties of PbTe–4Sr–2Na samples in detail. They proved that in addition to the correlation between the properties of the PbTe–4mol%SrTe sample and the thermoelectric properties, APT can also perform quantitative analysis in three dimensions.

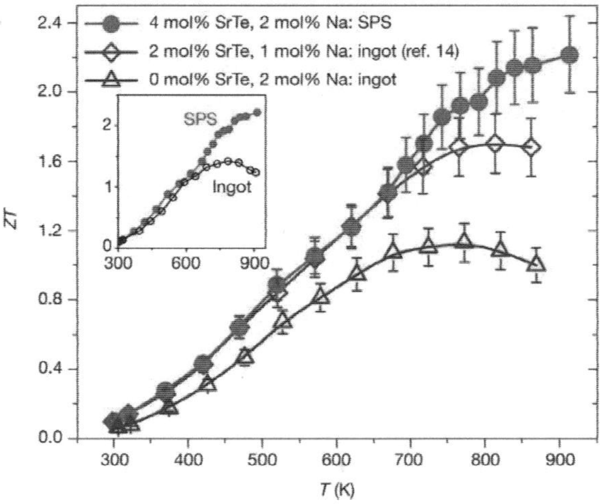

Figure 3 ZT of three methods in Biswas's work[10]

3.3. Effect of iodine doping on n-type PbTe based thermoelectric material

N-type iodine doped PbTe based thermoelectric materials are a hot direction in recent years as well. However, considering the intense sintering process, the sample in its final composition is often very different from the initial standard.

In 2011, Aaron et al. combined with the careful control of carrier density through iodine doping, revealed a significantly larger quality factor ZT of 1.4, which is higher than the previously reported n-type PbTe[14].

Later, the work of Juan et al. showed that in PbTe1-xIx, when x is 0.004 and 0.005, the ZT value is higher, and at this time it also has high conductivity, high power factor and low lattice thermal conductivity.They confirmed that iodine as valid donor in PbTe offers adequate carrier concentration with a handful of dopant. In addition, because the radii of I- and Te2- are similar, doping iodine has no significant effect on the lattice structure[15]. Figure 4 shows the ZT value and the comparison of others in their work.

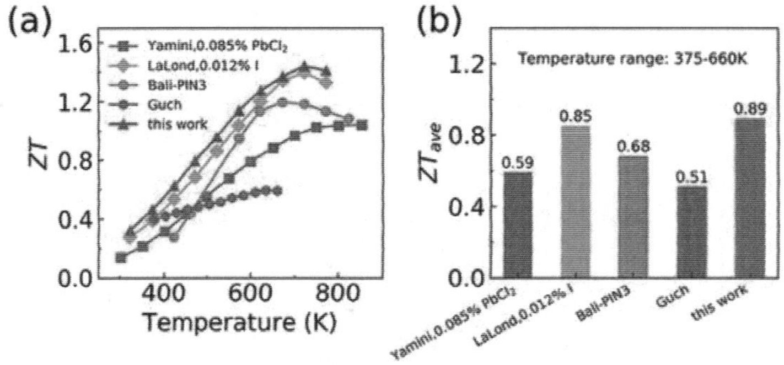

Figure 4 Comparison of the (a) ZT values and (b) average ZT values of Juan et al.' samples with other reported n-type PbTe samples doped with halides[15-18].

3.4. Balance the strong mechanical properties and high thermoelectric properties

Both p-type and n-type PbTe based thermoelectric materials enjoy perfect thermoelectric properties, although the mechanical properties are on the low side. In Fu's work, they chose to dope n-type PbTe-3%Sb material and improve the mechanical properties of the material by alloying it with PbS. According to their data, after alloying, the hardness of the sample increased by more than 60%, the microhardness increased from 54.2 kg mm-2 to about 88 kg mm-2 while the relative ZT value decreased by only 7.4%, still about 1.5(shown in Figure 5)[19]. It shows that a relatviely good balance is achieved between the mechanical properties and thermoelectric properties of PbTe materials. And this balance is conducive to the practical application in the thermoelectric power generation.The good balance between the mechanical properties and thermoelectric properties of PbTe materials is conducive to the practical application in the thermoelectric power generation field.

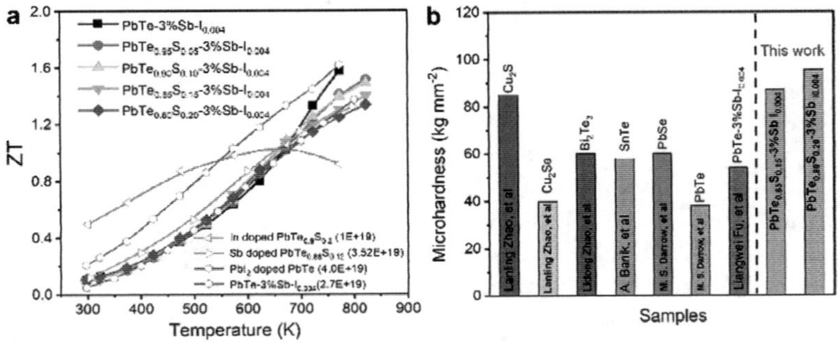

Figure 5 (a) Temperature dependence of the dimensionless figure of merit of PbTe1−xSx–3% Sb-I0.004 (x = 0.00, 0.05, 0.10, 0.15, and 0.20) samples. (b) Vickers microhardness of PbTe1−xSx–3% Sb-I0.004 (x = 0.00, 0.05, 0.10, 0.15, and 0.20) samples[19].

4. Conclusion

This paper discusses the changes of several doping methods on the properties of PbTe based thermoelectric materials. One method is Na-doped PbTe–PbS Thermoelectric Materials, in which way the ZT value of the material is improved by changing the electronic structure and reducing the lattice thermal conductivity. Other studies about Na-Doped PbTe–SrTe Thermoelectric Material showed that Na-rich precipitates can improve the power factor, resulting in a significantly high ZT value of PbTe – 4sr – 2Na system sample. Another method is iodine doping, which can greatly improve the carrier concentration of the sample without changing the crystal structure since the structure of I- and Te2- are similar. Additionally, alloying n-type PbTe-3%Sb with PbS can greatly improve the mechanical properties with little influence on the thermoelectric properties, which is beneficial to the practical application of thermoelectric devices in thermoelectric power generation.

As a new type of energy conversion material, thermoelectric materials can cleanly convert heat energy and electric energy into each other, and play an important role in reducing the consumption of fossil energy. In recent years, PbTe based thermoelectric materials have developed rapidly, and the ZT value has increased year by year. It is expected to achieve higher performance breakthroughs and large-scale commercial applications in the future.

Acknowledgement

First of all, I would like to thank my mentor professor He Jiaqing, who has provided me with a lot of help in guiding me into the field of thermoelectric materials and taught me a lot of knowledge and skills.

Then, much thanks i want to give to the seniors in our group for helping me a lot during my writing.

Thanks to the platform provided by South University of science and technology for my research in the field of thermoelectricity.

Thank mentor Han min for guiding me to write my paper.

References

[1] Jackson S. (2016) Thermoelectric n-type oxide materials for energy generation[D]. The University of Manchester (United Kingdom)

[2] Seebeck T J. (2010) Magnetische polarisation der metalle und erze durch temperatur-differenz[J]. Annalen Der Physik, 82(2): 133-160.

[3] Koumoto K, Wang Y, Zhang R, et al. (2010) Oxide thermoelectric materials: a nanostructuring approach[J]. Annual Review of Materials Research, 40:363-394.

[4] Peltier J C. (1834) Nouvelles expériences sur la caloricité des courants électrique[J]. Annales de chimie et de physique,56

[5] Shakouri A. (2011) Recent developments in semiconductor thermoelectric physics and materials[J]. Annual Review of Materials Research, 41.

[6] Shakouri A, Zebarjadi M. (2019) Nanoengineered materials for thermoelectric energy conversion[M]. Thermal Nanosystems and Nanomaterials,118:225-299.

[7] Yu, Xiao; Li-Dong, Zhao. NPJ Quantum Materials;London Vol. 3, Iss. 1, (2018). DOI:10.1038/s41535-018-0127-y

[8] Steven N. Girard, Jiaqing He, Xiaoyuan Zhou, Daniel Shoemaker, Christopher M. Jaworski, Ctirad Uher, Vinayak P. Dravid, Joseph P. Heremans, and Mercouri G. (2011) Kanatzidis Journal of the American Chemical Society 133 (41), 16588-16597 DOI: 10.1021/ja206380h

[9] He, J.; Zhao, L.-D.; Zheng, J.-C.; Doak, J. W.; Wu, H.; Wang, H.-Q.; Lee, Y.; Wolverton, C.; Kanatzidis, M. G.; Dravid, V. P. (2013) Role of Sodium Doping in Lead Chalcogenide Thermoelectrics J. Am. Chem. Soc. 135, 4624– 4627 DOI: 10.1021/ja312562d

[10] Biswas, K.; He, J. Q.; Blum, I. D.; Wu, C. I.; Hogan, T. P.; Seidman, D. N.; Dravid, V. P.; Kanatzidis, M. G. (2012) High-Performance Bulk Thermoelectrics with All-Scale Hierarchical Architectures Nature, 489, 414– 418 DOI: 10.1038/nature11439

[11] Pei, Y.; Tan, G.; Feng, D.; Zheng, L.; Tan, Q.; Xie, X.; Gong, S.; Chen, Y.; Li, J.-F.; He, J.; Kanatzidis, M. G.; Zhao, L.-D. (2017) Integrating Band Structure Engineering with All-Scale Hierarchical Structuring for High Thermoelectric Performance in PbTe System Adv. Energy Mater. 71601450 DOI: 10.1002/aenm.201601450

[12] Tan, G.; Shi, F.; Hao, S.; Zhao, L.-D.; Chi, H.; Zhang, X.; Uher, C.; Wolverton, C.; Dravid, V. P.; Kanatzidis, M. G.(2016) Non-Equilibrium Processing Leads to Record High Thermoelectric Figure of Merit in PbTe–SrTe Nat. Commun. 7, 12167 DOI: 10.1038/ncomms12167

[13] Yoon-Jun Kim, Li-Dong Zhao, Mercouri G. Kanatzidis, and David N. (2017) Seidman ACS Applied Materials & Interfaces 9 (26), 21791-21797 DOI: 10.1021/acsami.7b04098

[14] A. D. LaLonde, Y. Pei, and G. J. Snyder, Energy Environ. (2011)Sci. 4, 2090. https://doi.org/10.1039/c1ee01314a

[15] Juan Cui, Meimei Wang, Xiao Xu, Yue Chen, and Jiaqing He. (2019) Journal of Applied Physics 126, 025108; https://doi.org/10.1063/1.5101034

[16] A. Bali et al. (2016) J. Appl. Phys. 120, 175101. https://doi.org/10.1063/1.4965865

[17] S. Aminorroaya Yamini, H. Wang, D. Ginting, D. R. G. Mitchell, S. X. Dou, and G. J. Snyder. (2014) ACS Appl. Mater. Interfaces 6, 11476. https://doi.org/10.1021/am502140h

[18] M. Guch, C. R. Sankar, J. R. Salvador, G. P. Meisner, and H. Kleinke, J. Appl. (2012) Phys. 111, 063706. https://doi.org/10.1063/1.3694742

[19] L. W. Fu , J. Cui , Y. Yu , Y. Huang , Y. F. Wang , Y. Chen and J. Q. He , J. Mater. (2019) Chem. A, 7 , 6304—6311 DOI:10.1039/C9TA00400A

The formation mechanism of the line segregation in twin-roll casting (TRC) 3003 aluminium alloy

Xiaoming Qian[1], Longzhou Meng[1], Xue Sheng[2], Yong Li[2, 3, *], Zhaodong Wang[2, **]

[1]National Engineering Research Center for Equipment and Technology of Cold Strip Rolling, Yanshan University, Qinhuangdao 066004, China

[2]State Key Laboratory of Rolling and Automation, Northeastern University, Shenyang 110819, China

[3]Guangxi Advanced Aluminum Processing Innovation Center Co. LTD, Nanning 530007, China

*Corresponding author's e-mail: liyong@smm.neu.edu.cn, zhdwang@mail.neu.edu.cn

Abstract. The formation mechanism of the line segregation in twin-roll casting (TRC) 3003 aluminium alloy was investigated. Results revealed that line segregations exhibited in the as-cast microstructures of 3003 TRC strips, depending on the different electromagnetic parameters, the position of the line segregations appeared at the center, lower, upper layers of the strips. The formation of the line segregations took place during the solidification, alloying elements were pushed towards and gathered at the center line and the resulted in the formation of the center line segregation. The dual-wave interference was the reason the changes of the position of the line segregation, when the weakened zone of the dual-wave interference appeared at the lower or upper layer of the melts, the position of the line segregation changed accordingly, when the weakened zone of the dual-wave interference appeared at the center line, the line segregation diminished.

1. Introduction

Twin-roll casting (TRC) is a technique of which cast strips are produced by a combined process of casting and rolling [1, 2, 3]. Despite TRC possesses advantages of low production costs [4, 5] and high cooling rates that refines the microstructures [6, 7], macro segregations are often inevitable during TRC [8, 9, 10]. Center-line segregation appears in the TRC strip and bring negative influences on the mechanical properties and corrosion resistance of the sheet [11].

In the study carried out by You et al. [12], 3003 aluminium alloy was produced via TRC, in which typical center-line segregation was observed. By generating electromagnetic on the solidification field, the position of the center-line segregation moved toward upper-surface or bottom surface. However, no detailed explanation was provided.

In this study, the phenomena of center-line segregation in the study of You is demonstrated. Based on the electromagnetic control of TRC process, the detailed explanation of the formation of center-line segregation and the position change of center-line segregation was provided. The emphasis of this study was placed on the effect of wave interference on the center-line segregation in TRC aluminium alloys.

2. Experimental

3003 alloy was prepared by You et al. [14], the parameter in detail was as follow. The casting was conduct using a horizontal twin roll casting equipment built by EPM of Northeast University, and the roll 500 in diameter and 500mm in length, the maximum rolling speed was 7m/min, maximum rolling force was 70kN. The pulse power supply used in the experiment is SPMD series pulse power supply, with the model of ys9000-151000, specification 1000A/15V, input voltage 380V, input current 15A AC, output voltage 0~15V and output current 0~1000A. During the experiment, the lead out wires were respectively connected to the aluminum liquid in the launder and the outlet plate. The required parameters are obtained by adjusting. The peak values of pulse current used in this experiment are 300A and 200A. The excitation current of this experiment is 250A and 200A. The casting temperature was 703~710℃, and strip was obtained with thickness of 6mm. Microstructure observation was performed using optical microscope.

3. Results and discussion

3.1. Microstructures of the 3003 strip produced by TRC

The microstructures of the TRC 3003 strip are shown in figure 1. In general, elongated grains in the strip was observed under all the electromagnetic conditions, this is due to the rolling forced performed during TRC, resulting deformation of the strip, and the deform structure was retained in the strip. Besides, a very unique feature, line segregations were observed along with the rolling direction, as is shown in figure 1 (a), (b) and (c). Under the condition of IE=300A and IM=250A (figure 1(a)), the line segregations was placed at the near center line position, the line segregation exhibited not perfectly straight, but with the curved and forked morphology. When the parameter switched to IE=200A and IM=250A, the line segregations was moved towards lower surface of the strip (figure 1(b)), and the line segregation shows the morphology with straighter and thinner shape, compare with that of figure 1(a). When the parameter changed as IE=200A and IM=200A, the line segregation appeared as the similar morphology as that of figure 1(b), however, the position of the line segregation moved to the upper layer near the surface (figure 1(c)). As the parameter was set as IE=200A and IM=15A, no line segregation appeared in the microstructures (figure 1(d)), which indicated that the line segregation had been diminished by the electromagnetic field in this condition.

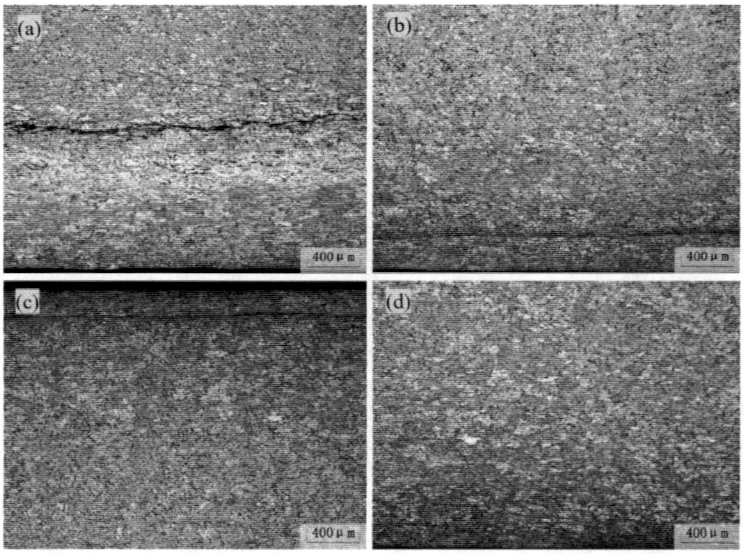

Figure 1. Microstructures of the 3003 strip of produced by TRC under different electromagnetic parameters conditions: (a) IE=300A IM=250A, (b) IE=200A IM=250A, (c)IE=200A IM=200A, (d)IE=200A IM=150A.

3.2. Formation mechanism of the line segregations

In general, the formation of eutectic and its size and fraction was closely related to the solidification conditions [12]. The line segregations was formed when solution atoms in the melt was gathered in the front of the solid/liquid interface, which resulted in more alloying elements reacted and formed eutectics. For the TRC process (figure 2), solidification took place at the roll surface as the melt encountered the rolls. Liquid zone, mushy zone and deformation zone were displayed in the TRC process. Along with the solidification process, alloying elements were pushed toward the center line of the solidification, resulting in the alloying elements were gathered in the front of liquid/solid interface. Consequently, the center line segregation formed and retained in the strip of 3003 alloy (figure 1(a)).

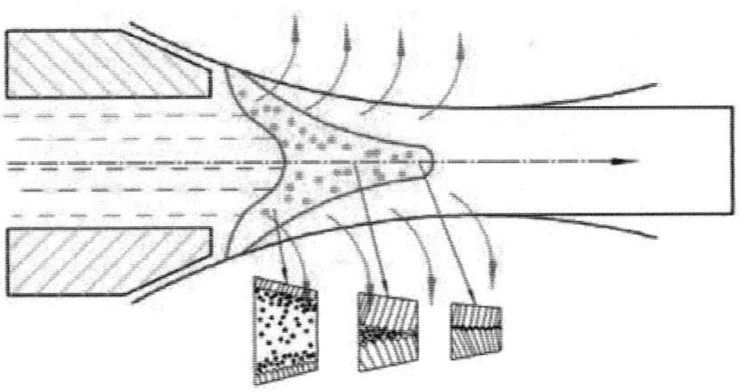

Figure 2. Line segregations interfered by the electromagnetic field.

In the experimental, the electromagnetic field was applied during TRC. Theoretically, one columns of simple harmonics (figure 3(a)) could be generated by the electric current in the melt of alloy. However, in the actual condition the dual-wave interference might occur (figure 3(b)).

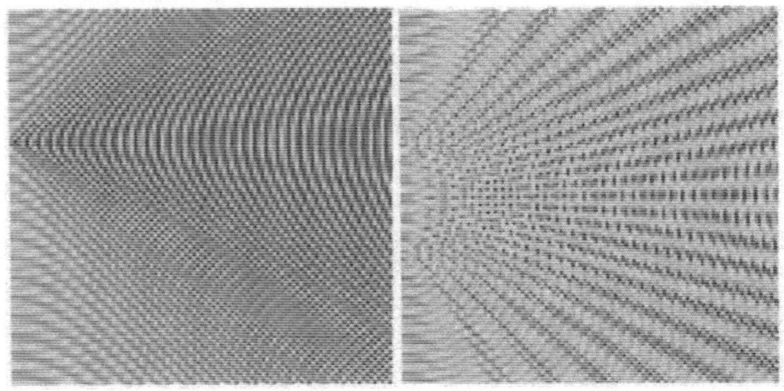

Figure 3. Schematic of (a) simple harmonics and (b) dual-wave interference.

When the dual-wave interference took place, strengthened area weakened area were arranged alternately. In the TRC solidification area, the alloying elements in the strengthened area were tend to be expelled to the weakened areas. For the condition of figure 1(b), in which the position of the line segregation moved to the lower layer near the surface, the weakened zone was at the lower position of the solidification region (figure 4(a)); while at the condition of figure 1(c), in which the position of the line segregation moved to the upper layer near the surface, the weakened zone was at the upper position of the solidification region (figure 4(b)). When the weakened zone was exactly at the center line of the solidification (figure 4(c)), the strengthened zone broken the center line segregation of the alloying elements, and resulted in the diminishment of the center line segregation in the as-cast microstructures (figure 1(d)).

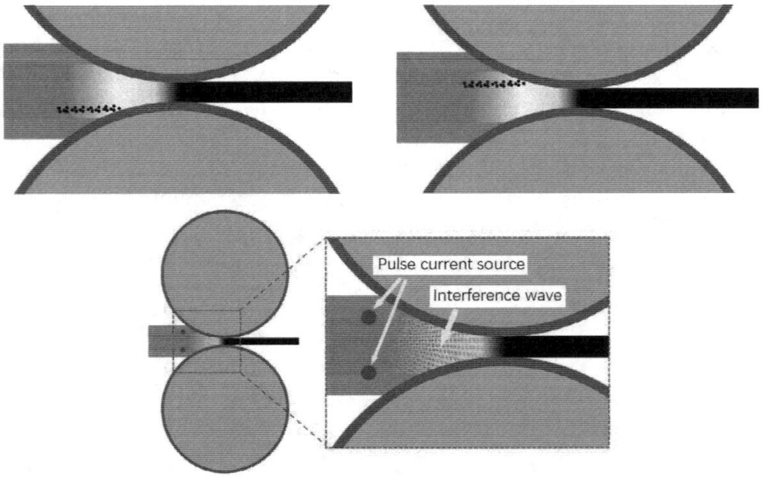

Figure 4. Schematic of the dual-wave interference and the distribution of the alloying elements.

4. Conclusion

1. Line segregations exhibited in the as-cast microstructures of 3003 TRC strips, and depending on the different electromagnetic parameters, the position of the line segregations appeared at the center, lower, upper layers of the strips.

2.The formation of the line segregations took place during the solidification, alloying elements were pushed towards and gathered at the center line and the resulted in the formation of the center line segregation.

3. The dual-wave interference was the reason the changes of the position of the line segregation, when the weakened zone of the dual-wave interference appeared at the lower or upper layer of the melts, the position of the line segregation changed accordingly, when the weakened zone of the dual-wave interference appeared at the center line, the line segregation diminished.

Acknowledgments

The authors gratefully acknowledge the support of the National Natural Science Foundation of China (Grant No. 51790485), the Fundamental Research Funds for the Central Universities (No. N2007008), Nanning science and technology major special projects (No. 20201041) and the Key research and development project of Shandong province (No. 2019JZZY010401).

References

[1] H. Zhao, P. Li, L. He, "Coupled analysis of temperature and flow during twin-roll casting of magnesium alloy strip," Journal of Materials Processing Technology, vol. 211, pp. 1197-1202, 2011.
[2] J. Zeng, R. Koitzsch, H. Pfeifer, B. Friedrich, "Numerical simulation of the twin-roll casting process of magnesium alloy strip," Journal of Materials Processing Technology, vol. 209, no. 5, pp. 2321-2328, 2009.
[3] H. Watari, T. Haga, N. Koga, K. Davey, "Feasibility study of twin roll casting process for magnesium alloys," Journal of Materials Processing Technology, vol. 192–193, p. 300–305, 2007.
[4] E. E. M. Luiten, K. Blok, "Stimulating R&D of Industrial Energy-Efficient Technology; The Effect of Government Intervention on the Development of Strip Casting Technology," Energy Policy, vol. 31, pp. 1339-1356, 2003.
[5] G. Kurz, J. Bohlen, D. Letzig, K. U. Kainer, "Influence of Process Parameters on Twin Roll Cast Strip of the Alloy AZ31," Materials Science Forum, vol. 765, pp. 205-209, 2013.

Journal of Physics: Conference Series **2194** (2022) 012007 doi:10.1088/1742-6596/2194/1/012007

[6] H. S. Di, Y. L. Li, Z. L. Ning, Z. Li, X. Liu, G. D. Wang, "New Processing Technology of Twin Roll Strip Casting of AZ31B Magnesium Strip," Materials Science Forum, Vols. 488-489 , pp. 615-618, 2005.
[7] M. Aljarrah, E. Essadiqi, D. H. Kang, I. H. Jung, "Solidification microstructure and mechanical properties of hot rolled and annealed Mg sheet produced through Twin Roll Casting route," Materials Science Forum, vol. 690, pp. 331-334, 2011.
[8] S. A. Lockyer, M. Yun, J. D. Hunt, D. V. Edmonds, "Micro- and Macrodefects in Thin Sheet Twin-Roll Cast Aluminum Alloys," Materials Characterization, vol. 37, pp. 301-310, 1996.
[9] C. Gras, M. Meredith, J. D. Hunt, "Micro Defects Formation During the Twin-Roll Casting of Al-Mg-Mn Aluminum Alloys," Journal of Materials Processing Technology, vol. 167, pp. 62-72, 2005.
[10] S. S. Park, Y. S. Park, N. J. Kim, "Microstructure and Properties of Strip Cast AZ91 Mg Alloy," Metals and Materials International, vol. 8, no. 6, pp. 551-554, 2002.
[11] S. S. Park, W. J. Park, C. H. Kim, B. S. You, N J. Kim, "The Twin-Roll Casting of Magnesium Alloys," JOM, vol. 61, no. 8, pp. 14-18, 2009.
[12] Tao You, "Study on electromagnetic roll casting and subsequent heat treatment of al-1.3% Mn and al-1.8% Mn aluminum alloys", Master's thesis, Northeastern University, PP 29-30, 2017.

Interfacial microstructures and Grinding performance of Synthetic Diamond Grits brazed by Cu-22Sn-3Cr Alloy

Lihua Gao[1], Yiqing Yu[1], Nian Duan[1], Dekui Mu[1, 2] and Mian Li [2*]

[1] School of Mechanical Engineering and Automation,

[2] Institute of Manufacturing Engineering, Huaqiao University, Xiamen 361021, China

*Corresponding author's e-mail: limian1758@stu.hqu.edu.cn

Abstract. To ensure the reliable bonding and low thermal damage of brazed diamond grits, a Ni-free Cu-Sn-Cr filler alloy was developed based on our previous study. Interfacial microstructure of diamond grits brazing used by Cu-22Sn-3Cr (in wt. %) filler alloy at 950 °C for 5, 10, 20, and 40 min were systematically investigated. Only Cr_7C_3 was generated at the interface between diamond and Cu-22Sn-3Cr alloy, and the thickness of Cr_7C_3 was increased after extended brazing time. The grinding performance of diamond tools brazed using Cu-22Sn-3Cr alloy has fewer ratios of diamond grits with breakage and fall-off failures and thus was found to be superior to that brazed by Ni-14Cr-10P filler alloy. In addition, alloy composition has a more significant influence than brazing time on the interfacial microstructure and mechanical behavior of brazed diamond grits; and the higher mechanical behavior of diamond grits brazed using Cu-Sn-Cr filler alloy is proved to be closely related to the low solubility of C in Cu-Sn-based alloy. Results obtained in this work are also crucial for developing a diamond brazing technique to fabricate diamond-related materials and devices.

1. Introduction

Owing to its superior physical and chemical properties, diamond abrasives are widely utilized for fabricating abrasive tools, which are essential for cutting, grinding, and polishing hard and/or brittle materials [1, 2]. Brazing has been regarded as a promising technique for fabricating abrasive tools due to its high bonding strength and large protrusion [3-5]. Up to date, two types of active alloys are widely utilized for brazing diamond: the Ti-containing Cu-Sn [6-8] or Ag-Cu [9-11] alloys and the Cr-containing Ni-Cr alloys [11-16]. The Ni-Cr-based filler alloy exhibits moderate activity, high mechanical strength, superior wear resistance and has been used for diamond brazing since the 1970s [12]. Although great research effort [14-16] has been made to optimize brazing strategies of Ni-Cr filler alloys, however, extensive formation of interfacial carbides and graphite or amorphous carbon remained to be a concern and the performance of diamond abrasive tools could be considerably impaired [11, 13]. Our previous studies indicated that the use of Ni-free low-temperature Sn-Cr [17], Sn-Cu-Cr [18] alloy can significantly ensure the integrity of brazed diamond grits. Nevertheless, these Sn-rich alloys exhibit undesirable grinding performance because of the low resistance to heat and wear. Obviously, the development of a novel Ni-free alloy is a possible option to overcome these shortcomings associated with Ni-Cr and Sn-rich alloys. Based on the Sn-Cu phase diagram, it is an efficient and effective means to adjust the mechanical properties of Cu-Sn alloy by manipulating Cu content. However, Lu et al. [19] found that the introduction of Cu-Sn elements to Ni-Cr-based filler alloy would affect the formation and morphology of interfacial chromium carbide. Therefore, a comprehensive understanding of the interface reaction and mechanical property of diamond grit brazed using Cu-rich Cu-22Sn-3Cr filler alloy is still required.

In this work, the interfacial microstructures of diamond grits brazed by a newly developed Cu-22Sn-3Cr alloy at 950 °C for 5, 10, 20, and 40 min were systematically investigated. Only Cr_7C_3 carbides were formed at the interface due to the low C dissolubility in Cu-Sn filler alloy, which was revealed to be responsible for the superior mechanical properties of diamond grits brazed by Cu-Sn-Cr alloy. In addition, diamond abrasive tools were fabricated for sapphire grinding. The grinding performance of diamond abrasive tools brazed by Cu-22Sn-3Cr filler alloy was significantly higher than that brazed by commercial Ni-14Cr-10P filler alloy. In addition, it is indicated that the composition of filler alloy played a more important role than brazing time in the interfacial microstructures and the impairment of diamond mechanical behavior.

2. Experimental details

In this study, Cu-22Sn-3Cr (in wt.%) filler alloys were prepared by mechanically mixing a certain amount of Sn, Cu-40Sn, and Cu-30Cr powders in a three-dimensional vortex mixer for 4 hours. The commercial Ni-14Cr-10P filler alloy was used for comparison. Before the sample was assembled, diamond grits (ISD-1650, ILJIN Co. Ltd., Korean) with grain sizes of 500 ~ 600 μm and the AISI 1045 steel substrate were cleaned in an ultrasonic acetone bath. As shown in Figure 1(a), the powder alloys and diamonds were placed on the substrate, and the assembled samples were placed in a quartz tube furnace with a vacuum better than 1×10^{-4} Pa and brazed at 950 °C for a certain period. After brazing, the samples were cooled to room temperature with the furnace. To explore the interface microstructure, the selected brazed specimens were installed in epoxy resin and then mechanically polished along the direction perpendicular to the brazing interface; some brazed specimens were deeply etched in aqua regia to remove the remaining filler alloy and steel matrix, to get the brazed diamond grits.

The sapphire grinding test was conducted on a HAAS OM-2A vertical machining center. As shown in Figure 1(b), a diamond abrasive milling tool with 20 mm outer diameter and 15 mm inner diameter, and 68 diamond grits were manually placed and brazed on each tool. Grinding parameters were n=15000 r/min, V_f=100 mm/min, a_p=0.2 mm. After grinding for 30 min, the number of failed diamond grits on the milling head was counted using optical microscopy, and each value was summed from 3 milling tools, i.e., 204 diamond grits were checked. Besides, in accordance with the JB/T 10987-2010 about 0.4 grams of brazed diamond grits after etching were tested in an impact tester (CYCJ-04A, Zhengzhou Research Institute of Abrasives & Grinding Co., Ltd., China) for 2000 times at the frequency of 2400 r/min. A Phenom ProX SEM installed with EDS, a Sigma 500 field emission SEM installed with EDS, an XPert Pro XRD, and a Raman microscope spectroscopy were used for microstructure detection and phase determination. A TEM (FEI TECNAI F20) was used to obtain interfacial microstructures of brazed diamond grits using samples prepared on an FEI Helios Nanolab-600i dual-beam focused ion beam (FIB) milling machine.

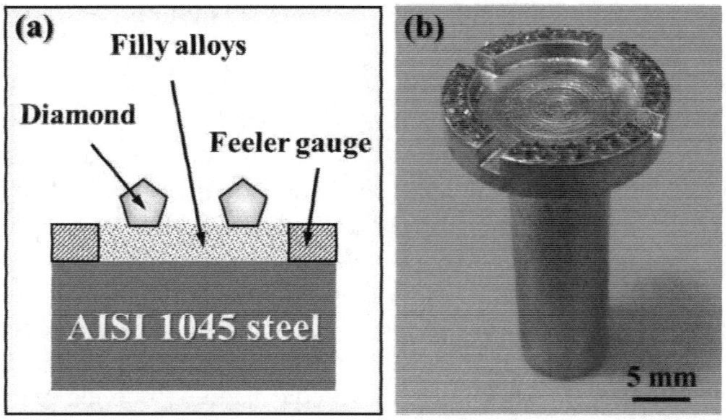

Figure 1: (a) Schematic diagram of assembled samples; (b) sample after brazing.

3. Results and discussion

3.1. Interfacial microstructures

Figure 2 (a)-(d) shows the cross-sectional interfacial microstructures of diamond grits brazed by Cu-22Sn-3Cr filler alloy at 950 °C for 5-40 minutes. After brazing for a short time of 5 min, a thin and discontinuous layer of the dark-grey interfacial product was generated at the interface between diamond grits and Cu-22Sn-3Cr filler alloy; when the brazing time was raised to 10 and 20 min, the interfacial product layer became continuous, but the thickness was slightly increased; a further elongation of brazing time to 40 min obviously increased the thickness of interfacial products. According to element distribution in Figure 2 (e), the concentration of Cr and C in the dark-grey interfacial phase was detected, indicating the formation of interfacial chromium carbides. Figure 3 (a)-(d) illustrates the SEM of diamond grits brazed by Cu-22Sn-3Cr filler alloy at 950 °C after the remaining filler alloy was removed by etching. It was found that as the brazing time was raised from 5 to 40 min, the coverage ratio of interfacial carbides was accordingly increased on the diamond surface. The morphologies of interfacial product in Figure 3 (a1)-(d1) revealed only a single layer of product generated at the reaction interface between diamond grits and Cu-22Sn-3Cr alloy, which was different from the formation of bilayers of chromium carbides on diamond brazed using Ni-Cr based filler alloy. Besides, the thickness of the interfacial product remained in a few microns, which is dramatically thinner than the thickness of chromium carbides at the reaction interface of diamond/Ni-Cr-P [13].

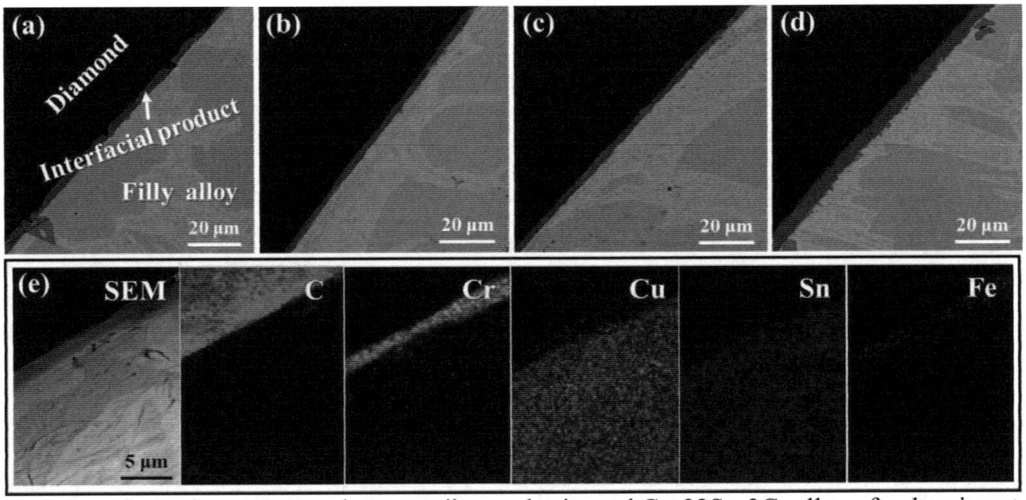

Figure 2: Interface microstructures between diamond grits and Cu-22Sn-3Cr alloy after brazing at 950 °C for (a) 5 min, (b) 10 min, (c) 20 min, and (d) 40 min; (f) microstructure and corresponded EPMA elemental mapping of interface brazed for 20 min.

Figure 4 (a) gives the XRD pattern obtained on brazed diamonds grits after remained alloy was removed by etching. As shown in the XRD pattern, only a Cr_7C_3 carbide was generated at the interface between diamond grits and Cu-22Sn-3Cr alloy, which was consistent with the monolayer interface product in Figure 3. To further understand the formation of chromium carbide, the TEM image of interfacial microstructures between diamond and Cu-22Sn-3Cr filler alloy after brazing at 950 °C for 20 min was obtained and shown in Figure 4 (b). The selected area HRTEM evidenced the formation of nano-sized Cr_7C_3 grains at the interface of diamond/Cu-22Sn-3Cr, which is consistent with the XRD result given in Figure 4 (a). The formation of interfacial Cr_7C_3 can be attributed to: unlike the carbon dissolving element Ni in Ni-Cr based alloy that catalyzes and dissolves diamond [11, 13], the solubility of C element in Cu-Sn alloy is limited [18]; in addition, the critical Cr concentration for Cr_7C_3 formation is easily reached in liquid Cu-Sn alloy [17, 18]. Consequently, these promote the formation of Cr-rich Cr_7C_3 chromium carbide.

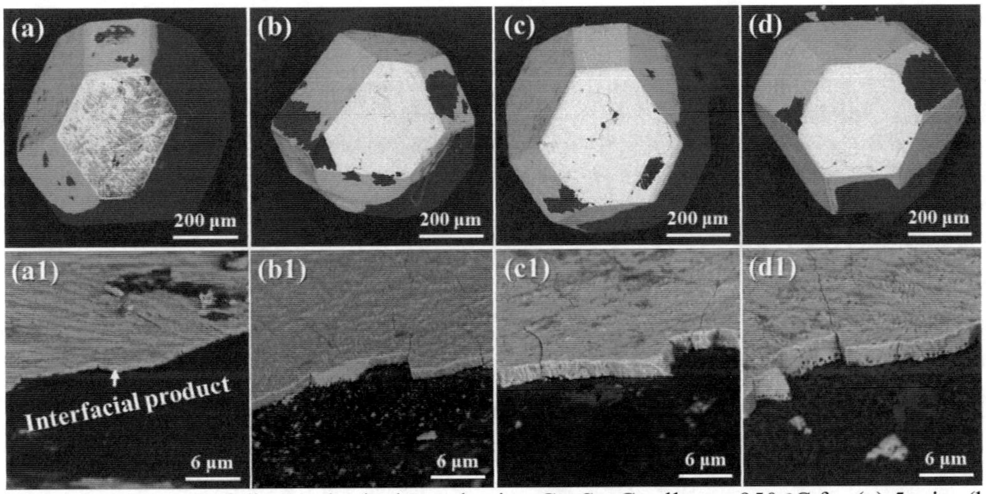

Figure 3: SEM images of diamond grits brazed using Cu-Sn-Cr alloy at 950 °C for (a) 5 min, (b) 10 min, (c) 20 min, and (d) 40 min; (a1-d1) morphologies of interfacial chromium carbides obtained from (a-d) diamond surface.

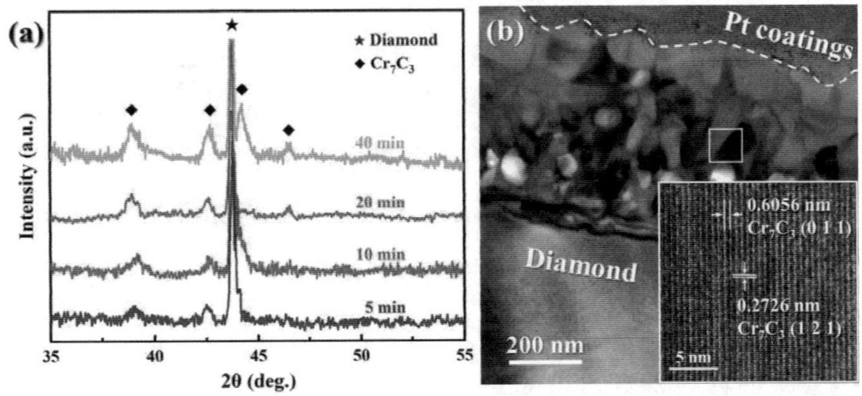

Figure 4: (a) XRD pattern of the interfacial product on diamond brazing used by Cu-22Sn-3Cr filler alloy at 950 °C for 5, 10, 20, and 40 min, (b) TEM and HRTEM image of interface brazed for 20 min.

3.2. Grinding performance

Improving the grinding performances of diamond abrasive tools is the fundamental motivation to develop novel filler alloys for diamond brazing. The breakage and fall-off in the grinding process are the primary failure modes of diamond grits and hence should be avoided as much as possible. Figure 5 is the statistics on breakage and fall-off failures and typical morphologies obtained from 204 brazed diamond grits after grinding sapphire for 30 min. As shown in Figure 5 (a), diamond grits brazed using Ni-14Cr-10P filler alloy for 20 min was found to have the highest percentage of 97.06 % in breakage failure, followed by 14.5 % of diamond grits brazing used by Ni-14Cr-10P filler alloy for 5 min. On the other hand, diamond abrasive brazed by Cu-Sn-Cr filler alloy was found to have a much lower probability of breakage failure than that of diamond abrasive brazed by Ni-Cr-P filler alloy, with the breakage frequency of 2.94 and 7.35 % for the diamond grits brazed using Cu-Sn-Cr alloy for 5 and 20 min. According to the typical morphologies of broken diamond grits brazed by Sn-Sn-Cr and Ni-Cr-P alloys in Figure 5 (b) and (c), respectively, brazed diamond abrasives were easily broken in the sapphire grinding process because of the ultra-high hardness of sapphire. However, the majority of diamond grits brazed by Cu-22Sn-3Cr filler alloy were intact, which indicates the excellent mechanical integrity of diamond grits. Moreover, as shown in Figure 5 (d), the highest percentage of fall-off failures was observed on diamond grits brazing used by Ni-14Cr-10P alloy at 950 °C for 20 min, and large diamond fragments are remaining in its fall-off pits as shown in Figure 5 (f). Therefore, it is reasonable to infer that the fall-off of diamond grits brazed using Ni-14Cr-10P filler alloy at 950 °C for 20 min was caused by grits fracture rather than the insufficient bonding strength. On the contrary, the percentage in the fall-off of diamond brazed by Cu-22Sn-3Cr alloy was decreased as the brazing time extended from 5 to 20 min because the interfacial reaction became more adequate. Last but not least, the diamond fragments on its pits were smaller, shown in Figure 5 (e).

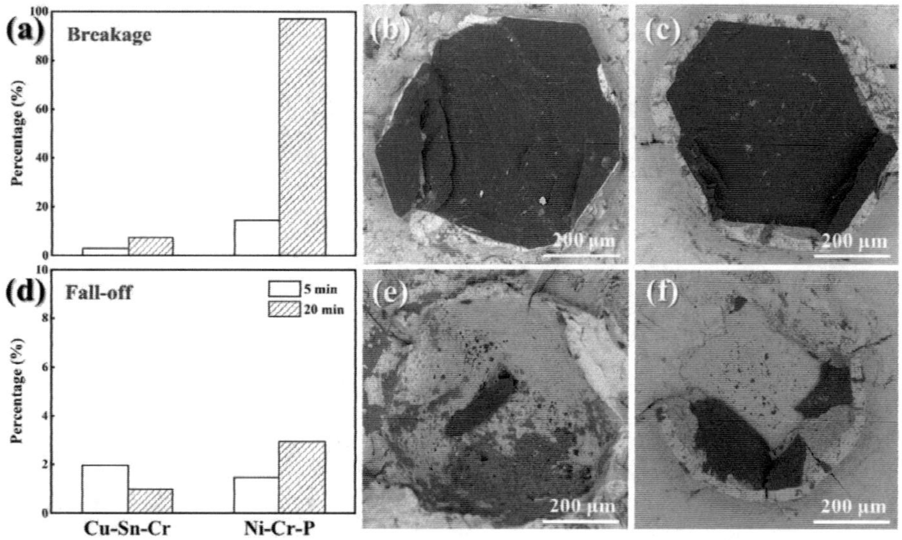

Figure 5: (a) Percentage of fracture failure of diamond grits brazing used by Cu-22Sn-3Cr and Ni-14Cr-10P alloy; morphology of abrasive fracture: brazed by (b) Cu-22Sn-3Cr and (c) Ni-14Cr-10P alloy at 950 °C for 20 min. (d) Percentage of fall-off failure of diamond grits brazing used by Cu-22Sn-3Cr and Ni-14Cr-10P alloy; morphology of abrasive fall-off: brazed by (e) Cu-22Sn-3Cr and (f) Ni-14Cr-10P alloy at 950 °C for 20 min.

Grinding is a complex process involving shearing, impacting, and squeezing [1, 2]. The brazing technology of diamond tools is therefore a comprehensive process that must balance the bonding strength and mechanical property of brazed diamond grits [3-5]. Among them, the occurrence of fall-off failure is due to the weak bonding strength, and the breakage of diamond grits is closely related to the mechanical property of brazed diamond. As shown in Figure 5 (d), the percentage in the fall-off of diamond grits brazed by Cu-22Sn-3Cr filler alloy for 5 and 20 min was as low as 1.96 % and 0.98 %, respectively, and that of diamond grits brazing used by Ni-14Cr-10P filler alloy for 20 min was the highest of 2.94 %. The ratios of diamond fall-off were much lower than that of diamond break, which demonstrates that the bonding strength between Cu-22Sn-3Cr alloy and diamond by generating Cr_7C_3 at the interface is reliable [6, 20, 21].

The impact toughness index (TI) of diamond grits under the corresponding brazed conditions was used to characterize the integrity of diamond grits brazed using the newly developed Cu-22Sn-3Cr and the commercial Ni-14Cr-10P filer alloys, and shown in Figure 6. The TI value of original diamond was 45.82 ± 0.5 %, which is reference data for evaluating the mechanical property of diamond grits brazed by different filler alloys in this study. As shown in Figure 6 (a), the TI value of diamond grits brazing used by Cu-22Sn-3Cr filler alloy at 950 °C for 5 and 20 min was slightly decreased to 39.42 ± 4.0 % and 33.31 ± 1.5 %, respectively. The variation of TI value can reasonably explain the slight worsening of the diamond break in Figure 5 (a). When Ni-14Cr-10P filler alloy was used, the TI value of diamond grits after brazing for a short time of 5 min was decreased to 24.81 ± 1.9 %. When brazed for 20 min, the TI value of diamond grits brazed using commercial Ni-14Cr-10P filler alloy was only 1.6 ± 0.6 %, and the diamond crushed material was pulverized entirely, as shown in Figure 6 (e). The almost disappeared mechanical integrity caused the severe breakage of diamond grits brazed by Ni-14Cr-10P filler alloy at 950 °C for 20 min, as shown in Figure 5 (a) and (c). Obviously, the influence of alloy composition is more remarkable than brazing time on diamond integrity. The above results fully indicate that the novel Cu-22Sn-3Cr alloy is superior to the commercial Ni-14Cr-10P alloy in terms of the mechanical property of diamond grits after brazing, which leads to the improvement of its grinding performance.

To further explore the variation of mechanical property of diamond grits after brazing used by different filler alloys, the phase information and corrosion of the diamond surface were examined. Figure 7 (a) is the Raman spectrums of original diamond and diamond grits brazed by the novel Cu-22Sn-3Cr and the commercial Ni-14Cr-10P filler alloys. Only the diamond peak was detected on diamond grits after brazing used using Cu-22Sn-3Cr at 950 °C for 5 and 20 min. However, when brazed using Ni-14Cr-10P filler alloy, a dome-shaped peak could be detected from 1355 cm^{-1} to 1530 cm^{-1} and became stronger as the brazing time was increased from 5 to 20 min. According to previous studies [8, 11, 15, 18], the existence of a dome-shaped peak probably corresponds to the formation of micro-graphite at 1355 cm^{-1} and amorphous carbon at 1530 cm^{-1}. In an ultrasonic alcohol bath, the interface product on the diamond surface after brazing was partially removed, as shown in Figure 7 (b)-(e). As shown in Figure 7 (b) and (c), when brazed using Cu-22Sn-3Cr alloy at 950 °C for 5 min, cavity morphologies were observed on the surface of the brazed diamond; when the brazing time was extended to 20 min, the cavity morphologies became denser. On the other hand, when brazed by Ni-14Cr-10P filler alloy, the corrosion surface of brazed diamond was relatively coarsened and denser than that brazed using Cu-22Sn-3Cr filler alloy under the same conditions as shown in Figure 7 (d) and (e). The corrosion morphology on the surface of diamond brazing used by Ni-Cr-based filler alloy was caused by the C dissolution from diamond into the liquid Ni, the formation of carbide, and the graphitization of diamond; meanwhile, the corrosion of diamond after brazing used by Cu-22Sn-3Cr filler alloy is mainly resulting from the formation of Cr_7C_3 [17, 18]. As a consequence, the variation of diamond integrity brazed by Cu-22Sn-3Cr and Ni-14Cr-10P filler alloys can be explained by the difference in phase diagram information and surface corrosion.

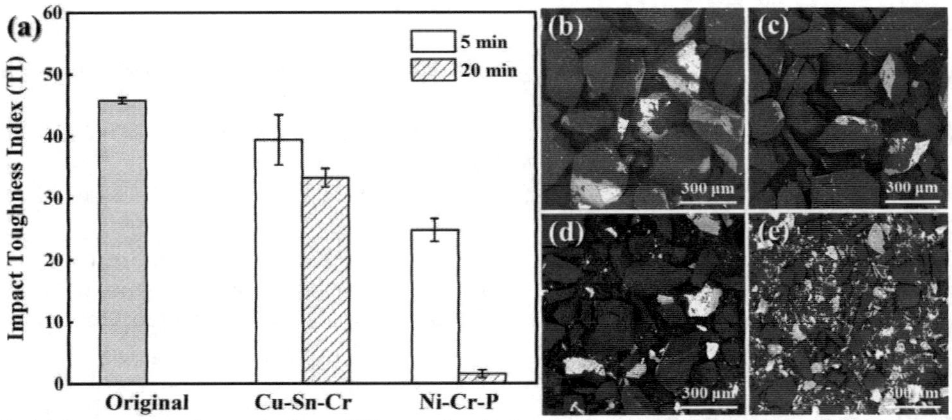

Figure 6: (a) Impact toughness index of diamond grits brazed by Cu-22Sn-3Cr and Ni-14Cr-10P alloy; morphology of diamond grits after impacting: brazed using Cu-22Sn-3Cr at 950 °C for (b) 5 min and (c) 20 min, brazed using Ni-14Cr-10P at 950 °C for (d) 5 min and (e) 20 min.

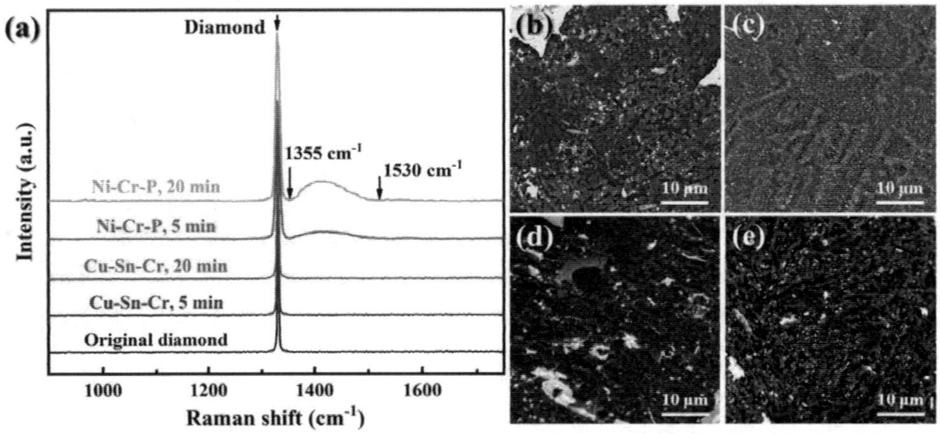

Figure 7: (a) Raman spectrums of diamond surface brazed by Cu-22Sn-3Cr and Ni-14Cr-10P alloy; corrosion surface of brazed diamond grits after deeply etching: brazed by Cu-22Sn-3Cr at 950 °C for (b) 5 min and (c) 20 min, brazed by Ni-14Cr-10P at 950 °C for (d) 5 min and (e) 20 min.

4. Conclusions

The synthetic diamond grits were successfully brazed using a newly developed Cu-22Sn-3Cr filler alloy in this work. After interface reaction at 950 °C for 5-40 minutes, only Cr_7C_3 carbide was detected at the interface between diamond grits and Cu-22Sn-3Cr filler alloy. Moreover, diamond grits brazed by the novel developed Cu-22Sn-3Cr filler alloys exhibited less breakage and fall-off failures than those brazed by the commercial Ni-14Cr-10P filler alloy in grinding sapphire. The high mechanical properties of diamond grits brazed using Cu-22Sn-3Cr filler alloy are due to the low dissolubility of C elements in Cu-Sn alloy; the formation of a thin and dense layer of interfacial Cr_7C_3 carbides is responsible for the improvement of bonding strength.

Acknowledgments

The authors would like to thank financial support by National Natural Science Foundation of China (Grant Nos. 51835004, 51805175) and Nation-guided Technology Development Program of Fujian Province (Grant No. 2021L3012).

References

[1] Malkin, S. (1989) Grinding Technology—Theory and Application of Machining with Abrasives. John Wiley & Sons, New York.
[2] Huang, H., Li, X., Mu, D., & Lawn, B. R. (2020) Science and art of ductile grinding of brittle solids. Int J Mach Tool Manu, 103675.
[3] Artini, C., Muolo, M. L., & Passerone, A. (2012) Diamond–metal interfaces in cutting tools: a review. J. Mater. Sci, 47: 3252-3264.
[4] Chattopadhyay, A. K., Chollet, L., & Hintermann, H. E. (1991) On performance of brazed bonded monolayer diamond grinding wheel. CIRP Ann. Manuf. Technol., 40: 347-350.
[5] Sung, J. C., & Sung, M. (2009) The brazing of diamond. Int. J. Refract. Met. Hard Mater., 27: 382-393.
[6] Ma, B., Pang, Q., & Lou, J. (2014) Rod-like brazed diamond tool fabricated by supersonic-frequency induction brazing with Cu-based brazing alloy. Int. J. Refract. Met. Hard Mater., 43: 25-29.
[7] Buhl, S., Leinenbach, C., Spolenak, R., & Wegener, K. (2012) Microstructure, residual stresses and shear strength of diamond–steel-joints brazed with a Cu–Sn-based active filler alloy. Int. J. Refract. Met. Hard Mater., 30: 16-24.

[8] Liao, X., Mu, D., Wang, J., Huang, G., Huang, H., Xu, X., & Huang, H. (2017) Formation of TiC via interface reaction between diamond grits and Sn-Ti alloys at relatively low temperatures. Int. J. Refract. Met. Hard Mater., 66: 252-257.

[9] Liu, Z., Liao, X., Fu, W., Mu, D., Xu, X., & Huang, H. (2020) Microstructures and bonding strength of synthetic diamond brazed by near-eutectic Ag–Cu–in–Ti filler alloy. Mat. Sci. Eng.: A, 790: 139711.

[10] Klotz, U. E., Khalid, F. A., & Elsener, H. R. (2006) Nanocrystalline phases and epitaxial interface reactions during brazing of diamond grits with silver based Incusil-ABA alloy. Diam. Relat. Mater., 15: 1520-1524.

[11] Chen, Y., Fu, Y., Su, H., Xu, J., & Xu, H. (2014) The effects of solder alloys on the morphologies and mechanical properties of brazed diamond grits. Int. J. Refract. Met. Hard Mater., 42: 23-29.

[12] Lowder, J. T., & Tausch, E. M. (1975) Method of Manufacturing Diamond Abrasive Tools, American Patents No. 3,894,673.

[13] Chen, J., Mu, D., Liao, X., Huang, G., Huang, H., Xu, X., & Huang, H. (2018) Interfacial microstructure and mechanical properties of synthetic diamond brazed by Ni-Cr-P filler alloy. Int. J. Refract. Met. Hard Mater., 74: 52-60.

[14] Sun, B. J., Xiao, B., & Liu, S. X. (2016) Microstructure and performance of diamond abrasive grains brazed in mesh belt furnace with ammonia dissociating atmosphere. Int. J. Refract. Met. Hard Mater., 60: 154-159.

[15] Mukhopadhyay, P., Simhan, D. R., & Ghosh, A. (2017) Challenges in brazing large synthetic diamond grit by Ni-based filler alloy. J. Mater. Process. Technol., 250: 390-400.

[16] Huang, G., Huang, J., Zhang, M., Mu, D., Zhou, G., & Xu, X. (2018) Fundamental aspects of ultrasonic assisted induction brazing of diamond onto 1045 steel. J. Mater. Process. Technol., 260: 123-136.

[17] Chen, J., Liao, X., Lin, Q., Mu, D., Huang, H., Xu, X., & Huang, H. (2019) Reactive wetting of binary Sn-Cr alloy on polycrystalline chemical vapour deposited diamond at relatively low temperatures. Diam. Relat. Mater., 92: 92-99.

[18] Li, M., Chen, J., Lin, Q., Wu, Y., & Mu, D. (2019) Interfacial microstructures and mechanical integrity of synthetic diamond brazed by a low-temperature Cu-Sn-Cr filler alloy. Diam. Relat. Mater., 97: 107440.

[19] Lu, J., Qi, W., Li, Y., Qi, F., Qian, M., & Qiu, X. (2017) Analysis on brazed diamond joints with modified Cu-based filler alloy. Int. J. Refract. Met. Hard Mater., 67: 141-146.

[20] Hwang, T. W., Evans, C. J., Whitenton, E. P., & Malkin, S. (2000) High speed grinding of silicon nitride with electroplated diamond wheels, part 1: wear and wheel life. J. Manuf. Sci. Eng., 122: 32-41.

[21] Hwang, T. W., Evans, C. J., & Malkin, S. (2000) High speed grinding of silicon nitride with electroplated diamond wheels, part 2: wheel topography and grinding mechanisms. J. Manuf. Sci. Eng., 122: 42-50.

Mn doped graphene aerogels for high capacity and long-life supercapacitor

Zhi Chen[1,2]*, Shuang Shan[2], Jiaxuan Liao[1,2]*

[1] Yangtze Delta Region Institute (Quzhou), University of Electronic Science and Technology of China, Quzhou, Zhejiang 324000, China

[2] School of Materials and Energy, University of Electronic Science and Technology of China, Chengdu 611731, People's Republic of China

[*] Corresponding author: drzhichen@163.com

Abstract. The pure graphene aerogel material is a double-layer capacitor material which provides capacitance by the surface area. Its conductivity and specific capacitance require further improvement. In this work, manganese doped graphene aerogels were synthesized by hydrothermal method using manganese chloride as a source of manganese. The effect of manganese content on the structure and properties of the material was studied. It is found that the specific surface area and pore volume of aerogels decrease with the increase of Mn content. The Mn doped graphene aerogel prepared at the current density of 0.1 A/g shows a capacitance of 180 F/g and exhibits pseudo capacitance performance in the three-electrode test. After assembling the button capacitor, the capacity retention rate is 93% after 1000 cycles at a current density of 0.5 A/g.

1. INTRODUCTION

Nitrogen doped carbon-based porous materials, such as nitrogen doped graphene aerogels have the characteristics of large surface area and rich pore structure [1-4]. This allows nitrogen doped graphene aerogels as a good carbon skeleton for carrying capacitive properties of the materials with pseudocapacitor materials [5-7]. As a common pseudocapacitor material, manganese oxide has attracted extensive attention because of its low price and rich content. In recent years, a lot of progress has been made in the research on the action mechanism and synthesis methods of manganese oxides with various valence states and crystal structures in the field of energy storage [8]. Manganese oxide, as a pseudocapacitor material, has made significant performance improvement in supercapacitors. However, manganese oxide is difficult to obtain good performance when used as electrode material alone because of its poor conductivity and the volume change caused by redox reaction in the cycle process.

Nitrogen doped graphene loaded manganese can make the two materials integrate their advantages and make up for their disadvantages, to improve the properties of the materials. Therefore, in this work, manganese doped graphene aerogels were prepared by hydrothermal method. The influence of different ratios of Manganese Sources on the structure and capacitance properties of aerogels is mainly studied.

2. Experimental methods

2.1 Preparation of Mn doped graphene aerogels

Firstly, graphene oxide dispersion is prepared. Weigh a certain amount of graphene oxide and add it into a beaker containing an appropriate amount of deionized water to prepare graphene oxide dispersion with a concentration of 4 mg/ml. After that, the solution was sonicated using a cell fragmentation instrument for 2 hours. Next, prepare additives and prepare 5 ml of ethylenediamine and ammonia solution in the ratio of 1:1 by volume. It was then added to 50 ml of graphene oxide dispersion and stirred with a magnetic stirrer for 30 minutes. Then add a certain amount of 0.2 M manganese chloride solution to the mixed liquid and stir for 15 minutes. Then transfer the solution to a hydrothermal kettle and heat it in an oven at 200 °C for 24 hours. Aging treatment shall be carried out after hydrothermal reaction. After aging, the hydrogel was put into the supercritical drying machine, and the carbon dioxide was used as a supercritical drying medium. Under the condition of 45 degree centigrade and a pressure of 7.5 MPa, the gel was dried for 24 hours, and the aerogels were obtained.

In the experiment, the amount of manganese chloride aqueous solution in hydrothermal reaction was taken as a variable to study the effect of the amount of manganese source on the material. The specific experimental setup is to add 5 ml, 7.5 ml and 10 ml of manganese chloride solution respectively. According to the different amount of manganese chloride added in the experimental group, they are named Mn5, Mn7.5 and Mn10 respectively.

2.2 Characterization and electrochemical test of materials

Then the materials were characterized by BET, XRD and SEM. The resistivity of the materials were tested by resistivity tester. The aerogels were then made into electrodes for supercapacitors. Cyclic voltammetry (CV), electrochemical impedance spectroscopy (EIS) and constant current charge discharge tests were carried out in a three-electrode system with 1m KOH solution as electrolyte. After completing the three-electrodes tests, the button capacitor was assembled. The electrochemical properties of the two electrodes were tested by CV, EIS and constant current charge discharge.

3. Results and discussion

The SEM test results of the material are shown in Figure 1. It can be seen from the figure that this material has a honeycomb like hole structure. Through comparison, it can be found that the structures of Mn5 and Mn10 materials are fluffy, while Mn7.5 presents a multi-layer structure with dense stacking. At the same time, the pore wall of Mn5 is thinner, and more aerogel sheets are attached to the Mn10 hole wall. Mn7. Although the stack is dense, there are holes between layers. In general, the material has a good pore network and can provide a path for ion transport in electrolyte.

Through the analysis of SEM images, we can see that Mn group materials can be divided into closely packed thick aerogels and relatively loose fluffy aerogels. In the loose structure, more graphene flakes are accumulated on the tube wall of mn10. These different structures are bound to have different effects on the working process of capacitors, so it is necessary to further explore them. To further understand these two structures, BET test, resistivity test and XRD test were carried out for the aerogels of materials.

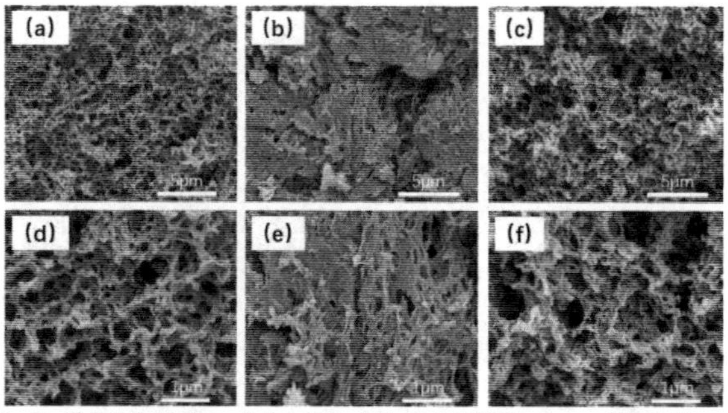

Figure 1 SEM of (a) Mn5, (b) Mn7.5and (c) Mn10 at 20000 magnification. SEM of (d) Mn5, (e) Mn7.5and (f) Mn10 at 50000 magnification.

The test results of bet and conductivity are shown in Table 1. It is obvious from the table that the pore volume, specific surface area and resistivity of the material decrease with the increase of Mn Addition. The specific surface area and pore volume decrease indicate that excessive manganese ions can inhibit the formation process of aerogels. At the same time, the reaction between manganese ions and graphene oxide will also destroy the existing pores, resulting in the decrease of pore volume. Although in general (compared to the three or four chapter), the addition of manganese ions will increase the resistivity of aerogels, but the resistivity of aerogels added with excess manganese ions is lower.

Table 1 BET and resistivity test results of aerogels

Sample	Specific surface area (m^2/g)	Pore volume (ml/g)	Resistivity (Ωcm)
Mn5	810.714	2.727	4.24
Mn7.5	766.172	2.559	2.20
Mn10	542.338	1.802	1.865

Figure 2a can be obtained by using the adsorption and desorption data obtained from bet test. It can be seen from the figure that the adsorption curves of Mn group materials are type IV curves and the hysteresis curve is type H3. The results show that the holes in the manganese doped graphene aerogels are mainly mesoporous. The results show that there is no significant difference in the void structure between the three aerogels. The material was tested by XRD, and the results are shown in Fig. 2b. According to the diagram, $Mn(CO_3)$ is found in the Mn materials, indicating that Mn is successfully incorporated into the aerogel structure. In addition, according to the obvious degree of peak shape, it can be known that the doping degree of manganese is directly proportional to the addition amount.

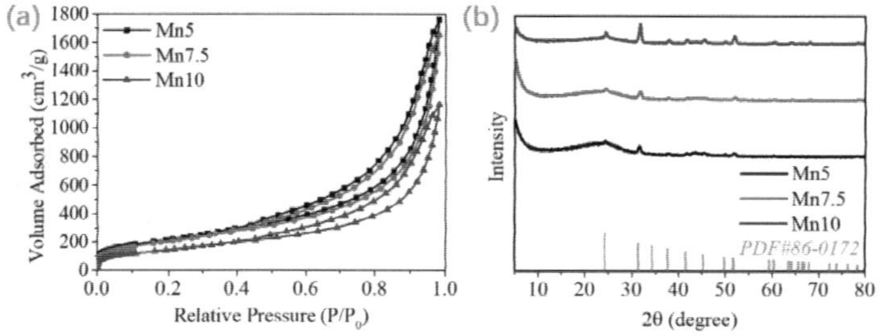

Figure 2 (a) Adsorption desorption curve of materials. (b) XRD test results of materials

In order to further understand the material properties, three electrode tests were carried out. The impedance test results are shown in figure 3a-b. The oblique line with high inclination in the low-frequency part and the almost invisible semicircle in the high-frequency part show that the material has good capacitive properties. Cyclic voltammetry was carried out on the three electrodes, and the results are shown in Fig. 3c-f. It can be clearly seen from the CV diagram that the material has peaks at -0.1 V and -0.32 v. This peak is caused by the reaction of manganese. In addition, according to the Figure 3f, the pseudo capacitance reaction of the dense stacked aerogel Mn10 is more intense, indicating that although the manganese content is not as good as Mn10, this stacked and connected structure helps to pseudo capacitance reaction. In addition, from the overlap of CV curves, it can be seen that the loose porous structure aerogels of Mn10 and Mn5 have stronger cycling performance.

In order to explore the capacitance performance of the material, the constant current charge discharge test was carried out for the three-electrode system, and the specific capacitance values under different current densities (0.2 A/g, 0.5 A/g, 1 A/g, 2 A/g, 5 A/g, 10 A/g) were calculated. The charge discharge curve and rate performance of the three-electrode system of Mn group material under the current density of 1 A/g are shown in Figure 4. According to the charge discharge curve, Mn5 is a charge discharge curve similar to a triangle, and its pseudo capacitance is not obvious. The charge discharge curves of Mn7.5 and Mn10 deviate from the triangle, and there are obvious turning points in the charge and discharge curves, showing obvious pseudo capacitance characteristics.

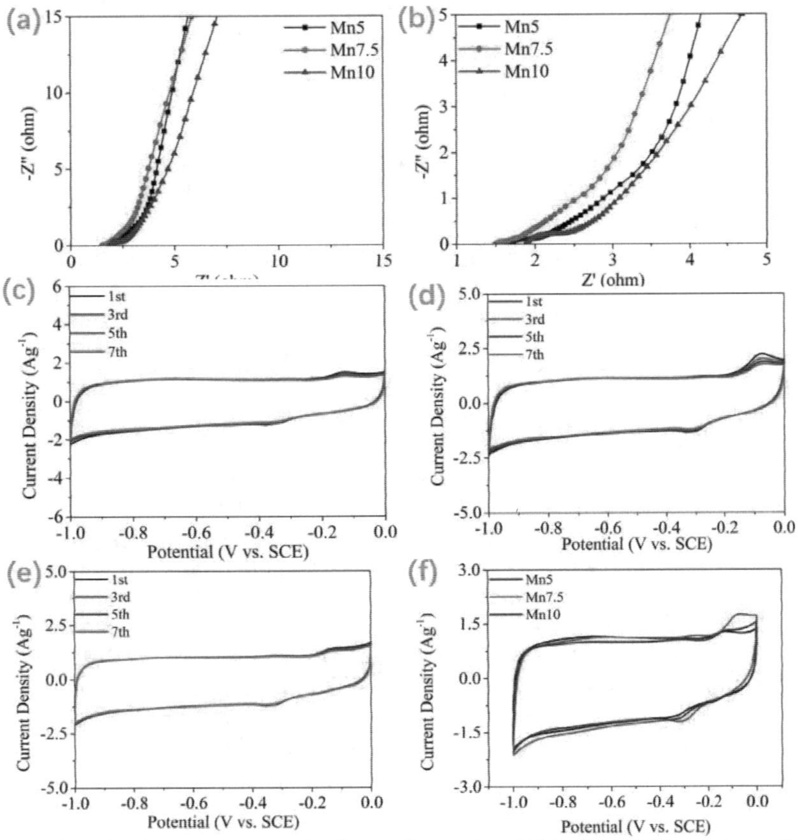

Figure 3 Electrochemical measurement in three electrodes. EIS of (a) full spectrum (b) and high frequency area. CVs of (c) Mn5, (d) Mn7.5 and (e) Mn10 at the scan rate of 10 mV/s. (f) CVs of all samples at the scan rate of 10 mV/s.

According to the rate performance, it can be seen that two materials Mn10 and Mn7 with pseudo capacitance characteristics The specific capacitance of Mn5 at low current density is much larger than that of Mn5 without obvious pseudo capacitance. And Mn7.5 aerogels with dense packing characteristics of aerogels exposed to electrolyte in a loose structure with an effective area greater than Mn10. Therefore, when the manganese content is low, Mn7.5 shows higher specific capacitance at low current density. However, with the increase of current density, the more abundant ion channels provided by loose structure play a role. The specific surface area of Mn10 is lower than Mn7.5. In the case, the capacitance of Mn10 exceeds Mn7.5. And the specific capacitance of mn5 with loose porous structure is more stable. This shows that the loose porous structure formed when the amount of manganese is too high or too low is conducive to the performance of the material at high current density.

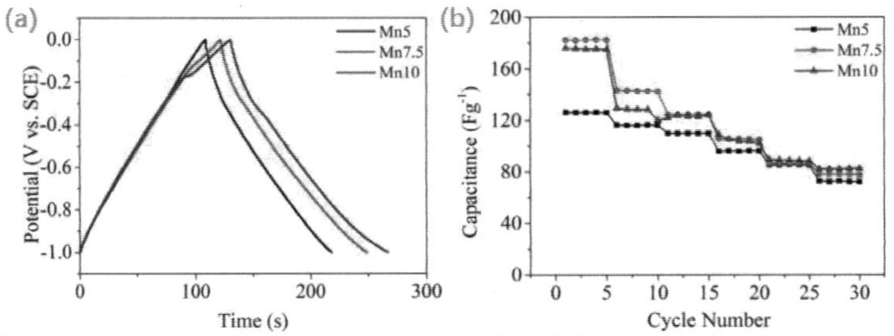

Figure 4 Electrochemical measurement in three electrodes. (a) Charge discharge curve of materials at current density of 1A/g. (b) Rate capacities of Materials

In order to study the practical performance of the material, the material was assembled into a symmetrical water buckle capacitor, and the impedance, cyclic voltammetry, constant current charge discharge and cycle life were tested. The results of the impedance test are shown in Figure 5a. It is obvious from the figure that the inclination of Mn5 and Mn10 low-frequency areas is greater than mn7.5. This shows that the diffusion rate of ions in Mn10 and Mn5 is faster than that in Mn7.5. The cyclic voltammetry test of button capacitor was carried out. The results are shown in Figure 5b-e. It can be seen from the figure that the capacitor presents a CV curve similar to a rectangle at this scanning rate. However, the overall capacitance is small, and the capacitance is completely proportional to the specific surface area of the material, and there is no corresponding peak on the CV Curve. It can be seen from the figure that the coincidence degree of the three is high, indicating that the structural change of the material is small during the cycle.

The cyclic charge discharge test of the capacitor and the magnification test results at current densities of 0.1 A/g, 0.2 A/g, 0.5 A/g, 1 A/g, 2 A/g, 5 A/g and 10 A/g are shown in Fig. 5f-g. It is obvious from the data in the figure that the performance of Mn10 is not reflected in the configuration of button capacitor, while Mn7 5 is only reflected at low magnification, while Mn5 is hardly affected. After that, the impedance test of Mn10 is carried out, and the results are shown in Figure 5h. It is found that the slope of the low frequency region of Mn10 impedance decreases after the test. This indicates that the structure of the material has changed. The structure of Mn10 is similar to that of Mn5, but there is no such change in Mn5. Therefore, this change may be due to the irreversible transformation of manganese in the process of charge and discharge, which destroys the structure of the material and makes the performance of mn10 unable to be reflected.

The cycle life of materials was tested. The results are shown in Figure 5I. It can be seen from the figure that the order of cyclic stability of devices from high to low is Mn10 (retention rate 100%), Mn5 (retention rate 97%) and Mn7.5 (retention rate 93%). This result shows that the internal structure of Mn10 has been destroyed and has become a device that completely relies on electric double layers to provide capacitance. For other devices, the higher the manganese content, the lower the retention rate. This is because the irreversible reaction of manganese in the cycle leads to the decrease of device capacity.

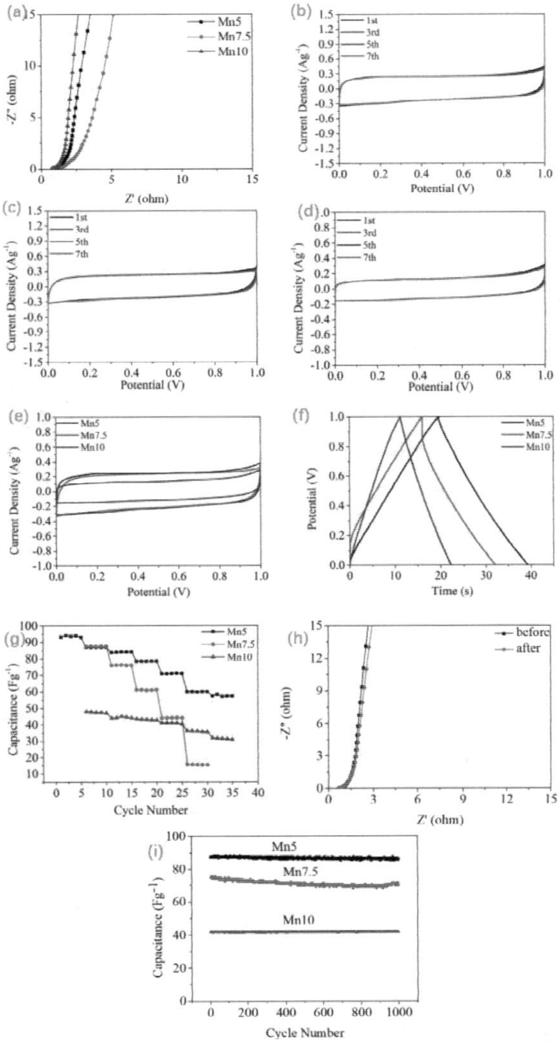

Figure 5 Electrochemical measurement in cell batteries. (a) EIS spectrum of cell capacitor. CVs of (b) Mn5, (c) Mn7.5 and (d) Mn10 at the scan rate of 10 mV/s. (e) CVs of all samples at the scan rate of 10 mV/s. (f) Charge discharge curve of materials at current density of 1A/g. (g) Rate capacities of Materials. (h) EIS comparison of Mn10 before and after test. (i) Long cycle performance test.

4. Conclusion

In conclusion, manganese doped graphene aerogels were successfully synthesized by hydrothermal method using MnCl2 as manganese source. The aerogel has higher specific surface area, resistivity, pore volume and specific capacitance. It is found that within the setting range, the higher the addition of manganese, the smaller the specific surface area and pore volume of the material, and the larger the specific capacitance at low magnification. It is also found that the different aeration of manganese will lead to different aerogels. A porous porous aerogel structure will be formed if the amount is too high or too low, and a stacked aerogel structure will be added. At the same time, the influence of these structures on the capacitance performance is studied. It is found that the stacked aerogel structure is advantageous to provide capacitance at low current density while loose porous type performs well under medium current density. Then the button symmetrical capacitor is assembled and the practicability of the material is studied.

Acknowledgements
The authors gratefully acknowledge the support by Sichuan Science and Technology Program of 2021YFH0126, Quzhou Science and Technology Bureau Project (2021D006) and the Fundamental Research Funds for the Central Universities (A030202063008029).

References
[1] F. Zhang, T. Liu, J. Zhang, E. Cui, L. Yue, R. Jiang, G. Hou, Carbon, 147 (2019) 451-459.
[2] Q. Yin, L. He, J. Lian, J. Sun, S. Xiao, J. Luo, D. Sun, A. Xie, B. Lin, Carbon, 155 (2019) 147-154.
[3] Z. Chen, J. Liao, W. Li, Y. Song, C. Chen, J. Yang, Z. Xu, T. Feng and M. Wu, Appl. Surf. Sci. 506, 144909 (2020)
[4] Z. Chen, W. Li, J. Yang, J. Liao, C. Chen, Y. Song, S. A. Ali Shah, Z. Xu and M. Wu, J Electrochem. Soc. 167, 050506 (2020)
[5] H. Zhuo, Y. Hu, Z. Chen, L. Zhong, Carbohydr Polym, 215 (2019) 322-329.
[6] M. Mirzaeian, Q. Abbas, D. Gibson, M. Mazur, Energy, 173 (2019) 809-819.
[7] Z. Song, W. Liu, N. Sun, W. Wei, Z. Zhang, H. Liu, G. Liu, Z. Zhao, Solid State Communications, 287 (2019) 27-30.
[8] Y. Guo, T. Feng, J. Yang, F. Gong, C. Chen, Z. Xu, C. Hu, S. Leng, J. Wang, M. Wu, Nanoscale, 11 (2019) 10763-10773.

Analysis of microstructure and protective performance of melt-blown materials for medical protective masks

Jin Zhang[1], Huipeng Wang[2], Chunliang Chen[2], An Gui[3], Xiaoyu Zhang[3], Yumin Wang[4], Yi Xiong[2,4,*], Wei Zeng[3,*]

[1] Department of Pharmacy, Zhongnan Hospital, Wuhan University, Wuhan 430071, Hubei, People's Republic of China

[2] Science and Technology Institute, Key Laboratory of Textile Fiber and Products, Ministry of Education, Laboratory for Electron Microscopy, Wuhan Textile University, Wuhan 430073, Hubei, People's Republic of China

[3] School of Electronics and Information Engineering, Anhui University, Hefei 230601, Anhui, People's Republic of China

[4] School of Physics and Technology, MOE Key Laboratory of Artificial Micro- and Nano-Structures, Wuhan University, Wuhan 430072, Hubei, People's Republic of China

* Corresponding author's e-mail: Yi Xiong, xiong@wtu.edu.cn; Wei Zeng, zengwei@ahu.edu.cn;

Abstract. Since the end of 2019, the Corona Virus Disease 2019 (COVID-19) pandemic has led to a surge in the use of all kinds of medical supplies, especially surgical masks. Based on the microstructure and anti-virus mechanism of melt-blown materials used for medical masks, this paper introduces the research status of nonwoven filter materials used for protective masks. At the same time, the surface interface structure of four disposable medical protective masks from different manufacturers was analyzed by scanning electron microscope, and the difference of melt-blown materials of these masks was studied. The results show that the fiber diameter of melt-blown mask with better protective effect is fine and compact, and the aperture formed between fibers is smaller. This reasearch provides new ideas for further research and development of non-woven materials for medical masks.

1. Introduction

Since December 2019, the outbreak and continued spread of Corona Virus Disease 2019 (COVID-19) has had a severe impact on the health and lives of people around the world, drawing wide attention to the issue of public health[1]. Diseases such as COVID-19, severe acute respiratory syndrome (SARS) and highly pathogenic avian influenza are transmitted mainly through the respiratory tract. And until specific drugs and vaccines for such diseases are fully available, wearing protective masks is an effective means of preventing the invasion and transmission of the virus for health workers[2-3].

Medical protective masks can be divided into four categories: disposable medical masks, medical surgical masks, medical particulate matter protective masks and medical protective masks, and their protective grade increases in turn[4-5]. Medical masks can be divided into planar structures and three-dimensional structures according to their appearance. Medical surgical masks and disposable medical masks are planar structures, and medical protective masks are three-dimensional arch structure. Medical masks take the form of multi-layer nonwoven fabric composite, including the outer layer, the filter layer and the inner layer, as shown in Figure 1. The outer layer is generally composed of spunbond or hot rolled nonwovens, which can not only work as the structural support but also intercept large particles and liquid droplets. The filter layer is the most crucial; it is usually made of charged melt-blown nonwovens, which can filter fine particulate matter[6-7]. The inner layer is composed of ordinary spunbonded, needled or spunned nonwovens, which is designed to ensure moisture absorption and improve skin affinity and comfort. The protective mechanism of masks is analyzed from the level of air filtration technology. According to the interception and interaction modes of particles, the protective mechanism is generally based on the following filtering mechanisms: interception effect, inertia effect, diffusion effect, electrostatic effect and gravity effect. The interception effect, inertia effect and electrostatic effect play a major role in medical mask materials. Among them, it is quite important to make full use of the electrostatic deposition effect. The core layer filter material of the mask is treated with electrostatic electret to make itself charged, and then increases the electrostatic effect and enhances the adsorption and filtration effect[8].

The main raw materials of medical masks are melt-blown cloth and spunbonded cloth made of thermoplastic polymer[9]. However, there are many shortcomings on these masks, such as poor moisture absorption, inadequate comfort, and environmental pollution. During the COVID-19 pandemic, the bridge of the nose and face of medical workers suffered varying degrees of compression and damage from prolonged wear of masks. Meanwhile, a large number of disposable masks cannot be disposed of in a timely and effective manner, which also puts pressure on environmental protection and public health[10-11]. Polypropylene (PP) masterbatch is the basic raw material of medical masks, which has the advantages of low cost, good heat resistance and stable physical and chemical properties[12]. It has been widely used in medical nonwovens as a high-quality raw material. In order to further improve the properties of PP nonwovens, it is often modified by various methods. For example, SiO_2 aerogel and PP are fused and blended to prepare SiO_2 aerogel/PP melt-blown nonwoven material. Compared with pure PP melt-blown material, its filtering effect is enhanced by about 50% [13]. Some researchers used titanium dioxide (TiO_2), a photocatalytic antibacterial material with strong antibacterial performance, to modify TiO_2 first to make its antibacterial performance more stable. Then the modified TiO_2 was loaded onto PP melt-blown nonwoven fabric in the online composite mode, and onto PE/PET spunbonded nonwoven fabric in the offline composite mode. Finally, the two materials were combined to prepare the modified TiO_2/spunbonded melt-blown antibacterial composite filter material[14].

2. Materials and Methods

2.1. Materials

Disposable medical masks from four different manufacturers were selected for this test, which were represented by numerical numbers. Their basic information is shown in Table 1, and their appearance is shown in Figure 2.

Table 1. The size and appearance color of the four masks.

Serial number	Size/ (cm × cm)	Color
1#	17.6×9.60	Sky blue
2#	17.7×9.05	Grass green
3#	17.5×9.25	Sky blue
4#	14.2×9.15	Sky blue

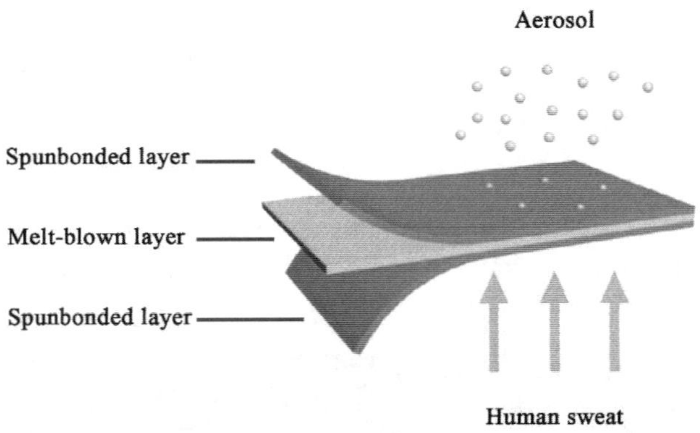

Figure 1. Schematic diagram of a mask composed of three layers of non-woven fabric.

2.2. *The experimental method*

The field-emission scanning electron microscope (SEM, JEOL-7800F) was used to observe the outer layer and the cross-section of the filter layer of the mask. Each mask was cut to a certain size and fixed on the sample table with conductive adhesive. Platinum was sprayed on the surface of the mask to increase the conductivity of the material, and the acceleration voltage was tested at 8 kV.

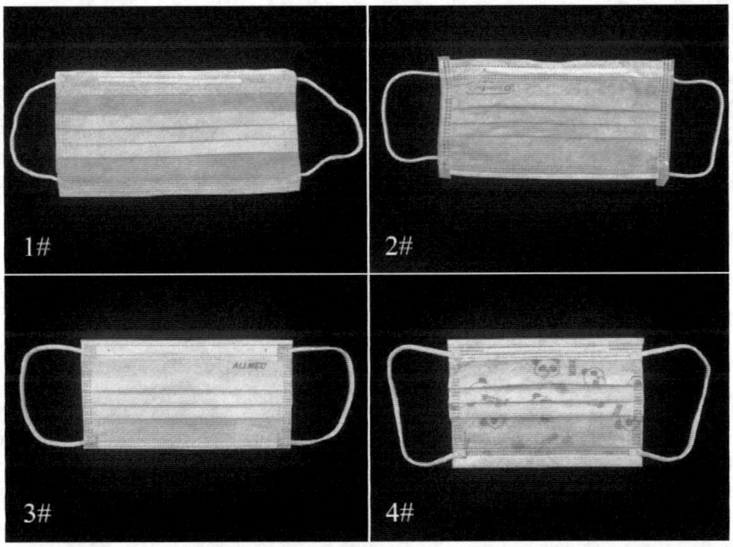

Figure 2. Appearance of masks from four different manufacturers.

3. Results and Discussion

In the production process of medical masks, the most critical material is the intermediate layer melt-blown cloth, mostly PP, which is made by melting, spinning, stretching, shaping and other steps. The PP fibers are ejected from the spinneret under high-speed and high-pressure hot gas flow, drawn into ultra-fine fibers with the diameter of 0.3~7.0 μm, evenly spread on the collection device. Then their own waste heat is used to bond into a network and reinforce into a cloth, finally the material got is electret treated to improve the filtration efficiency. Among them, the key process is electrostatic electret treatment, so that the filter material can increase the electrostatic adsorption of particles in the air, and greatly improve the filtering efficiency of the mask by using the electrostatic effect. The outer and inner layers are composed of spunbonded nonwoven fabric. The filter material of them has a large aperture scale and can filter larger particles. The filter layer is made of melt-blown nonwoven fabric to filter particles of even smaller size.

Figure 3 shows the SEM morphologies of the outer layers of the four masks and the enlarged images of single fiber. It can be seen that the fiber diameter of the outer melt-blown material is roughly the same. The outer layer is a single PP spunbond layer with an average fiber diameter of about 20 μm. The point nodes of the inner and outer layers are formed by hot rolling process, mainly to improve the strength of the spunbond material. The outer layers of masks 1# and 2# are relatively loose, while the outer layers of masks 3# and 4# are relatively dense.

Figure 3. SEM images of four masks. (a). 1#, (b). 2#, (c). 3#, (d). 4# SEM images of melt-blown material of mask outer layer; (e). 1#, (f). 2#, (g). 3#, (h). 4# SEM images of melt-blown material fibers on the outer layer of masks.

The intermediate layer melt-blown material is the most important factor affecting the filtering efficiency of masks. Figure 4 shows the morphologies of melt-blown mask intermediate layers from four different manufacturers. It can be seen from the figure that the microfibers in the middle layer of all masks are randomly distributed and have a certain disorder. This structure enables them to have a larger specific surface area, so that melt-blown materials have better filtering and shielding properties. The smaller the fiber diameter of the middle layer of the mask is, the higher the porosity of the fiber mat formed by the accumulation of fibers, the smaller the pressure drop resistance and the better the protective effect[15]. The smaller the average diameter of the melt-blown fiber in the middle layer of the mask, the better its density. It can be seen from the figure that the middle layer of mask 1# has poor workmanship and the fiber thickness is about 40-50 μm. The average diameter of the melt-blown material fiber in the middle layer of mask 2# is less than 10 μm, and the average diameter of the melt-blown material fiber in the middle layer of mask 3# and mask 4# is about 2 μm, and the fiber of the melt-blown material in the middle layer of mask 3# and mask 4# is thin and dense, with few voids, which is more conducive to filtration. There are many tiny voids between the fibers in the middle layer of 2# mask. However, the fiber diameter of the melt-blown material in the middle layer of mask 1# is thick and loose, with a large gap, which is not conducive to filtration. As can be seen from the SEM section of the mask as shown in figure 4(e-h), the structure of the mask is divided into three layers: outer layer, middle layer and inner layer. The interlayer is a melt-blown layer, and its microfibers are interlaced and randomly distributed. The above research indicates that the smaller the average diameter of the fibers of medical mask melt-blown material and the denser they are, the better their filtration performance will be.

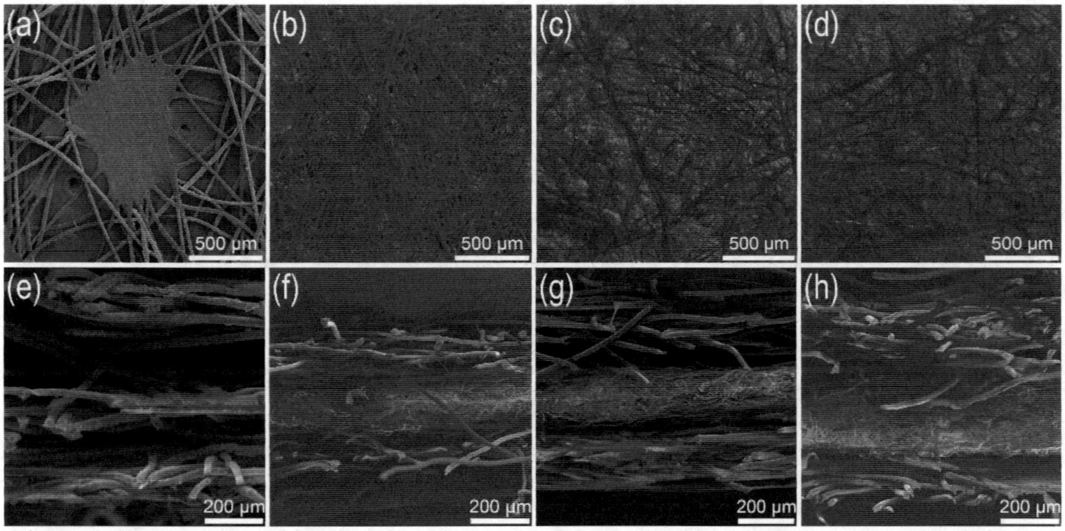

Figure 4. SEM images of four masks. (a). 1#, (b). 2#, (c). 3#, (d). 4# SEM images of melt-blown material in the middle layer of mask; (e). 1#, (f). 2#, (g). 3#, (h). 4# SEM images of cross sections of masks.

4. Conclusion

With the rapid growth of national economic level, the continuous improvement of medical and health level, and the continuous enhancement of people's awareness of life and health, the performance of protective medical masks is also constantly improved. From the single filtration efficiency to the comfort, biodegradability and antibacterial, etc., the performance of masks is becoming more and more diversified. Scanning electron microscopy was used to characterize the melt-blown materials of four disposable medical masks from different manufacturers. It shows that the melt-blown materials of the middle layer have higher filtration efficiency, and the smaller pore diameter, the better protective effect. The above research has potential guiding significance for the development and improvement of medical protective masks.

Acknowledgements

This work was supported by the Anhui Collaborative Innovation Project (GXXT-2019-036), Open Research Fund of National Engineering Research Center for Agro-Ecological Big Data Analysis & Application, Anhui University (No. AE201910, AE202001), Open Research Fund of Key Laboratory of Textile Fiber and Products (Wuhan Textile University), Ministry of Education (No. Fzxw2021026).

References

[1] Wang, L.S., Wang, Y.R., Ye, D.W., Liu, Q.Q. (2020) Review of the 2019 novel coronavirus (SARS-CoV-2) based on current evidence. Int. J. Antimicrob. Ag., 55: 105948.

[2] Shiu, E.Y.C., Leung, N.H.L., Cowling, B.J. (2019) Controversy around airborne versus droplet transmission of respiratory viruses: implication for infection prevention. Curr. Opin. Infect. Dis, 32: 372-379.

[3] Leung, N.H.L., Chu, D. K.W., Shiu, E.Y.C., Chan, K.H., Mcdevitt, J.J., Hau, B.J.P., Yen, H.L., Li, Y.G., Ip, D.K.M., Peiris, J.S.M., Seto, W.H., Leung, G.M., Milton, D.K. Cowling, B.J. (2020) Respiratory virus shedding in exhaled breath and efficacy of face masks. Nat. Med., 26(5): 676-680.

[4] Han, H.S., Shin, S.H., Park, J.W., Li, K., Kim, B.J., Yoo, K.H. (2021) Changes in skin characteristics after using respiratory protective equipment (medical masks and respirators) in the COVID-19 pandemic among healthcare workers. Contact Dermatitis, 85: 225-232.

[5] Siegel, J.D., Rhinehart, E., Jackson, M., Chiarello, L., the Healthcare Infection Control Practices Advisory Committee. (2007) guideline for isolation precautions: preventing transmission of infectious agents in health care settings. Am. J. Infect. Control, 35: 65-164.

[6] Shi, J., Zou, Y.Z., Wang, J.X., Zeng, X.F., Chu, G.W., Sun, B.C., Wang, D., Chen J.F. (2021) Investigation on Designing Meltblown Fibers for the Filtering Layer of a Mask by Cross-Scale Simulations. Ind. Eng. Chem. Res., 60(4): 1962-1971.

[7] Wang, D., Sun, B.C., Wang, J.X., Zhou, Y.Y., Chen, Z.W., Fang, Y., Yue, W.H., Liu, S.M., Liu, K.Y., Zeng, X.F., Chu, G.W., Chen, J.F. (2020) Can Masks Be Reused After Hot Water Decontamination During the COVID-19 Pandemic? Engineering, 6(10): 1115-1121.

[8] Lin, S.Z., Wang, S.X., Yang, W., Chen, S.W., Xu, Z.S., Mo, X.W., Zhou, H., Duan, J.J., Hu, B., Huang, L. (2021) Trap-Induced Dense Monocharged Perfluorinated Electret Nanofibers for Recyclable Multifunctional Healthcare Mask. ACS Nano, 15(3): 5486-5494.

[9] Desai A.N., Mehrotra P. (2020) Medical Masks. JAMA-J. Am. Med. Assoc., 323(15): 1517-1518.

[10] Wang, Z., An, C.J., Chen, X.J., Lee, K., Zhang, B.Y., Feng, Q. (2021) Disposable masks release microplastics to the aqueous environment with exacerbation by natural weathering. J. Hazard. Mater., 417: 126036.

[11] Kumar, P., Roy, S., Sarkar, A., Jaiswal, A. (2021) Reusable MoS_2 Modified Antibacterial Fabrics with Photothermal Disinfection Properties for Repurposing of Personal Protective Masks[J]. ACS Appl. Mater. Inter., 13(11): 12912-12927.

[12] Yim, W., Cheng, D., Patel, S.H., Kou, R., Shirley Meng, Y., Jokerst J.V. (2020) KN95 and N95 Respirators Retain Filtration Efficiency despite a Loss of Dipole Charge during Decontamination. ACS Appl. Mater. Inter., 12(49): 54473-54480.

[13] Zhang, H., Zhen, Q., Liu, Y., Song, W.M., Liu, R.T., Zhang, Y.F. (2019) Air filtration performance and morphological features of polyethylene glycol/polypropylene composite fibrous materials with embedded structure. Journal of Textile Research, 40(9): 28-34.

[14] Zhu, X.M., Dai, Z.J., Zhao, Y., Xu, H., Ke, Q.F. (2019) Fabrication and properties of modified TiO_2/Spun-bonded and melt-blown nonwoven antibacterial composite filter. Journal of Donghua University, 45(2): 196-203.

[15] Uppal, R., Bhat, G., Eash, C., Akato, K. (2013) Melt blown nanofiber media for enhanced quality factor. Fiber. Polym., 14(4): 660-668.

Numerical study of nano-particle composite paraffin phase change heat storage capsule

Fude Wang[1a*], Rusen Hou[1b*]

Shandong Huayu Univeisity of Technology, Dezhou City, Shandong Province. Shand ong, 253000, China

[a]*wfd87120@126.com, [b]*2463321404@qq.com

Abstract—The heat exchange characteristics of traditional heat exchange fluids such as water and oil in phase change heat storage can no longer meet the ever-increasing heat exchange requirements. The researchers found that by adding nano-scale particles to the basic phase change material (pure PCM) in a certain proportion to form a composite phase change material, the heat transfer characteristics of the material can be improved. In this paper, a model study of the phase change heat storage characteristics of the nano-particle composite paraffin wax phase change capsule with holes is carried out, and the effects of thermal disturbance, natural convection heat transfer and external convection heat transfer on the phase change heat storage are mainly studied. It is found that when nanoparticles are added to the phase change heat storage capsule, the composite phase change heat storage capsule formed by the phase change heat storage, the internal natural convection intensity increases, and with the increase of the particle share, the heat storage rate increases, and the heat exchange intensity increases.

1. Introduction

As researchers have further matured the research on the energy that can be continuously generated by nature such as the sun, everyone is more and more concerned about how to actually use these continuously generated energy. Yongqiang Liu[1] qualitatively analyzed the influence of the temperature and velocity of the heat exchange fluid on the phase change heat storage through the cloud diagram obtained by numerical simulation. It was found that when the flow rate and temperature of the heat exchange fluid are increased, the natural convection and heat storage inside the phase change material can be strengthened to a great extent.Wei Lu[2] et al focused on the changes in the density, viscosity, specific heat, etc. of composite materials with the particle share, and made detailed descriptions of the classic thermal conductivity model, the main production methods of composite nanomaterials and their existing problems. Qiang Li[3] further studied the Brownian motion of nanoparticles in composite materials based on the classic model. Wongwises[4] and others used experimental hot-wire methods to conduct experiments and measured the thermal conductivity and viscosity of composite materials. As a result, it is found that the thermal conductivity of the composite material is greatly increased relative to the basic material, and the viscous resistance of the particles will also increase.

So far, although many scholars have done in-depth research on the overall phase change thermal storage capsule. However, there are few researches on adding high thermal conductivity nanoparticles to the basic phase change material heat storage capsule to form a composite phase change material to change the performance of the phase change material, thereby enhancing the phase change heat storage.

2. Composite material phase change capsule model

This model uses the organic phase change material paraffin wax (RT44) as the heat storage material, and its melting temperature is 44 °C. It is known from the literature[2] that the carbon nano-particles at a lower proportion have little effect on the phase change temperature and can be ignored. . The latent heat of phase change is 250,000 J/kg. For the dynamic viscosity of water at the flow temperature, the value is 0.00035635 kg·(m3·s)$^{-1}$ according to reference. In this model, the flow rate of the heat exchange fluid is 1.2 m/s, the initial temperature is 350 °C, and the initial temperature of the phase change material is 313 °C. It only compares the difference in the phase change rate caused by the addition of nanoparticles with different physical properties. The phase change heat storage model used in this paper is designed based on the visual experiment of the foreign researcher GROY T[5] and others with the heat storage packaging container with a cavity above. The heat storage capsule radius is 38.1 mm (The void volume accounts for about 10.2%).As shown in Figure 1:

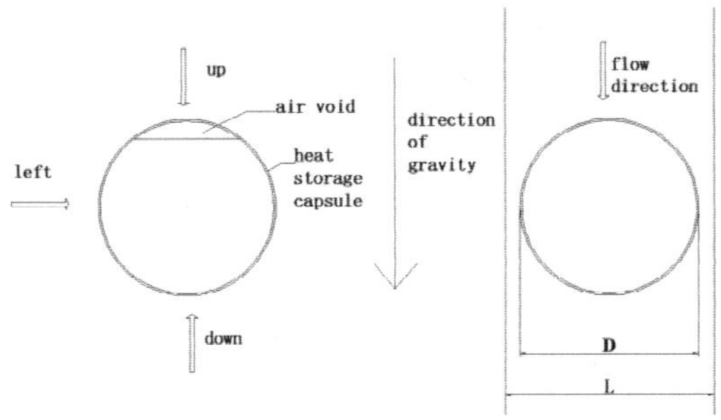

Fig.1 Physical model of a composite phase change capsule with holes

3. Selection of basic phase change materials

The basic phase change material used in this article is RT44. In order to simulate the phase change process of melting and solidification, the physical parameters are provided by its manufacturer. As shown in Table 1, its melting temperature is between 42 and 45 °C, and its volume expansion coefficient is 0.008 (1/k). This model selects stainless steel as the material of the outer wall of the phase change heat storage capsule, and the heat exchange fluid selected is water.

Table 1 Physical parameters of related materials

Material	ρ /kg·m–3	k /W·（m·K）–1	Cp /J·（kg·K）–1	L /KJ·kg–1
RT44	770（l）	0.2	-	250
	880（s）	0.2	-	
Water（350 K）	971.8	0.6	4195	
steel	8030	16.27	502.48	

4. Simulation results and analysis

4.1. Model argument

In order to verify the validity of the model and the feasibility of setting the material properties, the model is set up in accordance with the literature[6], and the simulated solidification process is analyzed. As shown in Figure 2, from the apparent point of view, at 600 s and 1500 s, the model simulation method in this paper can be well in line with the actual simulation results, so the phase change analysis method of the composite paraffin wax phase change capsule in this article has a certain degree of reliability. When setting in FLUENT, the residual error of the energy equation is less than 10^{-6}, and the residual error of the continuity and momentum equations is less than 10^{-3}. The selected time step is 0.001 s, and the number of grids is 243059.

Journal of Physics: Conference Series **2194** (2022) 012011 doi:10.1088/1742-6596/2194/1/012011

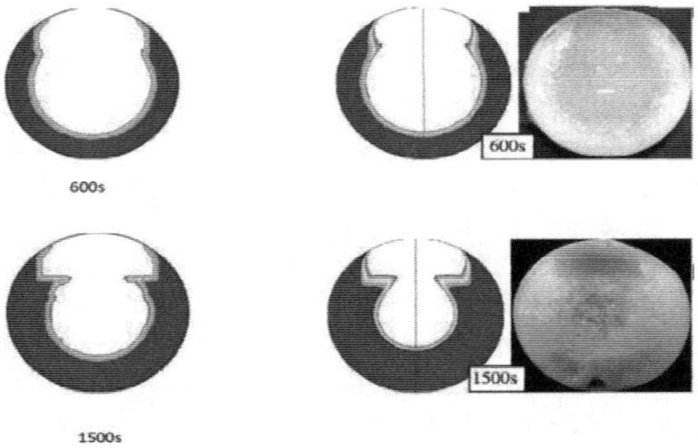

Fig.2 Comparison of model simulation and experiment in this paper

4.2. Analysis of flow field inside graphite/paraffin composite phase change capsule

Figure 3 shows the streamline diagrams of the graphite particle composite paraffin phase change capsules. The figure shows the intensity of thermal disturbance in the melted material during the melting of the composite phase-change heat storage material inside the capsule. At first, due to the high wall temperature, under the combined action of gravity and buoyancy, the melted liquid of the composite phase change material close to the wall rises along the wall, and then under the action of its own gravity, it rises again. When it falls, it will convectively exchange heat with the unmelted composite phase change material, thereby increasing the melting speed. When graphite nanoparticles are added, the flow lines of the internal flow field of the composite phase change material increase during the melting process, and the disturbance is strengthened. It shows that the addition of graphite particles makes the internal structure of the composite phase change material easier to melt, and during the melting process, the Brownian motion of the nanoparticles enhances the internal flow disturbance. The particle motion further strengthens the thermal disturbance of the melting process, which increases the streamline in the same time.

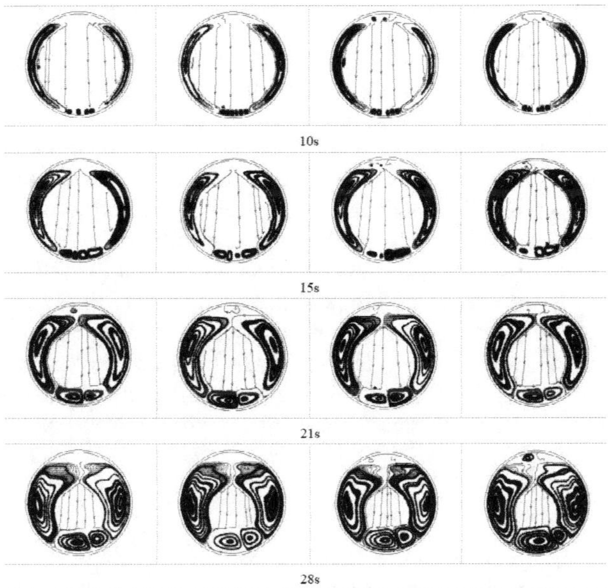

Fig.3 Streamline diagrams of graphite/paraffin composite phase change materials with different proportions of 0, 0.85%, 1.25%, 1.75% at different times

4.3. Analysis of heat transfer inside graphite/paraffin composite phase change capsule

Figure 4 shows the velocity vector diagrams of the composite paraffin phase change capsules . It can be seen from the figure that as the graphite nanoparticles are added to the basic phase change material and the volume share of the nanoparticles increases, the unsteady natural convection heat transfer increases and the melting process is accelerated. The addition of graphite particles changes the structure of the composite phase change material and strengthens its thermophysical properties. When the composite phase change material melts, more graphite particles can perform random Brownian motion, which strengthens the interaction between the graphite particles in the composite phase change material and the surrounding molecules. As time increases, the viscosity of the basic phase change material decreases and its temperature increases, which makes the Brownian motion of the graphite particles more intense and strengthens natural convection.

4.4. Analysis of Convection Heat Transfer Outside Graphite/Paraffin Composite Phase Change Capsule

Figure 5 shows that the convective heat transfer coefficient of the graphite/paraffin composite phase change heat storage material can be expressed as the composite phase change material in different proportions . The relationship between Nusselt number (Nu) and Reynolds number (Re). The Nusselt number in this paper is calculated at the 0°point of the capsule. The Nusselt number (Nu) and Reynolds number (Re) of composite phase change materials can be obtained according to the literature[7,8].

The convective heat transfer coefficient of graphite/paraffin composite phase change heat storage material is:

$$h_\theta = -k_w \frac{dT}{dr} \cdot \frac{1}{T_w - T_\infty} \tag{1}$$

Where: h_θ— local convective heat transfer coefficient, W/(m2·K)

θ is the sphericity measured from the stagnation point:

$$Nu_\theta = \frac{h_\theta d}{k_f} = -\frac{k_w D}{k_f\left(T_w - T_\infty\right)} \cdot \frac{dT}{dr} \tag{2}$$

It can be seen from the figure that the Nusselt number of the composite phase change material after adding graphite particles will increase compared to the basic phase change material at the same Reynolds number, and as the share of graphite particles gradually increases, the Nusselt number becomes relatively higher. This article uses formulas (1) and (2) to calculate the Nusselt number when the composite phase change thermal storage capsule melts under the corresponding Reynolds number.

Since the addition of graphite particles changes the overall structure of the composite phase change material, the overall characteristics of the capsule are more conducive to heat transfer, the melting of the material within a certain period of time is accelerated, and the temperature change is more obvious. Moreover, during the melting process, the temperature of the basic phase change material gradually increases, and the viscous resistance will also be reduced to a certain extent, which makes the melting process faster. Moreover, the dynamic irregular movement of graphite particles becomes more violent as the melting progresses, which further improves the overall thermal conductivity, reduces viscous resistance, strengthens convective heat transfer, and makes the time to reach the melting temperature shorter and shorter.

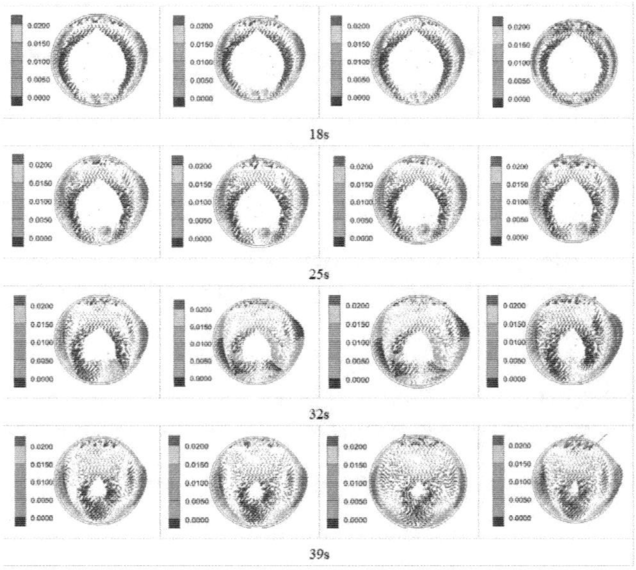

Fig.4 Velocity vector diagrams of composite phase change materials with different proportions of 0, 0.85%, 1.25%, 1.75% at different times

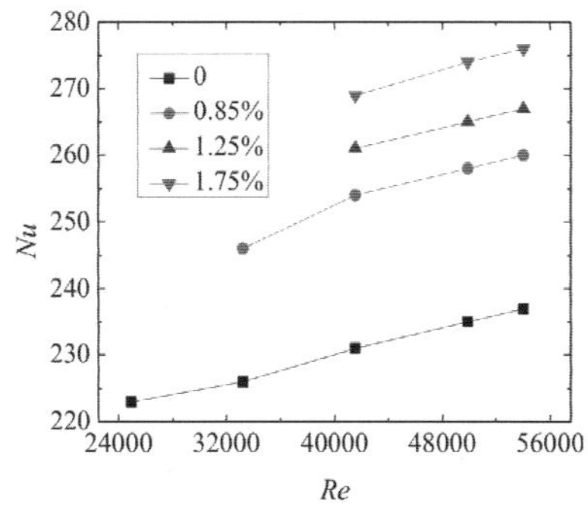

Fig.5 The relationship between Nu and Re under different composite phase change materials

5. Conclusion

The addition of graphite nanoparticles to the composite material is easier to melt. The movement of the nanoparticles further enhances the thermal disturbance and strengthens the natural convection. However, to a certain extent, because the viscosity increases, the strengthening is not obvious, and the movement speeds up and strengthens the convective heat transfer.

Acknowledgments

This work was funded by the 2020 Dezhou Municipal R&D Project.

References

[1] Yongqiang Liu. Numerical study of phase-change heat storage capsule and its accumulation heat storage characteristics [D]. Jilin: Northeast Dianli University, 2018.

[2] Wei Lu, Yongwei Wu. Preparation and performance study of nano-aluminum powder/paraffin/expanded graphite composite phase change material[J]. Thermal Energy and Power Engineering, 2017, 32(2): 101-105+140-141.

[3] Qiang Li. Research on enhanced heat transfer mechanism of nanofluids [D]. Jiangsu: Nanjing

University of Science and Technology, 2004.
[4] Duangthongsuk W, Wongwises S. Measurement of Temperature-dependent Thermal Conductivity and Viscosity of TiO2-water Nanofluids[J]. Experimental Thermal & Fluid Science, 2009, 33(4): 706-714.
[5] Revankar S, Croy T. Visualization study of the shrinkage void distribution in thermal energy storage capsules of different geometry[J]. Experimental Thermal and Fluid Science, 2007, 31: 181-189.
[6] Assis E, Zisking K, Letan R. Numerical and Experimental Study of Melting in a Spherical Shell[J], International Journal Heat and Mass Transfer, 2007, 50(9-10): 1790-1804.
[7] Buyruk E. Heat Transfer and Flow Structure Around Circular Cylinder in Cross-Flow[J]. Tr. J. of Engineering and Environmental Science, 1999, 23: 299-325.
[8] Ashjaee M, Amiri S, Habibi K. Slot Jet Impingement Heat Transfer from an Isothermal Circular Cylinder[J]. Thermal Issues in Emerging Technologies, 2008, 2: 399-404.

Analysis of engineering application requirements of nano coating materials

Wenan Zhong[1*], Dongjiu Zhang[1], Yuchao Wang[1], Yu Zhao[1], Jianfeng Liu[1]

[1]Xichang Satellite Launch Center

Corresponding author's email: wangyuchao@mail.sysu.edu.cn

Abstract: Coatings are the primary means to control the corrosion of marine engineering structures. It is the frontier and hot spot of nano coating materials research to use nano materials to modify anti-corrosion coatings, realize the complementary advantages between different materials, and prepare high-performance, environmental protection, energy saving and functional anti-corrosion coatings. However, with the application of nano coating materials in the atmospheric subtropical coastal environment, there are many problems of poor weatherability. Based on the environmental corrosion characteristics of the subtropical coastal zone, this paper evaluates the environmental adaptability of several nano coating materials in the coastal atmospheric environment, analyzes the engineering application requirements of nano coating materials, and puts forward the measures and suggestions for the development of nano coating materials in marine engineering in view of the main problems existing in the engineering application of nano coating materials in China.

1. Introduction

With the development of global marine resources, the major infrastructure and important industrial facilities are gradually pushed to the deep blue sea. However, the harsh service environment of the ocean has caused extremely serious corrosion of the infrastructure in the marine climate environment, which endangers the service life and safety of the marine infrastructure [1-2]. Moreover, the marine corrosion situation is complex and different service areas have different corrosion rates, bringing severe challenges in the development of corrosion-resistant materials and corrosion protection [3]. Energy chemistry in extreme environments is an important part of the national energy science and technology strategy. China is one of the few countries that have conducted deep-sea corrosion material tests [4], which provide a reliable basis for anti-corrosion research in marine atmospheric environment.

Coating is a primary means of corrosion control of marine engineering structures. Nanomaterials have become ideal materials for corrosion protection due to their excellent properties. In recent years, the research and application of new nano-coating materials have attracted a lot of attention [6]- [10]. Because of their convenient construction and widespread application, nano-coating materials have become one of the important factors that affect the service safety, life span and reliability of ships, offshore engineering, and ocean-going facilities and have attracted great attention from governments and industries all over the world [11]. Therefore, it is of great significance to develop the anti-corrosion coatings and coating technology in marine engineering and to establish the coating requirements of protective coating, which can effectively reduce the corrosion of exposed carbon steel, stainless steel, and aluminum components. In this way, the service safety and reliability of marine engineering and ships can be ensured, the occurrence of major disasters can be reduced, and the service life of marine structures can be prolonged.

Based on the characteristics of environmental corrosion in the coastal zone of the South China Sea, this paper evaluated the environmental adaptability of several nano-coating materials in the nearshore atmospheric environment and analyzed the engineering application requirements of nano-coating materials. In addition, the measures and suggestions for the development of nano-coating materials in marine engineering were put forward in view of the main problems existing in the application of nano-coating materials in marine engineering in China.

2. Characteristics of environmental corrosion in the coastal zone of the South China Sea

The high salt concentration (generally around 3.5%), rich oxygen, and numerous marine microorganisms and macroorganisms in seawater, together with the impact of waves and sunlight, resulted in severe marine corrosion environment [3]. The corrosion severity of seawater in typical sea areas of China was evaluated based on the metal corrosion rate, which showed that the seawater severity of Nanhai island was higher [12]. Atmospheric corrosion is classified into 6 categories in ISO 12944, with the highest corrosion grade in the marine environment. Due to the influence of sodium chloride particles contained in seawater droplets, the corrosion in the coastal land environment within 200 m also belongs to the category of corrosion in the marine environment [13].

Hainan Island is located in the coastal zone of the northern edge of the tropics. The whole island is surrounded by the coastline. It has the climate characteristics of high temperature, high humidity, thunderstorm, heavy precipitation, tropical cyclone landing, high salt fog and strong radiation [14]. The characteristics of tropical marine atmospheric environment are obvious. The whole island of Hainan, especially the Wenchang area, has high temperature all year round. The annual average temperature is 24.1 ℃, the annual average relative humidity is 86.4%, the annual average dew point temperature is 21.8 ℃, and the annual average sunshine hours is 1803.6 h/a [15].

In order to evaluate the environmental corrosion law of the northern coastal zone of the South China Sea more effectively, two environmental monitoring sites with different distances from the coastline in Wenchang area were selected. Based on the results of long-term environmental monitoring, environmental data from the southeastern China Sea region were analyzed to determine the characteristics of environmental corrosion.

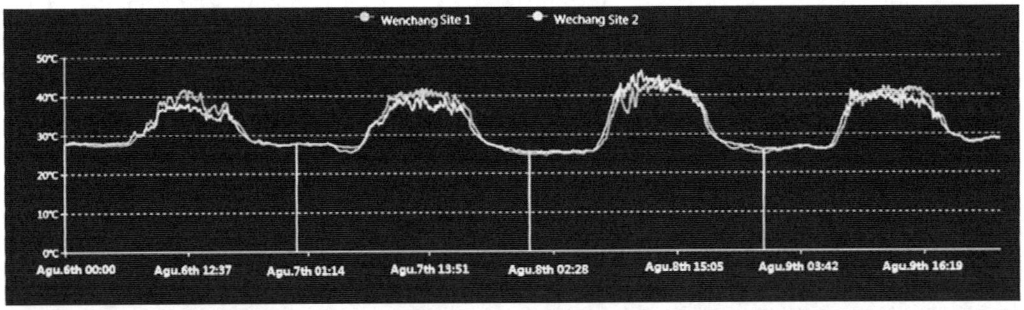

Figure 1. Temperature change curves of the two monitoring sites

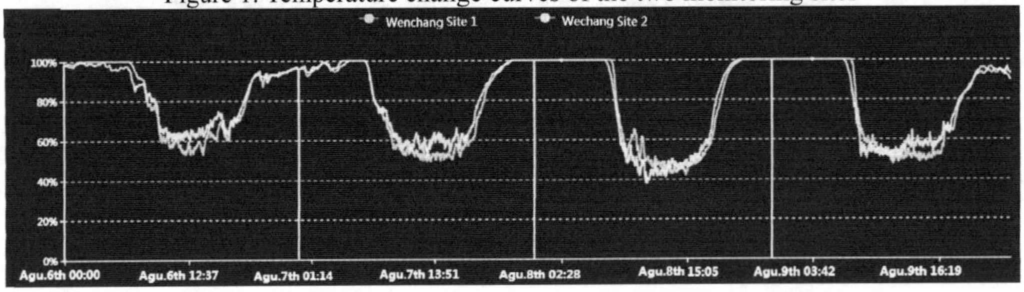

Figure 2. Material surface humidity change curves of the two monitoring sites

The changes in temperature and humidity of the two environmental monitoring sites at different distances from the coastline are basically the same. Figure 1 and Figure 2 show the temperature change curves and the material surface humidity change curves of the two sites, respectively.

The environment of the two sites was monitored for a long time by an online corrosion monitor, as shown in Figure 3. According to the ISO9223-9226 atmospheric corrosion classification standard, the environmental corrosion level of the two sites both reached level 5. The annual wetting time exceeded 6,500 hours, and the level was ζ5. As can be seen from Fig. 3, the atmospheric environment of the two sites was typical of high temperature, high humidity, high irradiation, and high chloride ion deposition environment. In this environment, the materials were continuously subjected to a drying-wetting cycle. The corrosion process was directly related to the surface wetting time.

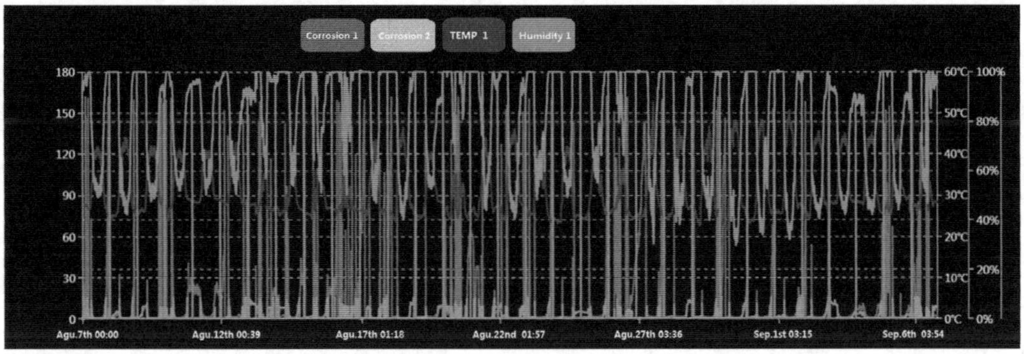

Figure 3. Long-term instantaneous corrosion rate curves of carbon steel and zinc alloy

A comprehensive analysis of the surface wetting status of each time period in a single day shows that the temperature and humidity were basically the same in the time period of 00:00–08:00, during which the characteristics of high temperature and high humidity were maintained, and the sample surface was in a state of condensation. In the time period of 08:00–12:00, the humidity decreased rapidly and the temperature rose rapidly, which corresponded to the drying process of the sample surface. During the period from 12:00 to 16:00, the temperature and humidity were stable again and the sample surface was further dried. During the time periods of 16:00–20:00 and 20:00–24:00, the surface of the sample was wetted, which was the reverse process of the first two stages.

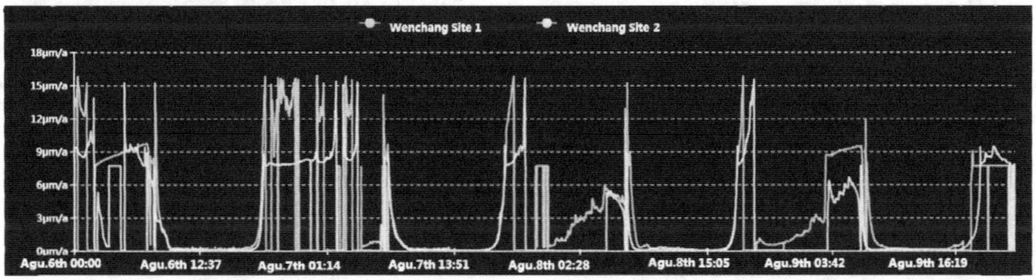

Figure 4. Corrosion behaviors of zinc alloy at the two sites

An online corrosion monitor based on electrochemical reaction mechanism of metal corrosion was used to analyze the instantaneous environmental corrosion degree. The corrosion law of zinc alloy at the two sites was compared, as shown in Figure 4. The annual corrosion rate of the two sites was quite different. The instantaneous corrosion rate and the average annual corrosion rate of site 1 were both half of those of site 2. It was found that the main factor affecting the corrosion rate was the deposition rate of sea salt particles at the two sites. The deposition rate of sea salt particles in site 1 was 0.1841 mg/

(100 $cm^2 \cdot d$) at site 1 and 0.3947 mg/ (100 $cm^2 \cdot d$) at site 2, which was proportional to the corrosion rate.

Wenchang launch site is located in low latitude area and close to the sea, with high annual average temperature, long sunshine time, high relative humidity, and extremely bad corrosion environment. The corrosion rate of carbon steel in the offshore field (site 1) is 0.3274 mm/a. According to the definition of corrosion environment in ISO 12944 and ISO 9223, the corrosion environment in Wenchang coastal zone belongs to the typical C5-M~Cx harsh corrosion environment grade.

3. Comparative analysis of environmental adaptability between nano-coating materials and traditional heavy-duty anti-corrosion coatings

Anti-corrosion coatings are applied to the surface of base course. After a polymer film is formed and cured, a mysterious film is formed, which isolates the protected substrate from the external corrosive medium, so that the corrosion reaction does not occur or the corrosion rate is delayed. In recent years, more and more researchers in our country began to pay attention to the research of nanoparticles to improve the corrosion resistance of coatings, some of which have entered the engineering application stage.

3.1 Materials and experimental methods

The base sample material is Q345B steel. The specific chemical composition is shown in Table 1. The size of the sample sheet is 100 mm × 50 mm × 3 mm.

Table 1. Chemical composition of Q345 steel as base material

Element	C	Si	Mn	P	S	Fe
Content（%）	0.16	0.43	1.56	0.016	0.007	97.827

Coating samples were selected based on environmental adaptability evaluation and corrosion resistance test of 4 kinds of coating materials and traditional heavy-duty coatings. The coating systems were shown in Table 2.

Table 2 Coating system to be tested

Coating System Number	Coating Systems	Film Thickness
1	Metal polymer nano-primer (90 μm, 3 coats) + metal polymer nano-intermediate paint (80 μm, 2 coats) + weather resistant metal polymer topcoat (30 μm)	200 μm
2	Nano composite emulsion coating, film thickness about 400 μm, primer (140 μm, 2 coats) + intermediate paint (120 μm, 2 coats) + weather resistant nano-composite emulsion topcoat (120 μm, 2 coats)	400 μm
3	SLF solvent-free nano-composite polymer epoxy coating (250 μm, 2 coats) + SLF fluorocarbon weather resistant topcoat (100 μm, 2 coats)	350 μm
4	KDF-111: EP108 modified epoxy primer (100 μm) + SHE-53-100 high solid epoxy primer (50 μm) + NM-01 nano-varnish (100 μm)	370 μm
5	725-H52-77 thick build epoxy anticorrosive paint (80 μm, 1 coat) + 725-H45-ZF101G heavy anticorrosive paint (300 μm, 2 coats) + 725-XF43-26 fluorocarbon topcoat (50 μm, 1 coat)	430 μm

Sea salt particles are chlorides, which can penetrate into the coating material and cause corrosion. Therefore, the cyclic salt spray resistance performance test was carried out referring to the cyclic A test procedure in the standard GB/T 31588.1-2015 "Paints and varnishes - Determination of resistance to cyclic corrosion conditions - Part 1: Wet (salt fog) / dry / humidity". The test was carried out in accordance with the cycle A of artificial weathering (method 1) in GB/T 1865-2009 "Paints and varnishes - Artificial weathering and exposure to artificial radiation - Filtered xenon-arc radiation".

3.2 Analysis of coating resistance to circulating salt spray

Table 3 showed the appearance of samples at different assessment stages during the cyclic salt spray test. According to the table, after 0-720 h cyclic salt spray test, there was basically no corrosion on the surface of the sample. After 960 h cyclic salt spray test, slight stain and rust began to appear on the surface of the sample surface. Some of them even bubbled and peeled off.

Table 3 Analysis of surface corrosion macro morphology in the cyclic salt spray resistance test

Number	Corrosion Morphology with Time in Cyclic Corrosion Environment Test					Conclusion
	120	240	480	960	1440	
1	no corrosion	no corrosion	no corrosion	the surface is dirty	the surface is dirty	the surface is dirty
2	no corrosion	no corrosion	no corrosion	slight discoloration of the surface	discoloration of the surface, 1 bubble	discoloration of the surface, 1 bubble, a small amount of rust and peeling on the edge, slight staining of rust
3	no corrosion	no corrosion	no corrosion	no corrosion	no corrosion	no corrosion, the surface is slightly stained
4	no corrosion	no corrosion	no corrosion	the edge is slightly stained	the edge is slightly rusted, peeled, bubbled, and slightly stained	the edge is slightly rusted, peeled, bubbled and slightly stained
5	no corrosion	no corrosion	no corrosion	no corrosion	no corrosion	no corrosion

Table 4 shows the comprehensive rating results of the surface state of the test sample after 1440 h cyclic salt spray test. According to GB/T 1766-2008 "Paints and varnishes - Rating schemes of degradation of coats", the comprehensive ratings of the five types of test samples for cyclic salt spray resistance are level 1, level 4, level 1, level 3, and level 0 respectively. The cyclic salt spray corrosion resistance of the whole set of nano-coating materials is the worst, followed by the micro nano-coating materials. When nano-coating materials are used as primer, their performance is slightly better, but all of them are worse than the traditional heavy-duty anti-corrosion coating.

Table 4 Comprehensive rating of cyclic salt spray resistance test

Number	Test Time	Light Loss Rate%	ΔE	Loss of light	Discoloration	Pulverization	Pangold	Spot	Stain	Crack	Blistering	Mold	Peel Off	Rust	Comprehensive grade
1	1440 h	25.5	2.9	2	1	0	0	0	0	0	0	0	0	0	1
2		4.2	6.0	1	2	0	0	0	2	0	1	0	2	1	4
3		16.9	2.5	0	1	0	0	0	1	0	0	0	0	0	1
4		9.7	3.9	0	2	0	0	0	2	0	1	0	1	1	3
5		0.2	1.4	0	0	0	0	0	0	0	0	0	0	0	0

3.3 Analysis of aging resistance of coating to xenon lamp

Table 5 shows the comprehensive test results of the surface state of the five types of test samples after 720 h xenon lamp aging test. Xenon lamp aging mainly assesses the ultraviolet (UV) aging resistance of the topcoat. According to GB/T 1766-2008, the comprehensive ratings of the five types of test samples for resistance to cyclic salt spray are level 5, level 3, level 0, level 3 and level 0 respectively. The fluorocarbon topcoat exhibits the best performance and the nano-coating system shows the worst performance. The light loss rate has reached 59.7%.

Table 5. Comprehensive rating of xenon lamp aging resistance test

Number	Test Time	Light Loss Rate%	ΔE	Loss of Light	Discoloration	Pulverization	Pangold	Spot	Stain	Crack	Blistering	Mold	Peel Off	Rust	Comprehensive grade
1	648 h	59.7	17.3	4	5	2	0	0	0	0	0	0	0	0	5
2	720 h	71.3	4.2	4	2	0	0	0	0	0	1	0	0	0	3
3		10.5	0.3	1	0	0	0	0	0	0	0	0	0	0	0
4		31.4	0.3	3	0	3	0	0	0	0	0	0	0	0	3
5		13.1	0.1	1	0	0	0	0	0	0	0	0	0	0	0

3.4 On-site adaptability assessment of coating system

Table 6 shows the on-site inspection results of in-service coatings of 5 types of coating systems, which evaluates the protective performance and environmental adaptability of the coating systems.

Table 6 Comprehensive rating of on-site adaptability assessment of coating systems

Coating system number	Coating thickness (μm)	Comprehensive Appearance Rating	Adhesion	Electric spark leak detection	Evaluation of coating system
1	238	Level 1	5A(Crossing method)	There is a leak at the weld	The construction of the project was completed in September 2016. After 2 years of service, the surface of the coating was slightly stained and there were anti-rust and leakage spots at the weld.
2	697	Level 2	5.32MP(Drawing method)	No leakage	The mailing sample was tested on site after 6 months of service. The coating thickness exceeded the standard and the coating surface was blistered.
3	637	Level 1	5A(Crossing method)	No leakage	The test construction was completed in May 2018. After 4 months of service, the on-site test showed that the coating thickness exceeded the standard and the coating surface was slightly blistered.
4	371	Level 3	5A(Crossing method)	No leakage	The test construction was completed in April 2018. After 5 months of service, the on-site test showed that the coating was obviously stained and blistered and the coating performance remained relatively good.
5	535	Level 1	4A(Crossing method)	No leakage	The test construction was completed in April 2018. After 5 months of service, the on-site test showed that the coating performance remained good.

Based on the analysis of laboratory results and on-site test results, it is found that some of the domestically produced nano-coating materials have relatively poor environmental adaptability in the Wenchang coastal zone. In particular, the aging resistance of the topcoat and the barrier performance of the intermediate paint both showed a gap with traditional heavy-duty anti-corrosion coating materials.

4. Analysis of engineering application requirements of nano-coating materials

China's marine coatings market is almost completely occupied by foreign countries. In particular, the coatings for ocean-going vessels, offshore platforms paint, and antifouling coatings are all foreign coatings. As far as the technical level is concerned, some domestic coating technologies have reached an applicable level, but there is a lack of opportunities for practical engineering applications. This not only affects the development of relevant domestic key technologies, but also affects the application of offshore platforms built in China in foreign countries [3].

Traditional anti-corrosion coatings contain heavy metals and some difficult-to-degrade organics, which will harm the environment no matter in the production process or in the use process. Therefore, the current field engineering applications requires nano-coating materials with performance comparable to traditional heavy-duty anti-corrosion coating materials, focusing on environmental protection, energy saving, resource saving, high performance, and functional development [17]. In order to expand the engineering application of nano coating materials in the South China Sea, several requirements need to be improved:

(1) Improving the barrier properties of nano-coating materials

By adjusting the composition of the nano-coating material, improving the microstructure of the coating, adjusting the combination of nanoparticles and organic solvents, a dense coating structure can be formed, which can significantly reduce the porosity of the coating and effectively block the penetration of water, oxygen, and corrosive ions such as chloride ions.

(2) Improving the weather resistance of nano-coating materials

The introduction of nanoparticles [18]- [19] can improve the rheology, adhesion, hardness, smoothness, and anti-aging properties of coatings. However, most of the nano-coating materials on the market are sensitive to UV radiation and are prone to light loss and pulverization. The research of nano-modified topcoat is highly desired. Modified materials such as super weather-resistant topcoat-fluorocarbon resin and fluorine-containing polyurethane are excellent choices for topcoat base materials. In addition to being used for hull paint, they can also be used for interior coatings in contact with strong corrosive media.

(3) Developing environmental friendly nano-coating materials

Environmentally-friendly high-performance heavy-corrosion protective coatings and environmentally-friendly long-life marine antifouling coatings are one of the important development directions for the sustainable development of "Green Hills and Clear Waters" and "the Belt and Road Initiative". How to effectively use the characteristics of nanomaterials and improve their environmental protection is one of the practical engineering application problems that need to be solved.

(4) Investigating the compatibility between nano-coating materials and traditional coatings

5. Conclusion

In view of the slight corrosion on the surface of the in-service anti-corrosion coating of some carbon steel structures in the marine atmospheric environment, the surface paint can be polished off and directly applied to the paint brand that has passed the recoatability test to renovate the coating. Solving the compatibility between nano-coating materials and traditional coatings is an important means to improve the efficiency of material application.

In addition, promoting the formulation of advanced anti-corrosion materials and technical standards and specifications, deepening the basic theoretical research on nano-coating materials, and increasing the anti-corrosion design before use are all key steps to further promote the application of advanced nanomaterials in anti-corrosion engineering.

References

[1] Duan T, Xu L, Ding K, Peng W, Cheng W. Corrosion behaviour investigation of 460 low alloy steels exposed in the natural deep-sea environment. Corros. Eng. Sci. Technol. 2019;

54(6244):1–8.
[2] Sun F, Li X, Lu L, Wan H, Du C, Liu Z. Corrosion behavior of copper alloys in deep ocean environment of south china sea. Acta Metall. Sin. 2013; 54(10):1211-1218.
[3] Zhou L. Consultation Report on Development & Strategy of Engineering Materials for Ocean Applications in China. Beijing: Chemical Industry Press Co., Ltd.; 2014. 3-5.
[4] Liang Z, Jiang W, Zhang G. Strategic development and discipline layout of energy chemistry of National Natural Science Foundation of China. Chin. Sci. Bull. 2021; 66(9):974-979
[5] Peng W, Duan T, Hou J, Guo W, Ding K, Cheng W, Xu L. Long-term corrosion behaviour of 1060 aluminium in deep-sea environment of South China Sea. Corros. Eng., Sci. Technol. 2021; 56(4):327-340.
[6] Zhao Y, Shen Z, Zhang X, Zhao X. Research progress on application of two-dimensional nano-materials in corrosion protection. China Powder Sci. Technol. 2021; 27(01):11-21.
[7] Research and application of nano-spraying materials. Inner Mongolia Electric Power. 2020; 38(04):3.
[8] Zhou Y, Liu X, Kang J, Yue W, Qin W, Ma G, Fu Z, Zhu L, Ding S, Wang H, Liang J, Weng W, Wang C. Corrosion behavior of HVOF sprayed WC-10Co4Cr coatings in the simulated seawater drilling fluid under the high pressure. Eng. Failure Anal.109:104338 ;2020 .
[9] Wang W, Wang H, Zhao J, Wang X, Xiong C, Song L, Ding R, Han P, Li W. Self-healing performance and corrosion resistance of graphene oxide-mesoporous silicon layer-nanosphere structure coating under marine alternating hydrostatic pressure. Chem. Eng. J. 2019; 361:792-804
[10] Mu X, Hu Y, Jin T, Liu C, Wang H. Corrosion Behavior of New Nano-coating in accelerated corrosion Environment. Equip. Environ. Eng. 2020; 17(02):31-40
[11] Han E, Chen J, Su Y. Corrosion and protection for marine offshore and coastal structures. Chemical Industry Press Co., Ltd. 2016. 169-171.
[12] Ding K, Zhang P, Liu S, Fan L, Guo W, Hou J. Study on the classification of seawater corrosivity of typical sea areas in China. Corros. Rev. 2020; 38(4):243-256.
[13] Ke W. Corrosion Investigation in China. Beijing: Chemical Industry Press Co., Ltd. 2003.
[14] Huang T. Research on Corrosion Behavior of weathering Steel in Marine atmosphere of the South China Sea. Central Iron & Steel Research Institute. 2018.
[15] Mu S. Analysis of the influence of high temperature high humidity and strong solar radiation on Launch Site Facilities and TT&C Systems. Journal of Spacecraft TT&C Technology. 2009; 28(5):24-29.
[16] Guo X, Ge S. Research progress in application of nanomaterials in waterborne anticorrosive coatings. Paint Coat. Ind. 2017; 47(8): 83-87.
[17] Xu B, Ou Z, Ma S, Qian Y, Zhang W. Nano Surface Engineering. China Mech. Eng. 2000; 11(6):708.
[18] Zhu X, Yan Q, Yu Y, Zhao H, Xue Q, Wang L. Orientation of Ultrathin α-ZrP Nanosheets in Aqueous Epoxy Resin for Anticorrosive Coatings. ACS Appl. Nano Mater. 2021; 4(5):5413-5424.
[19] Wang W. Preparation of new polymer nanocomposites and analysis of their superhydrophobic properties. Earth Environ. Sci. 2021; 714(3):032072.

Journal of Physics: Conference Series **2194** (2022) 012013 doi:10.1088/1742-6596/2194/1/012013

Impact Damage Characteristics of Quartz Fiber and NOMEX Paper Honeycomb Sandwich Panel composite materials

Xiang Zhou[1a*], Zhandong MA[1b], Xiaofei Liu[1c]

[1]AVIC Research Institute for Special Structures of Aeronautical Composites, Aviation Key Lab of Science and Technology on High Performance Electromagnetic Windows Ji'nan, Shandong, China

[a*]zhouxiang217@163.com, [b]eastma2002@126.com, [c]m13097205891@163.com

Abstract—To provide a reference for practical engineering application of fiber composite materials, this work researches on the impact resistance of type-B quartz fiber/modified cyanate ester composite material laminates and NOMEX paper honeycomb sandwich panels. Firstly, low-speed impact tests are carried out on the quartz fiber laminates and the NOMEX paper honeycomb sandwich panels by setting different parameters to analyze the impact energy and punch mass on impact damage; secondly, based on the experimental results, impact responses are analyzed; and the damage morphology of the two composite materials is observed through visual inspection and with the aid of the CT scanning technology. The results show that according to time-load curves, the NOMEX paper honeycomb sandwich panels have two peak load moments, respectively corresponding to the moments when the upper and lower panels are penetrated; the quartz fiber laminates show three different types of failures, namely substrate cracking, fiber fracture and delamination. As for the NOMEX paper honeycomb sandwich panels, two types of damage are found: fiber fracture and honeycomb core extrusion fracture. When the punch mass is constant, the greater the impact energy is, the more serious the damage is; and when the impact energy is constant, the greater the punch mass is, the more serious the damage is.

1. Introduction

Different from the anisotropic characteristics of metal materials, fiber composite materials with such excellent properties as high modulus and high strength have been extensively applied to aircraft structures, such as aircraft radomes, antenna housing, and leading edges of wings [1,2]. As a structural/functional component, it plays an important role in the aircraft structures to ensure the proper functioning of antennas and microwave systems in harsh environments. However, because the composite materials are often prone to shock loads due to external impacts during manufacturing, service and even maintenance (for example, bird strikes, hail strikes and tool drops), especially damages caused under low-speed impact conditions, its structure will be changed due to external impact and its service life will be shortened accordingly [3,4]. The "Composite Materials Handbook" (CMH-17) points out that low-velocity impact on the composite material structure refers to "the fall of object weighing 10-20 pounds from a height of several feet" [5,6]. In addition to visible damage on the surface, there are invisible impact damages in the composite materials, which will be more serious to jeopardize the mechanical properties and structural integrity of the composite materials [7]. Hence, this paper researches on on the impact response and damage types of the composite materials under low-speed impact loads through evaluation and analysis to lay a basis for the practical application of the composite materials.

Many scholars at home and abroad have made theoretical and experimental studies on the impact damage performance of the composite materials. Shen Zhen [8] made analysis on the impact damage state of composite materials and divided them into low-energy without damage, small-energy with non-obvious damage, obvious indentation and penetrating damage. Hou [9] conducted an impact test on the T300 grade CFRP laminate with a pneumatic gun. The test shows that the damage mainly acts on the substrate extrusion and delamination failure. In thin ply laminates, fiber breakage becomes more relevant while matrix cracking is reduced and the consequent induced delamination is practically suppressed. It is thought that interlaminar stresses and fiber breakage induce delamination [10,11]. Choi's research [12,13,14] reported that low velocity impact events generated matrix cracks, which lead to delamination damage. This paper presents results of experimental studies on the effects of different punch mass and energies for the impacts performed.

2. Experimental

2.1. Test materials

Material 1: The Type-B quartz fiber/modified cyanate ester composite material laminate is prepared by AVIC Research Institute for Special Structures of Aeronautical Composites; structure $[(+45/-45)/(0/90)]_{5s}$, thickness of the single-layer board of 0.2mm, resin content 38%, and the size of the specimen 150×100×4.0mm.

Material 2: The NOMEX paper honeycomb sandwich panel is made of upper and lower composite material panels and the honeycomb core in the middle. The panel material is made of the B-type quartz fiber/modified cyanate ester composite material. There are 4 layers in the upper and the lower panels respectively, each of which is 0.2mm thick; the honeycomb core is of a regular hexagonal NOMEX paper honeycomb with nominal density of 80 kg/m^3, hole size of 2mm, wall thickness of 0.05mm and height of 10mm; and the overall layering sequence of the sandwich panel is $[0]_4$/honeycomb/$[0]_4$, and the size of the specimen is 150×100mm.

2.2. Test scheme

The low-speed impact test is carried out according to ASTM D7136. The drop hammer impact test machine (equipment model ZCJ9302) is manufactured by MTS Industrial Systems (China) Co., Ltd. The test equipment includes the main frame, the transmission system, the secondary impact prevention device, the hammer, impact head, and supporting fixture, etc. The impact testing machine and specimen fixture are shown in Fig.1(a) and 1(b) respectively.

(a) Drop hammer impact testing machine (b) Specimen fixture

Fig.1 Low-speed Impact Test Device

Two sets of experiments: 1. Set the punch mass unchanged, and change the impact energy by setting different punch speeds; 2. Change the combination of impact energy and different punch masses to perform the test with constant impact energy. The test schemes are shown in Table 1 and Table 2. The above two test schemes are applied to the impact tests on type-B quartz fiber composite material laminates and NOMEX paper honeycomb sandwich panels to analyze the impact on the impact properties of the two composite materials under the two schemes.

Table 1. Test scheme 1

specimen No.	Punch mass (kg)	Punch speed(m/s)	Height of drop hammer (m)	Impact energy(J)
1	5.5	4.2	0.976	52
2	5.5	5.4	1.519	82
3	5.5	6.2	1.971	106

Table 2. Test scheme 2

specimen No.	Punch mass (kg)	Punch speed(m/s)	Height of drop hammer (m)	Impact energy(J)
1	5.5	5.2	1.409	75
2	7.5	4.5	1.035	75
3	10.5	3.8	0.738	75

3. Analysis and Discussion on Test Results

3.1.Shock response analysis

3.1.1. Type-B quartz fiber/modified cyanate ester composite material laminate

According to test schemes 1 and 2, the time-load curves of laminates under six different test conditions are obtained, as shown in Fig.2 and Fig.3, respectively. It can be seen from the curve that according to test scheme 1, when with constant punch mass and the impact energy is 52J, 82J, and 106J, the maximum impact load is 12274N, 15517N, and 15796N, respectively; with the increase of the impact energy, the curve jitter amplitude is getting more and more severe, when the impact energy reaches 106 J, there is a sudden drop of load, indicating an initial delamination damage occurred at this moment, as shown in Fig.2. According to scheme 2: When the impact energy does not change to 75 J, and the punch masses are 5.5kg, 7.5kg and 10.5kg, the maximum impact loads are 14771N, 14918N and 15534N respectively, and the change of curve trend is roughly the same, as shown in Fig.3.

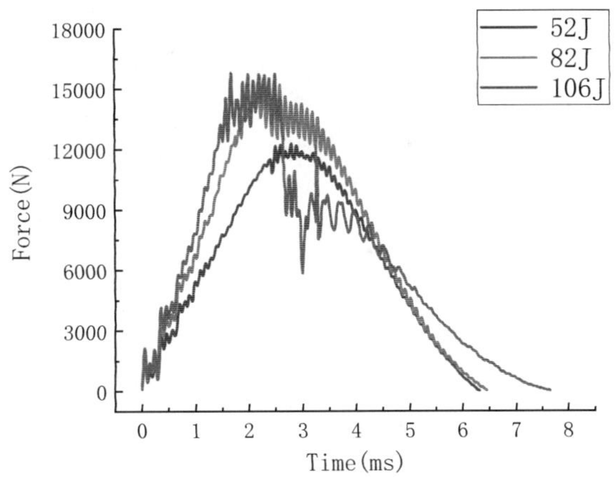

Fig.2 Time-load Curve of Type-B Quartz Fiber/Modified Cyanate Ester Laminate (Scheme 1)

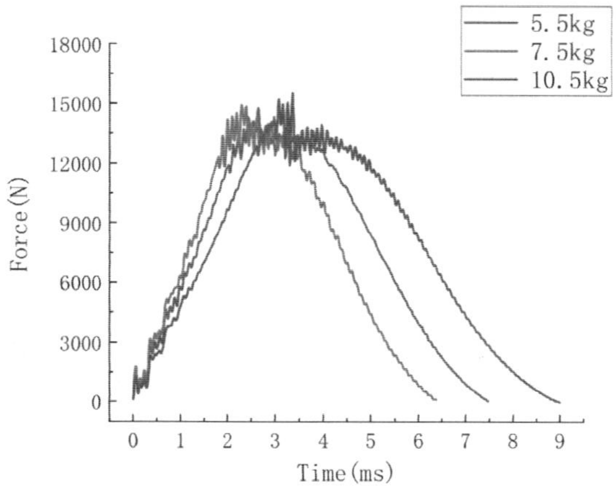

Fig.3 Time-load Curve of Type-B Quartz Fiber/Modified Cyanate Ester Laminate (Scheme 2)

3.1.2. NOMEX paper honeycomb sandwich panel

According to test schemes 1 and 2, the time-load curves of honeycomb sandwich panel under six different test conditions are obtained, as shown in Fig.4 and Fig.5, respectively. From Fig.4(a) and Fig.5(a), it can be seen that the time-load curve shows an upward trend and disappears near the maximum load with only one peak captured, and the corresponding limit peak loads of the upper panel and the honeycomb core are 4684N and 4817N respectively.

Under other test conditions, the time-load curve has two peak loads, as shown in Fig.4(b), 4(c), Fig.5(b), 5(c), the first peak load indicates the moment of penetration through the upper panel, and the appearance of the second peak load represents the moment of penetration of the lower panel. According to test scheme 1, the load values corresponding to the first peak of the time-load curve under impact energy 82 J and 106 J are 4612 N and 4986 N, respectively, and the second peak load is 4722N and 5431N; in scheme 2, when the punch masses are 7.5kg and 10.5kg, the corresponding first peak loads in the curve are 4638 N and 4907 N, respectively, and the second peak loads are 5036N and 5465N. It is found that the second peak load is always greater than the first peak load because the penetration of the upper panel absorbs the energy of the punch and slows down the speed of punching, resulting in a smaller impedance load on the upper panel than the lower panel.

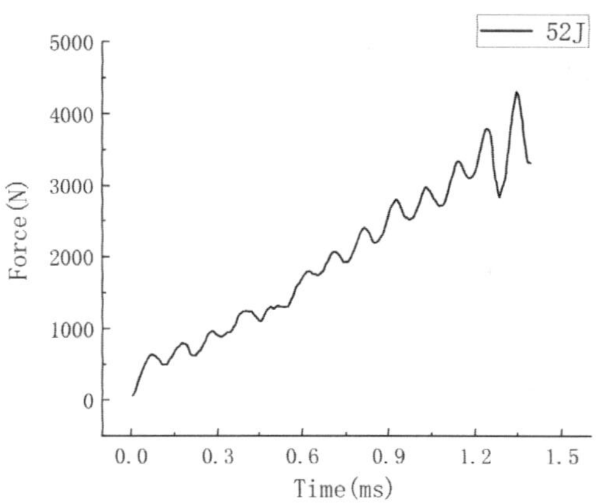

(a) 52J impact

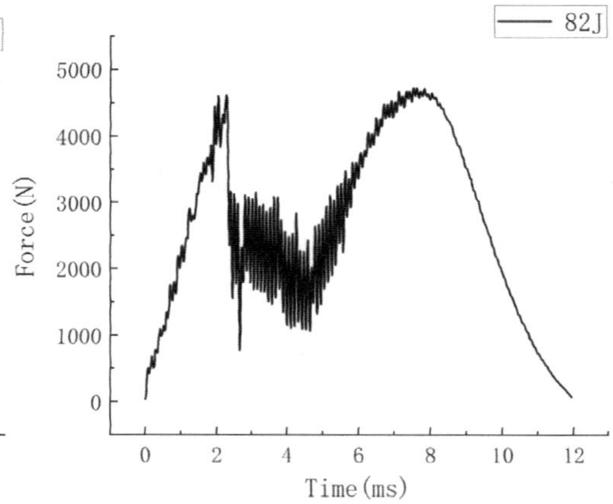

(b) 82J impact

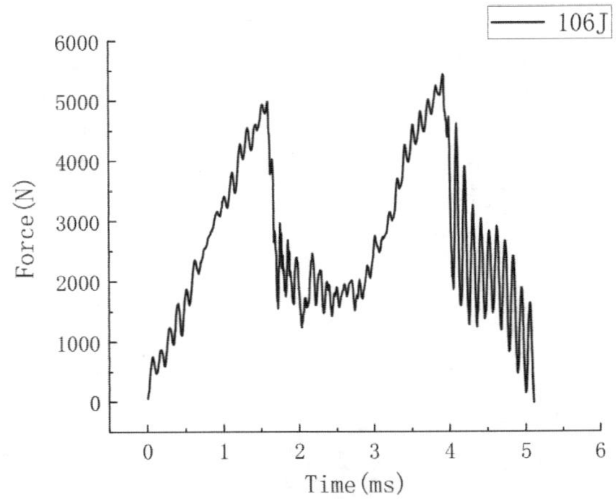

(c) 106J impact

Fig.4 Time-load Curve of NOMEX Paper Honeycomb Sandwich Panel (Scheme 1)

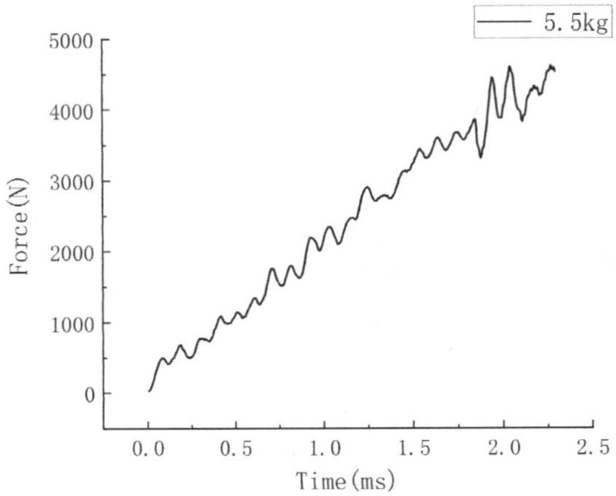

(a) 5.5kg impact

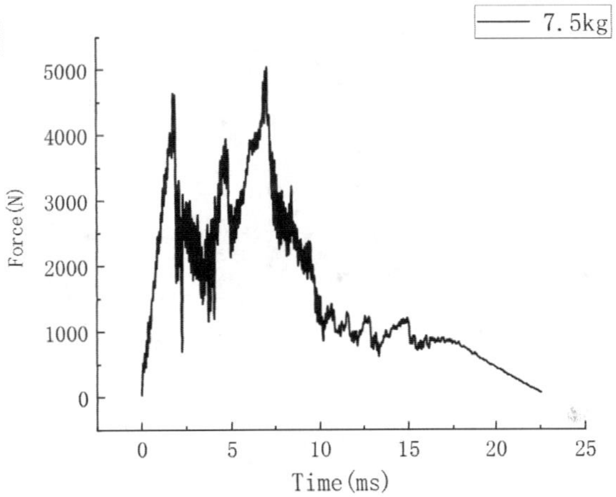

(b) 7.5kg impact

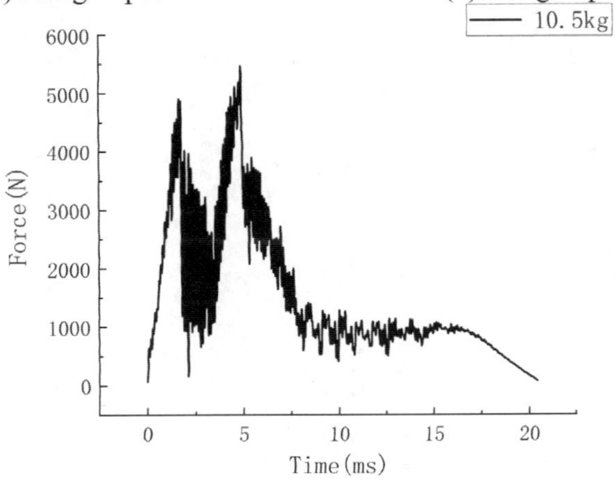

(c) 10.5kg impact

Fig.5 Time-load Curve of NOMEX Paper Honeycomb Sandwich Panel (Scheme 2)

3.2.Impact damage analysis

The damage degrees caused to the type-B quartz fiber/modified cyanate ester laminate as per test schemes 1 and 2 are shown in Fig.6 and Fig.7 respectively. It can be seen that the impact surface of the laminate (front side, the same below) has obvious pits, the impact front side damage is rhombic, and substrate cracking exists near the pits; while the reverse side of the laminate (reverse side, the same below) has obviously visible fiber breakage and substrate cracking; and as the impact energy increases, the damage becomes more serious.

(a)52J impact front side (b)82J impact front side (c)106J impact front side

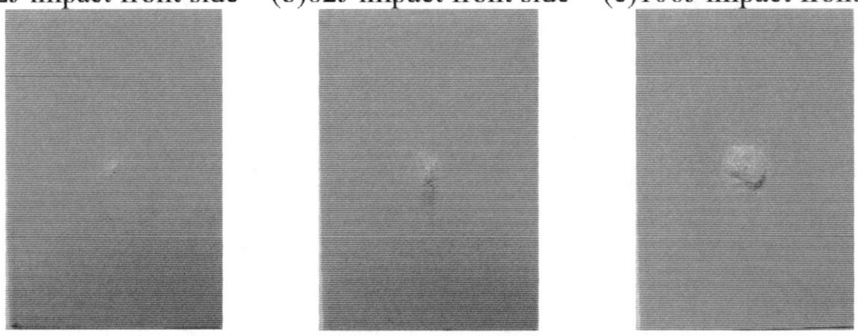

(d)52J impact reverse side (e)82J impact reverse side (f)106J impact reverse side

Fig.6 Impact Damage of Type-B Quartz Fiber/Modified Cyanate Ester Laminate (Scheme 1)

(a)5.5kg impact front side (b)7.5kg impact front side (c)10.5kg impact front side

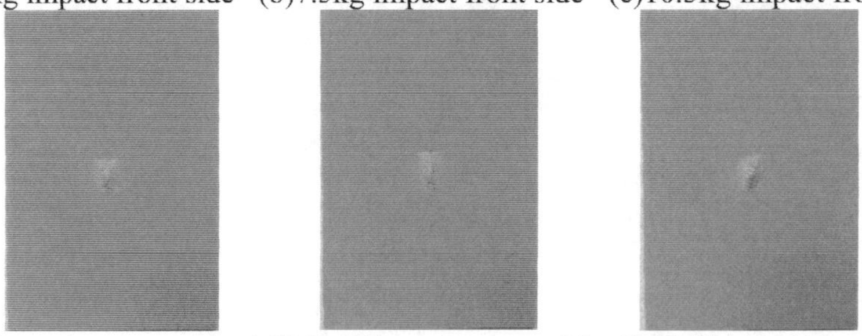

(d)5.5kg impact reverse side (e)7.5kg impact reverse side (f)10.5kg impact reverse side

Fig.7 Impact Damage of Type-B Quartz Fiber/Modified Cyanate Ester Laminate (Scheme 2)

The damage degrees of the NOMEX paper honeycomb sandwich panel as per test schemes 1 and 2 are shown in Fig.8 and Fig.9 respectively. It can be seen that the upper panel of the honeycomb sandwich panel is penetrated according to the two test schemes; the impact area on the front side of the specimen also has diamond-shaped damage, but the size of the crack near the pit is smaller than that of the quartz fiber. With the impact energy of 82J and 106J, the lower panels of the NOMEX paper honeycomb sandwich panel are also penetrated; and in the test scheme 2, the lower panels of the honeycomb sandwich panel under the impact of 7.5kg and 10.5kg punch masses are also penetrated.

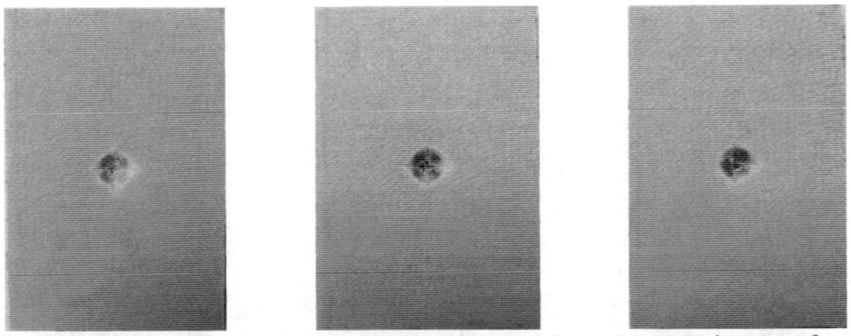

(a)52J impact front side (b)82J impact front side (c)106J impact front side

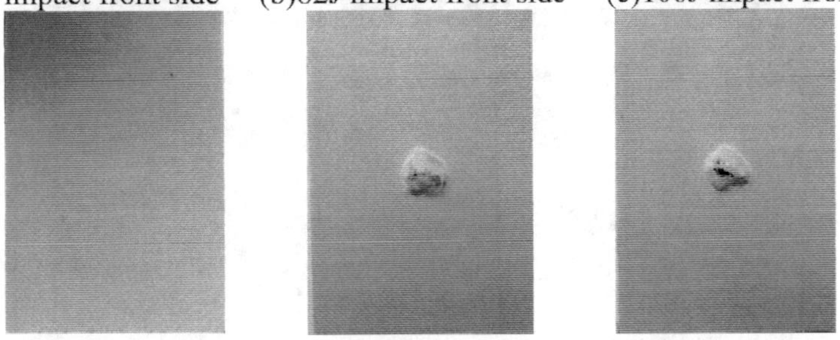

(d)52J impact reverse side (e)82J impact reverse side (f)106J impact reverse side

Fig.8 Impact Damage of NOMEX Paper Honeycomb Sandwich Panel (Scheme 1)

(a)5.5kg impact front side (b)7.5kg impact front side (c)10.5kg impact front side

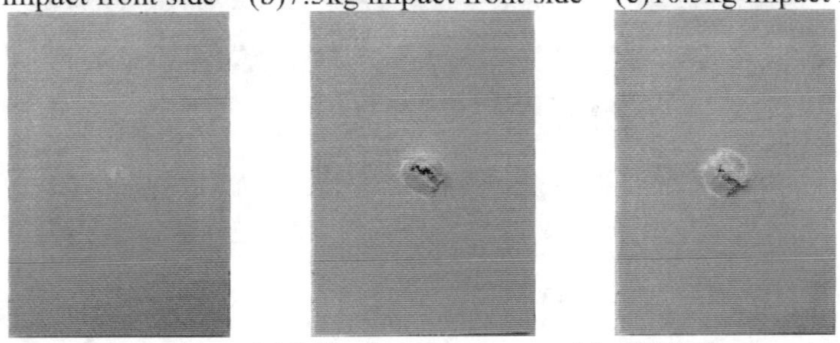

(d)5.5kg impact reverse side (e)7.5kg impact reverse side (f)10.5kg impact reverse side

Fig.9 Impact Damage of NOMEX Paper Honeycomb Sandwich Panel (Scheme 2)

To further examine the damage to the specimen near the impact damage area, computer tomography (CT) is used for specimen inspection. The MG452 CT inspection system manufactured by Germany YXLON is used for non-invasive analysis on the specimen. The equipment is characterized with density resolution ≤0.3%, spatial resolution ≥30 Lp/cm, and defect detection accuracy of 0.1mm. Damaged specimen is placed on a horizontal platform to collect and process the three-dimensional scanned image of the damaged laminate at the center line of the impact area.

The test scheme 1 is analyzed to understand the impact on the damage of the type-B quartz fiber laminate with different impact energies; the CT inspection image is shown in Fig.10. It can be seen from the Figure that the damage degree of the laminate increases with the increase of impact energy, and penetration is also formed. It can be seen from Fig.10(a) that under the condition of impact energy of 52J, the specimen is not fully penetrated, but local delamination occurs in the impact area. Delamination is a primary type of damage, accompanied by a small range of fiber breakage. It can be seen from Fig.10(b) and 10(c) that when the impact energy is 82J and 106J, the specimen will have a large area of substrate cracking and fiber fracture with the range of damage expanding along with the increase of impact energy.

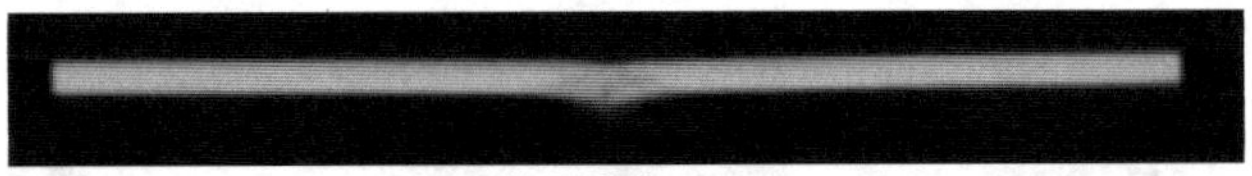

(a) 52J impact

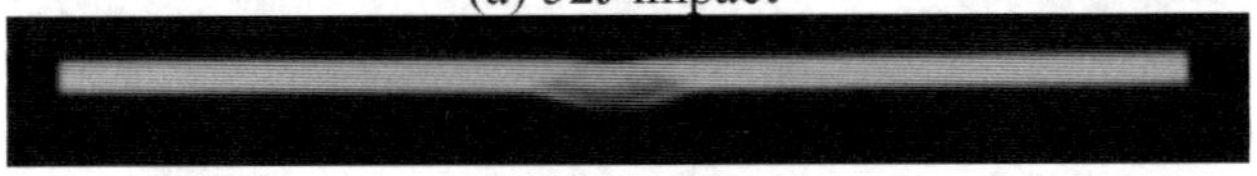

(b) 82J impact

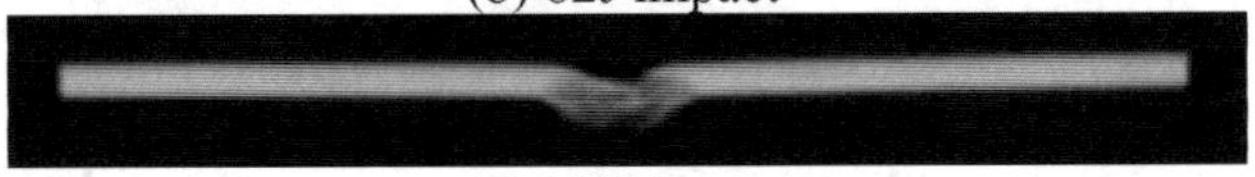

(c) 106J impact

Fig.10 CT inspection image of Type-B Quartz Fiber/Modified Cyanate Ester Laminate (Scheme 1)

The test scheme 2 is analyzed to understand the punch mass impact on the damage of the composite material laminate under constant energy. The CT inspection image is shown in Fig.11. It can be seen from this Figure that when the impact energy is constant, the damage degree of the laminate increases with the increase of the punch mass. Regarding the impact test with a punch mass of 5.5kg, two failures of substrate cracking and fiber fracture occur; for impacts with heavier punch masses of 7.5kg and 10.5kg, as the punch mass increases, the damaged area of fiber in the impact zone is enlarged. As shown in Fig.11(c), the heaviest punch mass of 10.5kg, the greatest damage is caused with a larger range of fiber breakage produced in the impact area.

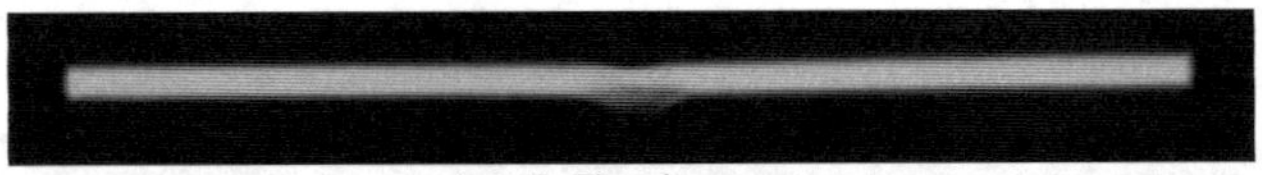

(a) 5.5kg impact

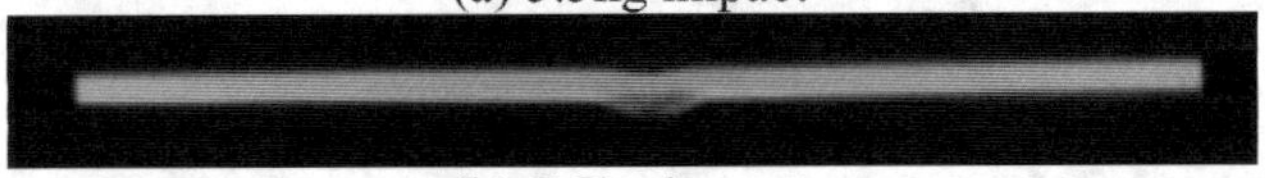

(b) 7.5kg impact

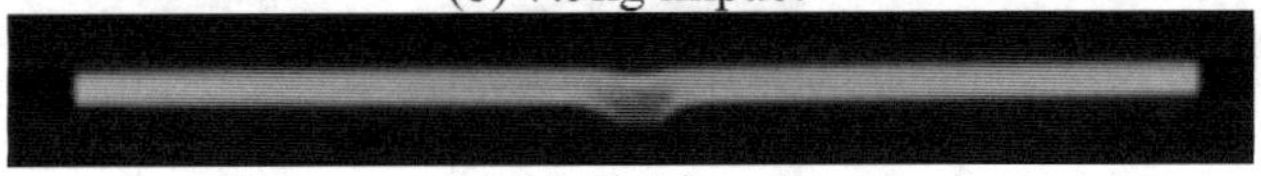

(c) 10.5kg impact

Fig.11 CT inspection image of Type-B Quartz Fiber/Modified Cyanate Ester Laminate (Scheme 2)

Observe the damage of the NOMEX paper honeycomb sandwich panel. The CT inspection image of Test scheme 1 is shown in Fig.12. It can be seen from Fig.12(a) that when the impact energy is 52J, the specimen is not completely penetrated, and the CT scan image shows that only the upper panel of the sandwich panel is penetrated; it can be seen from Fig.12(b) and 12(c) that when the impact energy is 82J and 106J, the specimen is penetrated completely. The damage covers panel fiber fracture and honeycomb core extrusion and fracture, and the damage area is enlarged from the impact center area to the surroundings.

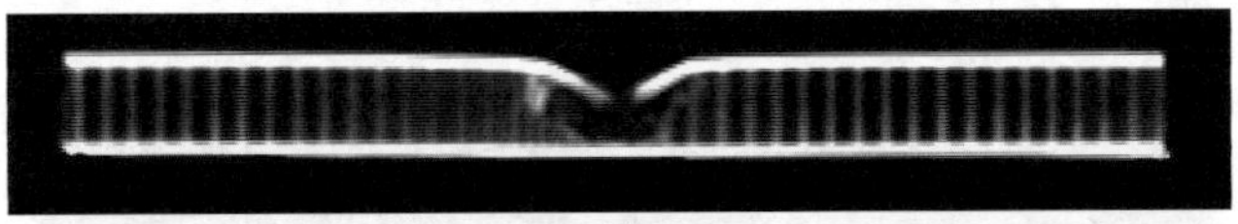

(a)52J impact

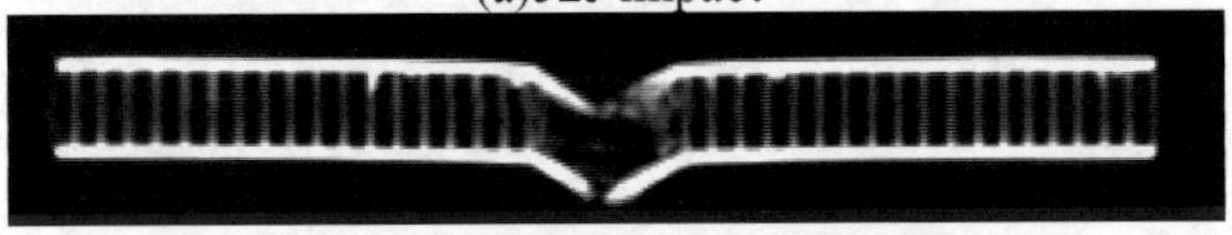

(b)82J impact

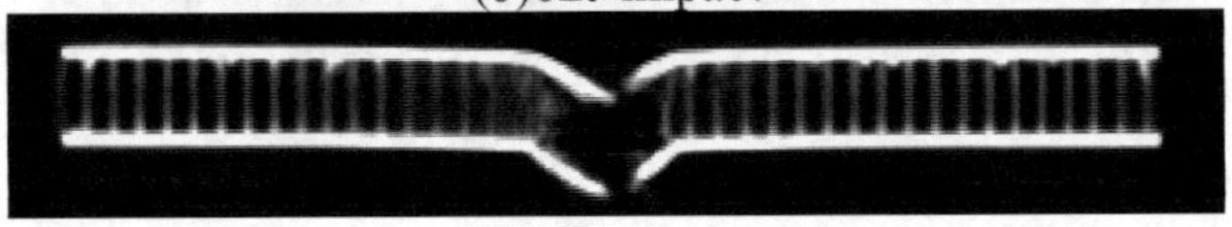

(c)106J impact

Fig.12 CT inspection image of NOMEX Paper Honeycomb Sandwich Panel (Scheme 1)

The CT inspection image of Test scheme 2 is shown in Fig.13. The punch mass impact on the damage of honeycomb sandwich panel under constant energy impact is analyzed. It can be seen from Fig.13(a) that when the punch mass is 5.5kg, the upper panel of the specimen is penetrated; from Fig.13(b) and 13(c), it can be observed that for the punch masses of 5.5kg and 7.5kg, the damages are the panel fiber fracture and core extrusion fracture in the impact zone and outward expansion of honeycomb core inside the panel from the impact center area. It can be seen that at a constant impact energy level, with the punch mass increases, the damage to the sandwich panel gets aggravated.

(a)5.5kg impact

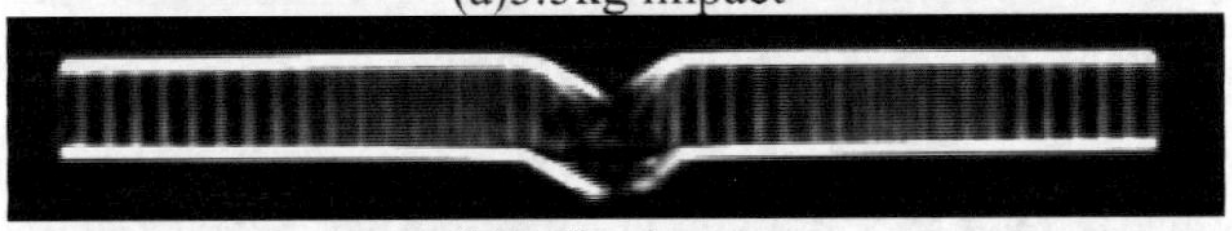

(b)7.5kg impact

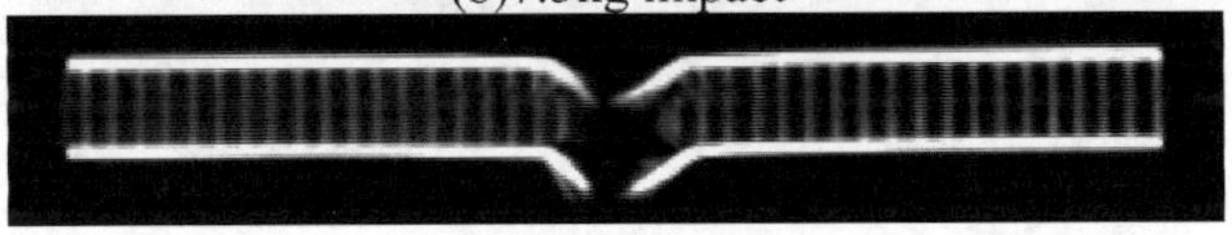

(c)10.5kg impact

Fig.13 CT inspection image of NOMEX Paper Honeycomb Sandwich Panel (Scheme 2)

4. Conclusion

(1) When the impact energy is 82J and 106J and the punch masses are 7.54kg and 10.5kg, two peak load moments will be generated when punching on NOMEX paper honeycomb sandwich panel, corresponding to the moments when both upper and lower panels are penetrated. The load on the

lower panel is greater than that of the upper panel because when the punch decelerates due to the resistance of the upper panel, the impact resistance of the lower panel will increase.

(2) For the type-B quartz fiber/modified cyanate ester laminate, with impact energy of 52J, delamination is the primary failure; with impact energies of 82J and 106J, the delamination further expands, resulting in substrate cracking and fiber fracture; for the NOMEX paper honeycomb sandwich panel, as the impact energy increases, the damage to the honeycomb core inside the sandwich panel will expand outwards along the impact center area.

(3) It is found that the type-B quartz fiber/modified cyanate ester laminate has three different types of failure: substrate cracking, fiber fracture and delamination. For NOMEX paper honeycomb sandwich panels, panel fiber fracture and honeycomb core extrusion are observed. When the punch mass is constant, the greater the impact energy is, the more serious the damage is, and when the impact energy is constant, the greater the punch mass is, the more serious the damage is.

(4) Research on impact damage of a quartz fiber composite and an NOMEX paper honeycomb sandwich panel composite material is a valued and challenging project, and research on impact damage and damage influence factors of the composite materials is beneficial to improving the efficiency of structural design, improving the applicability of composite material structures and fully exerting the advantages of the composite materials. This text has made a lot of research work, but still has a lot of incomplete contents needing to continue to be improved, and related deep research on the influences of punch materials, punch shapes, etc. on impact damage of the composite materials will be carried out in the follow-up process.

References

[1] Wang J, Wang H, Chen B, et al. A failure mechanism based model for numerical modeling the compression-after-impact of foam-core sandwich panels [J]. Composites Science and Technology, 2017, 151: 258-267.
[2] Guinard S, Allix O, Guédra-Degeorges D, et al. A 3D damage analysis of low-velocity impacts on laminated composites [J]. Composites Science and Technology, 2002, 62: 585-589.
[3] Léonard F, Stein J, Soutis C, et al. The quantifcation of impact damage distribution in composite laminates by analysis of X-ray computed tomograms [J]. Composites Science and Technology, 2017, 152: 139-148.
[4] Liu H, Nie H, Zhang C, et al. Loading rate dependency of Mode I interlaminar fracture toughness for unidirectional composite laminates [J]. Composites Science and Technology, 2018, 167: 215-223.
[5] Bogenfeld R, Kreikemeier J, Wille T. Review and benchmark study on the analysis of low-velocity impact on composite laminates [J]. Engineering Failure Analysis, 2018, 86: 72-99.
[6] Polymer matrix composites: guidelines for characterization of structural materials [M]. Washington, DC: Military Handbook MIL-HDBK-17-1F, June 2002. 16.
[7] Zhang X, Hounslow L, Grassi M. Improvement of low-velocity impact and compression-after-impact performance by z-fbre pinning [J]. Composites Science and Technology, 2006, 66: 2785-2794.
[8] Shen Z, Chen P H, Liu J S, et al. Compression failure mechanism of composite material laminates with defects [J]. Acta Aeronautica Sinica, 1991, 12 (3): 105-113.
[9] Hou J P, Petrinic N, Ruiz C, et al. Prediction of impact damage in composite plates [J]. Composites Science and Technology, 2000, 60: 273-281.
[10] Wagih A, Maimí P, González EV, Blanco N, de Aja JRS, de la Escalera FM, et al. Damage sequence in thin-ply composite laminates under out-of-plane loading. Compos Part A: Appl Sci Manuf 2016;87:66–77.
[11] Saito H, Morita M, Kawabe K, Kanesaki M, Takeuchi H, Tanaka M, et al. Effect of ply-thickness on impact damage morphology in CFRP laminates. J Reinf Plast Compos 2011;30(13):1097–106.
[12] Choi HY, Wang HS, Chang F. Damage of graphite/epoxy composite plates due to line-loading

Journal of Physics: Conference Series **2194** (2022) 012013 doi:10.1088/1742-6596/2194/1/012013

impact. J Compos Mater 1992;26:804–27.
[13] Choi HY, Wang HS, Chang F. A new approach toward understanding damage mechanisms and mechanics of laminated composites due to low-velocity impact: PartII-analysis. J Compos Mater 1991;25:1012–38.
[14] Choi HY, Wang HS, Chang F. A new approach toward understanding damage mechanisms and mechanics of laminated composites due to low-velocity impact:Part Iexperiments. J Compos Mater 1991;25:992–1011.

Journal of Physics: Conference Series **2194** (2022) 012014 doi:10.1088/1742-6596/2194/1/012014

Applications of Metal-Organic Framework Materials

Wenkui Chai[1, †], Youhong Shen[2, †], Jiahe Wang[3, †], Geling Zhang[4, *, †]

[1]CATS College Cambridge 1 High Street, Chesterton, Cambridge CB4 1NQ, UK.

[2]Dulwich International High School, Suzhou, 215000, China.

[3]International Department of the High School Affiliated to Shaanxi Normal University, Xi'an, 710000, China.

[4]Guangdong Country Garden School, Guangdong, 528000, China.

[*]guanghua.ren@gecacademy.cn

[†]These authors contributed equally.

Abstract-Metal-organic framework (MOFs) is a new group of porous crystal materials with ultra-high porosity. Its extraordinary surface area, fine and adjustable void surface characteristics and potential industrial-scale scalability make it an attractive target for further research. By searching for the data, we have integrated and summarized the important applications of MOF materials in many fields, such as catalysis, and its enclosed single metal or bimetallic ions provide efficient and synergistic catalytic ability; In adsorption, the special ionic liquid absorbent overcomes the shortcomings of other materials, greatly improves the chemical stability, reduces the corrosivity, and makes the process more environment-friendly and controllable. However, there are still many challenges, and the high cost and complicated preparation process are urgent problems to be solved at present. We need to have a deeper understanding of the growth mechanism of MOF films, improve their quality and thermoelectric stability, and combine them into low-cost polymer films through more reliable theoretical models to eliminate the defects of MOF materials as much as possible. We will continue to pay attention to the development of MOF materials, summarize the advantages of MOF materials, find out the shortcomings, and propose solutions to provide ideas and references for future research.

1. Introduction

Eighty years ago, few people in the world had heard of plastic, but in 1939, nylon, a kind of plastic, became a household name nearly one year after its debut at the New York World Expo. Nylon has been all the rage since the stockings industry started, and then it started to innovate the application of plastics in various fields. Clothing, household goods, electronic products, the pharmaceutical industry and other fields have been impacted by plastics unprecedentedly. After more than ten years' development, plastic has become a necessity and has been defined as the material of the 20th century. But now, in the 21st century, earth-shaking changes are taking place in the field of materials. New material is following the pace of plastics, and it is expected to become the decisive material in the 21st century in the future, that is, metal-organic framework materials (MOFs).

Plastic has become so popular because of its versatility, which consists of long and repeated monomers. Therefore, this polymer's physical and chemical characteristics, such as size, hardness, elasticity, and conductivity, can be determined by simply changing its monomers directly. Like plastic, MOF is a multifunctional application material, and it is also composed of repeating units with very small structures. It is connected by organic molecules to form a micron-sized high-rise structure, which can free divergence in all directions and forms special crystals through the rigid, uniform and precise atomic arrangement. This unique uniform structure enables scientists to study, design and process MOF materials from a new perspective. By accurately locating the position of each atom in the polymer and using modern technology, they can quickly simulate different crystal structures by modeling before laboratory synthesis to meet the different requirements of MOF materials in different applications.

So far, the research and application of MOF materials are still relatively few. Therefore, this paper introduces in detail the characteristics, main application fields and working mechanism of MOFs materials in various fields and puts forward what it should be.

2. Application of MOF materials

2.1.Catalysis

In recent years, metal organic framework materials (MOFs) have shown potential application prospects in many fields, aroused widespread research interest and achieved rapid development. Our research group mainly focuses on the following two aspects to carry out research and has obtained some preliminary research results.

Design, synthesis and pore tuning/modification of high-porosity stable MOFs. By applying a simple hydrophobic PDMS coating to the classic water/moisture unstable various MOFs, the water/moisture stability of these MOFs is greatly enhanced; more importantly, this hydrophobicity The thin layer of PDMS will not affect the intrinsic pore characteristics of MOFs and the use of catalytic sites [1, 2].

The synthesis of nanocomposites based on MOF pore confinement and MOFs as hard templates and their application in catalysis. Based on the pore confinement effect of MOFs, through a rational synthesis method, metal precursors are introduced into highly pore stable MOFs and reduced to obtain monometallic or bimetallic nanoparticles confined in the pores/cages of MOFs and provide high-efficiency/cooperative catalytic performance. It is particularly important that our host-guest synergy is based on the catalytic behavior of MOF and the catalytic performance of confined metal nanoparticles and the synergy of catalytic activity and selectivity between bimetallic nanoparticles together achieve more in one pot. Based on the high order and porosity characteristics of MOFs, the doped porous carbon nanocomposite is obtained by decomposing it, which exhibits excellent catalytic reaction performance in heterogeneous catalysis and fuel cell oxygen reduction reactions.

CO oxidation has generated considerable interest because of its importance in terms of emission control and its relative simplicity as an ideal model reaction in heterogeneous catalysis.

Noble metal nanoparticles (NPs) dispersed on solid support could be active catalysts for CO oxidation. However, it is well known that these metal particles tend to agglomerate and grow, which decreases the overall active surface area and causes the loss of catalytic activity. Metal-organic frameworks (MOFs), as a new class of functional porous materials, have been shown to be suitable for embedding and supporting metal NPs similar to zeolites, It is expected that the ordered crystalline structures of MOFs with the large and tunable cavity, high surface area, and controllable porosity could confine the migration and aggregation of metal NPs. Gao *et al.* [3] found that Pd/MOF-5 was an active catalyst for the Sonographer coupling reaction between aryl iodides and terminal acetylenes without the assistance of ligand and copper [4, 5, 6]. Ishida *et al.* reported Au clusters supported onto several kinds of porous coordination polymers (PCPs), including MOF-5 by solid grinding, and they investigated their catalytic properties for the liquid phase alcohol oxidation with molecular oxygen. Metal organic framework MIL-101(Cr) was employed to support preparing Au NPs catalysts by liquid-phase synthesis method, and Au/MIL-101(Cr) catalyst exhibited high catalytic activities in the liquid phase aerobic alcohol oxidation. However, most of the reported works with MOFs based catalysts are for catalytic applications in the liquid phase. There are few reports about gaseous reactions over MOFs supported catalysts. Jiang *et al.* pioneeringly reported the preparation of Au NPs supported on ZIF-8 (ZIF: zeolite imidazolate frameworks) catalysts by solid grinding method and used them as the catalysts for CO oxidation. The catalyst gave the medium activity for CO oxidation, and a full CO conversion was obtained at 200 °C for 5.0% (*w*) Au/ZIF-8.

Herein, we report the uniformly deposited Pd NPs on a microporous aluminium terephthalate MIL-53 to obtain a series of heterogeneous catalytic composite materials. In addition, the catalytic activities and reused stability of the as-prepared Pd/MIL- 53(Al) catalyst for CO oxidation were detailly investigated.

The silver-functionalized UiO-66 MOF [7] is prepared by a simple one-pot solvothermal reaction and characterized by X-ray diffraction, nitrogen adsorption-desorption, electron microscopy, X-ray spectroscopy, and thermogravimetric analysis. The functionalized MOF prepared by the test removes radioactive iodine from water. Furthermore, due to the excellent affinity of Ag particles and iodide, an excellent extraction capacity of ^{131}I (~1 MBq/g) is obtained. In practical applications, the obtained porous UiO-66-SO_3H@Ag material is combined into a 3D printing device using polyvinylidene fluoride as a binder by a simple method. The developed UiO-66-SO_3H@Ag 3D device shows the high efficiency of ^{131}I removal, excellent reusability and good performance of extracting ^{131}I from sewage and hospital waste samples. In all cases, more than 90% recovery is obtained rate.

Nowadays, the actual application of organometallic frameworks is also used for coating other materials most of the time, such as drug transportation, coating metal elements or coating catalysts—because of the use of MOF materials. As a result, the surface area is increased. Larger, fewer catalysts can catalyze more reactants, and the reaction rate will also increase. Luo's team [8] used ZIF-8/Pt/SiO_2 catalyst was prepared by combining atomic layer deposition technology and vapor phase conversion method. The Pt metal nanoparticles were deposited on SiO_2 nanowires, and the ZIF-8 film covers Pt and SiO_2 nanowires to form a sandwich structure. The structure of the catalysts was characterized by XRD, TEM, BET, IC-MS, XPS and CO–DRIFT, and the function of ZIF-8 film was studied by using the semi-hydrogenation of 1-heptyne as the probe reaction. The Pt particles are highly dispersed on the SiO_2 nanowires before and after the ZIF-8 overcoating, and the ZIF-8 film coats continuously on the entire catalyst with high conformity. The ZIF-8 film induces an electron density increase of Pt surface and increases the olefin selectivity from 14% to 70% in the 1- heptane hydrogenation reaction. A thinner ZIF-8 film increases the catalytic activity and cannot change the selectivity of olefin.

2.2.Adsorption

2.2.1.For Gas

The adsorption of CO_2 by traditional liquid alkanolamine adsorbent can not meet the requirements of green development in industrial applications. NH_2-β-CD-MOF [9] was first constructed by an American company by amino-functionalizing β-CD, the cheapest, readily available and biocompatible form. The selective adsorption of NH2-β-CD-MOF on CO_2/N_2(947.52) is also better than that of existing materials. The adsorption mechanism of CO_2 was analyzed by X-ray photoelectron spectroscopy and Fourier transform infrared spectroscopy.

The adsorption capacity of amino-functionalized β -cyclodextrin-MOF (NH_2-β-CD-MOF) can reach 12.3 cm^3/g, which is 10 times that of β-CD. It is noteworthy that NH_2-β-CD-MOF has a good adsorption selectivity for CO_2/N_2 compared to reported materials (947.52). In addition, in practical application, the ideal MOF adsorption material should still have good adsorption capacity and selectivity in the presence of water. The introduction of the amino group as a polar functional group can improve the selective adsorption of MOF and show that the amino group can competitively bind foreign water molecules and improve the water stability of NH_2-β-CD-MOF. NH_2-β-CD-MOF is not only cheap, green and non-toxic but also can be recovered by secondary ultrasound. After repeated recycling tests, NH_2-β-CD-MOF has a high carbon dioxide adsorption capacity.

2.2.2.For Liquid

Researchers at Lawrence Livermore National Laboratory have developed a cleaner, more stable and more efficient way to capture carbon dioxide using ionic liquids as carbon dioxide absorbents.

The current carbon capture technology mainly uses the chemical adsorption method. Carbon dioxide reacts with amines, which combine at low temperatures and separate at high temperatures. It is generally possible to separate the carbon dioxide from the waste gas containing carbon dioxide by passing it through an amine solution and then releasing the carbon dioxide by heating it in an appropriate place [10]. Monoethanolamide is currently used as a CO_2 absorber in a few commercial carbon capture coal-fired power plants. But monoethanolamides are corrosive, require large equipment, and are only effective at mild to moderate CO_2 pressures. Therefore, its cost and efficiency are not very ideal.

Amities Medi of the lab has been working to find new carbon dioxide absorbers. He tested several ionic liquids that can effectively dissolve carbon dioxide and got a lot of useful data. Unlike typical organic solvents, ionic liquids generally do not become vapors, so they are not prone to harmful gases and are easy to use. Medi found that using ionic liquids as CO_2 absorbers could overcome many of the drawbacks of monoethanolamides, making them cleaner and easier to use than the methods currently used. Its chemical stability is good, low corrosion, vapor pressure is almost zero, can be made into a film. Ionic liquids are of a wide variety, and there are many ions to choose from with potentially high carbon dioxide solubility.

2.3. Gas Separation

The separation of gas molecules with similar physical and chemical properties is a very important process closely associated with the manufacture of fuels, polymers and plastics. In the gas separation market and in small-scale enterprises, membrane separations are the most popular due to their low cost and the facile large-scale production. Furthermore, the emergence of metal-organic frameworks is ideal adsorbents for gas separation due to their large surface area, ultra-high porosity and modularity, and acceptable thermal stability. In existing studies, MOF membranes for gas separation can be divided into two main categories, pure MOF membranes and mixed matrix membranes (MMMs) [11, 12].

2.3.1. Pure MOF Films

Compared to conventional polymer and molecular sieve membranes, MOF film materials have advantages such as structural diversity, ultra-high porosity, and high orientation [13]. Metal-organic framework films are synthesized on inorganic or polymer supports by seeded secondary growth or in-situ growth synthesis methods. The following MOFs are commonly used to construct MOF films: IRMOF-1, Cu-BTC and ZIFs. In 2008, Lai and co-workers [14] first reported the *in situ* solvothermal syntheses of continuous, well-intergrown MOF-5 membranes prepared on the porous α-alumina substrate. Then they used MOF-5 membranes for gas separation research. The discovery that the diffusion of simple gas molecules through this material follows the Knudsen diffusion behavior has opened up a whole new field of research into the use of MOF membranes for gas separation. Zeolitic imidazolate frameworks (ZIFs), a subfamily of microporous MOFs based on transition metals (Co, Fe, Zn, Cu) and imidazolates as ligands, became one of the most studied MOFs in the field of separation membranes due to the excellent stability and tunable pore size. The *in-situ* synthesis method is also applied to the ZIF-8. Caro and co-workers [15] prepared sodalite (SOD) topology ZIF-8 membranes with molecular sieving performance by microwave-assisted synthesis, which performs high thermal stability. They also succeed in the preparation of ZIF-22 membrane on the 3-aminopropyl-triethoxysilane (APTES) -modified TiO_2 by using *in situ growth* method, shown in Figure 1, which represents the selectivity of H_2/CO_2, H_2/O_2, H_2/N_2, and H_2/CH_4 are 7.2, 6.4, 6.4 and 5.2 respectively.

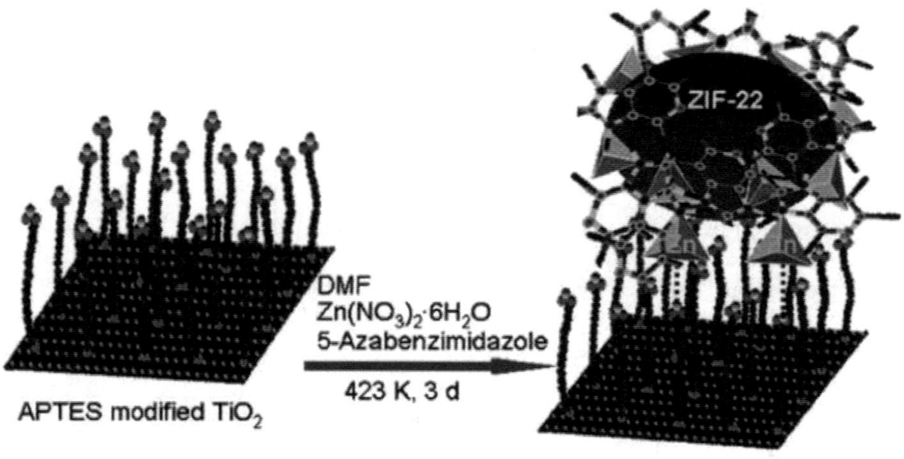

Fig. 1 Preparation of a ZIF-22 membrane by using 3-aminopropyl-triethoxysilane (APTES) as the covalent linker between the ZIF-22 membrane and the titania support [15].

2.3.2. Mixed Matrix Membranes

Compared to pure MOF membranes, mixed matrix membranes (MMMs) with a polymer matrix and MOF as filler have the advantages of lower cost, easy processing, good mechanical strength, better selectivity and permeability, which has a good prospect for commercial application.

HKUST-1 (Hong Kong University of Science and Technology $Cu_3(BTC)_2$) [16, 17] is a metal organic framework (MOF), which exhibits a cubic, paddle-wheel structure and is composed of copper metal sites and triangular 1,3,5-benzenetricarboxylate organic linkers. After an activation process, these water molecules can be removed, and the coordination site at the metal atoms is left unoccupied. Hence HKUST-1 shows high gas adsorption capacity. A variety of metal cations can be used to prepare the structure of HKUST-1 (M = Cr, Fe, Zn, Mo), and with the invention of M_2(dobdc), this suggests that metals are an important variable to consider when tuning material properties. In the case of metal ions possessing exposed coordination sites, adsorption can vary depending on the charge density on the metal coordination site [18].

Another well-studied family of MOFs bearing exposed metal cation sites is MOF-74 (M_2(DOT). Because of its high density of coordinatively unsaturated metal sites, it has high CO_2 adsorption capacities at 1 bar and 296 K. In experiments, Mg-MOF-74 shows an efficient separation of CO_2 (20%) from CH_4 (80%), which reflects attractive performance from a kinetic and thermodynamic point of view. In fact, the value of Mg_2 (dobdc) represents the highest low-pressure gravimetric and volumetric adsorption capacity of the metal-organic framework for CO_2. This shows the great potential of metal-organic framework materials in the field of gas separation.

3. Outlook

Pure MOF films have good thermal and chemical stability and are suitable for the physically and chemically complex environments of industrial applications, but the high cost, the complexity of the preparation process and processing difficulties limit their further application. In addition, the film growth mechanism needs to be studied in more depth, and the quality and thermoelectric properties of the films need to be further improved.

By combining the advantages of lower cost polymeric membranes, mature processing technology and uniform pore size of inorganic porous materials, mixed matrix membranes effectively overcome the inherent disadvantages of poor mechanical strength and difficulties in large-scale preparation of inorganic membranes, ultimately effectively improving the mechanical strength of the resulting films. At the same time, mixed matrix membranes require the use of only a very small amount of inorganic porous material compared to inorganic crystalline membranes, thereby substantially reducing the cost of separation membranes. However, there is an inherent problem of poor filler-substrate interfacial compatibility, which leads to defects in the MMMs and a reduction in selectivity. Therefore, more effective interfacial design strategies and more reliable theoretical models are needed to eliminate defects in membrane materials as far as possible.

4. Conclusion

As a new type of material, metal-organic framework (MOF) materials have good prospects for development and application in various fields such as the energy industry and drug delivery because of the characteristics such as high specific surface area, ultra-high porosity, as well as and high structural tunability. The research on MOFs synthesis, design and application has been discussed in depth. This review discusses the practical applications of MOF materials in catalysis, adsorption and gas separation, including porous carbon nanocomposites, amino-functionalized β-cyclodextrin-MOF and mixed matrix membranes, all of which perform well in their respective fields. However, there are still many challenges to overcome, such as high costs, complex preparation processes and processing difficulties, limiting the practical applications of MOF materials in production, leaving much room for the development of MOFs. Therefore, in the future, there is a need to rationalize the preparation methods to obtain higher performance, more economical and easier to fabricate MOF materials, and to investigate the influence of preparation methods on the performance of MOF materials, so that MOFs can be put into quantitative production for industrial applications.

References

[1] N.C. Burtch, H. Jasuja, K.S. Walton, Water stability and adsorption in metal-organic frameworks, Chemical reviews 114(20) (2014) 10575-612.
[2] W. Zhang, Y. Hu, J. Ge, H.L. Jiang, S.H. Yu, A facile and general coating approach to moisture/water-resistant metal-organic frameworks with intact porosity, Journal of the American Chemical Society 136(49) (2014) 16978-81.
[3] S.X. Gao, N. Zhao, M.H. Shu, S.N. Che, Palladium nanoparticles supported on MOF-5: A highly active catalyst for a ligand- and copper-free Sonogashira coupling reaction, Applied Catalysis A: General 388(196) (2010).
[4] N. Miyaura, A. Suzuki, Palladium-Catalyzed Cross-Coupling Reactions of Organoboron Compounds, Chemical reviews (1995) 95, 2457-2483.
[5] S. Stanforth, Catalytic cross-coupling reactions in biaryl synthesis, Tetrahedron (1998) 54, 263.
[6] J. Han, Y. Liu, R. Guo, Facile Synthesis of Highly Stable Gold Nanoparticles and Their Unexpected Excellent Catalytic Activity for Suzuki−Miyaura Cross-Coupling Reaction in Water, Journal of the American Chemical Society 131(6) (2009) 2060-2061.
[7] M. del Rio, M. Villar, S. Quesada, G. Turnes Palomino, L. Ferrer, C. Palomino Cabello, Silver-functionalized UiO-66 metal-organic framework-coated 3D printed device for the removal of radioactive iodine from wastewaters, Applied Materials Today 24 (2021) 101130.
[8] G. Luo, B, Zhang, X.-C. Yang, H.-B. Wu, F.-C. Meng, L.-M. Zhai, Y. Qin, Synthesis of ZIF-8 nano-film-coated Pt/SiO_2 by chemical vapor deposition for alkyne semi-hydrogenation, Journal of Fuel Chemistry and Technology 49(9) (2021) 60075-0.
[9] L. Xu, C.Y. Xing, D. Ke, L. Chen, Z.J. Qiu, S.L. Zeng, B.J. Li, S. Zhang, Amino-Functionalized beta-Cyclodextrin to Construct Green Metal-Organic Framework Materials for CO_2 Capture, ACS Appl Mater Interfaces 12(2) (2020) 3032-3041.
[10] Y. Hu, W.M. Verdegaal, S.-H. Yu, H.-L. Jiang, Alkylamine-Tethered Stable Metal-Organic Framework for CO_2 Capture from Flue Gas, ChemSusChem 7(3) (2014) 734-737.
[11] J.R. Li, R.J. Kuppler, H.C. Zhou, Selective gas adsorption and separation in metal-organic frameworks, Chemical Society reviews 38(5) (2009) 1477-504.
[12] Y. Shi, B. Liang, R.-B. Lin, C. Zhang, B. Chen, Gas Separation via Hybrid Metal–Organic Framework/Polymer Membranes, Trends in Chemistry 2(3) (2020) 254-269.

[13] Q. Qian, P.A. Asinger, M.J. Lee, G. Han, K. Mizrahi Rodriguez, S. Lin, F.M. Benedetti, A.X. Wu, W.S. Chi, Z.P. Smith, MOF-Based Membranes for Gas Separations, Chemical reviews 120(16) (2020) 8161-8266.

[14] Y. Liu, Z. Ng, E.A. Khan, H.-K. Jeong, C.-b. Ching, Z. Lai, Synthesis of continuous MOF-5 membranes on porous α-alumina substrates, Microporous and Mesoporous Materials 118(1-3) (2009) 296-301.

[15] A. Huang, H. Bux, F. Steinbach, J. Caro, Molecular-Sieve Membrane with Hydrogen Permselectivity: ZIF-22 in LTA Topology Prepared with 3-Aminopropyltriethoxysilane as Covalent Linker, Angewandte Chemie International Edition 49(29) (2010) 4958-4961.

[16] R.-B. Lin, S. Xiang, W. Zhou, B. Chen, Microporous Metal-Organic Framework Materials for Gas Separation, Chem 6(2) (2020) 337-363.

[17] S. Ketrat, T. Maihom, S. Wannakao, M. Probst, S. Nokbin, J. Limtrakul, Coordinatively Unsaturated Metal-Organic Frameworks $M_3(btc)_2$ (M = Cr, Fe, Co, Ni, Cu, and Zn) Catalyzing the Oxidation of CO by N_2O: Insight from DFT Calculations, Inorganic chemistry 56(22) (2017) 14005-14012.

[18] H. Aykac Ozen, B. Ozturk, Gas separation characteristic of mixed matrix membrane prepared by MOF-5 including different metals, Separation and Purification Technology 211 (2019) 514-521.

Numerical Simulation Research on Slow cook-off of Warhead of DNAN-based Melt-cast Explosive

Kebin Zhang, Wenbin Li*

ZNDY of Ministerial Key Laboratory, Nanjing University of Science and Technology, Nanjing 210094, China

*Corresponding author's e-mail: kb2018@njust.edu.cn

Abstract. RBOE explosive is a new type of DNAN-based insensitive melt-cast explosive with good thermal safety. In order to study the response characteristics of RBOE explosives under slow cook-off conditions, this study carried out a numerical simulation study of slow cook-off for the ϕ35mm×116mm RBOE cylindrical warhead. According to the thermal decomposition characteristics of each component of the explosive, the thermal reaction model of the RBOE explosive was established. Numerical simulation of RBOE warhead cook-off was carried out at a heating rate of 1K/min; the internal temperature rise curve, phase change cloud diagram and ignition cloud diagram of the warhead were obtained, and the thermal response characteristics of the RBOE explosive were analyzed.

1. Introduction

RBOE is a new type of DNAN-based insensitive melt-cast explosive (component mass ratio is 54wt% HMX, 20wt% NTO, 18wt% DNAN and 8wt% AL powder) [1]. There are few reports on the numerical simulation study of the explosive's cook-off. In order to study the cook-off characteristics of RBOE explosives, a theoretical model of RBOE explosive cook-off was established on the basis of the thermal reaction mechanism of the components in the mixed explosive. Based on the $\phi 35mm \times 116mm$ RBOE cylindrical warhead, a finite element model was established. Numerical simulation of RBOE warhead cook-off was carried out at a heating rate of 1K/min, and the response characteristics of the warhead under thermal stimulation were analyzed, which provides theoretical and data support for the thermal safety evaluation of RBOE warhead.

2. Establishment of theoretical model of RBOE warhead cook-off

2.1. Model assumptions

A numerical calculation model is established based on the thermal reaction mechanism of each component of RBOE explosives, and the following simplified assumptions are made:

(1) The heat transfer method in the slow cook-off process only considers heat conduction and convection heat transfer, ignoring the influence of radiant heat transfer on the temperature field.

(2) Ignoring the chemical reaction between the components in the mixed explosive, the self-heating reaction of each elemental explosive follows the Arrhenius law.

(3) After the explosive is melted, it is an incompressible flow; the density is only related to the temperature, and the change of density with pressure is not considered.

(4) The physical parameters of the warhead take the average value that changes with temperature, except for the explosive density.

(5) For the entire cook-off process before the response, it is assumed that the system never exchanges material with the outside world, and the mass, momentum and energy in the entire system are always conserved.

2.2. Cook-off model of RBOE explosive

RBOE is a mixed explosive made of 54wt% HMX, 20wt% NTO, 18wt% DNAN and 8wt% AL powder [1]. The numerical simulation of the cook-off process of the mixed explosive is mainly to accurately describe the thermal decomposition reaction of the mixed explosive [2]. In this study, a single-step + multi-step thermal decomposition reaction kinetic model is used to describe the decomposition reaction of each component, and the sum of the heat generated by their decomposition is used as the total reaction heat of the explosive.

2.2.1. Description of the thermal decomposition reaction of each component

(1) The thermal decomposition process of HMX is described by Tarver et al. [3] with a four-step chemical reaction:

Reaction 1: βHMX $\xrightarrow{r_1}$ δHMX (HMX endotherms and undergoes crystal form transformation);

Reaction 2: δHMX $\xrightarrow{r_2}$ solid intermediate product (endothermic reaction);

Reaction 3: solid intermediate product $\xrightarrow{r_3}$ gas intermediate product (a slight exothermic reaction produces intermediate gas products such as CH_2O、N_2O、HCN and HNO_2);

Reaction 4: Intermediate gas product $\xrightarrow{r_4}$ The final gaseous reaction product (secondary gas phase reaction produces gaseous products such as CO_2、CO、N_2 and H_2O).

(2) The thermal decomposition process of NTO is described by Kou et al. [4] with a one-step chemical reaction:

Reaction 5: NTO $\xrightarrow{r_5}$ product (exothermic reaction)

(3) The thermal decomposition process of DNAN is described with reference to the study of Chen Lang et al. [5] with a one-step chemical reaction:

Reaction 6: DNAN $\xrightarrow{r_6}$ product (exothermic reaction)

(4) This article believes that Al does not participate in the thermal decomposition reaction of explosives.

2.2.2. Thermal decomposition rate

From the above reaction process, it can be seen that except for HMX, other thermal decomposition reactions are single-step zero-order reactions. Therefore, the thermal decomposition process of NTO and DNAN is described by the traditional Frank-Kamenetskii model, and the thermal decomposition process of HMX is described by McGuire-Tarver model.

(1) HMX:

$$r_1 = Z_1 \exp(\frac{-E_1}{RT})\rho_a$$

$$r_2 = Z_2 \exp(\frac{-E_2}{RT})\rho_b$$

$$r_3 = Z_3 \exp(\frac{-E_3}{RT})\rho_c$$

$$r_4 = Z_4 \exp(\frac{-E_4}{RT}){\rho_d}^2$$

(2) NTO:

$$r_5 = Z_5 \exp(\frac{-E_5}{RT})\rho_e$$

(3) DNAN:

$$r_6 = Z_6 \exp(\frac{-E_6}{RT}) \rho_f$$

In the process, r_i is the mass reaction rate, Z_i is the pre-factor, E_i is the activation energy of the reaction, i is the reaction number, i=1, 2, 3, 4, 5, 6. ρ_m is the density of the reactants in each step, m=a, b, c, d, e, f. R is the universal gas constant, and T is the temperature of the reactant.

2.2.3. Thermal decomposition reaction heat

This article defines the reaction heat S generated by the thermal decomposition of RBOE as the sum of the heat absorbed and released by the decomposition of HMX, NTO and DNAN.

S(RBOE)= 0.54S(HMX)+ 0.2S(NTO)+ 0.18S(DNAN)

S(HMX)= $r_1Q_1 + r_2Q_2 + r_3Q_3 + r_4Q_4$

S(NTO)= r_5Q_5

S(DNAN)= r_6Q_6

Among them, Q_i is the heat of formation in each step of the reaction.

2.2.4. Thermal conductivity and specific heat capacity of explosives

Both the thermal conductivity λ and the specific heat capacity c of the explosive are functions of temperature, which are assumed to be constants here. According to the superposition principle, the thermal conductivity and specific heat capacity of RBOE can be expressed as [6]:

$$\lambda = 0.54\lambda_{HMX} + 0.2\lambda_{NTO} + 0.18\lambda_{DNAN} + 0.08\lambda_{Al}$$
$$c = 0.54c_{HMX} + 0.2c_{NTO} + 0.18c_{DNAN} + 0.08c_{Al}$$

3. Numerical simulation of RBOE warhead cook-off

3.1. Finite element model of RBOE warhead

This study selects the $\phi 35mm \times 116mm$ RBOE cylindrical warhead as the research object to carry out the numerical simulation study on the cook-off. Since the RBOE cylindrical warhead is an axisymmetric structure, a 1/2 model is established. The warhead structure diagram and the finite element model diagram are shown in Figure 1. The warhead is composed of an upper end cover, a case, explosives and a lower end cover. The case and other metal materials are all made of 45# steel. A hexahedral mesh is used when dividing the mesh, and the maximum size of the mesh element is 1mm. In the numerical simulation study of the warhead, the temperature boundary conditions were set on the outer surface of the warhead case, the upper end cover and the lower end cover: the heating rate was 1K/min, and the initial temperature was 293K.

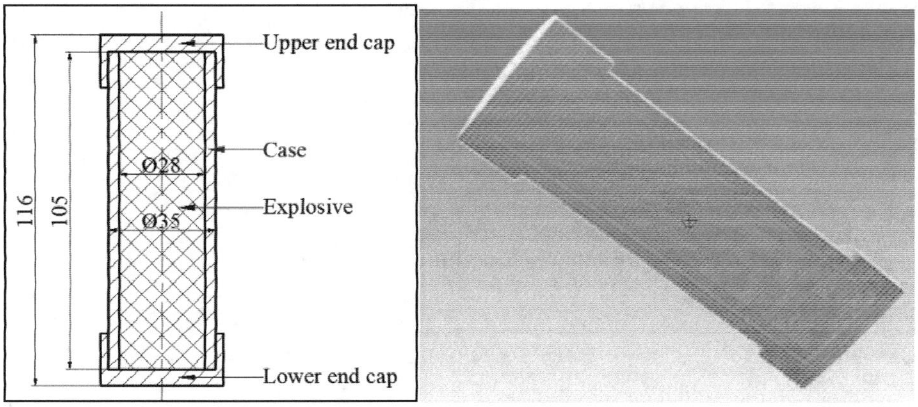

(A) Warhead structure diagram (b) Warhead finite element model diagram
Figure 1. RBOE cylindrical warhead

3.2. Solution parameter settings

Based on literature [2,7-10] and some physical property tests, the physical property parameters and reaction kinetic parameters of explosives are obtained in this study, as shown in Table 1~Table 3.

Table 1. Physical properties of each component of RBOE

	Density */kg·m-3*	Specific heat capacity */J·kg-1·K-1*	Thermal conductivity */w·m-1·K-1*
HMX (54wt%)	1850	1004.62	0.5258
NTO (20wt%)	1850	1088	0.27
DNAN (18wt%)	1450	1170	0.2
Al (8wt%)	2719	870	1.39

Table 2. Warhead physical parameters

	Density */kg·m-3*	Specific heat capacity */J·kg-1·K-1*	Thermal conductivity */w·m-1·K-1*	Viscosity /Pa·s	Melting point/K
RBOE	1850	1040.2948	0.4851	0.015	460.8
45 #steel	7850	460	32.6	-	-

Table 3. Reaction kinetic parameters of each component of RBOE

	i	*E/J·mol-1*	*Z/s-1*	*Q/ J·kg-1*
HMX (54wt%)	1	203574	7.9909E20	-42000
	2	221340	1.4130E21	-252000
	3	186060	2.0681E16	558600
	4	143220	1.5984E12	5615400
NTO (18wt%)	5	1.675E5	1.1E12	3631610
DNAN (20wt%)	6	1.72E5	1.2E11	4.96E6

The natural convection formed by the buoyancy force after the melting of cast explosives is a low-speed flow field. Therefore, Boussinesq approximation is adopted for explosives [11]. The density of explosives is only related to temperature without considering its relationship with pressure. It holds that the explosive fluid in the flow field is incompressible. The mathematical expression [12] is as follows:

$$\rho_f = \rho_0(1 - \alpha_v \Delta \mathrm{T})$$

Where: ρ_f is the explosive density; ρ_0 is the initial density of explosive; α_v is the coefficient of thermal expansion, $\alpha_v = 1.1 \times 10^{-3}$.

3.3. Analysis of simulation results

Figure 2 shows the temperature rise curve of the RBOE warhead case and the center point of the explosive, and the change curve of the liquid fraction of the explosive. From Figure 2, it can be seen that the RBOE explosive starts to melt in 4130s and ends at 5440s, and the mass fraction of the liquid phase reaches 1. In the process of explosive melting, the temperature difference between the center of the explosive and the case gradually increases due to the heat absorption of the explosive melting, and reaches the maximum at the end of the explosive melting. Due to the large temperature gradient, the explosives are completely melted, and the convective heat transfer intensity increases, resulting in an increase in the temperature of the explosives, and finally the temperature of the case tends to be the same. The RBOE warhead had an ignition reaction at 9883s, and the case temperature was 466K at this time.

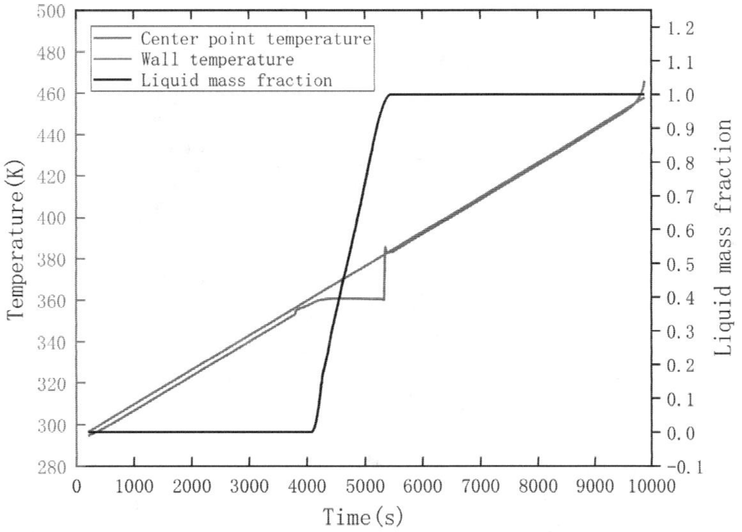

Figure 2. RBOE warhead temperature rise curve and liquid phase mass fraction change curve

Figure 3 is the cloud diagram of temperature distribution and liquid fraction distribution of rboe explosive before and after phase change. As can be seen from Fig. 3, the warhead temperature is concentric and symmetrically distributed before the explosive phase change. During the phase change process, the unmelted explosive is suspended under the warhead in a conical shape, and the temperature is distributed from high to low. After the phase transition, the internal temperature of the warhead also shows a trend of high and low. This further shows that after the explosive melts, convection occurs in the warhead due to the action of gravity and buoyancy.

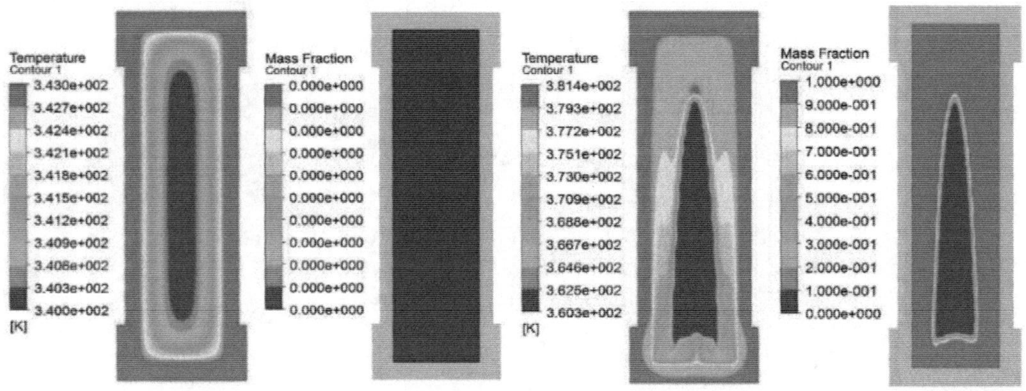

(a) Before the phase change (3000s) (b) Phase change (5200s)

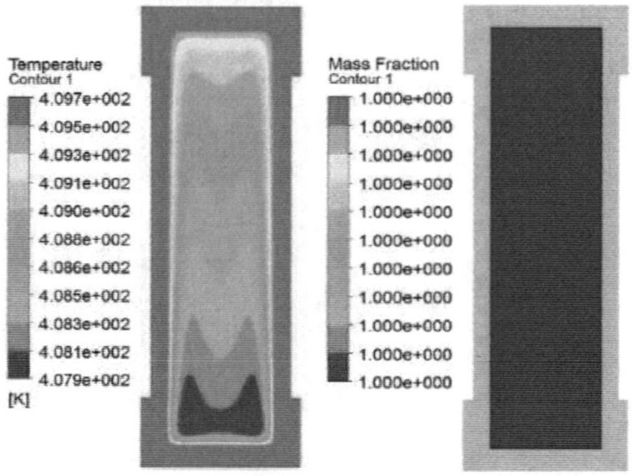

(C) After phase change (7000s)

Figure 3. Temperature cloud map and liquid phase mass fraction distribution cloud map before and after phase transition of explosive

Figure 4 shows the mass fraction distribution of each component of the RBOE explosive at different times and the temperature distribution inside the explosive at the time of ignition. At the moment of ignition of the explosive, βHMX has reacted completely, and solid intermediate products and gaseous intermediate products are successively formed. DNAN and NTO have good thermal stability, and there is basically no thermal decomposition reaction before the RBOE explosive is ignited. Due to the effect of convection, the ignition temperature of the warhead appears on the top of the warhead, and the ignition temperature of the explosive is about 492K.

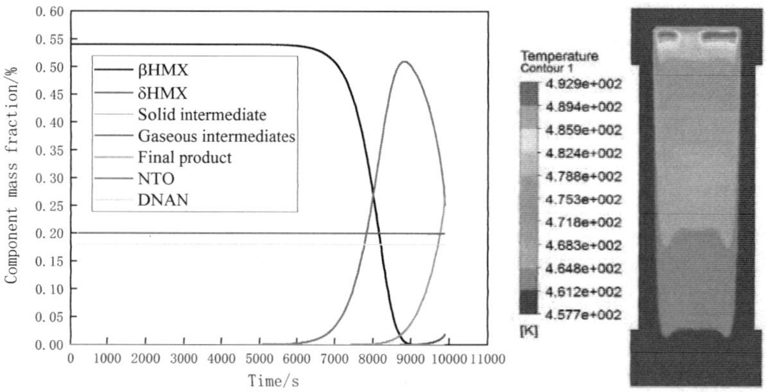

Figure 4. The mass fraction distribution of each component of the explosive at different times and the cloud diagram of the temperature distribution at the moment of ignition of the explosive

4. Conclusion

In this study, a single-step + multi-step thermal decomposition reaction kinetic model is used to describe the decomposition reaction of each component of RBOE explosive, and a RBOE explosive cookoff model is established to predict the distribution of the mass fraction of each component of the explosive over time. Through numerical simulation, it is found that the final ignition position of the warhead appears on the top of the warhead due to the convection effect after the explosive is melted. This research can provide theoretical and data support for the thermal safety evaluation of the RBOE warhead.

Reference

[1] Wang Shuai, Zhi Xiaoqi, Jia Qiulin, et al. Explosive burning test and numerical simulation based on mass_flux method[J]. Chinese Journal of Ordnance Equipment Engineering, 2020, 41(08): 1-6.

[2] Ma Xin, Chen Lang, Lu Feng, et al. Calculation of multi-step thermal decomposition reaction of HMX/TATB-based mixed explosive under firing conditions[J]34(1).

[3] Tarver C M, Koerner J G J J O E M. Effects of Endothermic Binders on Times to Explosion of HMX- and TATB-Based Plastic Bonded Explosives[J], 2007, 26(1): 1-28.

[4] Kou Yongfeng. Research on the severity of explosive reaction under firing conditions [D]. Beijing Institute of Technology.

[5] Chen Lang, Li Beibei, Energetic materials Ma J. Burning characteristics of DNAN explosives[J], 2016, 24(001): 27-32.

[6] Kou Yongfeng, Chen Lang, Ma Xin, et al. Baking experiment and numerical simulation of Hexogen-based aluminum-containing explosives[J], 2019, v.40;No.266(05): 85-96.

[7] Wang Ruijin, Zhang Kai, Wang Gang. Fluent technology foundation and application examples[M]. Fluent technology foundation and application examples, 2007.

[8] Dong Haishan, Zhou Fenfen. Performance of high-energy explosives and related substances [M]. Performance of high-energy explosives and related substances, 1989.

[9] Fang Hao. Study on the combustion calculation model and thermal characteristics of RBOL-2 explosive [D]. Beijing Institute of Technology, 2018.

[10] Wu Nana. Research on NTO/HMX-based Insensitive Press-charged Explosive[D]. North University of China, 2016.

[11] Zhang Ning, Li Guangzheng. Numerical study of natural convection in a rectangular enclosed cavity under special conditions [J]. Journal of Huazhong University of Science and Technology (Natural Science Edition), 2002, 30(007): 95-97.

[12] Song Shanshan, Guo Xueyan. Boussinesq approximation and numerical simulation of natural convection in a closed cavity[J]. Chinese Quarterly of Mechanics, 2012, 33(001): 60-67.

A Study on Forging Process of New Aluminum Alloy

ChaoFeng[1*], Weike Liu[1], Xianhui Cao[1], Yunlai Deng[2], Jun Wang[1], Yi Xie[1]

[1]State Grid Hunan Electric Power Company Limited Research Institute, Changsha 410007, China

[2]Changsha University of Science and Technology, School of Electrical and Information Engineering, Changsha 410007, China

*676459812@qq.com

Abstract: This paper employs the isothermal compression method to study the forging process of new aluminum alloy and analyzes its law of thermal deformation behavior at the temperature of 400~550°C and the strain rate of 1~-10-3s-1. It is found that flow stress is correlated with the temperature and strain rate and that the stress level decreases with the increase of temperature and the decrease of strain rate. This reveals the evolution law of the microstructure of the aluminum alloy in thermal deformation and lays a foundation for the subsequent studies and optimizationof the forging process.

1. Introduction

Because of its light weight, aluminum alloy can replace steel materials in many fields and achieve the goal of being lightweight. The strength of aluminum is low, generally 80~200MPa. By adding copper, zinc, manganese, magnesium, silicon and lithium to aluminum for alloying, the strength of aluminum alloy thus formed can reach 300~800MPa or even higher[1-2]. The main alloying elements added in the 6000 series aluminum alloy are Mg and Si. By adding them according to the proper proportion, the Mg2Si precipitation strengthening phase will be formed. This series of aluminum alloy has good comprehensive properties: high specific strength and corrosion resistance, good formability, low cost, etc. It has been widely used in the fields of national defense, military industry, rail transit, ground vehicles, surface ships, construction, electric power and electronics. The forged aluminum alloy can improve its microstructure and mechanical properties. While in terms of the cast aluminum alloy, after hot working deformation and subsequent treatment, the microstructure of castings undergo deformation, recrystallization and precipitation strengthening, so that the original coarse dendrites and columnar grains become microstructure of the forging flow line. The original metallurgical defects such as segregation, porosity, gas hole and slag inclusion in the ingot are compacted and welded. In this way, the strength, ductility, and toughness can be improved to obtain higher comprehensive properties[3-5].

2. Study plan

The first step is to determine the composition of the 6000 series aluminumalloy,whose performance index meets the requirements, find out its forging process performance and study the manufacturing process of the clamps for the transmission line. On this basis, the tests on the strength of aluminum alloy materials and the performance of clamps under simulated load conditions are then carried out.

3. Results and conclusions

3.1. Constitutive model for thermal deformation of 6000 series aluminum alloy

The constitutive relation of aluminum alloy can accurately describe the effect of material characteristics, such as strain rate, forming temperature, strain and work hardening behavior on deformation resistance. In accordance with the objectives of the project, an Al-Mg-Si-Mn alloy was designed and prepared, and the composition is shown in Table 1.

Table 1 Composition of the HD6X70 alloy (wt.%)

Element	Si	Mg	Mn	Fe	Ti	Zn	Cu	Al
Content	1.17	0.966	0.520	0.316	0.026	0.043	0.021	Bal

The isothermal compression method was adopted to study the thermal deformation behavior of this alloy and the constitutive equation of this new aluminum alloy was set up in accordance with the true stress-strain curve. Furthermore, the quasi-SEM-EBSD method was applied to discuss the recrystallization mechanism in the whole forging process, providing a basis for the formulation of the actual forging process.

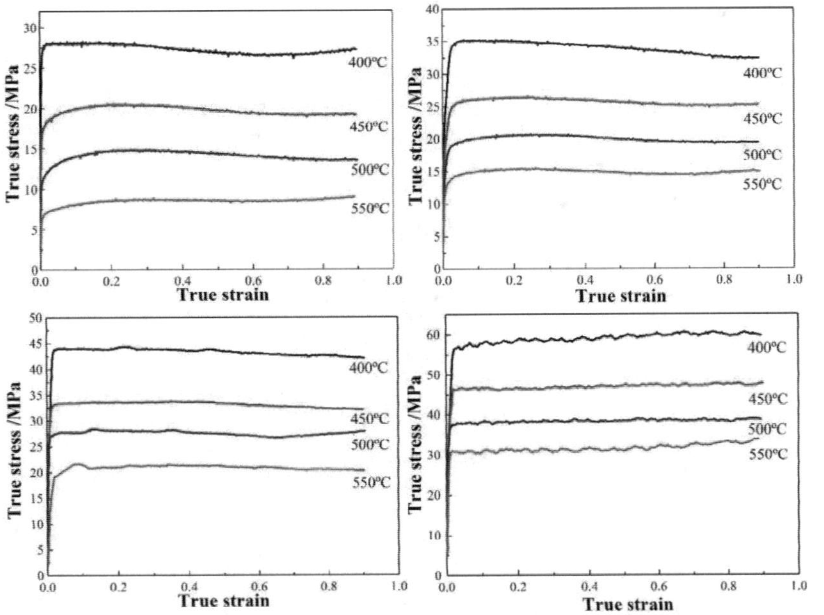

Figure 1 The true stress-strain curve at different temperatures
(a) $=10^{-3}s^{-1}$; (b)$=10^{-2}s^{-1}$; (c) $=10^{-1}s^{-1}$; (d)$=1s^{-1}$

Figure 1 shows the true stress-strain curves of the HB6X70 aluminum alloy at the temperature of 400°C, 450°C, 500°C and 550°C, strain rate of 0.001, 0.01, 0.1 and 1s-1 and true stress of 0.9 obtained by the isothermal compression experiments. The curves indicate that the thermal deformation temperature and strain rate significantly affect the flow behavior of the HB6X70 aluminum alloy. Generally speaking, at the given strain rate, the flow stress decreases with the increase of temperature. Meanwhile, at the given temperature, the flow stress tends to decline with the decrease of strain rate.

Figure 1 also suggests that, at the initial stage of deformation, the flow stress rapidly increases with the increase of stress and shows a tendency of yield decline under the condition of a high strain rate. The specific dynamic softening mechanism may be different. The typical performance under most experimental conditions is: the flow stress gradually decreases after the peak value until the high strain, showing dynamic flow softening. Deformation-induced flow softening is caused by different mechanisms such as deformation heating, strengtheneddynamic recovery (DRV), sub-grain coarsening, dynamic recrystallization (DRX), dynamic precipitation and dissolution, and precipitation phase coalescence/growth.

The effect of deformation heating on flow softening is very important. After the flow softening, flow stress near the steadystate occurred at the end of the experiment. This suggests that the flow stress remains relatively constant and does not increase with strain. Secondly, there is little flow softening when it is directly hardened to the steady state. This is the characteristic of DRV-dominated deformation behavior, which can be obviously seen in Figure 1-a and Figure 1-b. In addition, the flow stress continuously increases with the increase of strain, namely obvious continuous work hardening. The maximum flow stress is reached at the end. Under this condition, dynamic softening is insufficient enough to offset the work hardening effect in the thermal deformation process. This flow behavior can be obviously observed from Figure 1-c. The last one is relatively complicated. As shown in Figure 1-d, the peak stress is quickly reached after work hardening when the strain rate is 1s−1. Then dynamic softening occurs and subsequently, the flow curve goes up and down repeatedly, revealing that alternate hardening and softening continuously occur at this time.

Typically, the shape of the flow curve is decided by the competitive results of simultaneous work hardening and dynamic softening. What's more, the rate of the work hardening and the dynamic softening is decided by the temperature and the strain rate. At the beginning of the deformation, work hardening caused by the intense proliferation, accumulation and interaction of dislocations leads to the rapid increase of flow stress. At this stage, the softening caused by DRV is relatively less obvious. In other words, in the work hardening area, the dislocation generation rate is higher than the dislocation annihilation rate caused by DRV. With the increase of temperature or the decrease of strain rate, the work hardening amount decreases.

3.2. Constitutive model for thermal deformation of 6000 series aluminum alloy

The hot working diagram is widely applied to determine the conditions for the hot working of the alloy materials. The hot working diagram of the HB6X70 aluminum alloy can be established according to the dynamic material model. The total input power P is divided into plastic deformation dissipation power (G) and microstructure transformation dissipation power (J), as shown in formula (1).

$$P = \sigma\dot{\varepsilon} = G + J = \int_0^{\dot{\varepsilon}} \sigma d\dot{\varepsilon} + \int_0^{\sigma} \dot{\varepsilon} d\sigma \quad (1)$$

$$J = \int_0^{\sigma} \dot{\varepsilon} d\sigma = \frac{m}{m+1}\sigma\dot{\varepsilon} \quad (2)$$

$$\eta = \frac{J}{J_{max}} = \frac{m/(m+1)}{1/2} = \frac{2m}{m+1} \quad (3)$$

where, m denotes the strain rate sensitivity index of the alloy and is defined as:

$$m = \frac{dJ}{dG} = \frac{\dot{\varepsilon} d\sigma}{\sigma d\dot{\varepsilon}} \quad (4)$$

The dissipation efficiency parameter (η) can be expressed with the formula 3-4. *Jmax*represents the power consumption under the ideal linear conditions *(m=1, Jmax =σε/2)*. When j or η changes, defects such as voids and cracks may appear due to unstable deformation. These unstable areas can be marked on the hot working diagram. In a study on the hot working diagram, the value of the dimensionless parameter is adopted as the basis for the determination. The deformation instability occurs when such value is smaller than zero, as shown in the red area in Figure 2. When the temperature is lower than 480°C, the deformation conditions corresponding to the temperature in this area are prone to cause cracking. When the temperature is 500~550°C or the strain rate is low (for example, between 0.1~1s-1), the power consumption efficiency is the highest, indicating that the alloy has good hot working performance, and there are almost no unstable areas in the hot working diagram.

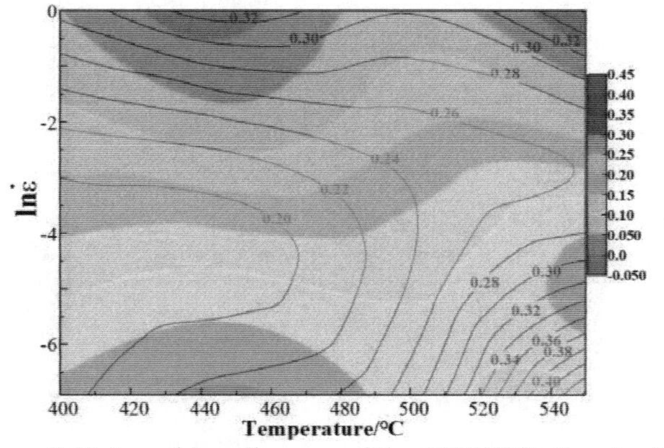

Figure 2 Hot working diagram of the HD6X70 alloy (ε=0.6)

3.3. Evolution law of microstructure of the 6000 series aluminum alloy in thermal deformation

3.3.1. Effect of the temperature on microstructure

Figure 3 shows the change of alloy microstructure with temperature. In the figure, the red zone (GOS>7.5) represents the deformed grain, the blue zone (GOS<1) represents the dynamically recrystallized grain and the yellow zone (1<GOS<7.5) represents the sub-grains. It can be seen from the IPF diagram that the grain structure after isothermal compression can be divided into two types. One is the large fiber particles which are about 20μm wide and has many sub-particles inside. The other is equiaxed recrystallized grains distributed along the grain boundary. The temperature has a great effect on the morphology of these grains. At a lower temperature of 400~450°C, the original banding grain is extruded into the fiber grain with a width of fewer than two sub-grains. The deformation is serious. As indicated by the arrows in the figure, many tiny equiaxed grains are formed in the fiber grain and distributed along the grain boundary. On the whole, at the temperature of 400~450°C, the average grain size after thermal deformation is relatively small. With the temperature rising to 500°C, the distribution of fiber grains becomes uneven and the size of both equiaxed grain and sub-grain grows. At 550°C, the grains completely recover, recrystallize and grow,with almost all the original banding microstructures transformed into equiaxed grains.

It can be seen from the GOS diagram that,within the range of 400~450°C, the deformation of the test samples is serious and the recrystallized grains, the sub-grains and the deformed microstructures are mixed together. Those marked by the white box in Figure 5-1 can clearly reflect these characteristics. With the increase of temperature, the number of recrystallized grains increases obviously and the size of the deformed microstructure decreases gradually. When the temperature is between 500°C and 550°C, due to dynamic recrystallization and recovery, there is almost no deformed microstructure, but there are a large number of recrystallized grains and some sub-grains.

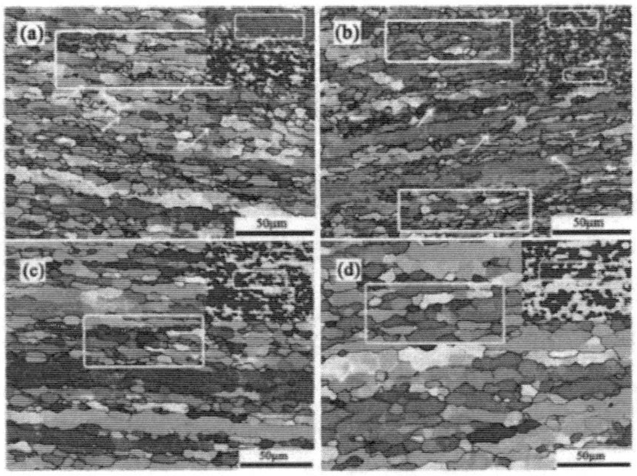

Figure 3 IPF diagram and intragranular orientation distribution (GOS) diagram of the sample in isothermal compression at the strain rate of 1s-1
(a) 400°C; (b) 450°C; (c)500°C; (d) 550°C

3.3.2.Effect of the deformation rate on microstructure

In addition to temperature, strain rate also has a significant effect on the grain morphology. As shown in Figure 4, when the strain rate is $10^{-3}s^{-1}$, almost all of them are equiaxed recrystallized grains, indicating a lower strain rate.The work hardening caused by deformation has enough time to be offset by the softening mechanism of the alloy, and the recrystallized grains also have enough time to grow. With the increase of the deformation rate, the recrystallization rate decreases and the grains are refined. When the strain rate is 10^{-2}~$10^{-1}s^{-1}$, the original grains are mixed with HAGBs and LAGBs, indicating that the recrystallization mechanism is CDRX. When the strain rate is $1s^{-1}$, as shown in Figure 4(b), the softening effect caused by recovery and recrystallization is not enough to offset the work hardening caused by deformation, resulting in inclusion deformation and recrystallized grains in the fine grain microstructure.

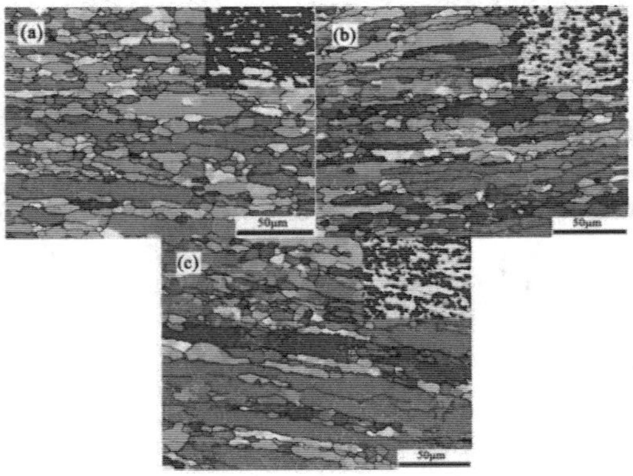

Figure 4 IPF diagram and GOS diagram of the sample at 450°C
(a) $\varepsilon=10^{-3}s^{-1}$; (b) $\varepsilon=10^{-2}s^{-1}$; (c) $\varepsilon=10^{-1}s^{-1}$

4. Conclusions

In this paper, we prepared the samples for studying the forging process principle of the 6000 series aluminum alloy with the isothermal compression method, and carried out studies on the thermal deformation behavior of such aluminum alloy at the temperature of 400~550°C and the strain rate of ~$10^{-3}s^{-1}$. We have the following main conclusions:

(1) The flow stress of the alloy is correlated with the temperature and strain rate. The flow stress reduces with the rise of temperature and the decrease of strain rate. Based on the experimental data, it is found that the apparent activation energy of thermal deformation of such alloy is 171.91kJ/mol. We then established the constitutive relation of thermal deformation of the alloy, laying a foundation for the subsequent studies.

(2) In the process of thermal deformation, the main recrystallization mechanism is CDRX, accompanied by GDRX at high lnZ. After heat treatment, the recrystallization mechanism is similar to that of heat treatment. Grain growth during discontinuous recrystallization occurs in the form of grain boundary migration driven by energy storage.

Acknowledgement
This work was supported by State Grid Hunan Electric Power Company Limited, Science project No.5216A521001H.

References

[1] Wesley W, Rudolf M, Ye T H, et al. (2019)The effect of incomplete solution treatment on the tensile behavior and mechanical anisotropy of aluminum lithium alloy.DiffusionFoundations.pp, 109-11.

[2] Eisaabadi B G, Yeom G Y, Murat T, et al. (2018)The effect of solution treatment time on the microstructure andductility of naturally-aged alloy A383 die casting.Material Science & Engineering A,722: pp.1-7.

[3] Zhang J,Li J Y, Tian S K, et al.(2019)Effects of solution treatment on microstructure transformation, tensile and exfoliation corrosion properties of aluminum alloy 7136. Journal of Materials Engineering and Performance.28(3):pp.1312-1323.

[4] Baghbani B A, Eivani A R, Jafarian H R,et al. (2020)Effects of hot forming cold die quenching and solution treatment on formability and pressing load during equal channel angular deformation of aluminum alloy AA2024. Journal of Materials Research and Technology. 9(3):pp.5599-5609.

[5] Hannes F, Benjamin M, Philipp W,et al. (2018)Influenceofsolution-annealing parameters on the continuous cooling precipitation of aluminum alloy 6082.Metals. 8 (4):pp.265-269.

Effect of reaction time on properties of TPEG type polycarboxylate superplasticizer

Zhuojun Jiang [1a*], Renliang You [1b], Yunhui Fang [1c], Yuanqiang Guo [1d], Tianxing Lin [1e]

KZJ New Materials Group Co., Ltd. Xiamen 361101, Fujian, China

[a*]jzj0407@126.com

Abstract. The effects of dropping time and holding time on the properties of TPEG type polycarboxylate superplasticizer paste and concrete were studied with 20000l stainless steel reactor as production equipment, ammonium persulfate as initiator, sodium hypophosphite as chain transfer agent and reaction temperature of 60 ℃. The experimental results show that within the experimental range, the dropping time has little effect on the initial dispersion effect of TPEG type polycarboxylate superplasticizer and has a great impact on the slump retention performance. The holding time has a great impact on the initial dispersion effect and slump retention effect of TPEG type polycarboxylate superplasticizer. The dropping time and holding time have little effect on the 3d, 7d and 28d concrete compressive strength of TPEG type polycarboxylate superplasticizer.

1. Introduction

Polycarboxylate superplasticizer has become the mainstream concrete superplasticizer product due to its good design, high water reduction rate and environmental protection [1-2]. TPEG (isoprenyl oxy polyoxyethylene ether) type polycarboxylate superplasticizer is the mainstream polycarboxylate superplasticizer in the market because of its outstanding water reducing performance [3].

The development of TPEG type polycarboxylate superplasticizer process generally passes the laboratory small-scale experiment first. Generally, 1L four port glass flask is used as the reaction vessel and equipped with laboratory mixer, while the production generally adopts enamel reactor or stainless steel reactor, with a volume ranging from 5000L to 50000l [4-7]. From small-scale test to production, the reaction amount varies greatly, and there are great differences in mass and heat transfer between the two. Therefore, the direct use of production equipment to study the impact of reaction time on the final product performance in the production of TPEG type polycarboxylate superplasticizer is of great significance to the industrial application of the product [8]. The reaction time in the production process of polycarboxylate superplasticizer includes dropping time and holding time. Therefore, this paper mainly studies the effects of dropping time and holding time on the properties of TPEG type polycarboxylate superplasticizer.

2. Experimental

2.1. Materials

2.1.1. The main synthetic experimental raw materials
TPEG (industrial grade), Acrylic acid (industrial grade), Ammonium persulfate (industrial grade), Sodium hypophosphite (industrial grade), Sodium hydroxide(30% aqueous solution, industrial grade).

2.1.2. Main performance test raw materials for experiment
（1）Cement (C): the "Minfu" brand 42.5R ordinary portland cement of Fujian Minfu building materials, and its performance indexes are shown in Table 1.

Table 1. Cement properties

80 u m square hole sieve Sieve residue /%	Standard consistency /%	Setting time /min		Flexural strength /MPa		Compressive strength /MPa	
		Initial setting	Final setting	3d	28d	3d	28d
0.8	28.4	160	245	4.7	8.8	28.5	50.2

（2）Gravel: Granite gravel produced by a stone factory in Zhangzhou, with particle sizes of 5 ~ 20mm (G1) and 16 ~ 31.5mm (G2). See Table 2 for specific performance indexes of gravel.

Table 2. Gravel properties

Particle size/mm	Apparent density /（kg/m3）	Bulk density /（kg/m3）	Void ratio /%	sediment percentage /%	needle and flake content /%
5～20	2660	1560	41	0.7	2
16～31.5	2640	1580	40	0.6	1

（3）Sand (S): Granite manufactured sand produced by a building materials Co., Ltd. in Xiamen. See Table 3 for its performance indexes.

Table 3. Sand properties

Type	mud block content /%	powder content /%	MB value	Crushing value index /%	Chloride ion content /%	Fineness modulus
Machine-made sand	0	4.4	0.6	12.8	0.00	2.8

2.2. Copolymerization
Add the measured water, TPEG and sodium hypophosphite into the 20000l stainless steel reactor, control the reaction temperature to 60 ℃, start dropping acrylic acid aqueous solution and ammonium persulfate aqueous solution, control it to drop at the time of experimental design, keep it warm for a period of time according to the time value of experimental design, add sodium hydroxide to adjust the pH value to 6.0 ~ 7.0, and then TPEG type polycarboxylate superplasticizer is obtained.

2.3. Performance test method

2.3.1. Net slurry fluidity test
The test shall be carried out according to the standard method in GB / T 8077-2012 test method for homogeneity of concrete admixtures.

2.3.2. Concrete test

(1) Concrete tests shall be carried out in accordance with GB/T50080-2016 Standard of Test Methods for Performance of Ordinary Concrete Mixtures.

(2) The formed concrete test block shall be subject to standard curing, and the 3d, 7d and 28d compressive strength of concrete shall be tested according to GB / T 50081-2002 standard for test methods of mechanical properties of ordinary concrete.

3. Experimental results and discussion

3.1. Effect of different dropping time on product performance

Adjust the dropping time of materials and keep the holding time after the reaction for 60 minutes. Investigate the effects of different dropping time on the net slurry fluidity and concrete performance of the product. The experimental results are shown in Fig. 1 and Fig. 2.

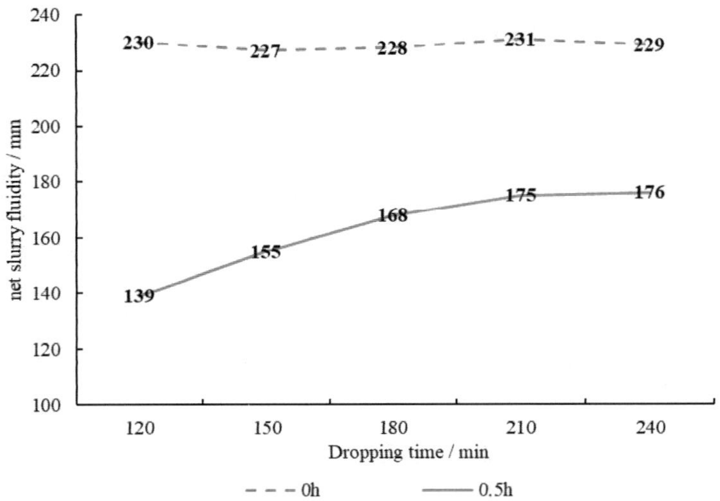

Fig. 1. Effect of different dropping time on fluidity of cement paste of the product

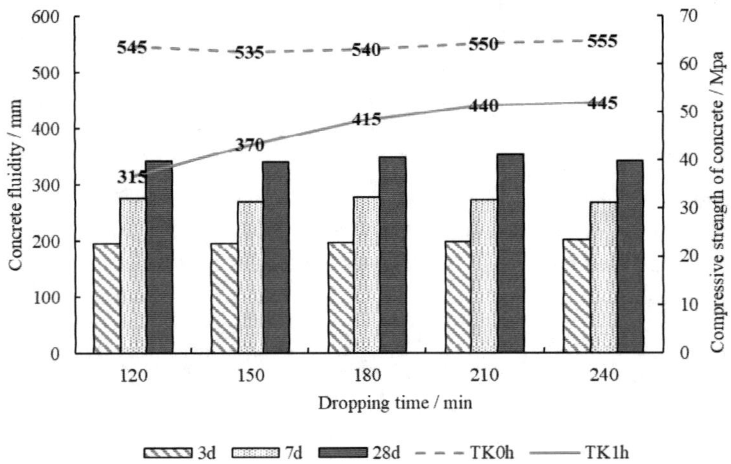

Fig. 2. Effect of different dropping time on properties of concrete of the product

As shown in Fig. 1 and Fig. 2, when the holding time during production is 60 minutes, with the extension of dropping time, the 0h net slurry fluidity and initial concrete fluidity (TK0h) of the production product change little, but the 0.5h net slurry fluidity and concrete 1h fluidity (TK1h) gradually increase, indicating that the slump retention performance of the product is gradually improved. In terms of concrete compressive strength, the products produced at different dropping times have little effect on the 3d, 7d and 28d compressive strength of concrete. Considering the production efficiency and product performance, the dropping time of 210 minutes is the best.

3.2. Effect of different holding time on product performance

When the dropping time is set to 210 minutes, the effects of heat preservation for 20 minutes, 40 minutes, 60 minutes and 80 minutes on the properties of product net slurry fluidity and concrete performance are investigated. The experimental results are shown in Fig. 3 and Fig. 4.

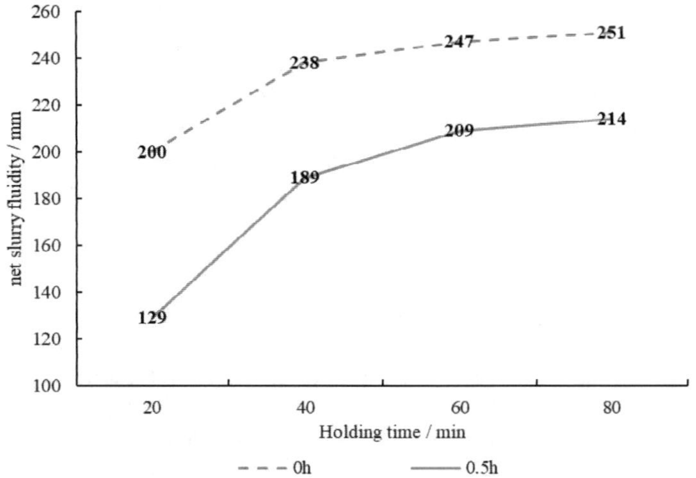

Fig. 3. Effect of different holding time on fluidity of cement paste of the product

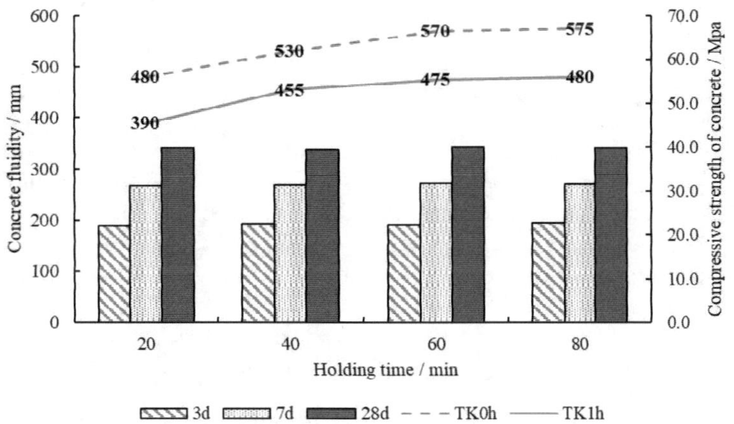

Fig. 4. Effect of different holding time on properties of concrete of the product

As shown in Fig. 3 and Fig. 4, when the dropping time is 210 minutes, with the extension of the holding time, the 0h net slurry fluidity and the initial concrete fluidity (TK0h) of the production products gradually increase and then tend to be stable, and the 0.5h net slurry fluidity and the 1h concrete fluidity (TK1h) also gradually increase and then tend to be stable, indicating that the slump retention performance of the products is gradually improved. In terms of concrete compressive strength, the products produced at different dropping times have little effect on the 3d, 7d and 28d compressive strength of concrete. Considering the production efficiency and product performance, the effect of holding time of 60 minutes is the best.

4. Conclusions

（1） Within the experimental range, the dropping time has little effect on the initial net slurry fluidity and initial concrete fluidity of the product, but has a great impact on the slump retention performance of the product. Considering the production efficiency and product performance, the dropping time of 210 minutes is the best.

（2） Within the experimental range, with the extension of holding time, the net slurry fluidity and concrete fluidity of the product gradually become better and tend to be stable. Considering the production efficiency and product performance, the effect of holding time of 60 minutes is the best.

（3） Within the experimental range, the changes of dropping time and holding time have little effect on the compressive strength of 3d, 7d and 28d concrete.

（4） From small-scale test to production, the dosage of reaction materials of TPEG type polycarboxylate superplasticizer varies greatly, there are great differences in mass and heat transfer between them, and there are also differences in reaction heat release under different initiation systems. Therefore, it is of great significance to directly use production equipment to study the impact of reaction time on final product performance during the production of TPEG type polycarboxylate superplasticizer to improve process effect and efficiency.

Acknowledgments
Thanks to Lin Tianxing and others for their guidance and help in the experiment. Thanks to them, my experiment can be successfully completed. Thanks to You Renliang, Fang Yunhui, Guo Yuanqiang for their guidance and help in the preparation of my thesis, which has greatly improved my study, scientific research and thinking. I would like to express my heartfelt thanks and deep respect to them.

References:

[1] Mo, X. Y., Jing, Y. J., Deng, M., et al. (2009) New research progress and summarization of polycarboxylate-type high performance superplasticizer. Concrete, (03):60-63.
[2] Lei L., Plank J. (2014) Synthesis and Properties of a Vinyl Ether-Based Polycarboxylate Superplasticizer for Concrete Possessing Clay Tolerance. Industrial & Engineering Chemistry Research, 53(3):1048-1055.
[3] Li, Y., Yang, C., Zhang, Y., et al. (2014). Study on dispersion, adsorption and flow retaining behaviors of cement mortars with tpeg-type polyether kind polycarboxylate superplasticizers. Construction and Building Materials, 64, 324-332.
[4] Liu, X., Wang, Z. M., Liang, X., et al. (2013). Preparation and characterization of polycarboxylate superplasticizer synthesized at non-aqueous condition. Applied Mechanics and Materials, 357-360, 1358-1361.
[5] Baoguo, M. A., Dai, Z., Xiao, J., Fang, C., et al. (2015). Design and implementation of functional regulation in production of polycarboxylate. New Building Materials, 42(2), 6-9.
[6] Tao, X. Design and application of intelligent controller for polycarboxylate superplasticizer automatic production line [D]. Beijing: Beijing University of technology, 2016.
[7] Fuying, Z. Research and design of polycarboxylic acid water reducer production process [D]. Beijing: Beijing University of technology, 2011.
[8] Su, Y., Yu, P., Huang, Z.J. (2003).Application of stirring technology in polymerization reactor [J]. Chemical propellant and polymer materials. 19(04), 19-23.

Effect of concrete viscosity on apparent quality of fair faced concrete

Renliang You [1a*], Pan Zhang [1b], Yuetao Qiu [1c], Qing Zheng [1d]

KZJ New Materials Group Fujian Co., Ltd. Zhangzhou 363000, Fujian, China

[a*]1561167808@qq.com

Abstract. In this paper, concrete with different viscosity was prepared by adjusting the water binder ratio of concrete. The floating time of table tennis was used to characterize the viscosity of concrete, and the effect of concrete viscosity on the apparent quality of fair faced concrete was studied. The test results show that with the increase of concrete viscosity, the apparent porosity of concrete decreases first and then increases. When the water binder ratio is 0.37 and the floating time of table tennis is 40s, the appearance quality of concrete components is the best.

1. Introduction

With the development and progress of China's construction industry, the requirements for concrete components are no longer limited to concrete strength, but also put forward higher requirements for the apparent quality of components. In recent years, the research and engineering application of fair faced concrete in China have developed rapidly [1]. The common appearance defects of fair faced concrete members refer to the "bubble hole defects" in the appearance of members [2]. At present, most of the evaluation methods for the appearance quality of concrete members in the industry are based on personal senses or manually counting the number of "bubble holes" in the unit area of concrete members. There is no specific quantifiable evaluation method and there is great subjective randomness.

Photoshop computer image processing technology is a digital way to analyze, recognize and process images, so as to display information more accurately and intuitively [3]. Through Photoshop Image processing technology, this paper accurately analyzes the apparent quality of concrete specimens made of concrete with different water binder ratio and different viscosity, and forms detailed basic data, which has certain reference value for the research of fair faced concrete engineering in the future.

2. Experimental

2.1. Materials

（1）Cement (C): the "Minfu" brand 42.5R ordinary portland cement of Fujian Minfu building materials, and its performance indexes are shown in Table 1.

Table 1. Cement properties

80 u m square hole sieve Sieve residue /%	Standard consistency /%	Setting time /min		Flexural strength /MPa		Compressive strength /MPa	
		Initial setting	Final setting	3d	28d	3d	28d
0.8	28.4	160	245	4.7	8.8	28.5	50.2

(2) Gravel: Granite gravel produced by a stone factory in Zhangzhou. See Table 2 for specific performance indexes of gravel.

Table 2. Gravel properties

Particle size/mm	Apparent density / (kg/m3)	Bulk density / (kg/m3)	sediment percentage /%	needle and flake content /%
5～25	2650	1460	0.7	2

(3) Sand (S): Granite manufactured sand produced by a building materials Co., Ltd. in Xiamen. See Table 3 for its performance indexes.

Table 3. Sand properties

Type	mud block content /%	powder content /%	MB value	Crushing value index /%	Chloride ion content /%	Fineness modulus
Machine-made sand	0	4.4	0.6	12.8	0.00	2.8

(4) Superplasticizer(A): KZJ point-QS special superplasticizer for fair faced concrete, which meets the requirements of GB 8076-2008 concrete admixtures.

(5) Water(W): Meet the requirements of JGJ 63-2006 water standard for concrete.

2.2. Performance test method

2.2.1. Test method for apparent porosity of concrete

When the quality defect of "bubble hole" appears in fair faced concrete components, take photos at an appropriate light angle. At this time, the defect part of "bubble hole" will show a shadow, which is very different from the color of surrounding normal concrete components [4]. After the image is preprocessed and cut, the area outside the concrete component is removed, and then imported into Photoshop software. Call out the histogram function in the software to display the pixel value of the whole image, and record the value as T1. Select the sampling color "shadow" through the color range function. At this time, the pixel value of the shadow area can be displayed on the square diagram, and record the value as T2, The calculation formula of concrete apparent porosity is shown in formula (1).

$$\gamma = T2/T1 \times 100\% \quad (1)$$

2.2.2. Test method for viscosity of concrete

The concrete viscosity test method uses the method described in ZL 201620627378.9 patent and uses the scheme of table tennis floating time to evaluate the concrete viscosity.

3. Experimental scheme design

The concrete with different viscosity is prepared by adjusting the water binder ratio, and the concrete slump is controlled within 155 ± 5mm through adjusting the dosage of water reducing agent, so as to avoid the impact of different concrete slump on the experimental results. The concrete mix is shown in Table 4, and the specific conditions of concrete with different viscosity are shown in Table 5.

Table 4. Concrete mix proportion kg/m³

C	S	G	W	A
480	780	950	158～288	0～6.24

Table 5. Concrete with different viscosities

Water binder ratio	Dosage of water reducing agent (kg / m3)	Slump(mm)	Table tennis floating time(s)
0.60	0	155	12
0.52	1.44	150	20
0.43	2.88	160	24
0.37	4.32	150	40
0.33	6.24	160	60

Place the prepared concrete in 150 mm × 150 mm × 150 mm test mold shall be vibrated for 50 s to form the test piece. The test piece shall be demoulded after curing in the laboratory for 24 h. The four sides of the test piece shall be photographed to form the appearance pictures of the test piece. The pictures shall be analyzed to obtain the apparent porosity of the concrete.

4. Experimental results and discussion

The effect of concrete with different viscosity (different table tennis floating time) on the apparent porosity of concrete is studied. The experimental results are shown in Fig. 1.

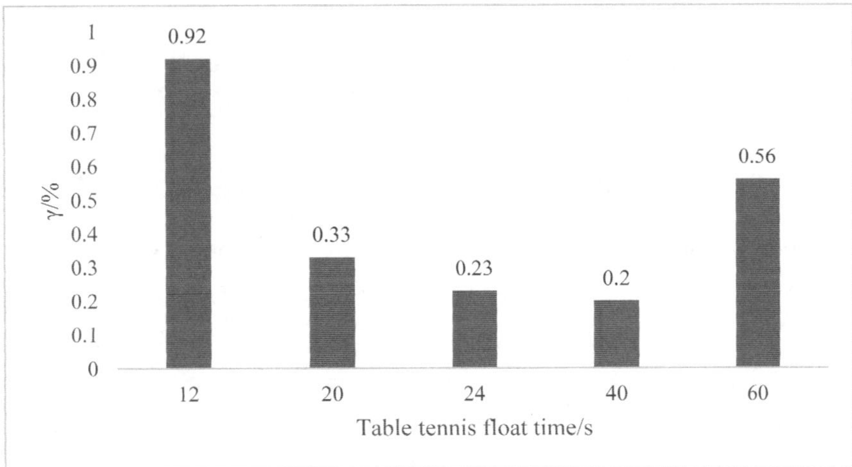

Fig. 1. Effect of different viscosity concrete (different table tennis floating time) on apparent porosity of concrete

The experimental results show that. With the increase of concrete viscosity (the increase of table tennis floating time), the apparent porosity of concrete decreases first and then increases. The water binder ratio is 0.37. When the table tennis floating time is 40s, the proportion of concrete apparent porosity is the lowest, which is 0.20%. The test group with water binder ratio of 0.60 and table tennis floating time of 12s has the highest proportion of concrete apparent porosity, reaching 0.92%. This may be because when the viscosity is too low, although the bubble discharge speed is fast, the concrete mixture is more likely to produce new bubbles during vibration, resulting in the high proportion of comprehensive concrete apparent porosity. The water binder ratio was 0.33, and the apparent porosity of the concrete in the test group with the table tennis floating time of 60s reached

0.56%, which may be caused by the excessive viscosity of the concrete and the difficulty of discharging bubbles from the concrete.

5. Conclusions

(1) Concrete viscosity or water binder ratio is one of the key factors affecting the appearance quality of concrete components. It is not blindly pursuing low viscosity to obtain better apparent quality. When the viscosity is too low, although the bubbles are discharged quickly during vibration, new bubbles are easy to be generated during vibration forming, resulting in the increase of the proportion of concrete apparent porosity and the deterioration of concrete apparent quality.

(2) When the water binder ratio is 0.37 and the table tennis floating time is 40s, the apparent quality of the concrete component is the best and the proportion of the apparent porosity of the concrete is the lowest.

(3) Through the research, it is found that the appearance quality of concrete components can be improved by controlling the concrete viscosity. Of course, factors such as concrete slump, concrete air content, the type of release agent and the type of formwork will affect the appearance quality of concrete components. Further research can be carried out in these directions.

Acknowledgments
Thanks to Qiu Yuetao and others for their guidance and help in the experiment. Thanks to them, my experiment can be successfully completed. Thanks to Zhang Pan, Zheng Qing for his guidance and help in the preparation of my thesis, which has greatly improved my study, scientific research and thinking. I would like to express my heartfelt thanks and deep respect to them.

References:

[1] Cui X., Xia W. J., Wang L. Z., et al. (2017) Research on Key Technologies of fair faced concrete preparation and construction. New building materials, 5:67-71.
[2] Wang L. Z., Lu L. H., Cui X., et al. (2016) Study on preparation technology of fair faced concrete. Concrete and cement products, 12:27-31.
[3] Liu Y. J. (2018) Research and implementation of Photoshop automatic evaluation system based on image analysis. Taiyuan University of technology
[4] Wang R. (2015) Application of light and shadow in modern photography. Popular literature and art, 18:189-190.

Research and application of polymer-surfactant binary combination flooding numerical simulation

Yanyu Zhang[1], Zhaowen Luo[1a*], Mengke Xie[2], Zhiyi Teng[1], Yue Wang[1]

[1]School of Petroleum Engineering in China University of Petroleum, Qingdao, China

[2]Southwest Oil & Gas Field Company, PetroChina, Chengdu, Sichuan, China

[a*]lzw15845857442@163.com

Abstract—At present, many oilfields in China have entered the stage of high water cut development, and the research on the methods of improving oil recovery has been paid more and more attention. Among them, the application of binary combination flooding is more and more extensive. In view of the target area, based on the fine geological modelling, this paper establishes a binary combination flooding model considering the influence of surfactant on oil-water interfacial tension, the adsorption of surfactant by rock, the shear and degradation of polymer, and the adsorption of polymer by rock. Based on the matching of reserves and production history, the development schemes under different development methods are designed and the development effect is predicted and evaluated, the influence of development mode, injection time and injection-production speed on development effect is also studied. The results show that the development effect of binary combination flooding is the best, which provides a theoretical basis for the subsequent efficient and reasonable development of the block.

1. Introduction

Chemical flooding is a kind of oil displacement method that improves the swept volume of injected water and oil displacement efficiency by injecting specific chemical agent solution into the formation, so as to further improve the recovery[1-3]. Chemical flooding mainly includes polymer flooding, surfactant flooding, alkali flooding, foam flooding and binary combination flooding. However, compared with conventional oil displacement methods, chemical oil displacement is a kind of oil displacement method with high risk and high production cost, which requires more appropriate oil displacement technology and development scheme in the development process.

The polymer-surfactant binary combination flooding technology is a new oil displacement technology developed by removing alkali on the basis of the traditional ASP flooding. This combination flooding technology has developed into one of the key technologies of chemical flooding to further improve oil recovery[4,5]. In this paper, Petrel software is used for geological modeling for the target block. The binary combination flooding model is established by STARS module of CMG reservoir numerical simulation software. Based on the matching of reserves and production history, the development effect of binary combination flooding is predicted, which provides a theoretical basis for the subsequent efficient and reasonable development of the block.

2. Numerical Simulation of Polymer-Surface Binary Combination Flooding

2.1. Establishment of reservoir numerical simulation mode

3D reservoir geological modeling is through the limited observation points, the integrated use of geological, seismic, logging and reservoir test data, the unknown reservoir between wells multidisciplinary, integrated, three-dimensional quantitative prediction. Based on the whole reservoir, the Petrel fine geological model is established. There are 31 layers in the model longitudinally. The data of block grid, well location and saturation are extracted from Petrel geological model and imported into STARS module of CMG reservoir numerical simulation software, as shown in Fig.1.

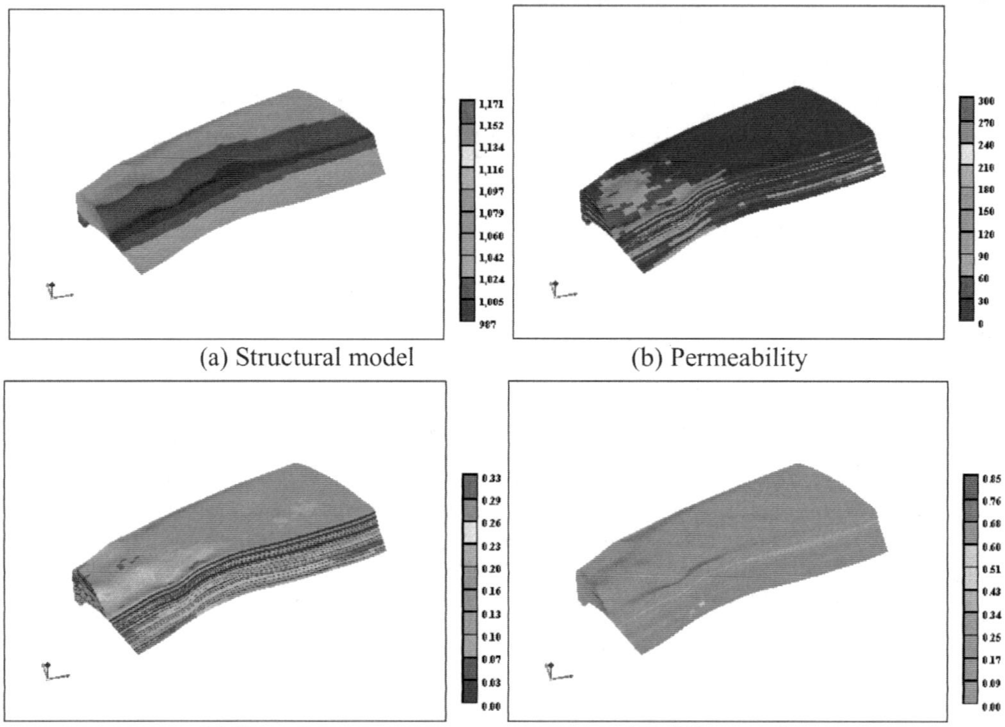

(a) Structural model (b) Permeability

(c) Porosity (d) Oil saturation

Fig.1 Block Numerical Simulation Model

In order to simulate the dual compound flooding process, four components of water, oil, TA2 and VES were set up in the STARS module of CMG reservoir numerical simulation software. The model not only considers the physical properties of oil and water, but also considers the specific properties of surfactants and polymers, such as the effect of VES on oil-water interfacial tension and the shear and degradation of TA2. The physical parameters of each component in the model are shown in Table 1.

Table 1. Table of physical parameters

Parameter name	Parameter value	Parameter name	Parameter value
Oil density, g/cm^3	0.9	VES density ,g/cm^3	3.791
Oil viscosity, mPa·s	8.5	VES viscosity ,mPa·s	1.5
Oil volume factor	1.052	TA2 density ,g/cm^3	0.76
Oil compressibility factor, 1/kPa	9.15E-07	TA2 viscosity ,mPa·s	27.9

2.2. Research on reserves and historical matching

In order to ensure that the established reservoir numerical simulation model can fully reflect the real reservoir development situation, it is necessary to match the reserves and production history of the model that is, to adjust the relevant parameters according to the actual geological data and development data of the reservoir within a certain range. After repeated comparison, the historical dynamic simulated by the model is closest to the actual dynamic to the maximum extent.

According to the reliability of the data obtained, the model saturation, effective thickness, relative permeability curve, rock compression coefficient and other parameters are properly adjusted, and the geological reserves are determined to be 1.67402 million tons after repeated fitting. According to the volume method in the reservoir engineering method, the geological reserves of the well group are calculated to be 1.5767 million tons. By comparing the two methods, the geological reserve error is 6.1 %, which proves the reliability of the geological model after fitting.

Well group production history is 18 years, using water flooding development. The whole block and single well production history are fitted with the main index parameters of cumulative oil production, cumulative water production and water content in the block and single well. According to the reliability of the data, the parameters such as permeability, relative permeability curve and rock compression coefficient in the model are appropriately adjusted. After repeated fitting, the historical fitting results are shown in Fig.2. It can be seen that the simulation results are in good agreement with the actual results, which provides a reliable reservoir numerical simulation model for the subsequent program prediction research and economic evaluation research.

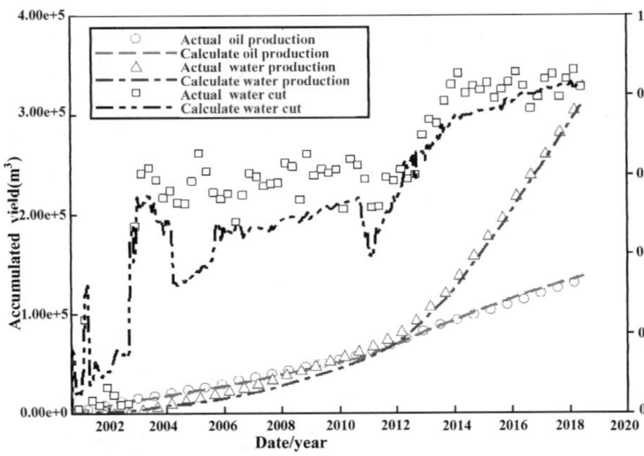

Fig.2 Historical matching results of whole area

3. Development scheme design and prediction

According to the laboratory experiment and field production, four kinds of development schemes are designed, including water flooding, surfactant flooding, binary compound flooding, polymer flooding and surfactant flooding. The influence of development mode, chemical agent injection time, injection speed and liquid production speed on the development effect is studied.

3.1. Prediction and results of water flooding scheme

The simulation time of water flooding scheme is 15 years, and the injection rate of well group injection well and the liquid production rate of production well are designed to be 0.7 times, 1 times and 1.3 times of the monthly speed at the end of the production history. The specific scheme design and simulation results are shown in Table 2.

Table 2. Water flooding scheme design and simulation results

Scheme number	Simulation time / year	Injection rate multiple	Recovery rate multiple	Cumulative oil production /m^3
1		0.7	0.7	130569
2	15	1	1	146046
3		1.3	1.3	163987

From Table 2, it can be seen that the injection-production speed has obvious influence on the oil production results of the whole block. With the increase of injection-production speed, the cumulative oil production of the block increases gradually, while the overall water content changes little. The low speed of water flooding injection and production leads to low recovery degree, which greatly increases the cost of site facilities, and the high speed of water flooding injection and production is easy to increase fingering phenomenon, advance water breakthrough time, resulting in a large number of residual oil. Therefore, increasing injection-production rate appropriately within the allowable range of conditions is conducive to efficient exploitation of reservoirs.

3.2. Prediction and results of surfactant flooding scheme

The concentration of VES in surfactant flooding scheme is 0.5 %, the injection time of VES is 1,3 and 5 years, and the remaining time is for subsequent water flooding. The specific scheme design and simulation results are shown in Table 3 and Fig.3.

Table 3 VES flooding scheme design and simulation results

Scheme number	VES injection time / year	Injection rate multiple	Recovery rate multiple	Cumulative oil production /m^3	Incremental oil production /m^3
4	1	1	1	146702	656
5	3	1	1	147049	1003
6	5	1	1	147125	1079
7	5	0.7	0.7	131078	792
8	5	1.3	1.3	166071	2084

The table shows that the oil production increases with the increase of VES injection time and injection-production speed. Surfactant flooding has fast effect and long duration. The daily oil production in the whole area began to increase significantly in 2019, and the water content decreased slightly. The surfactant flooding effect is better than water flooding.

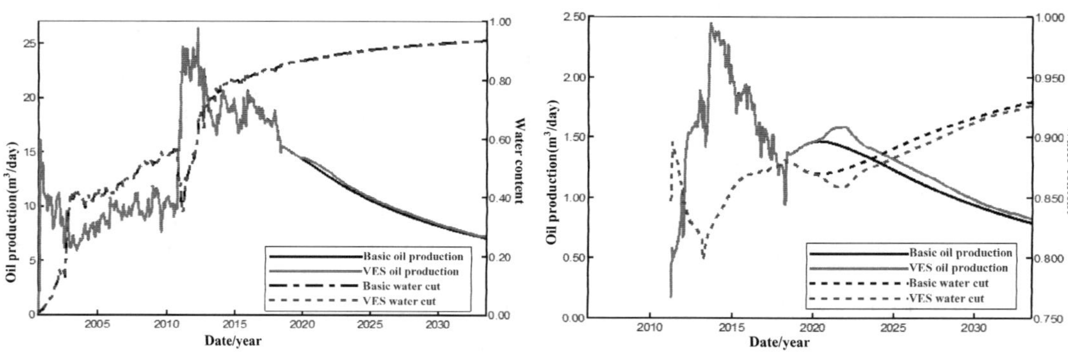

Fig.3 Simulation results of VES flooding whole area and typical wells

3.3. Prediction and results of binary combination flooding scheme

In the binary combination flooding scheme, the concentration of VES is 0.1 %, the concentration of TA2 is 0.05 %, the injection time of binary combination flooding is 1, 3, 5 years. The specific scheme design and simulation results are shown in Table 4 and Fig.4.

Table 4. Binary combination flooding scheme design and simulation results

Scheme number	Injection time / year	Injection rate multiple	Recovery rate multiple	Cumulative oil production /m^3	Incremental oil production /m^3
9	1	1	1	147623	1577
10	3	1	1	148577	2531
11	5	1	1	149180	3134
12	5	0.7	0.7	132347	1778
13	5	1.3	1.3	167784	3797

It can be seen from the table that the increase of oil production increases with the increase of chemical injection time and injection-production speed. Binary combination flooding began to take effect in 2020, with daily oil production significantly increased and water content decreased. The longer the injection time of binary combination flooding is, the slower the daily oil production decreases, the slower the water content increases, and the higher the cumulative oil production is.

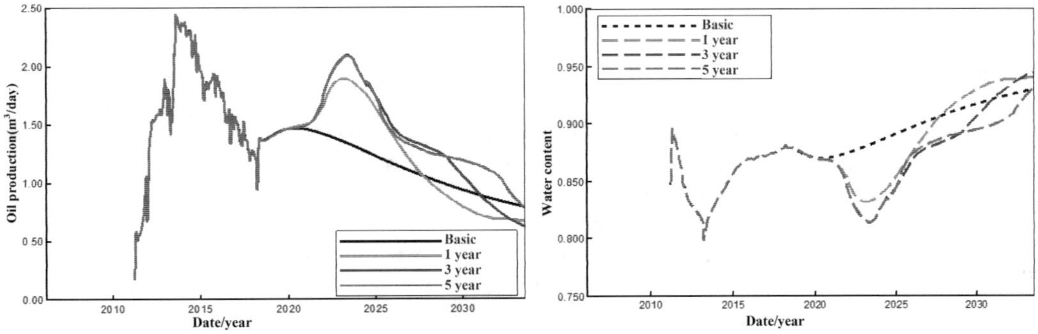

Fig.4 Daily oil production and water cut of typical wells with different injection times

3.4. Prediction and results of polymer flooding and surfactant flooding

The polymer flooding and surfactant flooding scheme first carried out TA2 flooding with 0.05 % concentration, and then carried out VES flooding with 0.1 % concentration. The injection time of TA2 is the same as that of VES, which is 1, 3 and 5 years. The specific scheme design and simulation results are shown in Table 5 and Fig.5.

Table 5. Polymer flooding and surfactant flooding scheme design and simulation results

Scheme number	Injection time / year	Injection rate multiple	Recovery rate multiple	Cumulative oil production /m^3	Incremental oil production /m^3
14	1	1	1	146888	842
15	3	1	1	148147	2101
16	5	1	1	147928	1882
17	5	0.7	0.7	131856	1287
18	5	1.3	1.3	166649	2662

It can be seen from table that the increase in oil production is the highest when the injection time is three years. With the increase of injection-production speed, cumulative oil production gradually increased. Fig.5 shows the daily oil production and water content of typical wells in polymer flooding and surfactant flooding at different injection times. After polymer flooding and surfactant flooding, the daily oil production gradually increases, and the water content gradually decreases. The effect in single well is obvious.

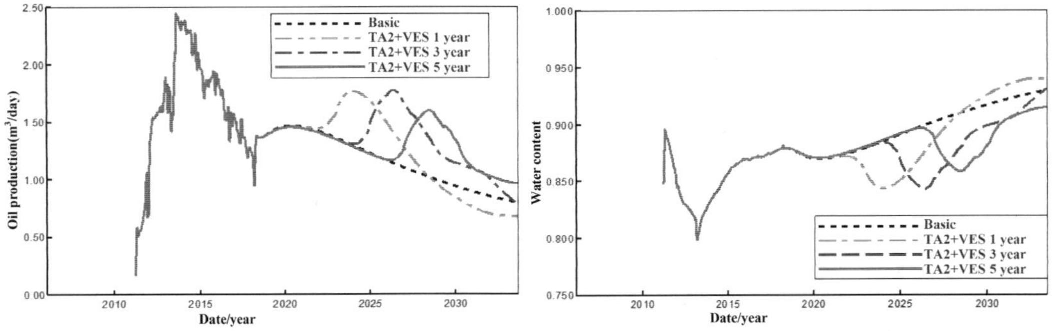

Fig.5 Daily oil production and water cut of typical wells with different injection times

However, compared with binary combination flooding, the increased oil production of polymer flooding and surfactant flooding is lower. It shows that the effect of binary compound flooding is the best.

4. Conclusion

Based on the establishment of three-dimensional fine geological model, the reservoir numerical simulation model is established through CMG. The matching of reserves and production history proves the reliability of the established model, which lays the foundation for subsequent program prediction and economic evaluation. Considering the special properties of surfactant and polymer, a numerical simulation model of binary compound flooding is established, which can accurately simulate the binary compound flooding scheme.

According to the actual situation, four development schemes of water flooding, surfactant flooding, binary compound flooding, polymer flooding and surfactant flooding are designed, and the influence of development mode, injection time and injection-production speed on the development effect is studied. The results show that under whatever injection time and injection-production rate, the increase in oil production of binary combination flooding is the highest, and the increase in oil production of the optimal scheme increases by 3797 m^3. In this paper, the optimal development scheme is proposed for the target block, and the development effect is predicted, which provides a theoretical basis for the subsequent efficient and reasonable development of the block.

References

[1] Shiran B S , Skauge A . Enhanced Oil Recovery (EOR) by Combined Low Salinity Water/Polymer Flooding[J]. Energy & Fuels, 2013, 27(mar.-apr.):1223-1235.
[2] Cristian S. Loaiza, Juan F. Patiño, Juan M. Mejía. Numerical evaluation of a combined chemical enhanced oil recovery process with polymer and nanoparticles based on experimental observations[J]. Journal of Petroleum Science and Engineering, 2020, 191
[3] Longde SUN, Xiaolin WU, Wanfu ZHOU, et al. Technologies of enhancing oil recovery by chemical flooding in Daqing Oilfield, NE China[J]. Petroleum Exploration and Development Online, 2018, 45(4):673-684.
[4] WANG Yefei, ZHAO Fulin, BAI Baojun, et al. Optimized surfactant IFT and polymer viscosity for surfactant-polymer flooding in heterogeneous formations. SPE 127391, 2010.
[5] Wei Y , Liu J , Luo L , et al. Performance of Polymer/Surfactant Binary Flooding in Enlarging Swept Volumes[J]. Journal of Southwest Petroleum University(Science & Technology Edition), 2018.

Microstructure characteristic of hot formed 22MnB5 steel and its relationship with delayed hydride cracking property

Hua Wang[1a*], Shiyu Hao[1b], Baolei Wang[1c], Chunfu Kuang[2d]

[1]Instrumental Analysis Center, Shanghai University, 200444, China

[2]PanGang Group Xi Chang Steel and Vanadium Co., Ltd, Si Chuan, 615000, China

[1a*]wh225@shu.edu.cn, [1b]shiyuhao@shu.edu.cn, [1c]wangbaolei@shu.edu.cn, [2d]kuangchunfu@126.com

Abstract——Hydrogen-induced delayed cracking is an important factor leading to the failure of ultra-high strength steels during service, thus in this study, the mechanical properties and hydrogen trapping of hot formed 22MnB5 steel considering different content of V (0, 0.05, 0.09 wt.%) and corresponding microstructure were investigated. Slow strain rate tensile tests (SSRT) were applied to obtain the tensile strength and elongation of samples under different medium conditions (air, HCl immersion and electrolytic hydrogen charging). Besides, microstructure and atomic distribution morphology of the samples were detected and analyzed by transmission electron microscope (TEM) and atomic probe tomography (APT) technique. Results showed that both resistances of hydrogen-induced delayed cracking in MnB-V5 and MnB-V9 samples were obviously better than that in MnB-V0 sample, besides, the delayed cracking resistances of MnB-V9 and MnB-V5 samples did not improve with the increase of V content, but basically remained the same. Lamellar residual austenite was found in all samples, but no hydrogen was grasped in the same area according to APT result. There was intragranular cementite in all samples, and it has grasped H in the same position. The carbides formed by V, Ti and other alloying elements were found in MnB-V5 and MnB-V9 and were considered to be an important factor in improving the anti-hydrogen delayed cracking performance of the steel.

1. Introduction

Hot-formed steel is an important kind of steel commonly used in manufacturing automobile structural parts. Hot stamping process, which combines the forming and quenching process of parts, can effectively solve the cracking and spring-back problems of parts under conventional stamping process, and is an important method to improve vehicle safety at present. However, the hydrogen-induced delayed cracking behavior of ultra-high strength steel plate becomes a more severe phenomenon that influences the application and service of steel. Therefore, with the wider application of ultra-high strength steel in automobile field, it is more necessary to study the delayed cracking behavior of ultra-high strength steel [1].

Hydrogen is reported to diffuse and penetrate in the steel substrate during both production and application processes, and the hydrogen adsorbed in micro cracks and cavities on metal surfaces reduced their surface energy and facilitated crack nucleation and extension. Studies showed that there were many factors, such as elements, precipitates, defects in substrates and steel microstructure that could influence the hydrogen embrittlement resistance in high strength steel [2]. Li et al [3]reported that adding appropriate amount of micro alloying element in steel is an effective way in improving the hydrogen embrittlement resistance of steel. V, as a common metal element among all the microalloying elements in industry, not only can improve the mechanical properties of steels, but also can improve the hydrogen embrittlement resistance of high strength steels.

In the present work, the mechanical properties and hydrogen-induced delayed cracking behavior of hot-formed 22MnB5 ultra-high strength steels containing different V contents with various hydrogen charging methods of air, HCl immersion and electrolytic H_2SO_4 were studied. The hydrogen-induced delayed cracking behavior mechanism was analyzed with employment of microanalysis technique of scanning electron microscope (SEM) and TEM. The study is expected to provide reference for the hydrogen-induced delayed cracking research in high strength steel plate.

2. Experimental Procedure

Three groups of 22MnB5 steel plates were provided from a domestic steel company, and the chemical compositions were listed in Table 1. The quenching processes of the steel plate are as follows: heating to temperature of 900-930 ℃ and holding for 3-5 min, then quenching to 800-850 ℃, holding for 5-10 s, finally cooling to room temperature at a rate about 30 ℃/s. The annealing process is as follows: heating to 720 ℃ at a rate of 5 ℃/s and holding for 120 s, then rapidly cooling to 280 ℃ at a rate of 40 ℃/s, over aging for 600 s, finally cooling to room temperature.

Table 1 Chemical compositions (wt.%) of 22MnB5 steels

	C	Si	Mn	P	S	Cr	Ti	B	V
MnB-V0	0.220	0.26	1.28	0.008	0.002	0.16	0.025	0.0029	0
MnB-V5	0.240	0.25	1.30	0.008	0.002	0.16	0.027	0.0029	0.052
MnB-V9	0.236	0.24	1.30	0.007	0.002	0.15	0.025	0.0027	0.09

The plate with a thickness of 1.2 mm was cut to a standard tensile sample as shown in Fig. 1. During experiment, low strain rate combined with three different hydrogen charging methods (air, HCl immersion and electrolytic H_2SO_4) were employed to simulate the various states of delayed cracking reaction. The first group of samples were tested in air; the second group was stretched in HCl solution at 0.1 mol/L; the third group was electrolytically charged with hydrogen and stretched in electrolyte (H_2SO_4 solution at 0.5 mol/L with 500 mg/L thiourea). The current density was 1 mA/cm^2. Strain rates of the three groups were same to be 10^{-6} s^{-1}.

The phase analysis of samples was carried out using SMART APEXII X-ray diffractometer. The phase was irradiated by CuKα (λ=0.1548). The working voltage was 50kV and the scanning speed was 2°/min. The microstructure and fracture morphology of the samples were observed by JSM-6700F emission scanning electron microscope (F-SEM). The microstructures of samples were observed and studied more deeply by JEM-2100F field emission transmission electron microscope (FE-TEM).

LEAP4000 atomic probe tomography (APT) equipment was used to detect the element distribution of steels with various compositions under different treatment conditions. Frist, the samples were cut into rod-shaped needles of 0.5 mm × 0.5 mm × 20 mm dimensions, and then the needle-like samples were electrolytic polished with 2 % perchloric acid 2-Butoxy ethanol. During detecting, samples were cooled to 50K, the pulse voltage frequency was 5kHz and the pulse fraction was 20 %. The obtained data was analyzed and processed using PoSAP software.

Hydrogen concentrations in samples were detected by hydrogen thermal analyzer, which is composed of a four-stage mass spectrometer, a heating device and a recording device controlled by computer, and the result date were processed by a professional software attached to the system. During test, the heating rate was 100 ℃/h and the holding temperature was 800 ℃.

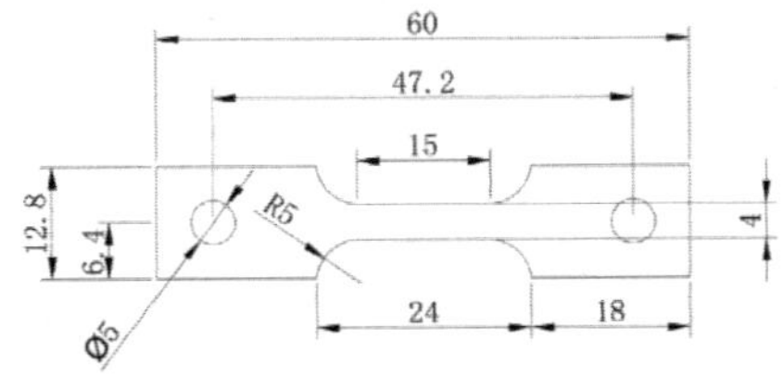

Fig. 1 Schematic diagram of samples using in tensile tests (mm)

3. Results and Discussion

The microstructure and morphology of the samples are shown in Fig. 2, which shows that the microstructures of 22MnB5 hot formed steel are fine martensite and fine carbides distributed around the martensite. It is known that the microstructure of hot formed steel is mainly composed of martensite with high dislocation density and a small amount of residual austenite. The property of hydrogen-induced delayed cracking is related to dislocation density, precipitation of the second phase and morphology and distribution of residual austenite. Besides, during the hot stamping process, the matrix of the material undergoes phase transformation and lattice distortion, which induces many defects and internal stresses and is highly susceptible to hydrogen-induced delayed cracking in this kind of steel.

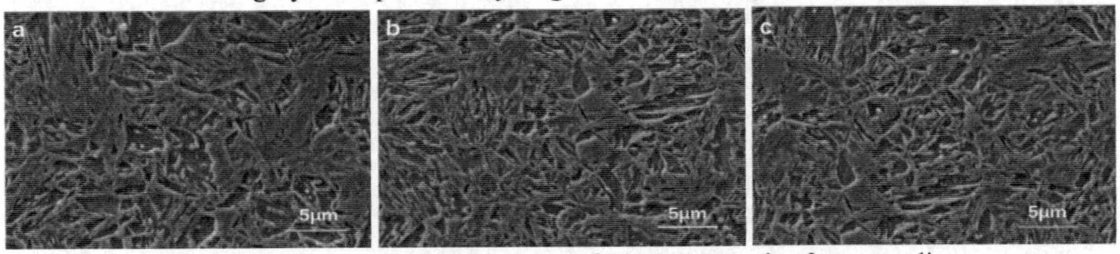

Fig. 2 Metallographic morphology of 22MnB5 steels after annealing;
(a) MnB-V0, (b) MnB-V0, (c) MnB-V9

The stress-strain curves of the three samples after SSRT test under three different conditions were shown in Fig. 3. The corresponding data were listed in Table 2. It can be found that there are elastic deformation, uniform plastic deformation and non-uniform plastic deformation, but no obvious yield stage in the stress-strain curves. Besides, compared Fig. 3a with b, it was obviously that the tensile strength and elongation of the samples decreased when the samples were in hydrogen environment, and the elongation was even only 1/3 to 1/6 of that in air environment. Thus, hydrogen can seriously influence the mechanical property of 22MnB5 steel especially when at a high concentration. As is known, the reason is that in the process of dynamic tension, when the plastic deformation occurs, the dislocation in the specimen multiply and move greatly, and the hydrogen atoms were carried by the dislocation to spread and gather in the three-dimensional stress zone. At the same time, the hydrogen atom reduced the bonding force between the matrix atoms, resulting in the sample even when it was subjected to lower external force.

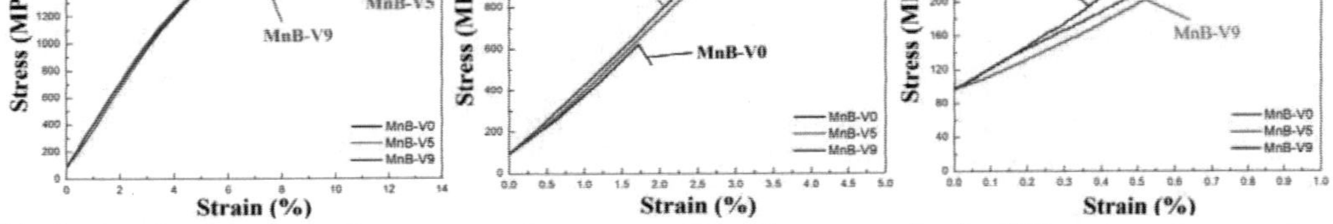

Fig. 3 Simultaneous tensile stress-strain curves of the three samples in different environments of (a) air, (b) HCl solution immersion and (c) H_2SO_4 solution electrolytic hydrogen charging

Table 2 Mechanical properties of the samples tested by SSRT under different medium conditions

Steels	Test conditions	TS/MPa	EL/%
MnB-V0	Air	1785	11.9
	HCl immersion	620	1.9
	Electrolytic hydrogen charging	240	0.5
MnB-V5	Air	1750	12.8
	HCl immersion	1215	3.9
	Electrolytic hydrogen charging	270	0.9
MnB-V9	Air	1585	10.9
	HCl immersion	1065	3
	Electrolytic hydrogen charging	300	0.9

Moreover, Fig. 3b and c showed the tensile elongation and strength of MnB-V0 samples under hydrogen atmosphere were obviously lower than MnB-V5 and MnB-V9 samples, that is, V addition in 22MnB5 steel is workable in improving resistance property of hydrogen embrittlement. But the mechanical property of the steel does not increase with the V content addition, on the contrary, the tensile strength of sample with 0.09 wt.% V showed serious decrease compared with the sample with 0.05 wt.% V in both air and HCl solution immersion conditions.

The residual hydrogen contents in the samples after tensile test were listed in Table 3. It shows that except for the electrolytic hydrogen charging samples, the hydrogen content in the other two groups of samples are at low levels. Combined with the changes of the mechanical properties shown in Table 2, we found that when a small amount of hydrogen is introduced into the sample, the performance of the sample decreases to a certain extent, but when the hydrogen reaches the order of magnitude of 1ppm, the performance of the sample will be seriously damaged, and it is no longer suitable for industrial applications.

Table 3 Hydrogen content test results in the sample after tensile test.

Hydrogen charging method	Steels	Hydrogen content (ppm)
Air	MnB-V0	0.003
	MnB-V5	0.002
	MnB-V9	0.003
HCl immersion	MnB-V0	0.04
	MnB-V5	0.026
	MnB-V9	0.031
Electrolytic hydrogen charging	MnB-V0	3.62
	MnB-V5	1.62
	MnB-V9	1.79

The fracture of SSRT of MnB-V5 in air (Fig. 4) that has many dimples on the fracture surface shows typical ductile fracture.

Fig. 5 shows the tensile fracture morphology of MnB-V5 immersed in HCl. Compared with the fracture surface of the air-stretched specimen, the tensile fracture dimple becomes larger after soaking in HCl, accompanied by obvious corrosion pits and microcrack areas, which means the plasticity of the specimen in this state decreases severely and tends to change to the cleavage fracture morphology gradually.

Fig. 6 shows the fracture morphology of MnB-V5 after electrolytic hydrogen-charged tensile. The ductile fracture characteristics of the specimens in this state almost disappear.

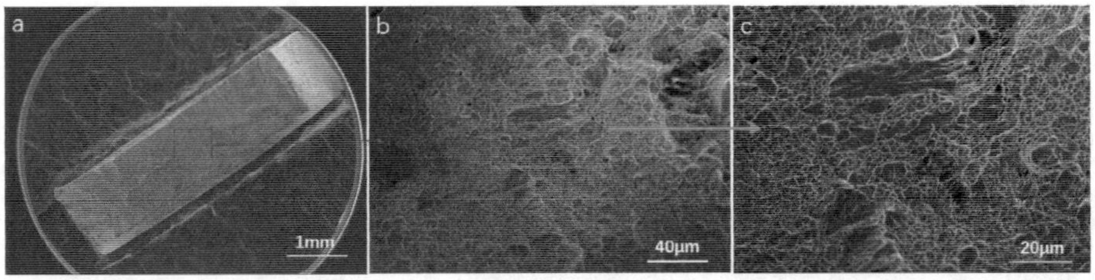

Fig. 4 Tensile fracture morphology in air under different magnification of (a)-(c) (MnB-V5)

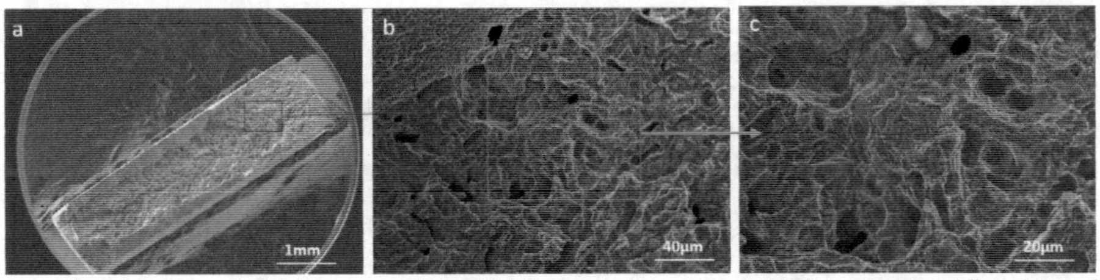

Fig. 5 Tensile fracture morphology after HCl immersion under different magnification of (a)-(c) (MnB-V5)

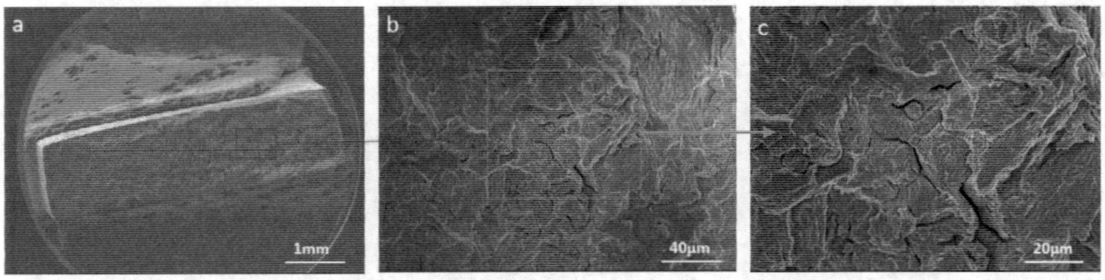

Fig. 6 Tensile fracture morphology of H_2SO_4 electrolytic hydrogen charging under different magnification of (a)-(c) (MnB-V5)

As reported in [4], a large amount of H generated by the cathodic reaction will be adsorbed on the surface of the sample, then the H atom will diffuse into the sample and enrich at the crack tip. The aggregated hydrogen atoms will reduce the binding force of the Fe-Fe bond and will further cause hydrogen embrittlement at the microscopic level. When the sample is loaded, the crack tip will become microcrack nucleation. Once the micro crack was formed, the hydrogen atom accelerates the crack growth by promoting the anodic dissolution of the crack tip or reducing the surface energy of the newly formed crystal plane.

Fig. 7 shows the XRD spectrum of MnB-V0, MnB-V5 and MnB-V9 sample. There is no or very low residual austenite, but only martensite in the samples.

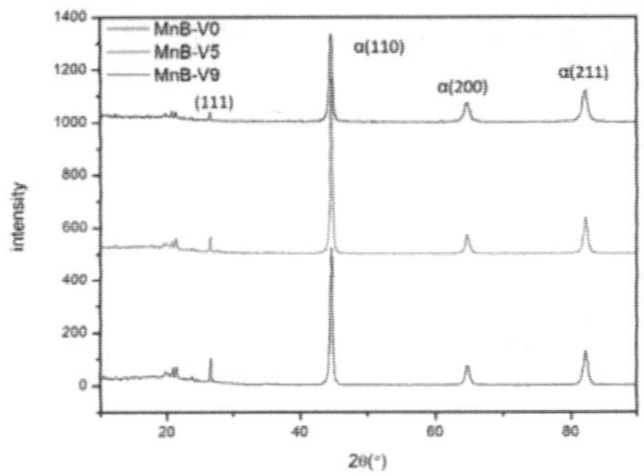

Fig. 7 XRD patterns of three sample with different V content

Figs. 8, 9 and 10 are TEM results of MnB-V0, MnB-V5 and MnB-V9 stretched in air, HCl immersion and electrolytic hydrogen charging, respectively. The lath size of air tensile sample in Fig. 8a is finer than hydrogen charged samples in Fig. 8b and c, which is in line with the tensile test results that the former sample has more deformation than the latter ones.

Fig. 8d shows the micro structure of MnB-V0. A thin film between the two lath interfaces at the red circle in Fig. 8d and e indicate that a fine residual austenite lamellar was remained between martensite lath after annealing. Studies have shown that residual austenite can be used as a hydrogen trap to improve the delayed cracking resistance of materials. Fan et [5] proposed that stable residual austenite is beneficial to improve hydrogen-induced delayed cracking, on the contrary, unstable residual austenite will aggravate the occurrence of hydrogen-induced delayed cracking.

Some intragranular phase were found precipitated in the martensite lath (Fig. 8f), and the SAD result proves them to be cementite. Cementite is formed in the process of slow phase transition, so its micro strain is small, and the resistance to hydrogen diffusion is weak. Generally, precipitation cementite on grain boundaries, especially thin film cementite will make delayed cracking sensitivity significantly enhanced [6]. From the tensile test data of the other two groups of samples, results can be obtained that the addition of V suppresses the formation of grain boundary cementite and affects the delayed cracking resistance of the three groups of samples.

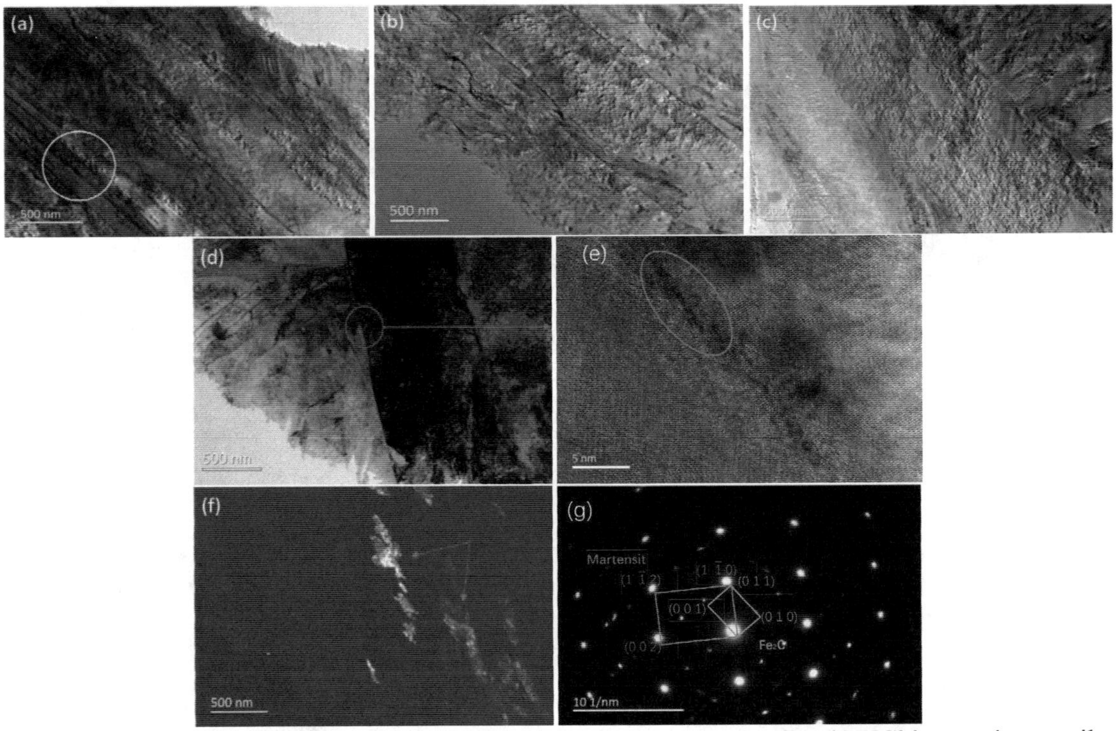

Fig. 8 TEM diagram of MnB-V0 sample microstructure after (a) air tensile, (b) HCl immersion tensile, (c) electrolytic hydrogen charging, and (d) martensite morphology, (e) martensite lath high resolution image, (f) cementite phase, (g) diffraction spot of f

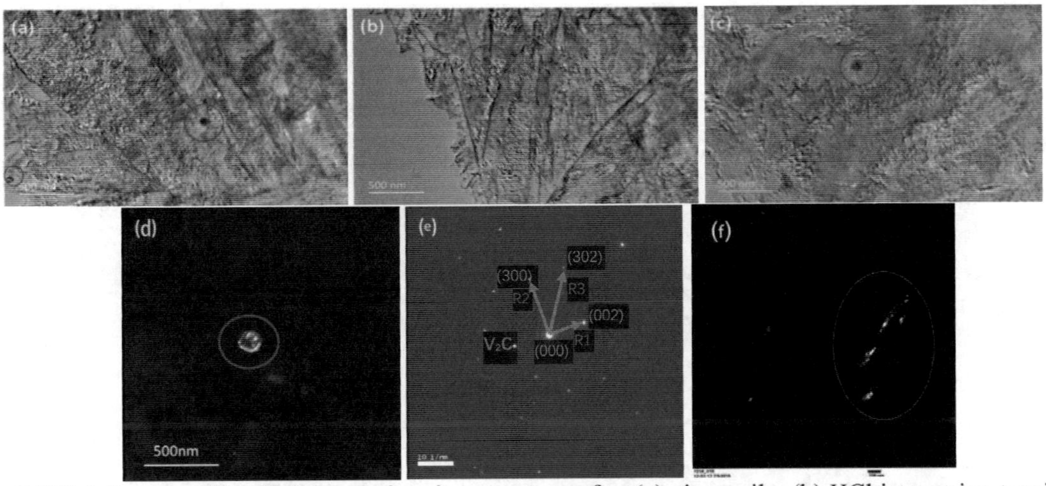

Fig. 9 TEM diagram of MnB-V5 sample microstructure after (a) air tensile, (b) HCl immersion tensile, (c) electrolytic hydrogen charging tensile, and (d) vanadium carbide precipitation phase, (e) diffraction spot of d, (f) dark field image of cementite

Fig. 10 TEM diagram of MnB-V9 sample microstructure after (a) air tensile, (b) HCl immersion tensile, and (c) electrolytic hydrogen charging tensile

Comparing Fig. 8, with 9 and 10, we found the lath of MnB-V0 is thicker than MnB-V5 and MnB-V9 because of the lower deformation. Similar residual austenite lamellar was found in MnB-V5 (Fig.9a) and MnB-V9 (Fig.10a). Thinner cementite was found in MnB-V5 (Fig.9i). In addition, dispersed precipitation of second phase particles was found in MnB-V5 and MnB-V9 (red circle zone in Fig. 9 and 10), and the structure of them were identified to be V_2C by calibration result of selected area diffraction spot.

Fig. 11 shows the scanning energy spectrum analysis of the selected area in MnB-V0 sample. The enrichment of Fe, Mn and Cr on the martensite lath boundary were proved to be in a same field, which confirms the existence of residual austenite lamellar.

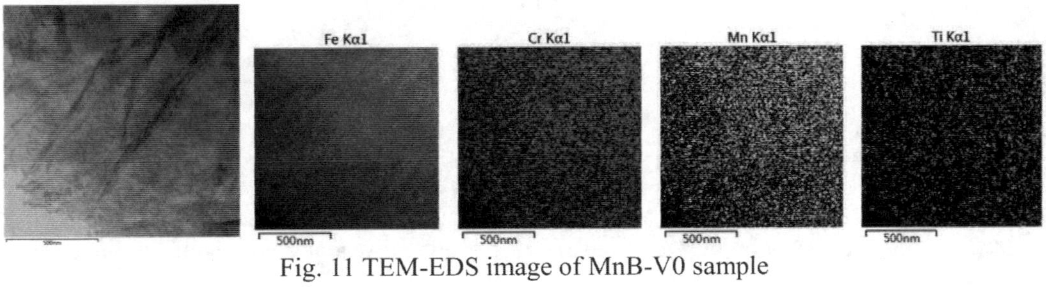

Fig. 11 TEM-EDS image of MnB-V0 sample

Fig. 12 shows the EDS result of MnB-V5 sample during TEM analysis. V and Ti elements were found to segregate in the precipitated particle area, which indicates the particle to be $(VTi)C_x$.

Fig. 12 TEM-EDS images of MnB-V5

Some researchers [7] have studied that coherent TiC precipitates can capture hydrogen when the cathode is charged at room temperature, while incoherent TiC particles in steel can not capture hydrogen at room temperature. Besides, the coherent and semi coherent TiC particles capture hydrogen at the precipitate/ferrite interface. However, some researchers [8] have reported that hydrogen is trapped inside the VC particles rather than at the interface of the precipitate matrix. Chen [9] found that deuterium is mainly precipitated at the interface. Combining the research results of previous scholars, with the microstructure characteristics of 22MnB5 steel in this study, we considered that different types, contents and sizes of carbides in each sample have captured hydrogen as hydrogen traps.

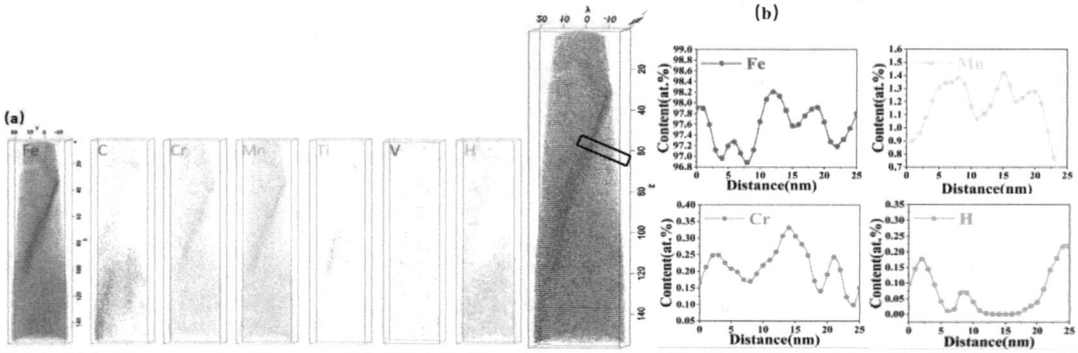

Fig. 13 (a) Element distribution of MnB-V0 (b) concentration distribution of elements (Fe, Mn, Cr, H) in the blue box

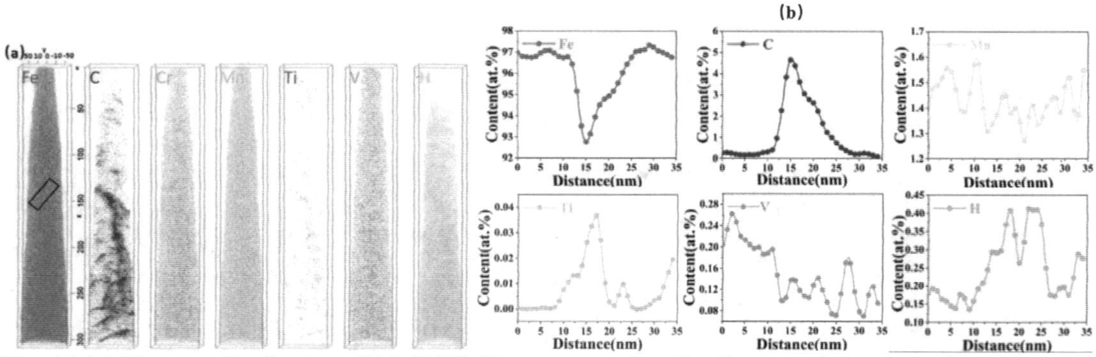

Fig. 14 (a) Element distribution of MnB-V9 (b) concentration distribution of elements (Fe, C, Mn, Ti, V, H) in the black box

Fig.13 and 14 are the element distribution diagrams of 22MnB5 samples measured by APT. The characteristics of the region shown in Fig.13 are consistent with the TEM results in Fig. 8, that is, the lamellar residual austenite structure do exists in 22MnB5 steels, and its thickness is less than 5 nm. As we known, H has high solubility in austenite and is not easy to diffuse, for example, Park et al [10] have proposed that residual austenite is an interphase hydrogen trap with high binding energy. However, unlike the reported theory above, the enrichment of H element does not occur in the lamellar residual austenite field in the present work (Fig. 13a). MnB-V9 sample has a character of cementite precipitation as shown in Fig. 14, and this is also consistent with the TEM results of MnB-V5 steel in Fig. 9f. Through the elements distribution results, we found that there was an obvious enrichment of H in the carbides position, accompanied by the accumulation of C and Ti elements.

The H segregation in cementite, can be used as a hydrogen trap, for example, Nagao et al [11] found that the (Ti, Mo) C precipitates in martensitic steel can effectively capture the free hydrogen atoms in the steel. Gutierrez et al [12] have proposed that the intergranular fracture started at the deformation twins formed on the coarse cementite at the grain boundary and the local shear stress concentration at the cementite interface. Still more, Dong [13] has found that the hydrogen diffusion rate of continuous annealing steel is obviously lower than that of batch annealing steel, and the reason is that many dispersed cementite particles in the ferrite grains of continuous annealing steel act as hydrogen traps, which restricts the diffusion of hydrogen atoms. It is generally believed that the fine and dispersed carbide in the grain but not the grain boundary will act as a hydrogen trap to slow down the diffusion of hydrogen in the material and to improve the delayed cracking resistance of the material. Unfortunately that no V was found in APT result, which was because the amount of V was low and the precipitates were dispersed.

According to all the tested results, the lamellar residual austenite in the three groups of samples didn't grasp hydrogen. There was cementite in all the three groups of samples though the sizes were different, and they can grasp hydrogen around them. Since the above two characters in 22MnB5 steel were similar in all the three groups of samples, they were considered to has no important effect on the hydrogen delayed cracking resistance in the steel. However, in MnB-V5 and MnB-V9 samples, the intragranular vanadium carbide was found, which was considered to be a hydrogen trap and has improved the hydrogen delayed cracking resistance obviously.

4. Conclusion

Based on the results and discussions presented above, the conclusions are obtained as below:

(1) The hydrogen-induced delayed cracking resistance of the samples containing V is obviously better than the samples without V. Comparing with the samples of MnB-V5 and MnB-V9, the hydrogen-induced delayed cracking resistance dose not improve with the increase of V content, but remains at a same level.

(2) There was lamellar residual austenite found in 22MnB5 steels with or without V addition, but no hydrogen was grasped in the same area according to APT result.

(3) There was intragranular cementite in all samples, and it has grasped H in the same position.

(4) Intragranular vanadium carbide was found in MnB-V5 and MnB-V9 samples, which was considered to be the hydrogen trap and was a key factor in improving the hydrogen delayed cracking resistance.

References

[1] Gianfranco Lovicu et al. Hydrogen Embrittlement of Automotive Advanced High-Strength Steels[J]. Metallurgical and Materials Transactions A, 2012, 43(11): 4075-4087.

[2] Chen Y S, Lu H Z, Liang J T, et al. Observation of hydrogen trapping at dislocations, grain boundaries, and precipitates. [J]. Science (New York, N.Y.),2020,367(6474): 171-175.

[3] Li J.X, Wang Y.B, Chu W.Y, et al. Effect of alloying element vanadium on hydrogen embrittlement sensitivity for rail steel[J]. Journal of Chinese Society for Corrosion and Protection, 1998(02): 35-40.

[4] V erpoort C, Duquette D J, Stoloff N S, et al. The influence of plastic deformation on the hydrogen embrittlement of nickel[J]. Material Science & Engineering, 1984, 64(1): 135-145.
[5] Fan Y H, Zhang B, Yi H L, et al. The role of reversed austenite in hydrogen embrittlement fracture of S41500 martensitic stainless steel[J]. Acta Materialia, 2017, 139: 188-195.
[6] Hong S, Lee J, Lee B J,et al. Effects of intergranular carbide precipitation on delayed fracture behavior in three Twinning Induced Plasticity (TWIP) steels[J]. Materials Science & Engineering A, 2013, 587: 85-99.
[7] Depover T, V erbeken K. The effect of TiC on the hydrogen induced ductility loss and trapping behavior of Fe-C-Ti alloys[J]. Corrosion Science, 2016, 112(11): 308-326.
[8] Wang Y, Hu S, Li Y, et al. Improved hydrogen embrittlement resistance after quenching–tempering treatment for a Cr-Mo-V high strength steel[J]. International Journal of Hydrogen Energy, 2019, 44(54): 1-10.
[9] Chen Y S, Haley D, Gerstl S S A, et al. Direct observation of individual hydrogen atoms at trapping sites in a ferritic steel[J]. Science, 2017, 355(6330): 1196-1199.
[10] Park Y D, Maroef I S, Olson D L. Residual austenite as a hydrogen trap in steel welds[J]. Welding Journal, 2002, 81(2): 27S-35S.
[11] Nagao A, Smith C D, Dadfarnia M, et al. The role of hydrogen in hydrogen embrittlement fracture of lath martensitic steel[J]. Acta Materialia, 2012, 60(13-14): 5182-5189.
[12] Gutierrez U, Raabe D, et al. Influence of Al content and precipitation state on the mechanical behavior of austenitic high-Mn low-density steel[J]. Scripta Materialia, 2013: 343-347.
[13] Dong F.B, Du L.X, Liu X.H, et al. Effect of annealing method on Microstructure and properties of boron bearing cold rolled deep drawing enamel steel[J]. Journal of Northeast University Natural Science Edition, 2013, 34(10): 1412-1415.

Applications and Future Developments of Flexible Organic Light-emitting Diode

Buyu Lei[1, *, †], Ruowei Li[2, †], Jingyao Wu[3, †], Ran Yan[4, †]

[1]RDFZ Chaoyang Branch School, Beijing, 100029, China.

[2]Fuzhou Overseas Chinese Middle School, Fuzhou, 350004, China.

[3]Cambridge Department, Tianjin Yinghua International School, Tianjin, 301725, China.

[4]Shenzhen College Of International Education, ShenZhen, 518000, China.

*Corresponding author. Email: leibuyu@rdfzcygj.cn

[†]These authors contributed equally

Abstract—With the rapid development of display technology, human beings have more and more stringent requirements for display products, e.g., their light weight, thinness, flexible performances, and versatility. Contemporarily, the flexible organic light-emitting diode (OLED) technology and its derived multi-layer structure tend to be wearable, light, and responsive in a short time. Although flexible OLED technology is popular, there are still many problems that needs to be addressed, e.g., new raw materials, rough flexible substrate, and low efficiency in workmanship, resulting in poor luminescence, rough and inflexible products. In this paper, a new type of flexible fiber display system is proposed by combining OLED with conductive optical fiber. This paper comprehensively introduces the mechanism, manufacturing technology and structure of this new type of flexible OLED. It covers various scopes like Internet information communication, real-time positioning, and medical assistance, etc. Despite some challenges and limitations, this product possesses a variety of advantages in terms of its convenience and flexibility, which makes the exploitation of a large range of applications probable. Therefore, it is highly expected that these promising prospects of development will stimulate the further progress in display fabrication, which helps to facilitate people's lives in the future. These results shed light on the significance of developments of flexible OLED technology.

1. Introduction

With the progress of societies and the improvement of people's living standards, one expects more and more high-tech display devices. Besides, it also makes the display technology develop by leaps and bounds in people's earnest expectations. OLEDs are favored plenty of consumers owing to their advantages e.g., self-emission, low driving voltage, high response speed, high contrast, wide viewing angle, good temperature characteristics, ultra-thin, impact resistance, flexible display etc. They are regarded as the third generation display technology after cathode ray tube (CRT) display and liquid crystal display (LCD) [1], and have gradually become a research hotspot in the display field.

Flexible OLED is also called flexible display, which is a flexible display device which is made of soft materials. At present, OLED technology is divided into flexible OLED and printed OLED.

Additionally, the curved screens and folding screens used by stacked mobile phones are all made of flexible OLED as display screens, which replaces the driving circuit thin film transistor (TFT), flexible OLED and flexible OLED packaging film [2]. The flexible OLED device consists of flexible substrates, flexible electrodes, transmission references and luminescent materials. Flexible OLED possesses the advantages of light weight, thin shape, folding characteristics and usage of new energy, etc. Moreover, flexible OLED also uses an innovative display technology, which develops rapidly and has broad development prospects. Many scholars have investigated flexible OLED and done related research, since this is a hot topic in contemporary times, which will bring a lot of convenience to researchers. In the display field, OLED devices can be widely used in watches, mobile phones, televisions, computers, and many other products. Because of its independent light-emitting function, no backlight is required, making the product lighter and thinner. Additionally, each pixel works independently, which makes the OLED display more power-saving, possess higher contrast and finer image quality [3]. In the field of lighting, OLED devices are widely used in indoor decorative lamps, eye protection table lamps, vehicle lamps, medical shadowless lamps etc. However, there are still lots of conundrums that have not been solved yet, e.g., new access to raw materials, rough flexible substrate, and low process efficiency, resulting in poor luminescence, rough products, and inflexibility [4]. Therefore, it is urgent to develop a new flexible fiber display system, which can meet the gradually growing demand.

In view of the above problems, a new flexible optical fiber display system combining OLED with conductive optical fiber is investigated in this paper. Besides, its preparation process, mechanism, advantages and disadvantages as well as prospects are discussed in depth. In addition, its application fields, advantages and limitations existing so far are also explored.

2. Flexible OLED

2.1. Mechanism of OLED

OLED is driven by semiconductor materials and organic light-emitting materials in an electric field, through carrier injection and recombination technology that leads to luminescence [5-7]. The principle is that indium tin oxide (ITO) glass transparent electrodes and metal electrodes are used as the anode and cathode of the device respectively. Under a certain voltage, electrons are injected into the electron transport layer through the cathode, and holes are injected into the hole transport layer from the anode, and then migrate to the luminescent layer, respectively. Excitons are formed after meeting the excite luminescent molecules, which emit visible light after radiation. The radiation light can be observed from the ITO side, and the metal electrode film also acts as a reflector [8].

2.2. Mechanism of flexible OLED

The polymer composite fiber loaded with luminescent active materials and transparent conductive polymer gel fiber are interwoven in the weaving process to form an electroluminescent unit (as shown in Fig.1), and a new flexible display fabric is realized by effective circuit control [9]. The display fabric presents a unique overlapping structure, which is formed by interlacing luminous warps and conductive wefts. Seen from the cross-sectional direction, one of them is a conductive yarn coated with luminescent material, and the other transparent conductive fiber is overlapped with its warp and weft by weaving. After alternating voltage is applied, the polymer composite luminescent active layer on the luminescent fiber is excited by the electric field in the lap joint area, forming luminous pixels one by one. Under the excitation of the electric field, the electrode and the light-emitting layer can emit light effectively by physical bonding. This method can unify the preparation of light-emitting devices and the weaving process of fabrics, and can initially meet the resolution requirements of some practical applications [10]. By changing luminescent materials, multi-color luminescent units can be realized and colorful display fabrics can be obtained.

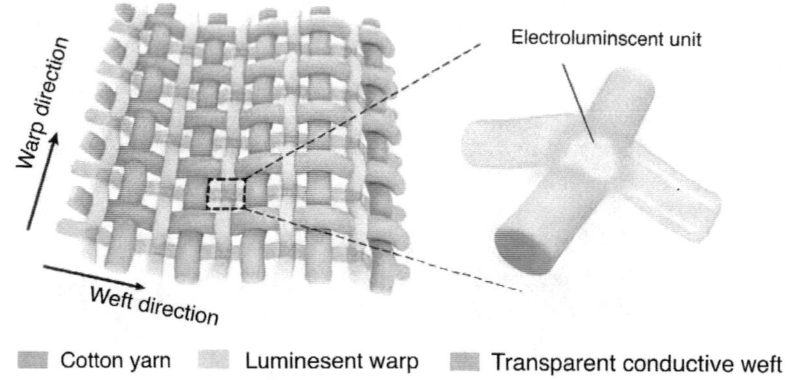

Fig.1 The unique lapping structure of flexible OLED fabric [7].

2.3. Manufacturing process of flexible OLED

The fabrication process of flexible fiber display system is mainly divided into three steps. They are the preparation of conductive fiber, OLED lattice and fibrotic grid [11]. The first is the preparation of conductive fiber. In order to make the display fabric adapt to the changes of the external environment, and even resist repeated friction, bending, stretching or other external forces, one needs to prepare a kind of polymer conductive fiber. The polymer conductive fiber can be selected from polyaniline, polypyrrole and other conductive polymers [12]. This kind of conductive fiber is used as the matrix of luminescent materials. The preparation method involved in the first step is melt spinning, which is a molding method using a melt spinning machine with polymer melt as raw material. The advantage of this method is that it can prepare fibers on a large scale. Fig. 2 presents schematic diagram of melt spinning principle. Using this method, even thousands of meters of fiber can be prepared.

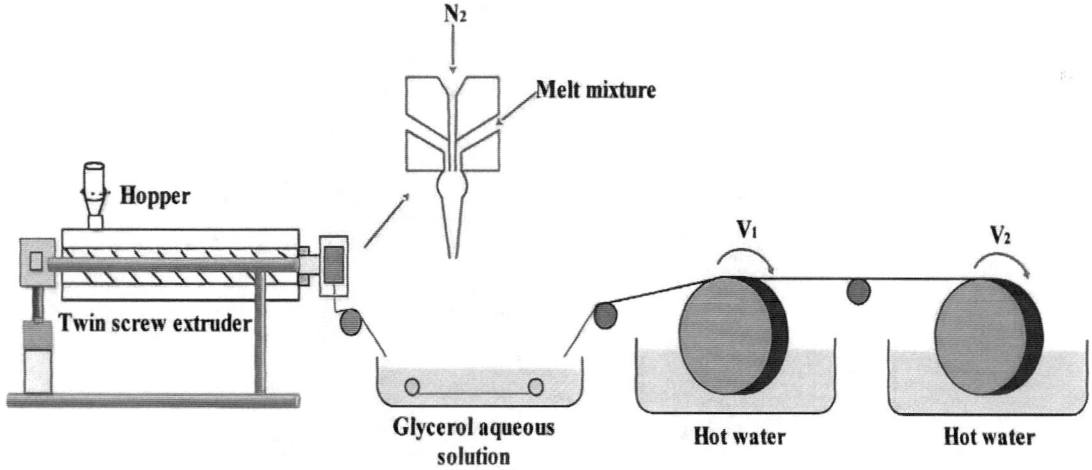

Fig. 2 Schematic diagram of melt extrusion [13].

The second part is the most critical step. We need to conduct some etching on the conductive fiber prepared in the previous step to obtain the dot matrix like pits as shown in Fig. 3. In the pits, one can fill some flexible and small OLED materials. These OLED array structures constitute the pixels of the display, which is the key to flexible display [14]. Small OLEDs combined with conductive fibers can give full play to their respective advantages and further improve the flexibility of light-emitting devices.

Fig. 3 Micro lattice structure on fibers [8].

The third step is to interconnect the functional fiber, the OLED loaded polymer composite fiber and the transparent conductive polymer gel fiber, and form the electroluminescence unit by the weft and warp. The collocation of the two fibers form the positive and negative poles of the fiber network. The principle of fiber luminescence (illustrated in Fig. 4) is divided into four steps. Firstly, the holes and electrons injected by anode and cathode. Then, the holes and electrons transport to emission layer (EML), need to keep high conductivity, block reverse transport [15]. Subsequently, the electron meets a hole in the EML, let exciton formation. In the last step, the electron falls to a lower energy level to fill the hole emitting a photon in the process, and tries to minimize defects to prevent loss of light. The power required by fiber display system can be supplied by battery, photovoltaic or piezoelectric materials [16]. With the designed circuit combination, one obtained the required text or pattern by programming, which is a unique lapping structure of interlaced weft and warp as exhibited in Fig.5.

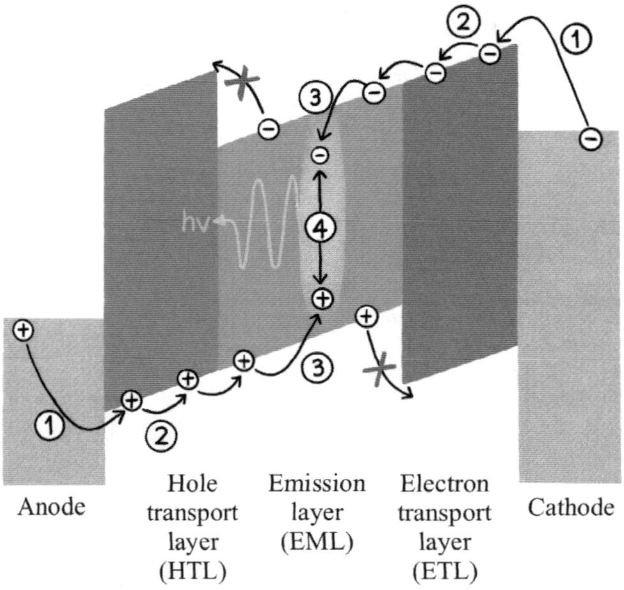

Fig. 4 Luminescence mechanism of OLED.

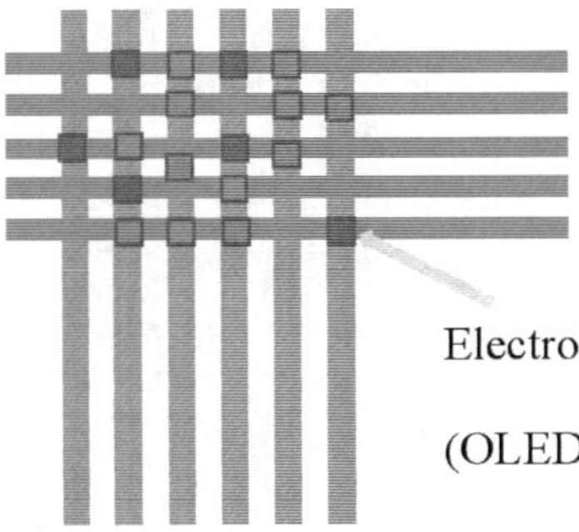

Fig. 5 Schematic diagram of flexible fiber display system.

2.4. Flexible OLED structure

For both multi-layer and a single-layer OLED, they are composed of a base layer, an anode, an organic layer, a conductive layer, an emission layer, and cathode composition. The structure of flexible OLED devices can be roughly divided into single layer structure, double layer structure, three layer structure and multi-layer structure [7].

Single-layer structure means that only one organic semiconductor thin film is prepared between the anode and the cathode as the light-emitting layer. Although this preparation is simple and low in cost, it is inefficient and often cannot meet the application criteria. Nevertheless, the double-layer structure solves this problem.

The double-layer structure consists of two specific ways: one is to insert an anode, a hole transport layer, a luminescent layer with electron transport capability and a metal cathode, and the other is to insert an anode, a luminescent layer with hole transport capability and a metal cathode in turn [13]. This double-layer structure greatly improves the working path and makes the device operate faster.

The three-layer structure is based on the single-layer structure, and hole transport layer and electron transport layer are added. With this structure, the carrier movement rate in the device can fluctuate within the controllable range, and the device will not be damaged and the material will not be lost. The multi-layer structure is based on the three-layer structure, with further hole injection layer, hole blocking layer or electron injection layer and electron blocking layer.

Multi-layer structure can not only improve the allocation and limitation of carriers, but also accelerate the flow and injection rate of carriers. Multi-layer structure is the most effective and efficient preparation method at present, but it also has the problem of high cost and unsuitable for mass production, i.e., it has not been widely used at present.

3. Applications

The fully flexible OLED fabric display system tends to show fantastic application prospects in a wide range of scopes. In light of its nature of high flexibility and good durability, it is highly possible that it can be put into use in lots of fields e.g., the internet of things and human-computer interaction, such as intelligent communication, real-time positioning and medical assistance, etc. [17].

3.1. Communications

This system can be used for intelligent communication, which offers an alternative way of instant communication with greater convenience in terms of saving time. On this basis, one can enter some letters by means of pressing the keyboards on the display and sending the message via the Enter button [18]. Based on setting, one can also input digits on the display. The sending message could be revealed on other's OLED fabric displays and also the mobile phones. Similarly, when a person sends a message via the phone, this message can be displayed on the clothes for people to read.

Indeed, this kind of communication which is based on OLED fabric display facilitates people's daily lives to a huge extent. For daily travelling excluding some long journeys, with the help of the OLED fabric display system on clothes, sending messages, ringing up and even making electronic payments are feasible. Therefore, one could deservedly go out without carrying phones, which reduces the risks of losing phones. However, there are someone who have gotten used to using their mobile phones and may be reluctant to change their habits. In this situation, the flexible OLED fabric display system can also be helpful. If the phones are out of power, the display system will provide people with timely support and avoidance of some troubles.

3.2. Navigation

This system can also be used for navigation and real-time positioning. A map can be clearly shown on the clothes with the exact bearings and overview of routes, which explicitly informs people of their precise real-time positions and helps them find the destination more quickly. With the guidance of the navigation, people can accurately know where they are and which direction they should go, i.e., they are unlikely to lose their way.

Additionally, besides daily travelling, in some work scenarios such as polar scientific expeditions or geological exploration, people may visit some dangerous places like oceans or snow mountains. Some hazardous natural disasters are inevitable, e.g., hurricanes, tsunamis and avalanches. For example, if somebody gets in a mortal accident and loses his way, his real-time location information can be displayed with just a few taps on the clothing, GPS and the map navigation can guide the way by the clothing. Furthermore, the distress signal can be easily sent so he can find a relatively safe place according to the map and wait for the rescue team. This is compared to wearing a monitor on your body, which provides safer conditions and greater security for exploration, due to the smaller probability of occurrence of accidents.

3.3. Medical assistance

Moreover, this system can be used in medical assistance as well. Because flexible OLEDs are used for luminescent light, the product will have similar properties compared with it. The colour can be displayed with high brightness and contrast with a fast, timely response [19]. By using electric currents, people are able to come into access to information about their health conditions. When the heart beats faster, there are higher waves with greater frequency displayed on the screen, and the world which expresses people's moods changes from 'relaxed' to 'anxious'. If possible, an electrocardiogram can also be shown on the screen using this technology, which could inform people of the potential existence of disease such as hypertension, risks of diseases attack and the need for treatment [20].

Besides, with the combination of more advanced apparatus, human being's brain waves could be collected and then decoded, finally displayed on the clothes in the form of words or colors. For example, via the process which is mentioned above, the desire of this man can be shown on the screen as a sentence of 'I need some food'. Certainly it can be much simpler, just using one word like 'hungry' or 'thirsty' to achieve exact meaning with maximum simplification. Moreover, different colors can be used in an emergency due to the fact that each colour can be endowed with a special meaning. For example, a red colour shown may represent that the person is in a dangerous situation and needs some assistance. This function helps to determine what people require in an easier way, which greatly brings convenience.

4. Advantages and limitations

4.1. Advantages

4.1.1 Multi-color light

The system can meet people's demands by combining light-emitting devices and silk fabric. With the use of the rational selection of dielectric polymer matrices and the semi-conductive organic material OLED, this product can release both light and colour. With the changing of the luminous material, it can create a multi-colour light fusion [21]. Flexible OLED is able to make the display brighter. Thus, with the combination of the materials, the resolution requirements of many practical applications are preliminary met, people are able to change the color according to their own preferences and get a colorful display fabric. These functions can be easily used in the aspect of decoration or in an emergency, which greatly brings convenience.

4.1.2. Fast response speed
Compared with other technologies, the use of OLED can accomplish a much higher response speed. Microsecond level can even be reached by the respond time. The faster response is able to promote greater speed of the present of light and colour, which helps to avoid long pause time. The response speed of OLED is about 1000 times faster than the LCD's response speed [22]. Therefore, the product becomes more efficient when people are using it.

4.1.3 Soft and comfortable
Displaying by OLED helps to achieve the property of soft and comfortable when wearing the display fabric (Fig.6). Firstly, OLED is produced on flexible substrates, i.e., it can achieve soft screens and exist external friction. Due to the fact that OLED is actively emitting light, it can accomplish a wider viewing angle. This means that the objects presented will not be distorted even in a large viewing angle. The vertical and horizontal viewing angle will reach more than 170 degrees. This property of long-term flexibility and toughness enables the display clothing to bend repeatedly and it is less easy to be damaged. When the human body moves and bends muscles and joints, the wearable display will also move with the body, acting like normal clothes people wear daily. Secondly, OLED has the property of ultra-thin and lightweight, people who wear it won't feel heavy. This enables everybody in a different age group, different height, and different situations to wear. The wearable display is portable. Therefore, it can achieve the significant property of clothing. Therefore, people can wear it without any discomfort [23].

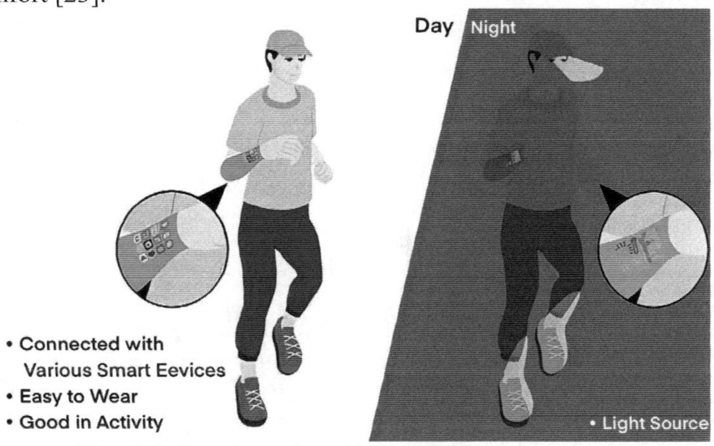

Fig. 6 Soft and comfortable bendable display [24].

4.1.4. Energy efficient
The use of OLED for the displayed fabric achieves sustainability of resources and energy efficiency. Due to the fact that OLED is mainly organic material and glass, it contains less toxic materials, which makes it easier for recycling. The fact that OLED does not require back-light reduces the consumption of power and fuel for energy, which then leads to lowering the emission of greenhouse gas. OLED provides design freedom in order to enhance the operating of the wearable display for the consumption of energy and greenhouse gas emission [25].

4.1.5. Wide temperature range
The system is able to work in a wide temperature range, which enables the wearable display to operate in almost any conditions. On this basis, it helps to reduce geographical restriction and the system can also be used in extremely cold regions.

4.2. Limitations

4.2.1. The challenge of packaging technology
The electronic components inside the OLED could be oxidized by contacting with air and water[26]. This would cause electronics short circuits. Therefore, flexible packaging is very important for the uses of flexible display fabrics. Now there are two primary encapsulation technologies, including cover plate packaging and Thin Film Encapsulation. Although the cover plate packaging is simple and effective, the biggest problem is that it cannot meet the needs of flexible display. Thus, the biggest advantage of OLED will not come true. The studies show that TFE provides good water and oxygen isolation for flexible OLED display[24]. However, the barrier layer of TFE is usually damaged due to large stress changes during repeated bending.

4.2.2. Complicated and costly production process
The flexible color tunable electroluminescent device composed of double-stacked emissive layers of ZnS phosphors dielectric polymer composite is developed by an all-solution processable method [26]. The conductive fiber and luminous point are arrayed respectively and interweave them into form Electroluminescent units. Integrity of multilayer structures. In the subsequent solution deposition of semiconductor, insulating and/or conductive layers, the underlying layer may be dissolved away or expanded by the solvent used to deposit the subsequent layers. If the solvent binds to the layer below, it expands, which results in a decrease in the performance of the layer [27].

4.2.3. Potential uncomfortableness
The flexible display fabrics are made from ZnS. The intrinsic stiffness of Zns might cause people to feel discomfort. Additionally, light emitting devices remain difficult to be used as wearable displays due to their low luminance and low efficiency. The first-generation OLED luminescence materials are fluorescent materials, mainly singlet exciton luminescence, because its maximum quantum efficiency is 25%, the luminescence efficiency is low, which is four times lower than the luminescence device made of phosphorescent materials. Phosphorescent materials can make use of singlet excitons and triplet excitons at the same time, and their external quantum efficiency (EQE) can reach 20%, which is higher than that of fluorescent materials [28].

4.2.4. Poor color purity and short life circle
Flexible OLED has the problem of insufficient color purity, which makes it difficult to display bright and rich colors. In the case of red, the COLOR presented by LCD can be pure red, but the same red on OLED is orange [29]. This may cause some display errors and lead to miscommunication. The organic polymer material in OLED has some defects in the life of blue light display, and will appear color difference after a long time of use. The data indicates that the average life of OLED is only 5000 hours under normal temperature and pressure, which is lower than that of LCD, which is at least 10,000 hours. Moreover, organic materials with low glass transfer temperature are susceptible to the temperature of the operating environment or the Joule Heating generated inside the OLED components.

4.2.5. The problem of Using in extreme environments
The high temperature caused screen damage, resulting in information communication barriers [30]. When the working temperature of LED display exceeds the bearing temperature of the chip, the luminous efficiency of LED display decreases rapidly, resulting in obvious light decay and damage.

LED display is mostly packaged with transparent epoxy resin. If the temperature exceeds the solid phase transition temperature (usually 125℃), the packaging material will turn into rubber and the thermal expansion coefficient will rise dramatically, leading to open circuit and failure of LED display [31].

5. Conclusion

In summary, starting from the development trend of display technology in the future, this paper introduces the flexible OLED fiber display and its working principle and development status. This functional, large-area display textile can form electroluminescence direct units with cotton yarn by weaving conductive and luminous fibers, so as to form a multifunctional integrated textile system for various applications. It is highly expected that it will soon cover varied fields, e.g., intelligent communication, real-time positioning and medical assistance, etc. With the improvement of flexible OLED and conductive fiber technology, the future display screen will be softer, lower cost and clearer. Although there are some problems remained unsolved (e.g., unwished poor luminescence and colour purity, short life span, temperature limitations etc.), flexible display is still the general trend of the development of display technology in the future. Obviously, it will occupy the most important market share of display technology in the future. In the wake of the integration of more functions, we expect these "smart textiles" to form future communication tools. These results offer a guideline for the great importance of flexible OLED based on its versatility of wide ranges of scope.

References

[1] J. Zhang, W. Huang, K.C. Chang, Y. Shi, C. Zhao, X. Wang, H. Meng, S. Zhang, M. Zhang, Performance Enhancement and Bending Restoration for Flexible Amorphous Indium Gallium Zinc Oxide Thin-Film Transistors by Low-Temperature Supercritical Dehydration Treatment, ACS Appl Mater Interfaces 13(7) (2021) 8584-8594.
[2] Y. Wu, Y. Yang, T. Li, S. Huang, H. Huang, S. Wen, Stretchable and foldable waveplate based on liquid crystal polymer, Applied Physics Letters 117(26) (2020).
[3] T. Wang, K. Lu, Z. Xu, Z. Lin, H. Ning, T. Qiu, Z. Yang, H. Zheng, R. Yao, J. Peng, Recent Developments in Flexible Transparent Electrode, Crystals 11(5) (2021).
[4] D. Wang, J. Hauptmann, C. May, Y.J. Hofstetter, Y. Vaynzof, T. Müller, Roll-to-roll fabrication of highly transparent Ca:Ag top-electrode towards flexible large-area OLED lighting application, Flexible and Printed Electronics 6(3) (2021).
[5] D.-H. Jang, K.-Y. Han, Multi-hardness hybrid silicone pad to laminate the bended cover window in a flexible OLED display, Journal of Mechanical Science and Technology 35(5) (2021) 1987-1992.
[6] S.Y. Jeong, H.R. Shim, Y. Na, K.S. Kang, Y. Jeon, S. Choi, E.G. Jeong, Y.C. Park, H.-E. Cho, J. Lee, J.H. Kwon, S.G. Im, K.C. Choi, Foldable and washable textile-based OLEDs with a multi-functional near-room-temperature encapsulation layer for smart e-textiles, npj Flexible Electronics 5(1) (2021).
[7] H.S. Kang, D.H. Kim, T.W. Kim, Organic light-emitting devices based on conducting polymer treated with benzoic acid, Sci Rep 11(1) (2021) 3885.
[8] Y.L. Wang, Overview of the development of flexible display technology, electronic testing No.413(08) (2019) 115-116.
[9] X. Shi, Y. Zuo, P. Zhai, J. Shen, Y. Yang, Z. Gao, M. Liao, J. Wu, J. Wang, X. Xu, Q. Tong, B. Zhang, B. Wang, X. Sun, L. Zhang, Q. Pei, D. Jin, P. Chen, H. Peng, Large-area display textiles integrated with functional systems, Nature 591(7849) (2021) 240-245.
[10] I. Verboven, W. Deferme, Printing of flexible light emitting devices: A review on different t echnologies and devices, printing technologies and state-of-the-art applications and future prospects, Progress in Materials Science 118 (2021).
[11] Y.-G. Kim, K.-Y. Han, Spread characteristics of OCR lamination for flexible OLED display process with flat and curved substrates, Japanese Journal of Applied Physics 60(5) (2021).

[12] C. Keum, C. Murawski, E. Archer, S. Kwon, A. Mischok, M.C. Gather, A substrateless, flexible, and water-resistant organic light-emitting diode, Nat Commun 11(1) (2020) 6250.
[13] D. Zhong, J. Zhou, Y. Wang, Hollow-fiber membranes of block copolymers by melt spinning and selective swelling, Journal of Membrane Science 632 (2021).
[14] H.S. Kang, D.H. Kim, T.W. Kim, Organic light-emitting devices based on conducting polymer treated with benzoic acid, Sci Rep 11(1) (2021) 3885.
[15] S.Y. Jeong, H.R. Shim, Y. Na, K.S. Kang, Y. Jeon, S. Choi, E.G. Jeong, Y.C. Park, H.-E. Cho, J. Lee, J.H. Kwon, S.G. Im, K.C. Choi, Foldable and washable textile-based OLEDs with a multi-functional near-room-temperature encapsulation layer for smart e-textiles, npj Flexible Electronics 5(1) (2021).
[16] J. Buhl, A. Cojocaru, M. Gerken, Customizing Tactile and Visual Perceptions of Organic Light–Emitting Diodes by Surface Coatings, Advanced Intelligent Systems (2021).
[17] D.-H. Jang, K.-Y. Han, Multi-hardness hybrid silicone pad to laminate the bended cover window in a flexible OLED display, Journal of Mechanical Science and Technology 35(5) (2021) 1987-1992.
[18] S.Y. Jeong, H.R. Shim, Y. Na, K.S. Kang, Y. Jeon, S. Choi, E.G. Jeong, Y.C. Park, H.-E. Cho, J. Lee, J.H. Kwon, S.G. Im, K.C. Choi, Foldable and washable textile-based OLEDs with a multi-functional near-room-temperature encapsulation layer for smart e-textiles, npj Flexible Electronics 5(1) (2021).
[19] H.S. Kang, D.H. Kim, T.W. Kim, Organic light-emitting devices based on conducting polymer treated with benzoic acid, Sci Rep 11(1) (2021) 3885.
[20] Y.L. Wang, Overview of the development of flexible display technology, electronic testing No.413(08) (2019) 115-116.
[21] X. Shi, Y. Zuo, P. Zhai, J. Shen, Y. Yang, Z. Gao, M. Liao, J. Wu, J. Wang, X. Xu, Q. Tong, B. Zhang, B. Wang, X. Sun, L. Zhang, Q. Pei, D. Jin, P. Chen, H. Peng, Large-area display textiles integrated with functional systems, Nature 591(7849) (2021) 240-245.
[22] D. Zhong, J. Zhou, Y. Wang, Hollow-fiber membranes of block copolymers by melt spinning and selective swelling, Journal of Membrane Science 632 (2021).
[23] D.X. Liu, J.Y. Zhong, B. Tang, X.H. Cao, W. Xu, S.X. Zhou, M.Y. Shi, R.H. Yao, H.L. Ning, J.B. Peng, Research progress of flexible and printed OLEDs, Liquid Crystal and Display 36(2) 12.
[24] PlasticsEurope, PlasticsEurope (2020) Plastics – the Facts 2020. An analysis of European plastics production, demand and waste data, (22 January 2021).
[25] S. Choi, S. Kwon, H. Kim, W. Kim, J.H. Kwon, M.S. Lim, H.S. Lee, K.C. Choi, Highly Flexible and Efficient Fabric-Based Organic Light-Emitting Devices for Clothing-Shaped Wearable Displays, Scientific Reports 7(1) (2017) 6424.
[26] X. Shi, Y. Zuo, P. Zhai, J. Shen, Y. Yang, Z. Gao, M. Liao, J. Wu, J. Wang, X. Xu, Q. Tong, B. Zhang, B. Wang, X. Sun, L. Zhang, Q. Pei, D. Jin, P. Chen, H. Peng, Large-area display textiles integrated with functional systems, Nature 591(7849) (2021) 240-245.
[27] S.Y. Jeong, H.R. Shim, Y. Na, K.S. Kang, Y. Jeon, S. Choi, E.G. Jeong, Y.C. Park, H.-E. Cho, J. Lee, J.H. Kwon, S.G. Im, K.C. Choi, Foldable and washable textile-based OLEDs with a multi-functional near-room-temperature encapsulation layer for smart e-textiles, npj Flexible Electronics 5(1) (2021) 15.
[28] M. Nakayama, N. Tanaka, T. Uchiyama, T. Ohkawauchi, Y. Tsuboko, K. Iwasaki, Y. Kawase, H. Matsuo, The stability of flow velocity and intracoronary resistance in the intracoronary electrocardiogram-triggered pressure ratio, Scientific Reports 11(1) (2021) 13824.
[29] Homer Antoniadis, Ph.D, Overview of OLED display technology, OSRAM, (2019).
[30] KYUNG CHEOL CHOI AND SEUNGYEOP CHOI, KAIST, OLED-on-fabric could be the next wearable wonder, (2021)
[31] ZheJiang violet, What is an OLED? All you need to know about OLED, OLED industry news, (9 April 2021).

[32] OLEDWorks, Sustanability: Imagine a world illuminated by organic lighting...OLED technology is good for the environment and people, (2017)
[33] S.-C. Kao, L.-J. Li, M.-C. Hsieh, S. Zhang, P.-M. Tsai, Z.-Y. Sun, D.-W. Wang, The Challenges of Flexible OLED Display Development, (2017).
[34] Lu Mengchen, Preparation of SnO_2 and ZnS:Mn quantum dot glasses with SPS and their optical properties, Donghua University, (2016).
[35] X. Xing, T. Lin, Y.-X. Hu, Y.-L. Sun, W.-Y. Mu, Z.-Z. Du, Y.-K. Liu, D.-F. Yang, S.-L. Shi, D.-Y. Zhang, Z. Cui, Inkjet printing high luminance phosphorescent OLED based on m-MTDATA:TPBi host, Modern Physics Letters B 33(12) (2019).
[36] Liu Dixuan, Zhong Jinyao, Tang Biao, Cao Xiuhua, Xu Wei, Zhou Shangxiong, Shi Muyang, Yao Ruihui, Ning Honglong, Peng Junbiao, Research progress of flexible and Printed OLED, (2020).
[37] Song Zhen, Study on ZnS based polychromatic force luminescent materials, (2018).

Combination of Photovoltaics and Organic Light-Emitting Diode Display

Jiarui Wang [1, †], **Xinyue Qiao** [2, †]

[1] Hefei No.6 High School, Anhui,230001, China, wangjiarui2019_8@163.com

[2] Ulink College of Shanghai, Shanghai 201600, xinyue.qiao@ulink.cn

[†]These authors contributed equally.

Abstract-Clean energy has the potential to deliver universal energy access in a way that is safe and powers bettering the environment for everyone. Photovoltaics, which transfer solar energy to electrical energy, can generate clean energy. This article introduces the internal structure, function, and working principle of each part of the system after combining photovoltaics and OLEDs. Then, this article concludes with some advantages: the system's energy source is clean energy and the system's other expanded applications. This new material-plastic electronic also faces some challenges. For example, solar energy received by the system is affected by factors such as latitude, which affects sunshine hours in a particular region, and weather, which affects light intensity. Moreover, there are several concerns like battery thermal management and battery life in practical application scenarios. So, this article needs to do more research and further experiments to improve the efficiency of solar power and maintain the health of the battery. In the future, this article suggests doing regular experiments to observe the system's efficiency and find the direction for improvement.

1. Introduction

Burning fossil fuels, such as coal, petroleum, and natural gas, for their economic development, can cause serious environmental pollutions: light pollution, air pollution, water pollution, etc. Those problems will also cause chain effects, affecting human health, global climate change, or species diversity endangered. Excessive exhaust emissions contribute to the earth's greenhouse effect while bringing mudslides and other catastrophic natural disasters. To deal with these problems, people are finding a more sustainable way of conflict between increasing demand in energy and economic development. Recently, the trend of using solar energy has gone well. Solar energy is considered a type of sustainable energy. Building a specifically designated solar power station, however, may cause light pollution to some extent.

In 2012, BASF and Philips made substantial breakthroughs in OLED technology which can be integrated into in-car roofs. There are tentative plans in the paper to switch to cleaner energy sources, but they do not set out to study their feasibility. Using photovoltaics is a better way of collecting clean energy, which is environmental-friendly. Solar energy is the energy supplied by nature – it is free and abundant. However, solar energy has not been widely used, and the stability and efficiency of its energy supply are still questioned. So far, no industry uses solar energy for power supply in a wide range of civil use. People have not found a suitable direction to realize energy supply and environmental protection at the same time. Also, the plastic, organic layers of an OLED are thinner, lighter, and more flexible than the crystalline layers in an LED or LCD. However, OLED's cost of production remains an intractable problem. The cost of producing OLED is too high, which also leads to high prices of final goods. Therefore, the article thinks of combining the two to make up for each other's defects. After the combination of the two, solar energy, which is clean and free energy, can be used more widely, and the cost of OLED products is greatly reduced by saving the cost of power supply.

In this article, this article provides a new idea, which is to install photovoltaics on glass buildings and cars to seek greater development and utilization of solar energy resources. The feasibility and specific design of the combination of solar power (photovoltaics) and OLED display are studied, and its advantages in energy, environment, and use are expounded. At the same time, the possible defects are pointed out, and the possible research directions are proposed to provide a reference for the combination of photovoltaic and OLED applications.

2. Using Solar Energy as a Source

2.1 Inside construction of the product

The panel of photovoltaics is designed to installed at the exterior part of the panel of Organic Light Emitting Diode. All panels should be put inside of the glass windows. The product can divide into four parts: photovoltaics, battery, wires, and OLED, and their functions are shown in Table 1. The working pattern is like using a charging phone. The detail of those parts will be presented below. Like a pump, the panel moves electrons through wires back into the battery and causes the recharge [1]. The detailed principle will be discussed in 2.2. Because the raindrop is a conductor, it will cause a short circuit and cause fire disaster if it touches the electric panel. Many natural phenomena like rain will damage the panel. Nowadays, many skyscrapers are made from glasses. To figure out this problem, installing the system inside the glasses can prevent outside erosions. Revealing photovoltaics could avoid the damage from rain or wind to reduce potential dangers, such as fire disasters, explosions, and installing problems. In addition, the position of the photovoltaics should be placed as Fig. 1, ensuring it won't shield inside vision of outside. As the figure is shown below, one glass can divide into two parts----one is the main body part, which OLEDs display can install here; another part is a small part above or below one's main body part, which can install solar panels to collect energy. Last but not least, to reach max energy, the size of the photovoltaic needs to be designed as large as possible (depending on the size of the glasses). Moreover, the size of OLED needs to fit the installation glass's size to ensure the display resolution (the clarity of images). Systems' revolution needs later scholars to test.

Table 1. Inside parts of the Combination of photovoltaics and OLEDs display and their main functions

Part of the product	Functions
Photovoltaics (solar panel)	Receive the solar energy and convert it to electrical energy in certain efficiency
Battery	Store the energy
Wires	Connect the photovoltaics to the battery, and battery to OLED to form a circuit
OLED	Use as display in multiple ways, such as provide people information or images

Fig. 1 The system of combination of OLEDs and photovoltaics

2.2 WorkingPrinciple

The glasses can be divided into two layers: the power layer(input) and the OLED displaying layer(output). The power layer is supported by some kinds of photovoltaic materials to gain energy. There are two typical kinds of photovoltaic materials: silicon class and organic polymer class, which enable the system to reach max energy input. To ensure the display effect, people need to reduce electroluminescence intensity by increasing the area of ITO transparent electrodes in organic LEDs [2].

2.2.1 Mechanism of Solar Power

There are four major steps during the mechanism over time, which eventually provide energy for the battery, which will be discussed below. The first step is light absorption. Some amount of light with a certain frequency will be absorbed, creating pairs of electrons and holes called excitons; Second step is charge separation over time. The excitons would make a separation, which means the electrons and holes will leave each other. The third step is diffusion, have been separated, the holes and electrons would diffuse towards different directions, creating an electromotive. In this step, electrical energy is generated. The final step is charge collection. The charge would get collected by the storing battery. The displaying time and energy collecting time are not happening simultaneously. To solve this, inserting a small battery at the edge of each glass to store the energy to release at the working time is good. To be elaborated, there are three features of the battery: to avoid the battery obstructing the light, the location of the battery is all around the edge [3-8]. The duration of the battery would be more than five years to keep the maintenance frequency down. The storing ability would be more than the daily consumption.

2.2.2 Mechanism of OLED Display

There are also four major steps of the mechanism of the OLED display. When OLEDs devices get electricity energy, holes and electrons will be injected by anode and cathode, as Fig. 2 shown below. Secondly, holes and electrons will transport to the emission layer(EML). Then, electrons meet a hole in the EML. Finally, electrons fall to a lower energy level to fill the hole emitting a photon because of the gap of layers' voltage. As the light emits, the OLED works.

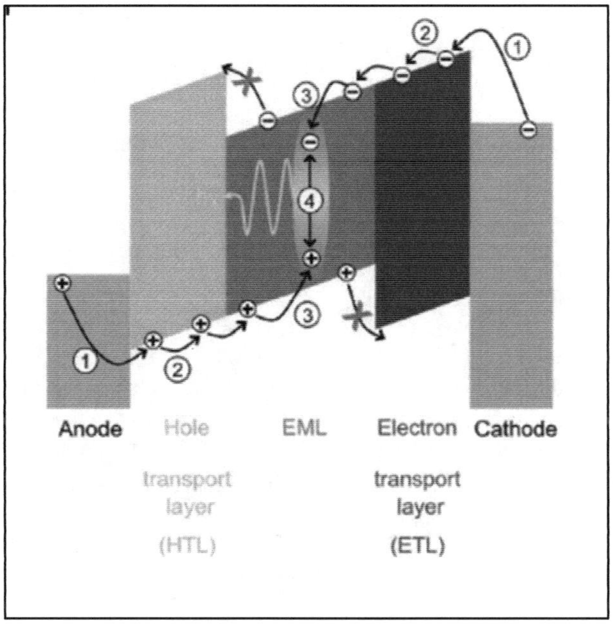

Fig. 2 Illustration of display mechanism.

3. Advantages

3.1 Clean Energy

Human beings are now facing several urgent problems that are the scarcely available resource and the resulting climate changes. As the major source for society's economic growth and development, fossil fuels are likely to cause sharp increases in temperature, wet-bulb temperature, and precipitation when burned. Extreme climate change will dramatically harm the health of adolescents and the elderly [9]. Apart from this, traditional energy like fossil fuels and oil is running out day by day. It is estimated that it would take about 1.5 piles of earth to supply the resources that humans are expected to consume [10]. The abuse of these carbon-containing energy sources has also caused many environmental hazards [11]. At the same time, residents living in resource-intensive areas are chronically exposed to potential pollution, and their incidence of endemic diseases such as pneumoconiosis is more than six times the average [12]. Over the long term, increases in resource extraction activities are associated with net increases in overall local mortality, cancer, and infant mortality. There is preliminary evidence that extraction activities are positively correlated with local pollution levels [13]. Compared with these resources, solar energy and environmentally friendly, recyclable and clean energy have a promising future due to their inexhaustible supply, universality, and high capacity [14]. The Sun provides the Earth with about 100,000 terawatts of energy, which is about 10,000 times the world's current energy consumption rate. Because of the many advantages of solar energy, photovoltaic cells are being used more and more widely. The article suspects that solar energy will play a key role in the sustainable energy system of the future [15].

3.2 Visibility

When OLED is not emitting light, it can be a transparent medium. Therefore, when OLED is applied to a transparent panel like the transparent solar panel, it can see the displayed information and see the scenery behind the panel. As our OLED used in the display is made of transparent OLEDs and transparent POVs, its visibility is very high, and it has different advantages in different application areas. When installed in the windows of subways, it will not affect the original lighting of subway cars. Also, passengers in the subway can simultaneously look out and touch the transparent display to learn more about current events and traffic conditions. This is the combination of new display technology and traditional Windows. The appearance of the OLED used in the display can improve the informatization of rail transit, accelerate its digital development, and offers travelers a better journey [16]. When installed in the windows of the car, it is different from the one installed in subways in functions. The OLED installed in the car's sunroof is a lighting system when switched on at night, and it turns back into a transparent panel that does not affect visibility when turned off during the day [17]. POVs can generate electricity in the daytime while the driver's visibility will not be affected.

3.3 Multiple Applications
The first one is displayed in different situations: usually, we would apply the glasses to display a big picture in the city. The big picture would be the advertisement of some public information that the government wants the screen to play. The advertisement part would be easy to imagine: the owner of the building would be paid if some other enterprises want to show something. The government information part is that when natural disasters are happening, the big screen would be a warning sign to warn the citizens to take cover or stay at a safe place. The performance part is that the screen takes the role of a live stream video display. Performing on display, it can provide information to passerby in this building. Then, separated glass could also do their jobs. They mainly act as an information passenger inside of the building. Managers inside the building could use it as a warning sign and messenger when there are accidents in the building or information to pass down. Also, using it as a meeting screen allows people to have a meeting without going upstairs or downstairs when it takes them a lot of time.

Another one is head-up visibility: when the OLED combines with a head-up display in the car, drivers don't have to look down to see speed, navigation distance, or navigation direction, which may avoid some traffic accidents to some extent. The OLED installed in the windshield has a similar working function to the one installed in the sunroof. They both can generate electricity through POVs when switched off, and a head-up display with OLED and POVS will not affect the sight of the eyes watching the outside lane. However, when the one installed in the windshield is switched on, it can show the real-time driving conditions of the car and nearby traffic conditions. Drivers can easily change routes or adjust the driving mode just through the head-up display.

4. Future Concerns

4.1 Energy Efficiency Limitation
Since the system will be installed in glass buildings, it is hard to make sure the building system could always receive the same amount of solar energy. Therefore, several efficiency-influencing factors need to be considered in future system's improvement. Because of the Earth's revolution, the daytime will be different in different seasons (the same reason with latitude impact). Therefore, it will cause energy inefficiency at a particular time and influence the systems' function. So, people need to install the system in a more useful way to raise the cost performance. For example, the high-ranking company administers always need to get in-time information; therefore, installing the system in their office is high cost-performance. Additionally, the technique of installing and cleaning the power layer, or photovoltaics, depends on real usage. We can see some solar power plants use mirrors to concentrate sunlight to gain higher energy.

4.2 Cell Thermal Management
Like phones, some devices with an electric circuit will be hot after a long-time of use. This is called the flow phenomenon, which may have a lot of potential risks. So, if a display is used intensively, that's dangerous. Its energy efficiency and lifetime will be reduced. Some solar power plants use mirrors to concentrate sunlight to gain higher energy; however, they always cause a temperature in the panel. With higher temperatures, the vibration of an atom will be increased; therefore, the bandgap size will be reduced. Consequently, the energy transfer efficiency will be lower. Conditions of high radiation, say 700–750W/m, modules can reach up to 40°C above ambient, 70°C, but this will obviously depend on

module design and building context. With the temperature higher, the voltage will be higher as well [18]. The high temperature and voltage may have potential risks, such as fire disasters and explosions. During the early time of the day, the temperature and sunlight's intensity is relatively low. Stop cooling solar panels in the early and last hours of the day can prolong the lifetime and increase the energy transfer efficiency [19]. Thus, in the future, the wire used in transport electricity needs to choose high-temperature resistant insulating materials to eliminate potential security risks. Besides, people need to control photovoltaics' use time depending on the intensity of sunlight to ensure the system's lifetime and energy-transfer efficiency.

4.3 Cell's Lifetime

What is excel of photovoltaics over conventional battery is its long life span and high-security level, but there are also some potential problems. Although the lifetime is longer than conventional one's, the efficiency cannot reach such a high after several years of usage. An electric product's lifetime depends on its battery. Therefore, prolong the cell's lifetime become an improvement in the future. In common cases, people choose to turn the e-product off to make sure the battery's healthy. But it is not a long-term plan. Table. 2 shown below, provides some influence factors of a cell's lifetime and use. Nevertheless, it is impossible to use mirror concentrating sunlight in metropolitan areas widely. Designing sunlight receiving device is an important future improvement plan. Because the life period of photovoltaic is around 15 years, the battery capacity and lifetime are needed to be tested every five years. After several years of usage, people can test the efficiency using time length, resistance, and highest working temperature to find out the highest efficiency factors in different stages. By doing so, users can control the photovoltaic and OLED combination system adjusting its position to get sufficient energy. Also, they can find the problems in time to prevent potential risks. Last but not least, as this technique mature, people can try to connect the internet to the display for a further application that needs internet access. For example, in the building, people inside can receive real-time news events around the world.

Table 2. Analyze of influencing factors and potential problems

Influence factors	Potential problems
High temperature	Lower the resolution of the OLED device, and high temperature might damage the circuit.
Low temperature	Lower the energy transfer efficiency
Frequently use	The current flow might burn the circuit and cause the operating system to break down.

5. Conclusion

This article provides ways to balance the conflict between increasing demand both in energy and economic development. This article designs a system that combines photovoltaics and OLEDs. Revealing photovoltaics in the glass could avoid damage from outside. Then, solar energy will be transferred to electrical energy for people to use by installing photovoltaics and OLEDs combination system in glass buildings or cars. This system has advantages in different aspects. In terms of energy supply, it has the benefits of free, environmentally friendly, sufficient energy resources. In terms of display, it has the advantages of a lighter material and higher visibility. This innovative system will have many expanded application. For example, it can be applied to the building wall for playing the news. It can also be installed in front of the car windshield to help drivers understand real-time traffic and car driving faster and more labor-saving. In the future, as this technique mature, people can try to connect the system with internet access to further its applications. At the same time, we also encountered some limitations and possible disadvantages in the research and design process, such as the impact of time and geographical location on efficiency. More deep research is needed in the future on the sustainability of energy supply and reducing the impact of external factors like climate change. Because this technology is immature, there are many challenges. We concern about the combination of photovoltaics and OLEDs' display effect, position influences, battery's thermal management and provide other possible applications and future improvements.

References

[1] Noel Kirkby, Barbara Kirkby, RVers guide to solar battery charging, 12 volt DC, 120 volt AC inverters, Aatec Publications (1987).

[2] Assaf Manor, Eugene A. Katz, Thomas Tromholt, Baruch Hirsch, and Frederik C. Krebs Origin of size effect on efficiency of organic photovoltaics （2011）

[3] The emergence of perovskite solar cells, Nature Photonics, 2014, 8, 506–514.

[4] Perovskite Solar Cells: From Materials to Devices, Small, doi: 10.1002/smll.201402767.

[5] Long-Range Balanced Electron and Hole Transport Lengths in Organic Inorganic CH3NH3PbI3, Science, 2013, 342, 344-347.

[6] 40.1% Record Low-Light Solar CellEfficiency by Holistic Trap Passivation using Micrometer Thick PerovskiteFilm[J]. Advanced Materials, 2021.

[7] Green M, Dunlop E, Hohl Ebinger J, etal. Solar cell efficiency tables (version 57). Prog Photovolt Res Appl. 2021,29(1):3–15.

[8] Ting H, Ni L, Ma S, et al. Progress inElectron Transport Materials for Solar cells[J]. ActaPhysica Sinica, 2015, 64(3):038802.

[9] A. Sommer, Burning Fossil Fuels: Impact of Climate Change on Health, Int J Health Serv 46(1) (2016) 48-52.

[10] ChengCheng Wang, Xinmin Wang, The resources consumed by humans need 1.5 earths to provide, Yangzhu (5) (2013) 85.

[11] M.A. Gulzar, H. Asghar, J. Hwang, W. Hassan, China's Pathway towards Solar Energy Utilization: Transition to a Low-Carbon Economy, Int J Environ Res Public Health 17(12) (2020).

[12] A. Biggeri, C. Lagazio, D. Catelan, R. Pirastu, F. Casson, B. Terracini, [Report on health status of residents in areas with industrial, mining or military sites in Sardinia, Italy], Epidemiol Prev 30(1 Suppl 1) (2006) 5-95.

[13] M. Matheis, Natural resource extraction and mortality in the United States, J Environ Manage 235 (2019) 112-123.

[14] J. Gong, C. Li, M.R. Wasielewski, Advances in solar energy conversion, Chem Soc Rev 48(7) (2019) 1862-1864.

[15] M. Gratzel, Photovoltaic and photoelectrochemical conversion of solar energy, Philos Trans A Math Phys Eng Sci 365(1853) (2007) 993-1005.

[16] Tao Jia, Peiyuan Li, Xueyuan Xu, Jinfa Liang, Gengze Lin, Ruhua Geng, Weijiu Zhang, Application of car windows based on transparent OLED display technology in urban rail transit, railway vehicles 59(2) (2021) 82-84.

[17] B. Plastics, R. Group, OLED lighting developed for use as a transparent car roof, British plastics & rubber (Feb.) (2012) 8.

[18] Randall Thomas, Photovoltaics and Architecture, Taylor & Francis (2001).

[19] Haskins H., Battaglia V., Christophersen J., Battery Technology Life Verification Test Manual (2005).

Research on the Progress of VOCs Adsorption by Biomass Nanocomposites

Qing Wei[1], Jie Yang[1a*], Changan Tian[2a*], Zhongmin Zhang[3], Xuan Zhang[3], Zhixin Zhang[3], Dong Li[1]

[1] School of Energy Materials and Chemical Engineering, Hefei University, Hefei, Anhui, 230601,China

[2] School of Chemistry and Civil Engineering, Shaoguan University,Shaoguan, Guangdong, 512005,China

[3]Anhui Tongyi Purification Technology Co., Ltd. Fuyang 236600

[*]Corresponding author's e-mail: yangjo@hfuu.edu.cn; tca2204@sgu.edu.cn

Abstract. The avoidance and control of the air is forthcoming, and its deterioration sources have a wide range of origin. Among them, the emission of VOCs is worthy of our attention. Its source is very wide, which seriously define the improvement of green chemical industry in our country. There are many types of VOCs, and when the concentration is too high, it will cause serious harm to the human body. Therefore, the avoidance and control of VOCs gas has become one of the current research tosspots. In this paper, several methods for alter VOCs gas are related, and it is found that adsorption is more economical and safer, and the core is the preparation of adsorbents. At the same time, the preparation of high-performance VOCs adsorbents based on TiO_2 as the matrix and manganese-based as the raw material of biomass waste is also prospected.

1. Introduction

Nowadays, the development of China's industrial economy has greatly improved our living standards, but it also brings many potential safety hazards to our life safety. Air pollutants mainly come from the exhaust emissions of key industries such as thermal power generation, iron and steel industry, refining tar and cement, petroleum industry, agricultural harvesting machinery, glass refining and motor vehicles, which directly lead to a significant decline in China's air quality. Nowadays, China strictly controls the emission of air pollution gases and strictly formulates relevant standards to ensure the effective prevention and improvement of China's environment [1]. VOCs not only seriously endanger our physical and mental health, but also our living environment [2-5], which is mainly reflected in the following aspects: once VOCs are discharged into the atmosphere without treatment, they will directly react with the light generated by sunlight to produce toxic organic compounds, thus forming highly oxidizing ozone, aldehydes and other harmful substances, which will lead to more serious pollution [6]. Since 2020, China has implemented new regulations on the prevention and control of VOCs Pollution to strictly control the emission of VOCs in China. Meanwhile, China strives to reduce VOCs emissions by more than 10% compared with the previous year in recent two years [7]. In the process of controlling VOCs emission, the task is still arduous. Therefore, it is urgent to develop efficient VOCs treatment technology. Therefore, VOCs adsorption technology is an environment-friendly and low-cost treatment and purification technology. It has a bright development prospect in our society and can promote the economic and social development of our country.

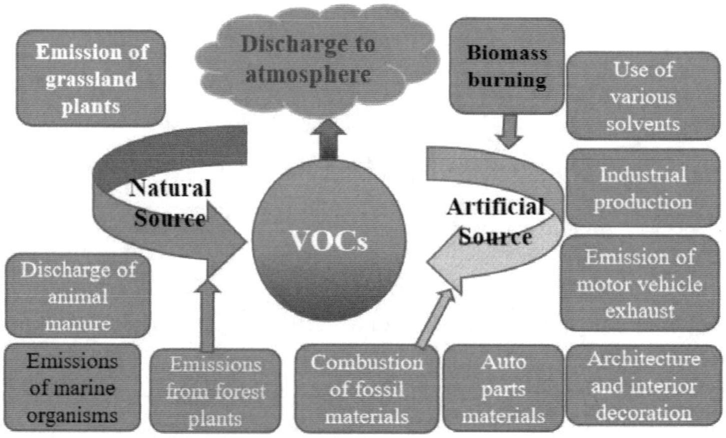

Figure.1 VOCs main source channels

2. Overview of VOCs treatment technology

The governance of VOCs mainly includes source reduction, strict control in the process and tail treatment [8]. Tail processing is the most important of the three links. In the process of tail treatment, it is usually implemented by physical or chemical means [9]. Physical means mainly include membrane material [10], adsorption and absorption method [11] and condensation method, which are collectively referred to as physical method, ionization method [12], combustion method [13], photocatalytic method and biological method [14].

(1) Membrane separation

Membrane separation method is to separate gas according to the rate difference of organic matter passing through membrane material. Therefore, it can be applied in the treatment of VOCs gas, and it can also be applied to the separation of aromatic and aliphatic compounds, esters, alcohols, aldehydes, phenols, ketones, etc [15-17].Therefore, this method can effectively solve the problem of recovery of organic pollutants, including the separation and recovery of volatile organic pollutants such as $HCHO$,C_7H_8,CH_3COCH_3, the recovery rate can reach more than 90%.

(2) Absorption method

Different gases have different solubility in the same solvent. At the same time, the gas will also produce chemical reaction with our selected absorbent. Using these characteristics, we can effectively use the solvent to absorb VOCs gas, so as to achieve the purpose of purifying the gas we need. At present, kerosene and water are mostly used as absorption solvents. Ozturk B found that fresh lubricating oil has excellent performance of removing VOCs gas, which is second only to the effect of water as absorbent [18].

(3) Adsorption method

The earliest method to deal with VOCs. Xi Jinying et al. Used this method to study a large number of VOCs gas treatment. In addition, adsorption method is widely used in high concentration VOCs gas treatment [19]. The main advantages of the adsorption method are high mature technical route, low cost and no secondary pollution. The disadvantage is that the treatment effect of the current adsorption method is limited by the adsorption capacity of the adsorbent itself. Therefore, it is of great significance to develop a composite and efficient green adsorbent.

(4) Biological method

After decades of development, many countries have begun to use biological methods for VOCs treatment on a large scale [20]. Microorganisms eliminate harmful gases generated in industrial production through respiration, which also provides a new idea and possibility for their elimination of VOCs gas. Its main device composition can be divided into biofilter, biofilter, biological washing tank[21-22], etc. This method has the excellent characteristics of low price, easy maintenance and strong treatment effect [23].

(5) Combustion method

Combustion method is also used on a large scale. This method is simple and effective. However, if VOCs gas is directly, it may cause other pollution, so attention should be paid during combustion. Thermal combustion requires fossil fuel as a carrier, which is expensive and easy to cause secondary pollution. Regenerative combustion needs to be carried out at high temperature, which is not very safe. Catalytic combustion reaction is a very classic gas-solid catalytic reaction, and its main essence is the oxidation reaction solution with the participation of active oxygen and deep oxygen [24-25]. The safety performance of this method is relatively good. Compared with other combustion technologies, the reaction temperature is low and there is no secondary pollution [26].

(6) Condensation method

Condensation method is widely used in the treatment of VOCs gas. It is implemented through the difference of saturated vapor pressure of volatile organic compounds at different temperatures and environmental conditions. By changing the temperature or pressure of the system, we can condense the volatile gaseous organic pollutants and turn them into gaseous pollutants, so as to separate the waste gas. This provides ideas and methods for the treatment of VOCs at room temperature, high temperature and high concentration. The removal degree is affected by two factors: cooling temperature and saturated steam pressure of the treatment object [27].

(7) Plasma technology method

At present, the research on the removal of VOCs by plasma technology is mostly in BDB reactor, using relevant catalysts. Usually, during high-frequency discharge, a large number of high-energy particles will be generated while plasma is generated. These high-energy particles can destroy the molecular structure, so as to promote the purification of VOCs gas and achieve the treatment of polluted gas [28].

To sum up, various research technologies for VOCs treatment, adsorption method is widely used in many VOCs treatment methods because of its low cost, high efficiency and no secondary pollution.

3. Preparation of biomass Nanocomposites

Researchers prepare composites by compounding a series of biomass nano materials with TiO_2, which has become a mainstream research. At the same time, the catalyst activity of TiO_2 as support reached the highest. TiO_2 is relatively cheap. It does not need other auxiliary reagents. It shows effective damage to toxic pollutants. It reacts at a certain ambient temperature and pressure and chlorinated organic compounds. The reaction products are usually carbon dioxide and water. For example, Lan [29] and others first demonstrated a simple hydrothermal induced solvent limited single micelle assembly method, which can synthesize monolayer two-dimensional ordered mesoporous TiO_2 nanosheets. The material has uniform size, thickness and high specific surface area. In addition to nano-TiO_2, Dinh et al. [30] prepared a series of TiO_2 nanocrystals by thermal reaction using oleic acid and oleamine as active agents.

Noble metal based catalysts (such as Pt, Pd and Zr) show high oxidation and selectivity to VOCs gas, improve their catalytic efficiency and enhance the progress of oxidation reaction, which has been widely recognized by experts from various countries [31]. Transition metal oxides show very good catalytic properties in oxidation reactions. In addition, they have lower cost, higher resistance to poisons, higher metal content and higher active surface area. Therefore, transition metals are widely used in the treatment of VOCs to cope with increasingly strict management regulations [32-33]. Manganese based catalyst is one of the low-cost active catalysts. It is active in catalytic reaction and has no toxic components. It shows good catalytic activity in a variety of volatile organic compounds, including n-hexane [34], acetone [35], benzene [36], ethanol [37], propane [38], etc. Wu et al. prepared a series of manganese series and titanium dioxide catalysts, studied the effect of the ratio of manganese group to titanium group on the catalytic activity, and found that when the ratio of manganese group to titanium group was 0.3, it had the highest catalytic activity. Biochar is widely used in water treatment and supercapacitors, and its unique structure deserves special attention in the research of VOCs gas treatment [39-40].

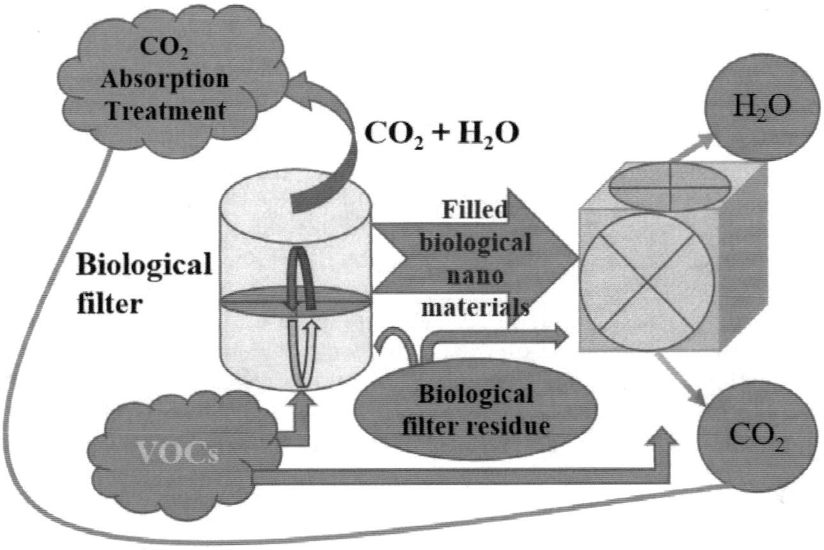

Figure 2. Process of processing VOCs with biomass nanomaterials

At the same time, the high adsorption properties of biomass activated carbon nanocomposites can also be widely used in daily tail gas and formaldehyde treatment. Banana peel is a potential huge raw material resource. At the same time, it also has the effects of clearing intestines and detoxifying. However, banana peel, as a waste of banana products, has not been fully developed and treated. Fresh banana peel accounts for 40% of the total banana mass; In addition, with the development of banana processing industry, its waste banana peel is directly discarded and cannot be fully treated and applied [41-43]. If this part of waste is fully utilized to prepare activated carbon, it can not only solve the problem that banana peel can not be treated due to large accumulation, but also achieve the purpose of making full use of waste[44-45].

4. Conclusion

The treatment of VOCs has always been a major research hotspot, and adsorption technology is a widely used method at present. In recent years, there have been relevant research reports on the preparation of activated carbon from banana peel, which is mainly used in supercapacitor, sewage treatment and other fields. However, there is no research on the preparation of biological activated carbon nanocomposites from banana peel to treat VOCs. Therefore, it is planned to explore biomass nanocomposites with better performance by designing synthesis routes and changing industrial conditions, and there are few reports on the research of composite manganese based and titanium dioxide materials. Our research group plans to design and synthesize a series of biomass nanocomposites, hoping to be used as air purification materials in daily life.

Acknowledgments

This work was supported by Hefei University Quality Engineering Demonstration Experimental Training Center Project(2020hfusxzx02), National Natural Science Foundation Project (61804039), Guangdong Basic and Applied Basic Research Foundation(2021A1515010671, 2020A1515011221), National College Students' innovation and entrepreneurship training program project (202010576005,202011059103,202011059100,202011059117,202011059114), Shaoguan University Education and Teaching Reform Project(syjy220201203), the Natural Science Foundation of Anhui Province of China under Grant(2108085ME152), the Characteristic innovation project of colleges and universities in Guangdong Province(2021KTSCX122), Natural Science Foundation of Shaoguan University(SZ2020KJ03).

Reference

[1]National Environmental protection standard. (2017) "13th five year" development plan. Tianjin paper,39 (2): 21-48
[2]C Gennequin, M Lamallem, R Cousin, et al. (2009) Total oxidation of volatile organic compounds on Au/Ce-Ti-O and Au/Ce-Ti-Zr-O mesoporous catalyst.Journal of Materials Science,44:6654-6662
[3]R Atkinson. (2000) Atmospheric chemistry of VOCs and NO_x.Atmospheric Environment,34:2063-2101
[4]S Scirè S,Minicò C,Crisafulli,et al.(2003) Catalytic combustion of volatile organic compounds on gold/cerium oxide catalysts.Applied Catalysis B: Environmental, 40: 43-49
[5]Y Liu, H Dai, J Deng, et al.(2014) Mesoporous Co_3O_4 supported gold nanocatalysts:Highly active for the oxidation of carbon monoxide, benzene, toluene, and oxylene. Journal of Catalysis, 309:408-418
[6]He Hong, Li Junhua, he Hong. (2008) Environmental catalysis. Science Press,
[7]Wang Haifeng, Qin song, Jiang Xi, et al. (2018) Research progress on VOCs emission reduction in sintering process of iron and steel industry. Iron and Steel, 1:1-7
[8]China Environmental Supervision. (2017)"Work plan for the prevention and control of volatile organic compounds pollution" in the 13th five year plan,9(22):15-16
[9]Torimoto T,Okawa Y, Takeda N,et al. (1997) Effect of Activated Carbon on Photodegradation Behaviors of Dichloromethane.J Photochem Photobio A:Chem, 103(1-2):153-157
[10]Huang Donglin. (2006) Membrane separation and recovery of volatile organic gases. Dalian University of technology
[11]Li Liqing, Liang shidun, Tang Rui, et al. (2013) Adsorption performance of acid modified activated carbon for toluene and methanol. Journal of chemical engineering, 64(3): 970-979
[12]Fan Hongyu, Li Xiaosong, Liu Yanxia, et al. (2011) Removal of toluene from indoor air by a new process catalyzed by storage discharge plasma with integrated cycle. Journal of chemical industry, 62(7): 1922-1926
[13]Jiang S, Song S. (2013) Enhancing the performance of Co_3O_4/CNTs for the catalytic combustion of toluene by tuning the surface structures of CNTs. Applied Catalysis B:Environmental, 140-141(8):1-8
[14]Go Mez Cerezomn, Munoz Batista MJ, Tudela D,et al. (2014) Composite Bi_2O_3-TiO_2 catalysts for toluene photo-degradation:Ultraviolet and visible light performances. Applied Catalysis B: Environmental, S156-157(3):307-313
[15]Zhang Lin,Chen huanlin,Chai Hong.(2002) Research progress on membrane treatment of volatile organic compounds waste gas. Chemical environmental protection,22(2):75-80
[16]Xu Zhikang,Liu Zhenmei,Dai Qingwen.(2002) Research progress of polymer membrane separation of olefins-alkanes based on solubility diffusion mechanism. Petrochemical industry,31(2):135-140
[17]ALBERICI RM, JARDIM WF.(1997) Photocatalytic destruction of VOCs in the gas-phase using titanium dioxide. Applied Catalysis B:Environmental,14(1):55-68
[18]OZTURK B, YILMAZD.(2006) Absorptive Removal of Volatile Organic Compounds from Flue Gas Streams. Process Safety & Environlnental Protection,84(5):391-398
[19]Xi Jinying,Wu Junliang,Wang can.(2012) Investigation and Analysis on the application of industrial VOCs gas treatment technology.China Environmental Science,32(11): 1955-1960
[20]VANLH,DEBI,JACOBS P,et a1.(2004) Amembrane bioreactor for the removal of dimethyl sulphide and toluene from waste air. Water Science Technology,50(4):215-224
[21]OTTENGRAF S P,VAN A H.(1983) Kinetics of organic compound removal from waste gases with a biological filter. Biotechnology Bioengineering,25(12):3089-3102
[22]Den W,Huang C,Li C H.(2004)Effects of cross-substrate interaction on biotrickling filtration for the control of VOC emissions. Chemosphere,57(7):697-709
[23]Yang Hao, Li YanXu, Lu Zi. (2009) Study on biological treatment of volatile organic waste gas

(VOCs). Guangdong chemical industry, 36(8): 125-127
[24]Xie Yunlong.(2006) Study on VOCs catalytic combustion performance of cordierite ceramic honeycomb coated non noble metal composite oxide catalyst. Hangzhou: Zhejiang University of technology,
[25]Qian Jun,Shen Meiqing,Wang Xiaoling,et al.(2001) Study on catalyst performance for combustion control of organic waste gas pollution. Combustion science and technology,7 (3):242-244
[26]Scire S,Minico S,Crisafulli C,et a1.(2003) Catalytic combustion of volatile organic compounds on gold cerium oxide catalysts.Appl Catal B:Environ,40(1):43-49
[27]Wu Bijun, Liu Xiaoqin. (2005) Research progress of VOC pollution control technology. Electric power technology and environmental protection, 21(4):39-42
[28]Li Jie, Li Jian, Jin Yukui, et al. (2006) Study on treatment of volatile organic compounds (VOCs) by low temperature plasma technology. Beijing University of technology.
[29]Lan K, Liu Y, Zhang W, et al. (2018) Uniform Ordered Two-Dimensional Mesoporous TiO2 Nanosheets from Hydrothermal Induced Solvent Confined Monomicelle Assembly. Journal of the American Chemical Society,140(11):4135-4143
[30]Dinh C T, Nguyen T D, Kleitz F, et al. (2009) Shape Controlled synthesis of Highly Crystalline Titania Nanocrystals. ACS Nano, 3(11):3737-3743
[31]Tidahy HL, Hosseni M, Siffert S, et al. (2008) Nanostructured macro-mesoporous zirconia impregnated by noble metal for catalytic total oxidation of toluene. Catalysis Today, 137(2-4):335-339
[32]Saqer S M, Kondarides D I, Verykios X E. (2011) Catalytic oxidation of toluene over binary mixtures of copper, manganese and cerium oxides supported on $A1_2O_3$. Appl Catal B:Environ, 103(3-4):275-286
[33]Kim S C, Shim W G. (2010) Catalytic combustion of VOCs over a series of manganese oxide catalysts. Appl Catal B: Environ, 98(3-4):180-185
[34]Zhou G, He X,Liu S, et al. (2015) Phenyl VOCs catalytic combustion on supported CoMn/AC oxide catalyst. Journal of Industrial Engineering Chemistry, 21:932-941
[35]Wang X, Zheng Y, Lin J. (2013) Highly dispersed Mn-Ce mixed oxides supported on carbon nanotubes for low-temperature NO reduction with NH_3. Catalysis Communications, 37:96-99
[36]Carabineiro S, Tavares P B, Figueiredo J L. (2012) Gold on oxide-doped alumina supports as catalysts for CO oxidation. Applied Nanoscience, 2(1):35-46
[37]Peluso M A, Pronsato E, Sambeth J E, et al. (2008) Catalytic combustion of ethanol on pure and alumina supported K-Mn oxides: An IR and flow reactor study. Applied Catalysis B: Environmental, 78(1-2):73-79
[38]A C L, A A B, A P G,et al. (1998)Evaluation of γ-MnO_2 as a VOC Removal Catalyst: Comparison with a Noble Metal Catalyst. Journal of Catalysis,178(1):214-225
[39]Wu Z, Tang N, Ling X, et al.(2010) MnOx/TiO_2 composite nanoxides synthesized by deposition-precipitation method as a superior catalyst for NO oxidation. Journal of Colloid and Interface Science, 352(1):143-148
[40]Qian K, Kumar A, Zhang H, et al.(2015) Recent advances in utilization of biochar. Renewable & Sustainable Energy Reviews, 42(1):1055-1064
[41]Y Zhang, Z Gao,N Song,et al.(2016) High-performance supercapacitors and batteries derived from activated banana-peel with porous structures.Electrochimica Acta, 222:1257-1266
[42]Ingole R S, Lataye D H, Dhorabe P T. (2017) Adsorption of phenol onto banana peels activated carbon. KSCE Journal of Civil Engineering,21(1):100-110
[43]D Yu, L Wang, M Wu. (2018) Simultaneous removal of dye and heavy metal by banana peels derived hierarchically porous carbons. Journal of the Taiwan Institute of Chemical Engineers, 93: 543-553
[44]O Fasakin,J K Dangbegnon,D Y Momodu,et al.(2018) Synthesis and characterization of porous carbon derived from activated banana peels with hierarchical porosity for improved electrochemical performance. Electrochimica Acta,262:187-196

[45]Gupta H, Gupta B. (2016) Adsorption of polycyclic aromatic hydrocarbons on banana peel activated carbon. Desalination and Water Treatment,57(20): 9498-9509

Molecular Dynamic Simulation in Organic Semiconductor Investigation

Junwen Gu[1*]

[1]The high school affiliated to Renmin University of China, Beijing, China

* Gujunwen0208@sohu.com

Abstract—The details of charge transfer events inside the organic semiconductor puzzled researchers for many years. This paper focuses on the simulation method's capability to investigate on polaron's properties and insights about charge transfer events inside the organic semiconductors. The paper adopts a molecular dynamic simulation method called LMD simulation from the journal of chemical science to model the charge transfer between pairs of fullerene molecules. By performing the simulation, the paper analyzes and evaluates the results it gained and concludes, by showing a large-scale investigation capability and accurate details for a single event, the molecular simulation is capable to study the charge transfer events of semiconductors.

1. Introduction

Photovoltaic is an electronic device that uses photoelectrical effect to produce electricity. The invention of organic semiconductors provides new facilitation of photovoltaic. The organic semiconductor has various advantages like much lower cost, ease of fabrication process and applicability of flexible elements. However, the inefficiency of charge transfers heavily interrupts its application on the photovoltaic. To investigate the charge movement inside the organic semiconductor, in a conventional way, scientists need to solve enormous complex quantum mechanical equations. Due to the variation of circumstances, solving equations is nearly impossible in many cases. Thus, the utilization of the molecular dynamic simulation method debuts. Molecular dynamics (MD)uses computational methods to measure and simulate the physical movements of atoms and molecules in a given system for a period of time. The molecular dynamics can help researchers to investigate the properties of complex systems, reveal the systems' evolution and simulate the condition that cannot carry out in the laboratory at a much lower cost than conventional methods and can simulate a quite large scale the experiment at the same time. In this paper, we research a special attempt that uses LMD simulation to investigate the properties of polaron, particles that formed by charges due to strong coupling with lattice degree with freedom and investigate the insights of charge transfer events between fullerene molecules to test the capability of LMD simulation in the investigation of micro molecular behaviors. Moreover, this paper also analyzes, evaluates, and compares the results of simulation with other investigation methods. By observing a great accuracy in a single event simulation and a large-scale simulation capability, this research confirms that we can utilize molecular dynamic simulation to investigate the micro properties of molecular behaviors. This outcome is beneficial because it means we can use a much lower cost and quicker time to simulate the charge transfer event inside the organic semiconductor. Thus, by applying the appropriate model, scientists can simulate the charge movement much quicker and with a significantly lower cost. Hence, researchers can advance the invention and modification of photovoltaic significantly and improve the solar power conversion rate. More importantly, this research also shows a new possibility to investigate charge transfer events from a simpler way which can help scientists to investigate the effects of multiple parameters on the charge transfer for lower cost both in financial and time. Besides, the validity of the molecular dynamic simulation method that is certified in the paper can apply in other areas that relates to electricity. By investigating the charge transfer event inside the devices microscopically, it can provide more insights for scientists for the origin of electricity and their electrical behaviors.

2. The molecular dynamic experiment in semiconductor

2.1. The purpose and methods used in the experiment

The development of organic semiconductors is an improvement in electric engineering. Organic semiconductors can be used to manufacture flexible devices and printing technology. Nowadays, scientists are still working on the revelation of the charge transportation inside organic semiconductors. The revelation of charge transportation can help researchers to deduce the mechanism and prevail the organic semiconductors in a wide range of applications. Yet, the vast number of atoms and complex conditions inside obstruct the research. Nevertheless, an article published in the journal Chemical Science issue 4 2017 provides a new way to investigate this sophisticated process. [1]

The article suggests that the development of theoretical methods that accurately models charge transport would represent a major contribution to the optimization of this promising renewable energy technology.[1] The scientist group used Langevin molecular dynamics (LMD) to carry out the entire experiment. Besides, the scientists assert that they present a multiscale approach grounded in constrained density functional theory(Density functional theory (DFT) is a type of electronic structure calculation that has rapidly gained popularity.[4] (CDFT)29 calculations.[1] They present that once a set of CDFT is conducted to a potential energy surface to the LMD transfer, the process can be modeled at a negligible cost which can be helpful to simulate the transferring process easier. The scientists apply their model to phenyl-C61-butyric acid methyl ester (PCBM) molecules [1] because the recent study exhibited that the excess charges will mostly locate on a single PCBM molecule.

2.2. The experiment process and results shown

The scientists' group use 7 steps to summarize and introduce the experiment's implementations: (1) Perform CDFT calculations, as well as unconstrained DFT calculations, to find the optimized geometries of the initial state, final state, and transition state, (2) Perform single-point CDFT calculations for a series of geometries along with the reaction coordinate. There are two sets of single-point CDFT calculations., (3) Use diabatic PESs to compute the adiabatic PESs. The lower adiabat is then used to create our PES for LMD, (4) This adiabatic PES, like the diabatic PESs, is a function of the reaction coordinate. However, we wish to obtain a pseudo-atom PES as a function of the spatial coordinate (5) Effective mass and friction are assigned, (6) The spatial-coordinate PES, the effective mass, and the friction are then used as input for our LMD simulations, (7) LMD results are then analyzed to compute charge transfer rates and examine the details of polaron movement through space. [1]

First, the optimization of CDFT and reaction coordinates can illustrate the initial state and the final state of moving charge by conducting an extra charge constrained to only one of the two molecules participating in charge transfer events. Notably, the charge can be constrained to the entire PCBM molecules or only the C60 molecules. The optimized geometric graph can be represented by R which each molecule contains coordinates XYZ in PCBM molecules. The geometric graph corresponding to the reaction coordinates R is extracted from the charge transporting event including initial state(R1), transition state (Rt) and final state(R2). This gradual change can be described by the following interpolation suggested by Wu et al.:[3]

$$R_{\mathrm{q}} = \frac{\mathrm{q}(q+1)}{2} R_1 - (q-1)(q+1) R_T + \frac{q(q-1)}{2} R_2 \tag{1}$$

Second, we create diabats by conducting CDFT single-point energy calculations for a series of PCBM geometries along with the reaction coordinate Rq. [1] The energy is shown by each point along the diabats. The set of nuclear positions corresponds to the charge constrained to the fullerene 1 and charge constrained to the fullerene 2.

Third, the diatbats and their coupling can be used to derive adiabats (a system that cannot exchange heat with its surroundings. [8]) PES in LMD simulation. Because the adiabats represent a charge that gradually moves along the reaction coordinate, it is the natural choice for a PES in LMD simulation. The shapes of these diabats capture the coupling between adiabats. [1]A low peak is produced by a high coupling while a low coupling will produce a strong peak. A full dimensional PES is essential in the construction of a full-dimensional LMD. The scientists suggest a simple measure to obtain the full-dimensional graph. They perform an LMD simulation to simulate a charge that transfers in a straight line from one molecule to another molecule. Thus, a 1-dimensional PES can be used.

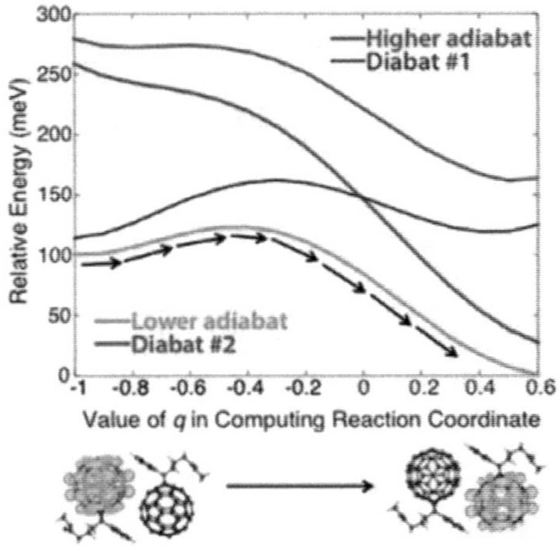

Figure 1 The diabatic and adiabatic PES obtained from CDFT calculation [1]

The graph shows that the initial adiabatic state energy is not equivalent to the final adiabatic state energy. This asymmetry is due to the removal of the outlying points mentions above. The researchers mention that asymmetry is the real state that happens inside organic semiconductors because the molecules are usually in disorder inside the system. Moreover, the disorder causes a great variation in PESs between different pairs of molecules and the scientists only want to find one reasonable PES for the improvement and testification for the LMD model.

Fourth, the scientists state that converting the adiabatic coordinates to spatial coordinates is essential for an LMD simulation to indicate the movement of particles in space. First, researchers use an equation that is shown in Figure1 to derive the reaction coordinates Rq fromR1, RT and R2. Then, the researchers compute the spatial coordinates Xsp by using multiple equations: 1. Xsp= A1X1+A2X2 2. A1=C12/C12+C22 A2=C22/C12+C22. X1 and X2 are the positions of fullerenes, C1 and C2 describe the contribution of each diabat to the adiabat at a given point on the reaction coordinate and A1 and A2 give the contributions of diabats #1 and #2 to the adiabat at each point on the reaction coordinate [1]. Afterwards, the adiabats PES shows as a function of space. The curve is mirrored and expanded in one-dimensional space to create a full PES for LMD simulation. The bellowing graph show specific steps to derive the full PES from optimized geometries.

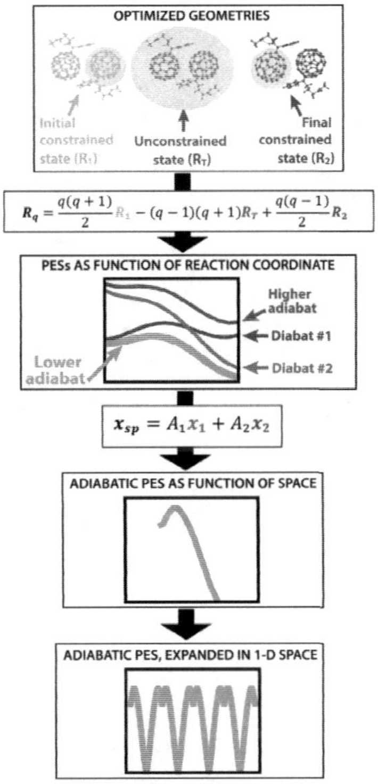

Figure 2 The sequence of producing a PES graph suitable for LMD simulation [1]

Fifth, the authors propose that calculations of effective mass or friction are more challenging tasks because there are no methods for deriving mass that widely apply in organic semiconductors. Thus, scientists investigate the relationship that each variable has on the transfer rate and adopt the value of effective mass and friction that show consistency with experiment observation. Therefore, scientists could formulate the properties of quasi-particle (describes a physical concept, which treats elementary excitations in solids, like spin waves, as particles. As the particles do not consist of matter, they are called quasiparticles. [5]) from this method. The scientists use values of meff from 7.22 × 10−2 to 7.22 × 103 amu [1] The effective mass is not the mass of electron nor the mass of nuclei. The scientists state that exploring the wide range of effective mass is essential to the development of the model and comparison to the benchmark. The friction is much more complex in this case, the scientists conclude the formula to calculate the friction γ = pγ D, the magnitude of p is varied from3.6,0.36 and 0.036[1] and the γ D is proportional to the square root of meff.

Sixth, due to the random nature of the simulation, the ensemble of orbits is essential for an accurate prediction. The researchers use the average time to obtain the rate of transfer. The velocity of molecules is distributed randomly that has a mean of 0 and standard deviation of (KT/meff) ^ (0.5), each point in the meff/γ parameter will perform a 500 times calculation. The calculation that has an average time shorter than 5ns will not be adopted because it corresponds to charge mobility several orders of magnitude lower than that needed for a functional OPV device [1]

2.3. The analysis and conclusions of the experiment

Step 7 includes the interpretation, analysis of the data and results present in the experiment. The authors divide the analysis part of the LMD result into 3 sections.

First, the scientist states that using a defined estimate rate of transfer can modify the model. Researchers use the diffusion constant of material D=KL2/2 and mobility = eD/KBT derive from

Einstein relation to calculate the mobility range in PCBM is from 0.001–0.08 cm2 V−1 s−1[6]. The k value is from k = 5.14 × 109 s−1 to k = 4.11 × 1011 s−1.

Second, scientists compare the results of LMD simulation with other theories. First, the researchers compare the LMD simulation results with Kramers'theory.

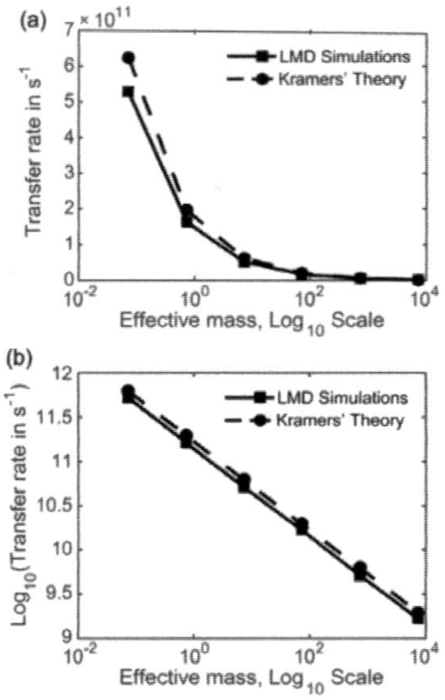

Figure 3 The results from LMD simulation and Kramers'theory [1]

The researchers suggest 2 possibilities that cause this discrepancy. First, the Kramers'theory relied on the theory of friction, but the LMD simulation does not make any simplification of the model. Second, Kramers'theory does not contain all information on the PES. On the other hand, LMD simulation incorporates the complete shape of all regions of the PES.

The second comparison is made between LMD simulation and Marcus theory (a method for calculating rates of electron transfer in outer-sphere processes [7]). The results show a big discrepancy that the outcome of Marcus's theory is 2 magnitudes larger than the LMD simulation. Scientists suggest that this difference is due to the rate of reorganization of polaron transfer. The Marcus theory relied on the assumption that the rate of reorganization is quicker than the rate of reaction. However, in the condition of a low reorganization rate, the accuracy of Marcus's theory decreases rapidly.

In conclusion, researchers state that the Marcus theory is appliable in low friction conditions while Kramers' theory is appliable in high friction circumstances. On the other hand, the LMD simulation is not constrained by the level of friction and can choose the right friction and effective mass for the experiment interest.

Last, the LMD simulation also provides an approach for scientists to investigate when and why the transfer occurs. For instance, the scientists show an experiment by using a different mass of pseudo-atom to show their discrepancy in transfer events.

The graphs exhibit below show that the pseudo-atoms with lower effective mass have a more frequent oscillation pattern and hop over the barrier slower than the pseudo-atoms with larger effective mass. In contrast, the pseudo-atoms with larger effective mass hop over the barrier quickly and have a relatively slow oscillation pattern.

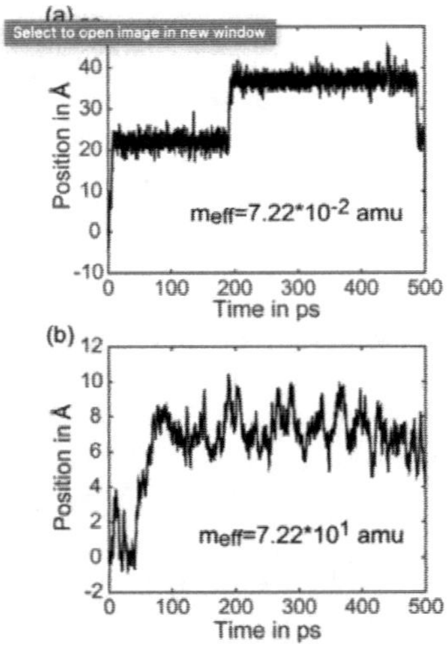

Figure 4 The movement of pseudo-atoms with different effective mass in 500ps time span [1]

3. The future of semiconductor and molecular dynamics

Molecular dynamics have a serious drawback. For example, the time span is too short for an experiment. The scale of molecular dynamics is usually in nanosecond and picoseconds. Though scientists are working on it to expand the scale to a higher time level, it will at least take researchers 15 years to expand the time limit to the micrometer level. Simulations of macromolecular assembly will also extend to include very large complexes [2] due to the rapid development of computer capability.

The future of semiconductors is also promising. The development of electronics facilitates the progress of semiconductors to be cheaper, efficient, and smaller. An organic conductor is a promising area in the future. They are flexible and widely used in printing technology. Yet, the low efficiency is the main obstacle to their further development. However, by using molecular dynamics to understand the mechanism in transporting charges through the organic semiconductor. The researchers can modify and improve the conductivity of organic semiconductors significantly. Besides, organic semiconductors are manufactured with multiple ingredients and compounds to minimize their production costs, so the organic semiconductors also have a huge potential in mass-scale production.

4. Conclusion

The scientists show that the main advantage of the LMD model was it can simulate the large scale of particles at a relatively low computational cost, allowing much larger volume simulation that is intractable in other simulation methods. The scientists patch a series of one-dimensional transportation events to simulate a three-dimensional transportation event. Though, there are multiple molecules adjacent to the molecules that charge reside. However, the simulation method can predict the results by utilizing the random numbers R(t) and accepting the particle that transfers most quickly to be the new position of pseudo-atom. The scientists show that morphology should be considered. The research group acknowledges that the PES was only performed by considering the orientation between 2 molecules. Yet, the overall morphology will have an effect on the molecules' friction and effective mass.

The scientists use the LMD model to derive the polaron motions in the organic semiconductor. They note that the interaction between charges and surrounding vibration is an active area for researchers to investigate the multiple properties of polaron in organic semiconductors. Besides, they also conclude that the low computational cost is the most important advantage of this model. At last, they admit that the transport rate of charges in organic semiconductors is the most concerning problem in the development of organic semiconductor manufacturing.

The investigation has more important meaning. The experiment successfully measures the multi parameters of the process of charge transfer inside the organic semiconductor. The high accuracy of a single event test and the capacity of large-scale experiments prove the capability of the molecular dynamic simulation in the research of semiconductor. The lower cost and more efficient and convenient operations will enable scientists to investigate the property and invent new organic semiconductor more quickly in the future.

References

[1] Kenley M. Pelzer (2017) Molecular dynamics and charge transport in organic semiconductors: a classical approach to modeling electron transfer
[2] David W.Borhani David E.Shaw(2011, Demcember 20)molecular dynamics simulations in drug discovery https://link.springer.com/article/10.1007%2Fs10822-011-9517-y
[3] Q. Wu and T. V. Voorhis, J. Phys. Chem. A, 2006, 110, 9212
[4] Kyle A. Baseden and Jessse W.Tye* (2014) Introduction to density functional theory: Calculation by Hand on the Helium Atom https://pubs.acs.org/doi/10.1021/ed5004788
[5] WHU (2012) what are quasi particles https://www.unimuenster.de/Physik.AP/Demokritov/en/Forschen/Forschungsschwerpunkte/mBECwaqp.html
[6] W. J. Grzegorczyk , T. J. Savenije , T. E. Dykstra , J. Piris , J. M. Schins and L. D. A. Siebbeles , J. Phys. Chem. C, 2010, 114 , 5182 —5186
[7] Valerie Shurtleff (2016 January 26) A very Brief introduction to the concepts of Marcus theory https://macmillan.princeton.edu/wp-content/uploads/VWS-Marcus.pdf thermo-bestehen.de(n.d.) Adiabat https://www.thermo-bestehen.de/adiabat

A study on the capture of a droplet impact on a fiber

Wenquan Gu[1a], Zhishan Bai[1b*]

[1]School of Mechanical and Power Engineering, East China University of Science and Technology, Shanghai 200237, China

[a]wenq_gu@mail.ecust.edu.cn, [b*]baizs@ecust.edu.cn

Abstract-The phenomenon of droplets captured by fibers is ubiquitous and plays an important role in natural and industrial applications. This study focused on the impact process during a droplet impact on a fiber and established a physical model to describe this process. The critical velocity of capture above which the droplet might not be captured after the impact was deduced according to this model and the influence factors were investigated. These findings will be helpful to elucidate the mechanism of the impact between droplets and fibers and optimize the design of coalescence applications.

1. Introduction

The phenomena of droplets impact on fibers are ubiquitous and commonly observed in natural and industrial applications: rain droplets impact on spider webs on rainy days [1], collect water through fiber meshes in deserts [2], filtrations applied to separate liquid in aerosol [3,4], and coalescence separation devices [5,6]. These impact phenomena attracted intensive research and the captured phenomenon has been a vibrant topic of engineering and science due to the extensive applications [7].

The study of droplets impact on fibers was derived from the droplets impact on surfaces, and the geometry of surfaces reduced from plane to fiber eventually [8-10]. The first research of droplets impact on fibers was performed by Hung et al [11], and they focused on the impact phenomena of large amounts of droplets impact on a fiber. The typical phenomena after the impact were classified as droplets disintegration and dripping, and a relationship between the phenomena and dimensionless numbers such as Weber number and Bound number was established. The influence of fiber wettability and the phenomena of droplets impact on fiber meshes were investigated in further research [12]. The dynamics of a droplet impact on a fiber were investigated by Lorenceau et al [13] and the study showed that the droplet could not be captured by the fiber when the impact velocity exceeded a critical value which was termed as the critical velocity of capture. This velocity was studied both experimentally and theoretically through forces analysis, and further studies showed that off-center impact and the inclined fibers could significantly increase this critical velocity which could enhance the capture ability of fibers [14, 15]. The impact between fibers and wetted fibers was investigated by Liang et al [16, 17], and some special phenomena such as rebound and spreading were observed. Comtet et al [18] and Dressaire et al [19] studied a droplet impact on a flexible fiber respectively and found that the capture ability of fibers could be improved by choosing an optimal fiber length. Kim et al [20] extended the impact phenomena by changing the impact velocity in a more wide range and the typical outcomes could be classified as capturing, single droplet falling, and splitting. Besides, the residual volume of droplets left on the fiber after falling was also investigated.

This research investigated the impact process of droplets impact on fibers and primarily focused on the capture phenomenon during the impact which is of vertical importance to the industrial applications.

Some models were established to describe the impact process and have good accuracy with the experiment results under certain conditions, however, they were not suitable for universal situations. The existing deviations could be quite obvious and some new models were needed to describe the impact process. Therefore, we studied the impact process based on the experiment phenomena and developed a new model aiming at calculating the critical velocity of capture more precisely. Innovation through the experiment data available in the literature and the influence of droplets and fiber properties on the critical velocity was researched through numerical calculations. These studies will be helpful to elucidate the mechanism of the impact between a droplet and a fiber and optimize the design of industrial applications.

2. Theoretical Model

The phenomena of a droplet impact on a fiber could be briefly classified as capture phenomenon and detach phenomenon shown in Fig. 1 according to previous studies [11, 20]. The droplet kept moving after impact on the fiber and the capture phenomenon occurred when the impact velocity was smaller than a critical value which was regarded as the critical velocity of capture v_c which was first proposed by *Lorenceau* et al [13] and many factors such as surface wettability [20], impact position (the eccentric of droplets and fibers [14]) and the impact angles [15] could have a significant influence on it. The impact process we investigated in this study began at the contact of the droplet and the fibers and ended at the status that the droplet was just about to leave the fiber and these two situations correspond to *Status A* and *Status B* in Fig. 1 respectively.

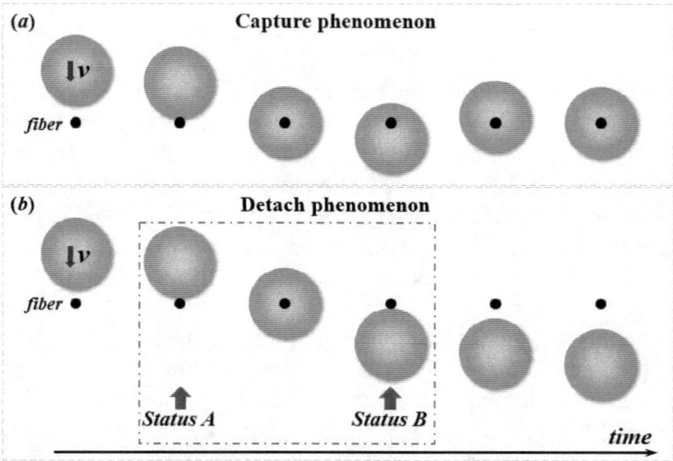

Fig. 1 The schematic of capture phenomenon and detach phenomenon of a droplet impact on a fiber

According to previous studies, a theoretical model was established to describe the impact process and this model was established through the droplet dynamics [13]: the movement of the droplet subjects to the *Newton's Second Law* and the forces acting on the moving droplet were gravity, drag force caused by the fiber and the capillary force caused by the surface tension. Hence, the critical velocity could be solved by the differential equation mathematically, and this model was a significant innovation and important research for subsequent studies. However, this model simplified the forces during the impact process by neglecting the change of drag force and capillary force which could bring in significant deviations. In this section, we analyze the forces during the impact process more preciously and establish a new model to describe the movement of the droplet after the impact.

When a droplet impact on a fiber, the droplet moves due to inertance and forces such as gravity, drag force, and capillary force. The gravity forces the droplet to leave the fiber while other forces resist the droplet from detaching. We established a coordinate system to investigate the impact process quantitatively, and the forces were shown in Fig. 2.

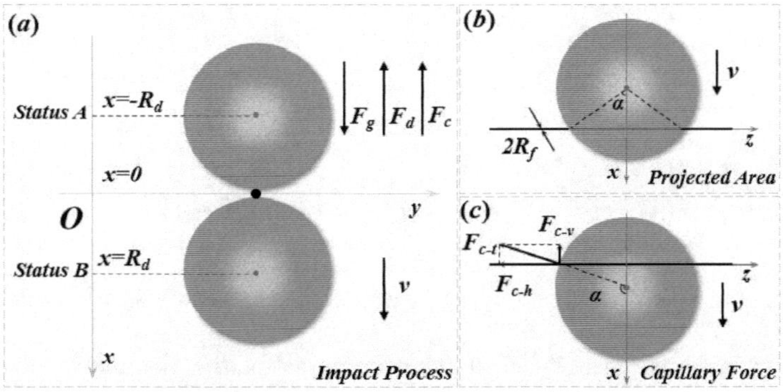

Fig. 2 The coordinate system and forces acting on the droplet

We donated x as the axis pointing downward and set the original point at the center of the fiber as shown in Fig. 2(a). The droplet was assumed spherical during the impact process and much larger than the fiber during this study and moved downwards. Hence, the impact starts at the initial position of $x=-R_d$ while the final position was $x=R_d$. Besides, the droplet was regarded as detached from the fiber entirely which meant the gravity of the droplet kept constant during the impact process and can be calculated as:

$$F_g = \frac{4}{3}\pi R_d^3 \rho g \tag{1}$$

Where R_d is the droplet radius; g is the local gravitational acceleration; ρ is the density of the droplet. However, the other two forces were difficult to calculate due to the movement of the droplet.

Drag force was primarily determined by velocity and the projected areas of contact area between droplet and fiber according to the previous study [13], and can be written as:

$$F_d = C_d \rho v^2 A_p \tag{2}$$

Where C_d is the drag force coefficient; v is the droplet velocity during the impact process; A_p is the projected area in the moving direction and is exhibited as the red line in Fig. 2(b). Due to v and A_p being ever-changing during the impact, the drag force is very difficult to calculate. To solve this problem, we import angle α shown in Fig. 2(b) to calculate the projected area. This angle changes from 0° to 180° during the impact process. Therefore, the projected area A_p can be calculated as:

$$A_p = 4R_d R_f \sin\alpha \tag{3}$$

Considering the geometrical relationship and the coordinate system in Fig. 2(b), the cosine of angle α can be described as $-x/R_d$. Hence, the sine of angle can be expressed as:

$$\sin\alpha = \sqrt{1-\left(\frac{-x}{R_d}\right)^2} \tag{4}$$

Where x is the coordinate of the droplet center, and R_d is the droplet radius. It is difficult to obtain an exact analytical solution by utilizing Eq. (4) and some simplifications are indispensable to get an explicit solution.

Considering $\sin\alpha$ increases from 0 to 1 and decreases to 0 again during the impact process, we assume $\sin\alpha$ changes linearly and we simplify the equation as:

$$\sin\alpha = 1-\left|\frac{-x}{R_d}\right| \tag{5}$$

The comparison between the simplified expression and the original relationship is shown in Fig. 3. The red line represents Eq. (4) while the blue line represents Eq. (5).

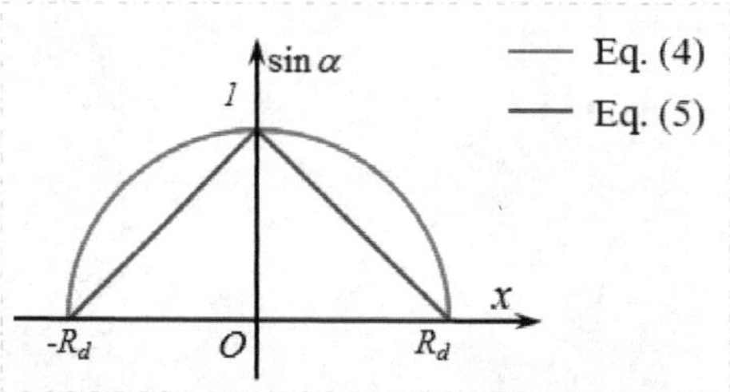

Fig. 3 The comparison between the simplified expression and the original relationship

Although there are differences between the two expressions, the changing tendency and the values at the threshold points are the same which means this simplification is quite acceptable. Hence, the drag force can be written as:

$$F_d = 2C_d \rho v^2 R_f R_d \left(1 - \left|\frac{-x}{R_d}\right|\right) \tag{6}$$

The capillary force is another crucial factor during the impact process and is caused by the surface tension due to the contact of the droplet and the fiber. The maximum value of this force is $4\pi\sigma R_f$ when the droplet is about to detach according to previous research [13]. As noted in Fig. 2(c), the capillary force can be decomposed to horizontal component and vertical component which has a significant influence on the droplet movement and the total capillary force in a vertical direction can be written as:

$$F_c = 4\pi\sigma R_f \cos\alpha \tag{7}$$

Besides, the cosine of α can be expressed as $\cos\alpha = -x/R_d$ according to the geometrical relationship between the droplet and the fiber, and this force:

$$F_c = -4\pi\sigma R_f x / R_d \tag{8}$$

During the impact, the droplet moves under the coefficient of forces above and the impulse cause the momentum to change:

$$M(v_B - v_A) = (F_g - F_d - F_c)\Delta t \tag{9}$$

Where v_B is the velocity at *Status B* and v_A is the velocity at *Status A*. For the differential form of Eq. (9):

$$\frac{dv^2}{dx} - \frac{2}{M}\left(\frac{4}{3}\pi R_d^3 \rho g - 2C_d \rho v^2 R_f \left(R_d - |x|\right) - 4\pi\sigma R_f x / R_d\right) = 0 \tag{10}$$

This equation is the description of the droplet velocity changes with displacement after impact. When the droplet exactly detaches from the fiber, the droplet velocity at the maximum displacement position $x=k$ reduces to $0 m/s$ which means $v(x=k) = 0$. Under this condition, the droplet velocity can be solved and can be written as:

$$v^2 = e^{-\frac{b(2R_d - x)x}{2}} \left(\frac{a e^{R_d b x}}{b\sqrt{e^{bx^2}}} + \frac{(c + aR_d)\sqrt{2\pi e^{bR_d^2}}\, erf\left(\frac{\sqrt{2b}}{2}(R_d - x)\right)}{2\sqrt{b}} - \frac{a}{b}\sqrt{e^{bR_d^2}} \right) \tag{11}$$

Where $a = \frac{6\sigma R_f}{\rho R_d^4}$, $b = \frac{6C_d R_f}{\pi R_d^3}$, $c = -2g$, are parameters related to the system properties and *erf(x)* is the *Gauss error function* and the expression is:

$$erf(x) = \frac{2}{\sqrt{\pi}} \int_0^x e^{-\eta^2} d\eta \tag{12}$$

The Eq. (12) is the droplet velocity change law that the droplet exactly detach from the fiber and the impact velocity under this condition can be calculated by calculating the velocity at the position $x=k$ and this is the critical velocity of capture v_c that can be calculated as:

$$v_c = \left[\frac{a}{b}\left(1-\sqrt{e^{4bR_d^2}}\right) + \frac{(c+aR_d)\sqrt{2\pi e^{4bR_d^2}}\, erf\left(\sqrt{2bR_d^2}\right)}{2\sqrt{b}} \right]^{1/2} \tag{13}$$

Thus, Eq. (13) is the critical impact velocity above which the droplet may not be captured by the fiber and can be obtained for the given droplet radius, fiber radius, droplet surface tension, and droplet density. This velocity is of significance to the design and optimization of industrial filter facilities.

3. Result and Discussion

3.1 Numerical validation

The accuracy of the above model is verified in this section by comparing the computational results with experimental data in the reference [13]. The experimental data were performed under a certain condition and the parameters were shown in Tab. 1.

Tab. 1 Parameters of experiment

Physical Parameters	Unit	Water
Density ρ	kg/m^3	990
Viscosity μ	Pa·s	0.897×10^{-3}
Surface tension σ	N/m	72.1×10^{-3}
Gravitational acceleration g	m/s^2	9.81

Based on the above experiment condition, we made a simplifying assumption that the surface morphologies of fiber were smooth and neglected the influence of fiber surface roughness to facilitate comparisons with the experiment data. The comparisons of numerical results with experimental data are shown in Fig. 4.

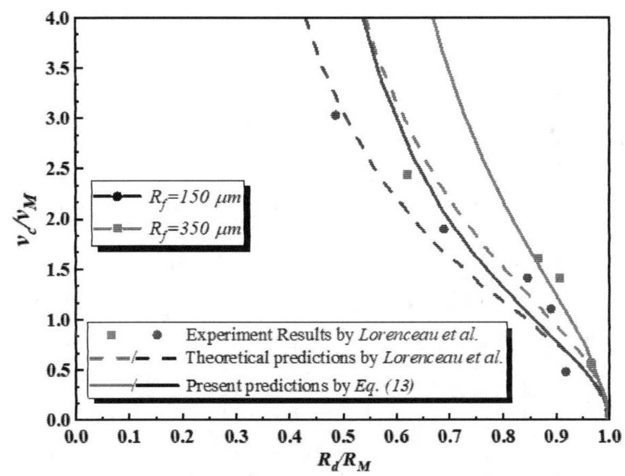

Fig. 4 The comparison of numerical results and experiment data

The abscissa of Fig. 4 is the ratio of droplet size R_d to the maximum droplet size R_M which is calculated by Eq. (13), and the ordinate is the ratio of the critical velocity of capture v_c to the capillary characteristic velocity v_M which is defined by $v_M = \sqrt{4gR_0}$ that R_0 is a characteristic length related to fiber radius R_f and the capillary length $\kappa^{-1} = \sqrt{\sigma/\rho g}$, and the relationship is $R_0 = 3^{1/3} R_f^{1/3} \kappa^{-2/3}$ [13].

As noted in Fig. 5, the prediction of this physical model (solid line) is consistent with the experimental data obviously and has better accuracy than the model in references (dash line) [13]. The scatters are the experimental data obtained from the reference. The predicted velocity decreases with the increase of droplet size which is in accord to the experiment data. The predicted value decreases sharply at first and then experiences a relatively smooth decrease stage and then decreases sharply again. The predicted results exhibit good agreement with the experimental data. Furthermore, with the increase of droplet size, the critical velocity gradually decreases to *0* which means when the droplet may not be captured by the fiber when the droplet size is larger than a critical size.

Besides, as shown in the graph, there still exist errors between the predicted values and experiment values. The main reason for the deviation may be our simplified assumptions in calculating the drag force and capillary force and the experiment data may be imprecise due to the experiment and test limit.

Although there are some deviations between the predicted results of the current model and the experimental data, the predictions are completely acceptable. In particular, the predictions under the situation $R_d/R_M>0.8$ are very consistent with the experimental results which means this model is much more suitable for larger droplets. Therefore, this model can be used to analyze the detaching behavior qualitatively and is very useful for many practical applications, such as providing a useful guideline for the design of coalescence separation, fibrous filters, and fog harvesting equipment.

3.2 The influence of Oh number on the behavior of droplet impact

The impact process is influenced by the droplet properties according to experimental results and theoretical analysis. Under this condition, the droplet properties such as viscosity, density, and surface tension, and size (radius) can be described through a dimensionless parameter *Ohnesorge Number* which is the description of the relative influence of viscous versus capillary-inertial effects.

To better understand how *Oh* influences the impact process, the variations in the critical velocity of capture v_c with *Oh* on four different fibers were investigated here through numerical calculation according to Eq. (13). In this work, *Oh* was changed by changing the droplet size R_d and varied from 0.003-0.010 and shown in Fig. 5.

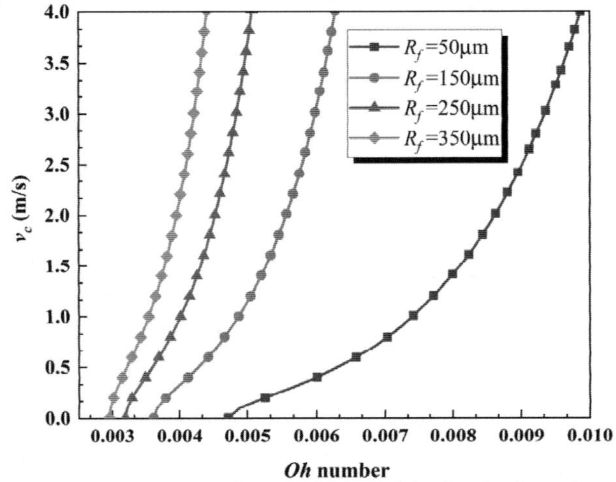

Fig. 5 Variations in *Oh number* with the critical velocity of capture v_c

The dependence of the critical velocity on *Oh* number is plotted in Fig. 5. The variations of critical velocity of capture v_c with *Oh number* on four different fibers are significantly different, and the critical velocity increases apparently with the increase of fiber size. Under the condition *Oh* is 0.004, the critical velocity changes from 0.3m/s to about 2.3m/s when the fiber changes from 50μm to 250μm. Besides, v_c increases more quickly when the fiber is larger. When *Oh* changes from 0.0035 to 0.004, v_c changes from 0.9m/s to 2.1m/s when R_f=350μm and changes from 0.4m/s to 1.0m/s when R_f=250μm.

Moreover, the critical velocity may decrease to 0 with the decrease of *Oh* which means the droplet may not be captured by fibers when *Oh* is less than a critical value, and this value increases with the decrease of fiber size. As *Oh* reflects the relative influence of viscous versus capillary-inertial effects, decreasing the droplet viscosity and increasing the surface tension or droplet size all will lead to the decrease of *Oh* and further influence the critical velocity.

In a word, the v_c increases with the increase of *Oh* and fiber size which means increasing the droplet viscosity and decreasing the surface tension, droplet size all will lead to v_c increase and benefit to the droplet captured by fibers. These will be useful to the design of filters and other industrial applications.

3.3 The influence of fiber properties

The fiber size R_f has a significant influence on the forces during the impact process and then affects the movement of the droplet. Accordingly, the change of fiber sizes needs to be studied and to lucubrate the influence of R_f, the variations in the v_c/v_M with R_d/R_f is also investigated under four different conditions and shown in Fig. 6.

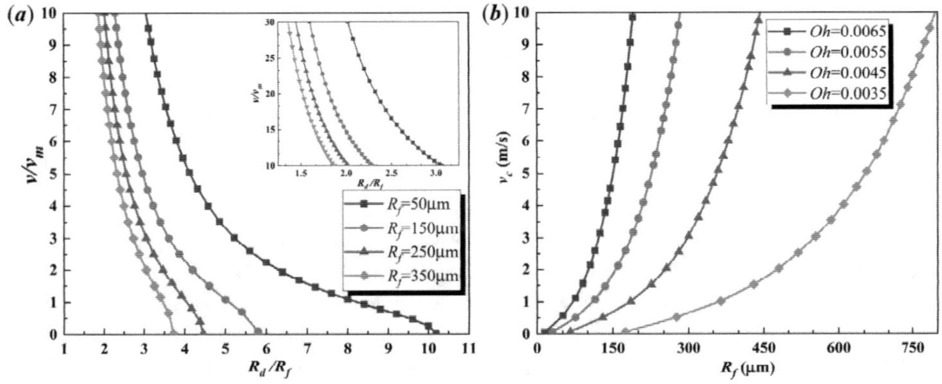

Fig. 6 (*a*) Variations in radius ratio with velocity ratio. (*b*) Variations in fiber radius with the critical velocity of capture

As shown in Fig. 6(a), the critical velocity increases with the increase of fiber size, and the increasing velocity is larger when *Oh* is larger. These findings attributed to the fact that larger fibers are more easily to capturer larger droplets and larger impact velocities. This means increasing the fiber size is good in enhancing the capture of a droplet.

As shown in Fig. 6(b), the dimensionless velocity decreases with the increase of the ratio of droplets to fiber, and eventually decreases to zero. Besides, the dimensionless velocity is much larger when the fiber is small under the same droplet ratio. All these indicate that the small fiber is a benefit to the capture of droplets, the droplets range is large and the relative velocity is small which is in favor of detaching after being captured by fiber in industrial applications.

To show the influences of *Oh* and R_f on the velocity more intuitively and understand these factors preferably, the variations of threshold capture velocity with *Oh* and R_f were plotted in Fig. 7.

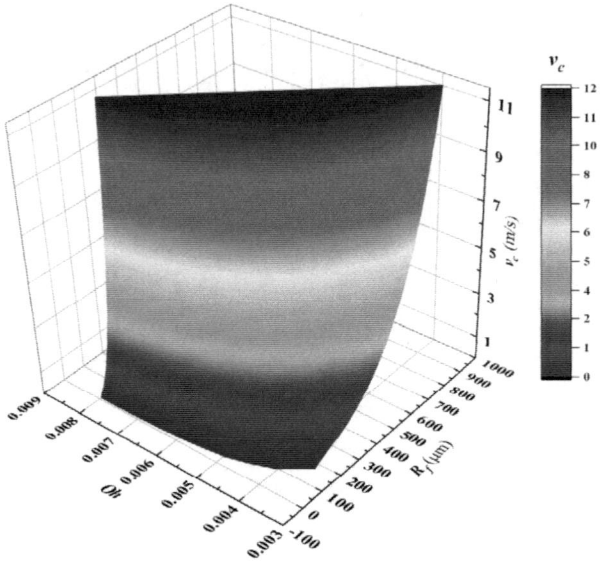

Fig. 7 The variations in fiber radius and Oh number in critical velocity of capture

As noted in Fig. 7, the changing law of the critical velocity and the fiber radius and *Oh* can be observed obviously. The velocity increases with the increase of fiber radius and *Oh*. The velocity decreases to zero when the fiber is less than a value or *Oh* is less than a value which means this droplet cannot be captured by fiber under this condition. Accordingly, to enhance the capture of a droplet, we can consider it from fiber and droplets respectively. Increasing the fiber radius properly may augment the critical velocity and then enhance the capture ability; increasing the droplet viscosity and decreasing the surface tension or droplet radius all will increase *Oh* and then enhance the capture ability of fiber to some degree. These are the theoretical directions and the specific operations during the industrial applications need to consider the real operation condition.

4. Conclusion

In this work, the impact process between a droplet and a fiber was investigated and a new theoretical model was established to study the critical velocity of capture. This model takes the specific change of forces during the process and makes reasonable simplifications that the theoretical values have good accuracy with the experiment results. This model will be helpful to investigate the mechanism of a droplet impact on a fiber.

In addition, the factors that influence the threshold velocity of capture were also discussed, such as the effects of fiber and droplets properties on the critical velocity of capture. The critical velocity increases with the increase of fiber size which could increase the capture of a droplet when impact. And the droplet properties influence the critical velocity very much: when increasing the droplet size and decreasing the droplet viscosity and increasing the droplet surface tension all will lead to the increase of critical velocity of capture which may be a new thought to increasing the critical velocity of capture. These investigations are of great importance to industrial applications.

Acknowledgments

This work was supported by the National Natural Science Foundation of China, China (51578239).

References

[1] Zheng Y, Bai H, Huang Z, et al. Directional water collection on wetted spider silk. Nature. 2010;463(7281):640-643.

[2] Shanyengana ES, Sanderson RD, Seely MK, Schemenauer RS. Testing greenhouse shade nets in collection of fog for water supply. Journal of Water Supply: Research and Technology-Aqua. 2003;52(3):237-241.
[3] Contal P, Simao J, Thomas D, et al. Clogging of fibre filters by submicron droplets. Phenomena and influence of operating conditions. Journal of Aerosol Science. 2004;35(2):263-278.
[4] Kampa D, Wurster S, Meyer J, Kasper G. Validation of a New Phenomenological "Jump-and-Channel" Model for the Wet Pressure Drop of Oil Mist Filters. Chemical Engineering Science 2015, 122, 150–160.
[5] Li J, Gu Y. Coalescence of Oil-in-Water Emulsions in Fibrous and Granular Beds. Separation and Purification Technology 2005, 42 (1), 1–13.
[6] Šećerov Sokolović R M, Govedarica D D, Sokolović D S. Separation of Oil-in-Water Emulsion Using Two Coalescers of Different Geometry. Journal of Hazardous Materials 2010, 175 (1), 1001–1006.
[7] Abishek S, Mead-Hunter R, King A, Mullins Benjamin. Capture and re-entrainment of microdroplets on fibers. Physical Review E 2019, 100.
[8] Levin Z, Hobbs PV. Splashing of Water Drops on Solid and Wetted Surfaces: Hydrodynamics and Charge Separation. *Philosophical Transactions of the Royal Society A: Mathematical, Physical and Engineering Sciences*. 1971; 269(1200):555-585.
[9] Dear JP, Field JE. High - speed photography of surface geometry effects in liquid/solid impact. *Journal of Applied Physics*. 1988; 63(4):1015-1021.
[10] Bussmann M, Chandra S, Mostaghimi J. Droplet Impact onto Arbitrary Surface Geometries. ASME Fluids Engineering Division Summer Meeting. 1997; FEDSM97-3073.
[11] Hung LS, Yao SC. Experimental investigation of the impaction of water droplets on cylindrical objects. *International Journal of Multiphase Flow*. 1999; 25(8):1545-1559.
[12] Hung LS, Yao SC. Dripping phenomena of water droplets impacted on horizontal wire screens. *International Journal of Multiphase Flow*. 2002; 28(1):93-104.
[13] Lorenceau É, Clanet C, Quéré D. Capturing drops with a thin fiber. *Journal of Colloid and Interface Science*. 2004; 279(1):192-197.
[14] Lorenceau E, Clanet C, Quéré D, Vignes-Adler M. Off-centre impact on a horizontal fibre. *Eur Phys J Spec Top*. 2009; 166(1):3-6.
[15] Piroird K, Clanet C, Lorenceau É, Quéré D. Drops impacting inclined fibers. *Journal of Colloid and Interface Science*. 2009; 334(1):70-74.
[16] Liang G, Guo Y, Yang Y, Guo S, Shen S. Special phenomena from a single liquid drop impact on wetted cylindrical surfaces. *Experimental Thermal and Fluid Science*. 2013; 51:18-27.
[17] Liang G, Yang Y, Guo Y, Zhen N, Shen S. Rebound and spreading during a drop impact on wetted cylinders. *Experimental Thermal and Fluid Science*. 2014; 52:97-103.
[18] Comtet J, Keshavarz B, Bush JWM. Drop impact and capture on a thin flexible fiber. *Soft Matter*. 2016; 12(1):149-156.
[19] Dressaire E, Sauret A, Boulogne F, Stone HA. Drop impact on a flexible fiber. *Soft Matter*. 2016; 12(1):200-208.
[20] Kim S-G, Kim W. Drop impact on a fiber. *Physics of Fluids*. 2016; 28(4):042001.

Study on Compatibility of Waterborne Surface Treatment Coatings

Wenbo Li [1], Yingfei Yang [2], Dongdong Song[2], Jiran Zhu[1], Yi Xie[1]

(1. State Grid Hunan Electric Power Company Limited Research Institute, 388 ShaoShan N Rd, Changsha 410000, China; liwb@alum.imr.ac.cn;

(2. Institute of Advanced Wear & Corrosion Resistant and Functional Material, Jinan University, 601 Huangpu W Ave, Tianhe District, Guangzhou 510632, China)

Abstract: the severe corrosion of power grid equipment has always been an important factor affecting the service life of equipment. Rust transfer technology is used to treat the corroded equipment to improve the corrosion resistance of transmission grid cabinet equipment in high temperature and high humidity environment. Taking acid modified acrylic acid as the main component, the composite conversion solution was obtained by adding phosphoric acid and phytic acid. Through the surface morphology, electrochemical test and adhesion test after rust conversion, the universality and corrosion resistance of the rust conversion solution were analyzed; Combined with zinc phosphate primer and zinc rich primer, the effects of rust conversion solution on the adhesion and salt spray corrosion resistance of the two commercial primers were studied. The compound conversion solution has good rust conversion effect on atmospheric corrosion rust layer, salt spray corrosion rust layer and damp heat corrosion rust layer; The corrosion resistance and adhesion of atmospheric corrosion rust layer are obviously improved after being treated with rust conversion solution. The adhesion of zinc phosphate primer and zinc rich primer on rust conversion solution is 2.1 times and 0.8 times higher than that of atmospheric corrosion rust layer without rust conversion, respectively. The compound rust conversion solution has strong universality and can convert atmospheric corrosion, salt spray corrosion and damp heat corrosion rust layers. At the same time, it has good corrosion resistance, which can significantly improve the corrosion potential of the corroded surface and reduce the corrosion current density. In addition, the composite rust conversion solution can significantly improve the adhesion and corrosion resistance of the primer coating.

1 Introduction

Infrastructure networks, such as transmission grid and communication network, are the blood of industrial production and people's life. Ensuring the safe and reliable operation of infrastructure networks such as power grid is a necessary prerequisite to ensure economic activities. When the power grid operates in high humid and hot natural environment and industrial pollution environment, important control components such as metal cabinet equipment are also faced with corrosion and accelerated damage caused by long-term condensation in humid atmosphere. Aiming at the problem of metal corrosion protection, organic anti-corrosion coatings are the most effective and feasible means at present because of their shielding effect [1], chemical protection effect [2], electrochemical protection [3]. However, research shows that it is necessary to ensure good coating / metal interface bonding in order to realize the function of anti-corrosion coating [4].

Common substrate pretreatment includes mechanical method and chemical derusting method [5], which can ensure good bonding of coating / metal interface. However, due to the limitations of equipment, site and environmental protection in the field, effective substrate treatment cannot be carried out. Therefore, low surface treatment coatings have become a research hotspot. Rust coating, also known as low surface tolerance coating, can directly stabilize, passivate or transform the rust layer, so that the active rust layer becomes a stable substance, so as to achieve the dual purpose of rust removal and protection of metal surface. At present, there are three types of rust coatings widely studied: permeable [6,7], stable [8,9] and conversion [10,11]. Some products developed for solvent based epoxy resin in China have excellent performance [12,13], but they are not conducive to popularization and use due to environmental protection restrictions.

Waterborne coatings are an important direction for the development of coatings industry in the future. Phosphoric acid [14], phytic acid [15] and phosphate functional components [16] are the most commonly used functional components in low surface treatment coatings, but most of these substances are acidic, resulting in hidden dangers in the stability of aqueous film-forming system, which limits the development of high-performance waterborne low surface treatment coatings. Based on this, this paper uses the new modified acrylic resin as the film-forming system, combined with the functional components of phosphoric acid, phytic acid and phosphate, to trial produce the waterborne low surface treatment coating. By studying the synergistic effect of each functional component and its compatibility with the resin system, and analyzing the matching with the subsequent coating, it provides a basis for the development of the new waterborne low surface treatment coating.

2 Experimental procedure

2.1 coating preparation

Chemical composition of Q235 steel used (calculated by mass fraction): C $\leqslant$ 0.22%, Mn $\leqslant$ 1.4%, Si $\leqslant$ 0.35%, s $\leqslant$ 0.05%, P $\leqslant$ 0.05%, Fe balance. The pretreatment process is: soak in degreasing solution for 40 min → clean with water → grind with sandpaper to No. 150 → wash with alcohol → dry in air for standby. The substrate is treated in three ways to form rust layer as coating substrate: 1 indoor damp and heat, 2 indoor salt spray, 3 natural exposure.

Film forming resin system: acid modified acrylic resin. The drugs used in this paper are phosphoric acid, phytic acid, zinc phosphate, propionic acid, hexamethylenetetramine and glycerol.

The sample preparation steps are as follows: 1) dissolve a certain amount of phosphoric acid, phytic acid and propionic acid mixture, a certain amount of zinc phosphate and hexamethylenetetramine mixture and a certain amount of glycerol in an appropriate amount of deionized water, stir at 1000 rpm for 1 h to obtain component A; 2) adding a certain amount of silicone modified defoamer in a certain amount of acid modified acrylic resin emulsion to obtain 0.5h B at low speed. 3) Mix components a and B, stir evenly at low speed and stand still for 20min; 4) Apply to the surface of rusted substrate.

2.2 performance test and organization observation

1) The coating was observed by phenom prox scanning electron microscope, digital camera and Leica dmi3000m optical microscope, and the elements and components were analyzed by EDS attached to the scanning electron microscope.

2) Use Ultima IV; RIGAKU corporationxrd was used to analyze the composition before and after rust transformation.

3) Chenhua e604 was used for electrochemical test, the solution was tested with 3.5% NaCl, and the corrosion properties before and after rust transformation were analyzed.

4) The coating adhesion was tested with defelko at-m adhesion tester.

5) Salt spray test shall be conducted for the coating according to the national standard GB t 1771-2007 determination of resistance to neutral salt spray of paints and varnishes.

3 Results and discussion

3.1 surface morphology of different rust layers after rust transformation

Atmospheric corrosion (outdoor exposure for one month), salt spray corrosion (neutral salt spray corrosion for 72 h) and damp heat corrosion respectively (humidity 85%, temperature 80 °C 200 h) the surface is coated with compound conversion solution. The micro morphology after rust conversion is shown in Figure 1. It can be seen from the figure that the rust layer of conversion solution has good film-forming effect under three conditions, and there are a small amount of cracks and bulges on the surface. The existence of microcracks can increase the permeability of the coating, and bulges can increase the surface roughness and improve the adhesion of the coating The composite conversion solution has good conversion function for the surface of a variety of corrosion products.

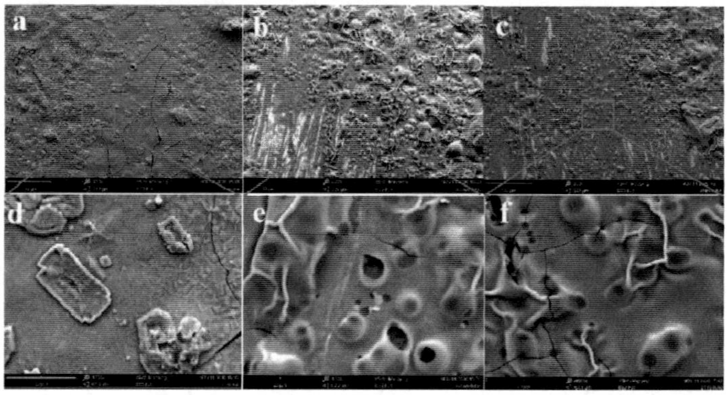

Fig. 1 surface morphology of different rust layers after rust Transformation: (a, d) atmospheric corrosion rust layer, (B, e) salt spray corrosion rust layer and (C, f) damp heat corrosion rust layer

3.2 binding performance of rust conversion solution after treatment

In order to analyze the adhesion between the rust conversion layer and the substrate, the adhesion of the rust conversion layer was tested. First, a pull-out test was carried out. However, after the rust conversion layer completely reacts and is stable, the surface energy of the film layer is low and the adhesion with the adhesive is poor. The pull-out test is due to the fracture between the adhesive and the film surface, and the real bonding condition of the film layer cannot be measured, as shown in Fig. 2 (a) and Fig. 2 (b), indicating that the conversion layer has a good bonding with the substrate.

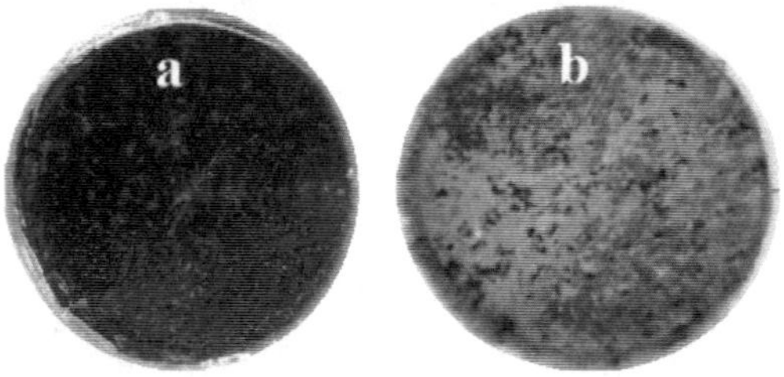

Fig. 2 drawing test results of atmospheric corrosion rust layer after rust Transformation: (a) sample surface, (b) test column surface

3.3 electrochemical performance after rust conversion solution treatment

After atmospheric corrosion The polarization curve and impedance spectrum of the (actual exposure) rust layer before and after rust transformation are shown in Fig. 3. After the treatment of the composite conversion solution, the corrosion potential of the sample increases significantly and the corrosion current decreases slightly. The specific electrochemical results are shown in Table 1. The results of the planned curve show that the corrosion resistance of the sample is significantly improved after the treatment of the composite conversion solution. The composite corrosion conversion solution is used for conversion The corrosion potential increased obviously and the corrosion current decreased obviously, indicating that the corrosion conversion layer has a good protective effect. The results of electrochemical impedance spectroscopy showed that the composite corrosion conversion solution improved the corrosion resistance of the corroded substrate.

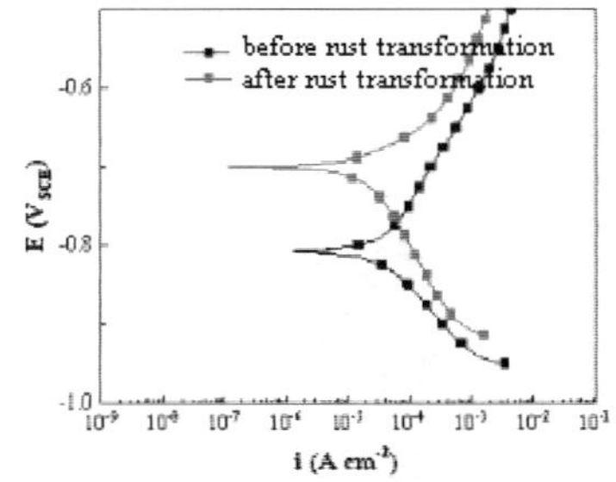

Fig. 3 polarization curve of atmospheric corrosion rust layer after rust transformation

Table 1 electrochemical parameters of actual exposed samples and rust transformation

number	corrosion potential (mV)	Corrosion current (10-5A/cm²)
Actual exposure sample	-806	3.54
After rust transformation	-698	0.89

3.4 effect of rust conversion solution on coating adhesion

In order to verify the practical application performance of the rust conversion solution, commercial zinc phosphate and zinc rich primer were selected as reference materials and coated on the surface of the test piece after rust conversion. It is compared with coating two kinds of commercial anti-corrosion paint on the rusted surface after three days of salt spray. As shown in Fig. 4, it can be seen that when zinc phosphate antirust coating is directly applied on the rusted surface, the rusted part of the substrate penetrates into the coating surface, forming unevenly distributed yellow corrosion points on the coating surface; After rust conversion, the surface is coated with zinc phosphate antirust coating with uniform and complete color distribution. When zinc rich primer is directly applied on the corroded surface, the color distribution of the sample surface is uniform after the paint film is cured, but the surface roughness is uneven. In contrast, the surface after rust conversion is uniform and complete after zinc rich primer is applied.

Fig. 4 surface morphology of atmospheric corrosion rust layer coated with primer and rust conversion solution plus primer, surface morphology with zinc phosphate antirust primer (a), with zinc rich primer (b), with rust conversion solution and zinc phosphate antirust primer (c), with rust conversion solution and zinc rich primer (d).

The results of pull-out test on the samples coated with primer on the surface of atmospheric corrosion rust layer and coated with primer after rust conversion are shown in Figure 5. It can be seen from the figure that when zinc phosphate primer is coated on the surface of the sample after rust conversion, its adhesion is more than twice that of the sample without rust conversion. The adhesion of zinc rich primer coated on the surface after rust conversion is nearly twice that of the sample without rust conversion. The above results show that rust conversion solution can significantly improve the adhesion of primer.

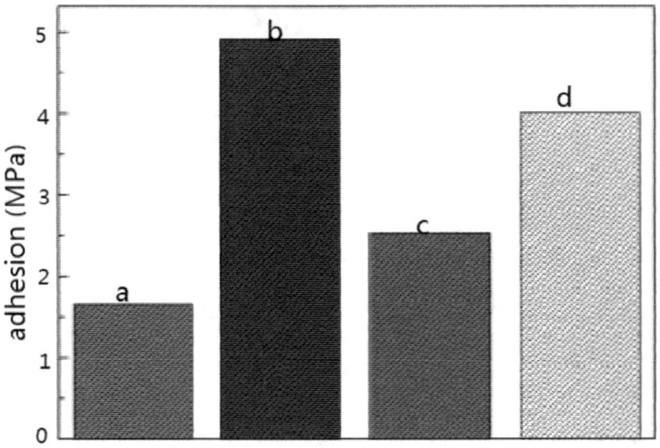

Fig. 5 adhesion of atmospheric corrosion rust layer coated with primer and rust transfer solution plus primer, adhesion with zinc phosphate antirust primer (a), with rust conversion solution and zinc phosphate antirust primer (b), with zinc rich primer (c), with rust conversion solution and zinc rich primer (d).

In view of the excellent conversion ability of the rust conversion solution, it was applied on site in combination with zinc rich anticorrosive paint . It is applied to the side cabinet plate and lower cabinet plate of distribution network cabinet equipment.

Fig 6. field application of rust conversion solution and zinc rich anticorrosive paint in rust resistance of cabinet equipment

4 Conclusion

In this paper, the rust conversion solution with acrylic resin as the main component is developed, and the universality, corrosion resistance and adhesion to primer of the rust conversion solution are tested. The following conclusions are drawn:

(1) Rust conversion solution has strong universality, which can convert atmospheric corrosion, salt spray corrosion and damp heat corrosion rust layer;

(2) Rust conversion solution has good corrosion resistance, which can significantly improve the corrosion potential of corroded surface and reduce the corrosion current density;

(3) Rust conversion solution can significantly improve the adhesion and corrosion resistance of primer coating.

References

[1] YH. Wu. Correlation Between Steel Corrosion and Coating Wet Adhesion[J]. Synthetic Materials Aging and Application, 31 (2002) 28-31.

[2] X. Lu, Y. Zuo, X. Zhao, et al. The influence of aluminum tri-polyphosphate on the protective behavior of Mg-rich epoxy coating on AZ91D magnesium alloy[J]. Electrochimica acta, 93 (2013) 53-64.

[3] D. Xie, J. Hu, S, Tong, et al. The development of Zinc-Rich paints[J]. Journal of Chinese Society for Corrosion and Protection, 24 (2004) 314-320.

[4] I. Gonzalez, D. Mestach, J.R. Leiza, et al. Adhesion enhancement in waterborne acrylic latex binders synthesized with phosphate methacrylate monomers[J]. Progress in Organic Coatings, 61 (2008) 38-44.

[5] F. Deflorian, L. Fedrizzi. Adhesion characterization of protective organic coatings by electrochemical impedance spectroscopy[J]. Journal of adhesion science and technology, 13 (1999) 629-645.

[6] C. Peebles, J. Garcia, A.P. Tornheim, et al. Chemical "Pickling" of phosphite additives mitigates impedance rise in Li ion batteries[J]. The Journal of Physical Chemistry C, 122 (2018) 9811-9824.
[7] V. Guillaumin, D. Landolt, Effect of dispersion solution on the degradation of a water borne paint on steel studied by scanning acoustic microscopy and impedance[J]. Corrosion science, 44 (2002) 179-189.
[8] X. ZU, J. HU, F. WANG, et al. Research progress of environment-friendly anti-rust coatings [J]. Chemical Industry and Engineering Progress, 9 (2008).
[9] V. Vetere, M. Deya, R. Romagnoli, et al. Calcium tripolyphosphate: An anticorrosive pigment for paint[J]. Journal of Coatings Technology, 73 (2001) 57-63.
[10] N. Li, H. Lv, W. Fu, et al. Preparation and Performance of a New Waterborne Rust Converting Agent[J]. Contemporary Chemical Industry, 47 (2018) 56-59.
[11] J.-M. Yeh, K.-C. Chang, Polymer/layered silicate nanocomposite anticorrosive coatings[J]. Journal of Industrial and Engineering Chemistry, 14 (2008) 275-291.
[12] L. Ning, Y. Chen, K. Yang, et al. Research and Application of a Highly Adaptable Anticorrosion Coating for Transmission Tower[J]. Corrosion Science and Protection Technology, 2019(2) 149-154.
[13] K. Xu, J. Fang, B. Lian. Study on Preparation and Properties of High Performance Surface-Tolerant Epoxy Coating [J]. Paint & Coatings Industry, 2019(4):27-33.
[14] Singh D D N, Bhattacharya D. Performance and mechanism of action of self-priming organic coating on oxide covered steel surface[J]. Progress in Organic Coatings, 2010, 67(2):129-136.
[15] Li J, Ge S, Wang J, et al. Water-based Rust Converter and its Polymer Composites for Surface Anticorrosion[J]. Colloids and Surfaces A Physicochemical and Engineering Aspects, 2017:S0927775717309408.
[16] Feng L, Yuan P. Corrosion protection mechanism of aluminum triphosphate modified by organic acids as a rust converter[J]. Progress in Organic Coatings, 140(2020): 105508.

Research of Water Vapor Barrier Property Testing Methods for Hollow Containers

Meng Li [1, 1st], Lina Liu [1, 1st], Jinxiu Wang [2, 1st], Shifeng Wang [1, *], Weichang Mao [1., 1st], Xin Chen[3,1st]

[1.] PLA Naval Medical Center

[2.] PLA Naval Medical University

[3.]Jinan Guoke Medical Technology Development Co., Ltd.; Shandong Engineering Research Office, Suzhou Institute of Biomedical Engineering and Technology, Chinese Academy of Sciences

[1st]: Joint author

[*]: Corresponding author

156362518@qq.com

Abstract: Barrier property is a characteristic of hollow containers and can be used to meet the need for shelf life of the contents; the hollow container shape is variable, which makes it difficult for overall barrier property testing. Based on this, this paper discusses the common testing methods of water vapor barrier property of hollow containers. Firstly, the advantages and disadvantages of the traditional weighing method and sensor method in testing the water vapor barrier property of containers are compared, typical applications are taken as an example, and finally two new methods, namely electrolytic sensor method and infrared sensor method, which can quickly and accurately test the moisture permeability of hollow containers, are introduced in detail, so as to achieve good and accurate test for various hollow containers, which can provide reference for related industries in selecting testing technologies for container barrier property.

1. Introduction

Hollow containers are mainly plastic containers processed by the hollow molding method, they often exist in the form of bottles, barrels and cans, and can be used to hold various liquid or solid foods, medicines, chemical products and so on. Because the raw materials go through high temperature, cooling, inflation and extension and other technological steps in the container processing process, it is impossible to reflect the final performance of the produced container by testing the related performance of the raw materials [1]. , The testing process of barrier property, as an important performance of hollow containers, is difficult due to the irregular shape of the container. In the past, sheets at some positions of the container wall were often cut for barrier estimation test. However, due to the uneven wall thickness of the container, it could not represent the barrier property at the junction at the bottom of the container or the mouth of the container. Therefore, this estimation test method could not accurately characterize the overall barrier property of the container [2]. Taking the water vapor barrier property (i.e. moisture permeability) of the container as an example, the traditional weighing method is often used nowadays to measure the moisture permeability (also called water vapor permeability) of the whole container, but this method has some disadvantages in use. Therefore, the fast and accurate sensor methods, including electrolytic sensor method and infrared sensor method, are increasingly recommended, especially, the infrared sensor method has the best applicability and repeatability.

2. Test with Weighing Method for Hollow Containers (Traditional Method)

2.1. Test standards and principles of the method

The moisture permeability test method of hollow containers commonly used in the previous period is evolved on the basis of film and sheet moisture permeability test method - weighing method, domestic and foreign standards used for the weighing method test of containers are GB/T 6981-2003 "Test Methods for Moisture Permeability of Rigid Packaging Containers", GB/T 6982-2003 "Test Methods for Moisture Permeability of Flexible Packaging Containers", ASTM D3079-94 (2009) "Standard Test Method for Water Vapor Permeability of Heat-sealed Flexible Packaging for Dry Articles", YBB00092003-2015 "Measurement Methods for Water Vapor Permeability". The above standards can be used for moisture permeability tests of flexible and rigid plastic packaging containers. Although there are differences between various methods, the basic method principle is the same.

In this method, the hollow container sample with desiccant in it is sealed for sample preheating treatment, and then the sample is transferred into the test chamber (room) with constant moisture heat conditions for moisture heat test, according to the moisture permeability of the hollow container sample, the appropriate interval is selected for regular weighing of sample mass, when the amount of moisture permeability is stable, the increase in sample mass and the rate of change of time are used to calculate the amount of moisture permeability of the container.

2.2. Defects of the method

The method is formed on the basis of the more widely used and long-standing film weighing method, it appears to be simple to operate, but it is more difficult to achieve automatic instrument testing, and the test efficiency and accuracy of the results are poor, and there are mainly the following shortcomings:

(1) The test process is time-consuming: according to the weighing interval recommended by relevant standards, for a container with high moisture permeability (low moisture resistance), the recommended minimum weighing interval is at least 3 days; for a container with low moisture permeability (high moisture resistance), the weighing interval of 15 days to 30 days is recommended. Usually the container moisture permeability needs to test at least three points of quality data, so even hollow container samples with high moisture permeability need 7 days to 9 days to complete the test, and testing of samples with low moisture permeability takes several months.

(2) Weighing and test environment differences affect the accuracy of the results: the standards require the sample to be removed from the test environment to the weighing environment for weighing, and this environmental difference is easy to destroy the state of osmotic equilibrium of water vapor, affecting the accuracy of the final experimental results. In particular, the shorter the weighing interval, the greater the number of weighing, the more obvious this impact is.

(3) The water vapor pressure difference between the two sides of the sample is difficult to keep stable for a long time: because the desiccant used in the weighing method is limited (existing standards require not more than 100 g), the moisture absorption capacity of the desiccant will gradually decrease as the test proceeds, and even adsorption saturation will occur, and accordingly the water vapor pressure difference between the two sides of the sample will also change. When encountering this problem, desiccant can be replaced to continue the test, but the replacement process will cause quality deviation and further affect the accuracy and repeatability of test results.

(4) Poor repeatability: due to the differences in the sealing preparation and weighing practices of different test operators for container samples, the repeatability of the test results will be affected to a certain extent.

(5) Limited accuracy of the weighing system: since the accuracy of the balance used in the weighing method is closely related to its range, and the amount of moisture permeability of the container is greater

than that of the film test, the accuracy of the test derived from the weighing method in conducting the container test is significantly worse than that in the film test.

3. Test with Sensor Methods for Hollow Containers (New Methods)

3.1. Test standards and principles of the methods

At present, the sensor methods used to test the moisture permeability of hollow containers are mainly electrolytic sensor method and infrared sensor method, the current method of testing containers using electrolytic sensor method is based on GB/T 21529-2008 "Measurement of Water Vapor Transmission Rate of Plastic Films and Sheets - Electrolytic Sensor Method", and test containers in the infrared sensor method and other packaging components have relevant standards that can be implemented. At present, the referable standard is GB/T 31355-2014 "Water Vapor Permeability Test Method for Packaging and Containers - Infrared Sensor Method", and there are no referable foreign standards.

These two sensor methods are basically of the same method principle (as shown in Figure 1), the container sample is used to separate the permeation chamber into two independent airflow systems, a humidity chamber with specified relative humidity is formed at the container said, a flowing dry carrier gas test chamber is formed at the other said. As a result of the water vapor concentration difference at the both sides of the sample, the dry carrier gas carries the water vapor through the wall of the sample to the sensor, the sensor changes the amount of humidity into an electrical signal or current value output, calculates the amount of water vapor through a certain volume of the specimen per unit of time (i.e., water vapor transmission rate). The only difference is that the electrolytic method has a chemical electrolytic tank sensor containing phosphorus pentoxide to test electrolytic current value, while the infrared method uses a sensor that applies the physical principle of water vapor absorption of infrared energy to test electrical signal value.

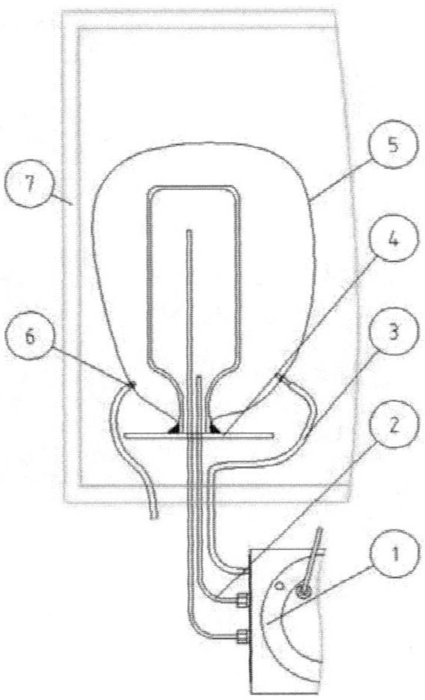

1. Test chamber integrated block; 2. Air pipe; 3. Inlet pipe; 4. Container specimen; 5. Outer cover; 6. Outlet pipe; 7. Temperature control device

Figure 1: Schematic diagram of hollow container testing by sensor method

Usually, the sensor method can realize the moisture permeability test of various shapes of hollow containers, the test accuracy can reach 0.0001g/pkg-d, and the test cycle is greatly shortened (generally only 3 to 4 days), which can ensure a stable, fast and accurate test of container moisture permeability under different temperature and humidity environment conditions.

3.2. Test methods

The electrolytic sensor method and the infrared sensor method are similar in relevant test steps, and different only in the processing data results of the sensor. Taking the infrared sensor method as an example, the specific test methods are as follows:

(1) Sample preparation and connection: moisture permeability testing of the container as a whole usually includes the moisture permeability testing of bottles (barrels, jars) without a cover and those with a cover. When testing the moisture permeability of bottles (barrels, jars), the container needs to be inverted and fixed in the packaging test bracket of the instrument with container mouth down, and then the container and bracket connection is sealed with a good special sealant (such as epoxy resin) (shown in Figure 2). When testing the overall moisture permeability of bottles (barrels, jars) with a cover, a small hole is gently drilled on the container bottom or the container wall to allow the passage of the carrier gas inlet and outlet pipe, the small hole is fixed to the bracket, and then sealant will be sealed at the connection, it is needed to ensure that the recessed part at the container bottom and the pipes are filled with sealant to ensure tightness.

Figure 2: Schematic diagram of sample preparation and connection method

(2) Installation of the specimen into the instrument: the sample to be tested is installed in a device with constant known temperature and humidity inside the test instrument, the upper chamber of the test is replaced with a specific outer cover (see Figure 3, the instrument shown is the W3/230 water vapor transmission rate tester), and the sample separates the permeation chamber into a separate airflow system (see the principle of the test method).

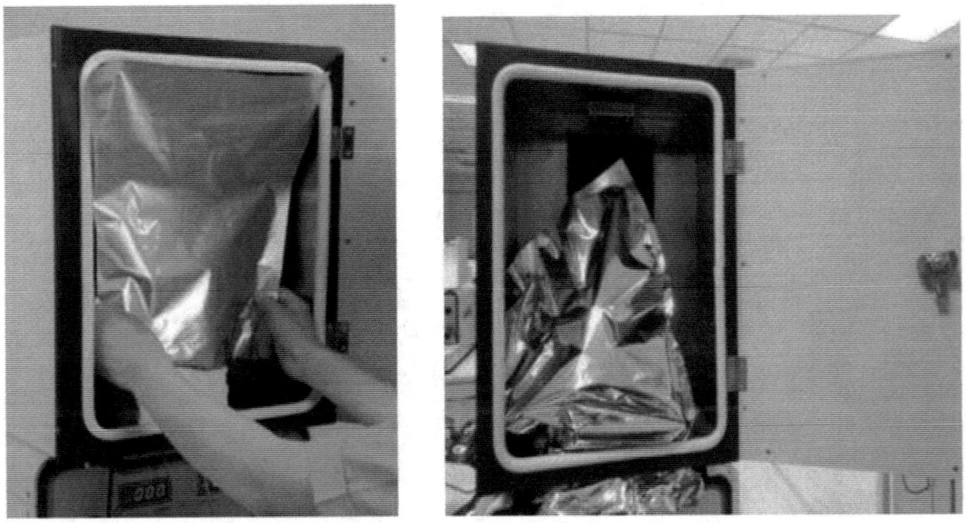

Figure 3: Schematic diagram of specimen installation

(3) Carrier gas purging and setting: set the test temperature, humidity and other test parameters, pass the dry carrier gas into the specimen, set the purging time according to the volume of the container specimen, and blow the air inside the specimen at a certain flow rate. After blowing, adjust the carrier gas flow to the set value and keep it stable.

(4) Test and record: the instrument starts the test until the change of voltage output value is within the set ratio, the test reaches a stable state and the test is finished. The instrument automatically records the voltage value and calculates the water vapor transmission rate of the final specimen.

4. Typical Applications of Sensor Methods

When the sensor methods are used to test hollow containers with high moisture resistance (low moisture permeability), the advantages of high test accuracy and fast test efficiency are highlighted, so they have a high utilization rate in food, medicine packaging materials and other industries. Generally, the hollow containers for barrier testing are mainly bottles, barrels, jars and other packages, but special attention shall also be paid to the testing of hollow containers such as soft bags, hoses and bubble caps.

(1) Moisture permeability test of soft bags

When using the weighing method to test infusion bags, doypacks and other soft bags, it is difficult to maintain the space inside the bag, especially when the bag is filled with distilled water or desiccant, making it difficult to realize effective sealing of the bag. Therefore, the use of sensor methods can effectively solve the above-mentioned problems.

When adopting sensor methods for testing, attention shall be paid to the control of the pressure inside and outside the soft bag, ensure that the actual gas through the area is the entire surface area of the soft bag to prevent errors in the experimental results. Therefore, the expansion of the bag can be achieved by passing a certain amount of carrier gas into the bag to maintain sufficient space inside the bag, as shown in Figure 4: Schematic diagram of infusion bag sample preparation.

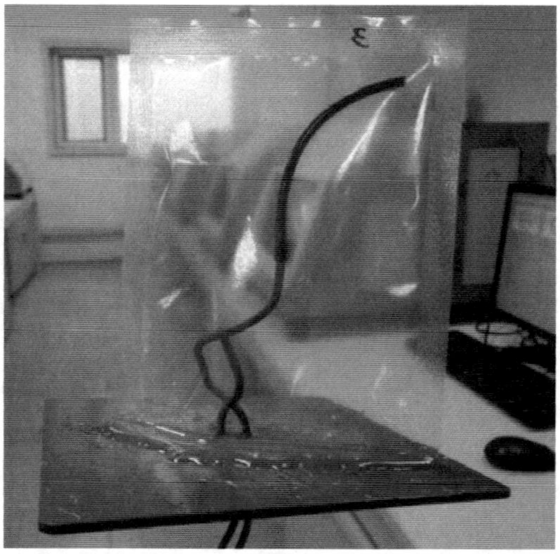

Figure 4: Schematic diagram of infusion bag sample preparation

(2) Moisture permeability test of plastic hoses

It is difficult to realize effective sealing when the weighing method is practiced in moisture permeability test of hoses due to the internal filling of distilled water or desiccant, especially the small size of the hose increases the difficulty of sealing. Therefore, sensor methods are mostly considered in hose test.

In the hose testing process, it is needed to choose the right location for holing, otherwise the sealant cannot seal effectively, usually the middle of the plastic pipe is holed, and it is necessary to ensure that the length of the gas line within the hose can meet adequate blowing without causing hose breakage.

(3) Moisture permeability test of small hollow containers

Bubble caps, capsules and other small hollow containers are small, so the weighing method is used only to test the moisture permeability of their substrates, this method cannot be used to test the moisture permeability of the containers [3], only sensor methods can be used to test those containers, and special custom test brackets and other instrument accessories are needed to complete the sample preparation and connection (such as the schematic diagram of capsule sample preparation shown in Figure 5). In addition, the carrier gas flow rate needs to be strictly controlled throughout the test process to prevent sudden pressure changes due to instantaneous and significant changes in the gas volume inside the bubble cap or capsule, resulting in abnormal sample status and affecting the test results. Therefore, sensor methods have a broader application prospect in the field of testing the moisture permeability of small specimens.

Figure 5: Schematic diagram of capsule sample preparation

5. Conclusions

The defects of long testing period, poor accuracy of test results and limited applicability of the traditional weighing method, which has been used for many years to test the moisture permeability of hollow containers, have become increasingly prominent, which in turn has led to the rise of the highly accurate, fast and efficient sensor methods in the field of moisture permeability testing of hollow containers. The sensor methods can be used to test the moisture permeability of various types of hollow containers, such as flexible bags, hoses, plastic bottles, drums, etc., providing more rigorous and reliable test data for monitoring the impact of moisture permeability of hollow containers on the shelf life of internal products.

References

[1] Wu Donglei, Zhang Lili, Xu Liang, et al. Oxygen Permeability Testing of PET Barrier Containers and Its Applications in Beer Packaging [J]. Packaging and Food Machinery, 2017, 35(003):62-64.
[2] Zhao Jiang. Development Status of Overall Barrier Property Testing of Hollow Containers [J]. China Packaging, 2018(10):46-47.
[3] Anonymous. Special Discussions on Test Methods of Water Vapor Barrier Property for Hollow Containers and Applications in a QQ Group [J]. Packaging Frontiers 2019(5):4.
[4] Hideo Fukushima, Noboru Takahashi. Hollow Containers and Methods for Manufacturing Hollow Containers:, CN106687280B[P]. 2019.
[5] Yao Xuerong, Liu Lizhi. A Polyethylene Composition with Improved Water Vapor Barrier Properties and Method of Preparation:, CN110713647A[P]. 2020.

Synthesis and performance study of a highly dispersed HPEG-type polycarboxylate superplasticizer

Zeyu Lin

KZJ NEW MATERIALS CO., LTD, Xiamen 3631010, China

linzeyu350628@qq.com

Abstract—HPEG has long been widely used as an alternative to TPEG. In this study, a new high-performance HPEG-type polycarboxylate superplasticizer PCE-A was synthesized by using H_2O_2 and sodium hypophosphite/Vc redox initiation system to provide free radicals under aqueous polymerization conditions with HPEG and acrylic acid as the main raw materials. The results showed that the best performance of the synthetic polycarboxylate superplasticizer was achieved when the acid to ether ratio was 4.50, Vc was 0.25%, H2O2 was 0.80% and sodium hypophosphite was 1.7%, which could effectively reduce the amount of admixture without affecting the working performance of concrete, and the concrete strength was higher.

1. Introduction

With the continuous development of science and technology, people's requirements for the performance of concrete are also increasing, so the types of concrete applications continue to increase [1-3]. As the third generation of polycarboxylate superplasticizer, polycarboxylate superplasticizer is the most widely used polycarboxylate superplasticizer in the market due to its low admixture, high water reduction rate, good enhancement effect, strong molecular adjustability, and no formaldehyde used in the synthesis process [4].

At present, the polyether macromonomers used in the production process of polycarboxylate superplasticizer mainly include MPEG, APEG, TPEG and HPEG, among which TPEG and HPEG occupy most of the market [5]. In this study, a highly dispersed HPEG-type polycarboxylate superplasticizer was synthesized using HPEG and acrylic acid as the copolymer monomer, hydrogen peroxide and Vc as the initiating system and sodium hypophosphite as the chain transfer agent. The effects of different synthesis factors on the dispersion performance of polycarboxylate superplasticizer were investigated, and its application performance was compared with that of similar polycarboxylate superplasticizer with better performance in the market.

2. Experiment parts

2.1. Raw materials

Methyl allyl polyethylene glycol ether (HPEG, Mn=2400), Acrylic acid (AA), Hydrogen peroxide solution (H2O2), Sodium hypophosphite, Vc, Sodium hydroxide solution (NaOH), 30 wt%.

2.2. Synthesis of polycarboxylate superplasticizer (PCE)

Add HPEG, sodium hypophosphite and water to the four flasks with accurate weighing and start stirring. After the HPEG is completely dissolved in the four flasks, turn on the peristaltic constant flow pump and add acrylic acid solution, H2O2 solution and Vc material solution drop by drop for 2~2.5h. After the drop, keep warm for 1.0~1.5 h. After the reaction is finished, adjust the pH value to neutral with NaOH to finally produce the polycarboxylate superplasticizer PCE-A with 50% solid content.

2.3. *Performance test methods*

2.3.1. *Gel Permeation Chromatography (GPC)*

The samples were examined by gel permeation chromatography (GPC, Waters 1515/2414, Waters, USA) using 0.10 mol/L aqueous sodium nitrate as the mobile phase.

2.3.2. *fluidity of cement paste*

According to GB/T 8077-2012 "concrete admixture homogeneity test method", the amount of water reducing agent is 0.18% of its solid content.

2.3.3. *Concrete Performance Testing*

According to GB 8076-2008 "Concrete Admixtures", PCE was tested for compressive strength ratio and slump of concrete. The concrete mix ratio (kg/m^3) used to evaluate the workability and strength of the concrete in this paper is: m (cement):m (sand):m (stone):m (water) = 360:830:1000:175.

3. Results and discussion

3.1. *Orthogonal experimental design and analysis*

A 3-level 4-factor ($L_9 3_4$) orthogonal experiment was designed with four factors: acid to ether ratio (A), mass percentage of Vc (B), mass percentage of H_2O_2 (C) and mass percentage of sodium hypophosphite (D). The results of the horizontal factor table and orthogonal test are shown in Tables 1 and 2.

Table.1 The orthogonal experiment design factor

Factors / Levels	A	B/ wt%	C/ wt%	D/ wt%
1	3.50	0.10	0.70	1.60
2	4.00	0.20	0.80	1.75
3	4.50	0.30	0.90	1.90

Table.2 Results and analysis of the orthogonal experiment

Sample	A	B/wt%	C/wt%	D/wt%	Fluidity of cement slurry/mm
1#	3.50	0.10	0.70	1.60	170.00
2#	3.50	0.20	0.80	1.75	180.00
3#	3.50	0.30	0.90	1.90	155.00
4#	4.00	0.10	0.80	1.90	185.00
5#	4.00	0.20	0.90	1.60	185.00
6#	4.00	0.30	0.70	1.75	185.00
7#	4.50	0.10	0.90	1.75	200.00
8#	4.50	0.20	0.70	1.90	200.00
9#	4.50	0.30	0.80	1.60	195.00
k1j	168.33	185.00	185.00	183.33	-
k2j	185.00	188.33	186.67	188.33	-
k3j	198.33	178.33	180.00	180.00	-
R	30.00	10.00	6.67	8.33	-

Table 2 shows the results of the PCE orthogonal test, from the table, it can be seen that when the orthogonal test is carried out, the R value can reflect the influence of each factor, where A to D are 30.00, 10.00, 6.67 and 8.33 respectively, and the order of influence on PCE is A > B > D > C. Through the orthogonality, it can be seen that the best synthesis process is A3B2C2D2, that is, the acid to ether ratio is 4.50, B is 0.20%, C is 0.80% and D is 1.75%.

3.2. effect of acid ether ratio on properties of PCE

According to the results of the orthogonal test, the best process was selected and the rest of the conditions were kept constant to investigate the effect of acid to ether ratio on the performance of polycarboxylate superplasticizer, the experimental results are shown in Figure 1.

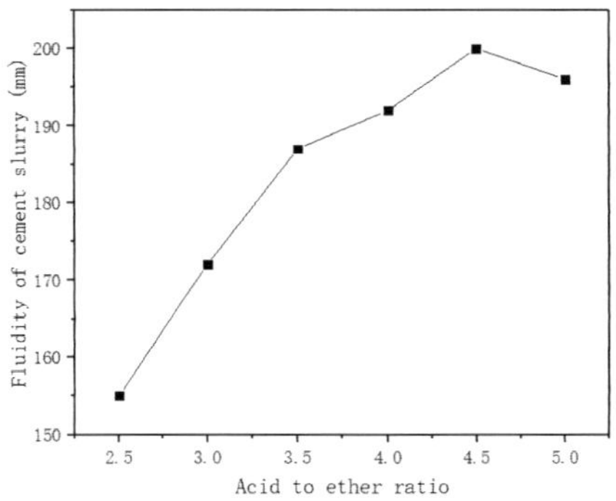

Fig. 1 Effect of acid ether ratio on the initial paste fluidity of PCE

As can be seen from Figure 1, when the acid to ether ratio is less than 4.5, the fluidity of cement paste increases with the increase of the acid to ether ratio; when the acid to ether ratio is 4.5, the the fluidity of cement paste increases with the increase of the acid to ether ratio reaches its peak. This is due to the fact that as the acid to ether ratio increases, the effective adsorption amount increases, and its adsorption on the surface of cement particles provides electrostatic repulsion to disperse the cement particles, thus making the dispersion performance of polycarboxylate superplasticizer improve; when the acid to ether ratio is too large, the number of monomers on the main chain is relatively reduced, which cannot provide sufficient electrostatic repulsion to reduce the spatial site resistance effect, thus leading to the reduction of fluidity of cement paste.

3.3. Effect of Vc on the initial paste fluidity of PCE

The best process was selected and the rest of the conditions were kept constant to investigate the effect of Vc dosage on the performance of polycarboxylate superplasticizer, the experimental results are shown in Figure 2.

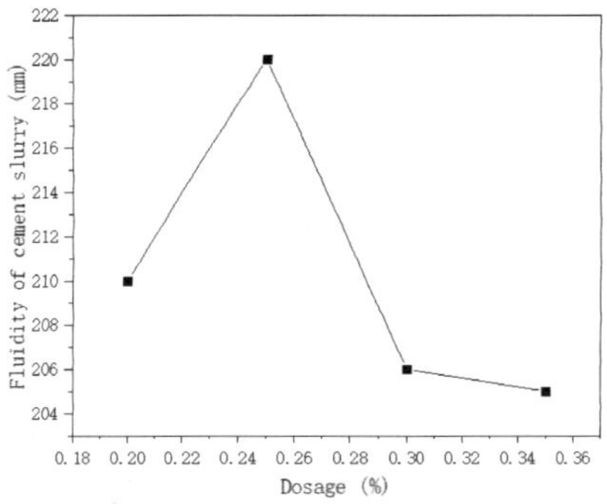

Fig. 2 Effect of Vc on the initial paste fluidity of PCE

As can be seen from Figure 2, with the increase of Vc dosage, the fluidity of cement paste first increases and then decreases. May be due to, with the increase of Vc dosage, the system due to H_2O_2 will Vc oxidation, the release of free radicals more, the reaction activity is higher, so the synthetic polycarboxylate superplasticizer dispersion performance is better; when the amount of Vc is greater than 0.25%, H_2O_2 gradually consumed, its release of carboxyl groups gradually reduced, so the synthetic polycarboxylate superplasticizer dispersion performance gradually reduced. Therefore, the best amount of Vc is 0.25% of the mass of large monomer.

3.4. Effect of sodium hypophosphite on the initial paste fluidity of PCE

The best process was selected and the rest of the conditions were kept constant to investigate the effect of sodium hypophosphite dosage on the performance of polycarboxylate superplasticizer, the experimental results are shown in Figure 3.

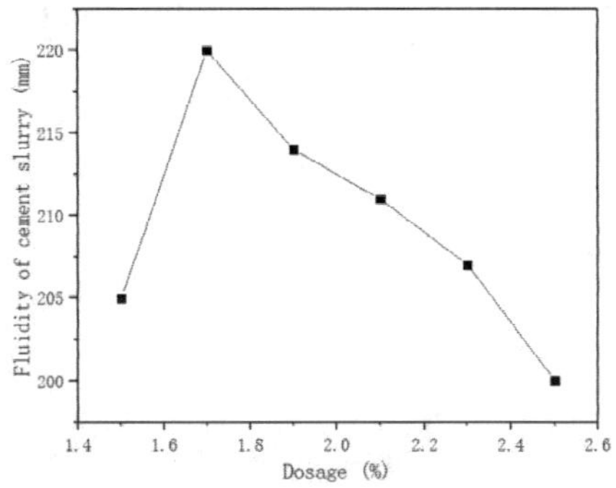

Fig. 3 Effect of sodium hypophosphite on the initial paste fluidity of PCE

From Fig. 3, it can be seen that with the increase of sodium hypophosphite dosage, the initial flow of fluidity of cement paste showed first increase and then decrease. The main reason may be that, at the beginning, as the dosage of sodium hypophosphite increases, the length of the main chain shortens, the proportion of long side chains increases, its spatial dispersion effect is enhanced and the initial dispersibility increases; however, when the dosage of sodium hypophosphite exceeds 1.7%, the main chain is too short, the speed of adsorption of cement slows down and the initial dispersibility decreases. Therefore, the optimum amount of sodium hypophosphite is 1.7% of the mass of the macromonomer.

3.5. GPC

The polycarboxylic acid water reducing agent (PCE-A) synthesised by the optimum process was subjected to GPC tests with a commercially available comparable product (PCE-S) with better performance. Table 3 shows the gel permeation chromatography data for PCE-S and PCE-A.

Table.3 Results of optimized ratio of the fluidity of cement paste and GPC

Sample	Mn	MW	MP	MW/Mn	Conversion rate/%
PCE-S	31730	72368	43147	2.280730	86.61
PCE-A	24220	45010	38773	1.858391	90.79

The data show that compared with PCE-S, PCE-A has lower number average molecular mass, heavier average molecular mass and peak molecular mass, more concentrated molecular distribution and higher conversion rate, which is conducive to obtaining water reducing agent products with excellent dispersibility.

3.6. Concrete application performance

The polycarboxylate superplasticizer (PCE-A) synthesised by the optimum process was tested against a commercially available product with better performance (PCE-S) for concrete and the test results are shown in Table 4.

Table.4 Concrete application performance test results

Sample	Dosage/%	(Slump/Slump expansion) /mm		Compressive strength/MPa		
		0 h	1 h	3 d	7 d	28 d
PCE-S	1.63	215/585	205/380	26.7	36	43.3
PCE-A	1.42	220/590	200/400	27.6	36.4	46

The results show that the PCE-A synthesised in this study can achieve the same working performance with reduced admixture and its concrete has a higher compressive strength, indicating the superior performance of PCE-A.

4. Conclusions

(1) A polycarboxylic acid high performance water reducing agent PCE-A was synthesised by free-radical polymerisation of the macromonomer HPEG at room temperature under the following conditions: acid to ether ratio of 4.50, Vc of 0.25%, H2O2 of 0.80% and sodium hypophosphite of 1.7%.

(2) The polycarboxylate superplasticizer reducing agent prepared under the best process conditions compared with similar commercially available products, to achieve the same work performance, the amount of admixture can be reduced by 13%, and the concrete strength is higher, has a broad application prospects.

(3) The adsorption mechanism of polycarboxylate superplasticizer reducing agents and their effect on cement hydration have not been systematically investigated and tested in this paper. Further research and optimisation will be carried out in future trials.

References

[1] Xiaobin Qian,Shilin Zhao,Xiaobing Zhang,Xiaobin Huang. Research on properties and acting mechanisms of concrete superplasticizers [J]. Journal of Nanjing Tech University (Natural Science Edition),2002(02):61-64.

[2] J. Plank et al. Chemical admixtures — Chemistry, applications and their impact on concrete microstructure and durability[J]. Cement and Concrete Research, 2015, 78 : 81-99.

[3] Zhenping Sun. Six Urgent Problems in the Further Study on Polycarboxylate Based Plasticizers [J]. Journal of Building Materials,2020,23(01):128-129.

[4] Shanshan Qian et al. Synthesis, characterization and working mechanism of a novel polycarboxylate superplasticizer for concrete possessing reduced viscosity[J]. Construction and Building Materials, 2018, 169 : 452-461.

[5] Yexiang Wang,Wenjuan Wang. Research Progress of Naphthalene Series Highly Efficient Water Reducing Agent [J]. Chemical Enterprise Management,2018(04):199-200.

Preparation Process and Quality Characterization of High Density Fine Grain Osmium Target for Coated Cathode

Junda Ma[1,2,3 a], Yang Xia[2,3 b *], Yuanfeng Xie[1,2,3 c], Hong Lv[2,3, d] and Chao Zhang[2,3 e]

[1]State Key Laboratory of Advanced Materials for Smart Sensing, GRINM Group Co., Ltd., Beijing 101407 China

[2]GRIMAT Engineering Institute Co., Ltd., Beijing 101407, China

[3]General Research Institute for Nonferrous Metals, Beijing100088,China

[a]18811501186@163.com, [b]xiayang@grinm.com, [c]xieyuanfeng@grinm.com, [d]lvhong@grinm.com, [e]zhangchao@grinm.com,

Abstract—High quality osmium targets are the key influence in improving the emission performance, lifetime and reliability of M-type cathodes. In this experiment, oxidative distillation was used to prepare osmium powder with a purity greater than 99.99%. The effect of hydrogen sintering and hot pressing sintering processes on the densities and grain size of the targets was investigated. The results show that the oxidation distillation process can effectively separate Os from Si, Fe, Ni, Zn and other impurity elements, and the purity of osmium powder prepared by multiple oxidation distillation was not less than 99.99%. Both hydrogen sintering and hot pressing sintering processes can be used to obtain osmium targets with a density of not less than 95%. However, the sintering temperature and time used in the hot pressing sintering process under pressure assistance are significantly lower than those of hydrogen sintering, which is more conducive to inhibiting grain growth during the high densification sintering process. By optimizing the parameters of the hot pressing sintering process, the sintering temperature of 1500°C and the holding time of 1.5h resulted in an osmium target with a density of 99.11% and a grain size of ≤12. XPS analysis of the highly dense fine crystalline osmium targets prepared by hot pressing sintering showed that the C and O mass fractions on the surface of the targets were 0.85% and 4.43% respectively, and the C and O contents gradually decreased with increasing etching depth. The above highly dense and fine crystalline osmium targets can be used in the M-type cathode standard coating process to obtain high purity, dense and strong bonded osmium films.

1. Introduction

Millimeter-wave electric vacuum devices have the advantages of high resolution, high precision and high energy density, and are widely used in military and civil applications such as 5G communications, integrated air-space communications and high-resolution radar. The cathode is the "heart" of the electric vacuum device, its life and reliability directly determine the performance and reliability of the traveling wave tube[1,2]. With China's high detection accuracy, wide width, lightness and deep space exploration and other air and space science and technology development needs, there is an urgent need for centimeter wave and millimeter wave high-power and high-pulse space traveling wave tubes.

The M-type cathodes by depositing a thin metal or alloy film (Os, Ir, Re or Os-Ir) on the surface of B or S type impregnated cathodes to reduce cathode surface fugitive work, increase cathode emission current density, cathode life and reliability[3-5]. The actual work function of the surface osmium coated cathode is 0.2 eV lower than that of the conventional barium tungsten cathode, and the emission current densities of the osmium coated cathode and the tungsten coated cathode are 8.04 A/cm^2 and 3.02 A/cm^2 respectively at 1000°C. The emission performance of the Os film cathode is much better than that of the W film[6,7].

Osmium (Os) is the densest metal with a density of 22.59 g/cm^3. Pure osmium melts at 3027°C, second only to tungsten and rhenium. Its crystal structure is dense and hexagonal, hard and brittle, not easily machined, furthermore, osmium is easily oxidized, generating OsO_4 which is highly toxic.

Osmium targets for magnetron sputtering are mostly prepared by powder metallurgy due to their high melting point and difficulty in shaping. The quality of the osmium targets for coating directly limits the performance of the M-type cathodes. A large number of domestic and abroad studies have shown that the factors that have the greatest impact on the quality of sputtering targets are purity, density, grain size, dimensional accuracy, weave and so on. For a long time, most of the domestic demand for osmium targets relies on imports, the product performance of the target purity, grain size, density and other important information requirements of the index is basically lacking. In recent years, the domestic demand for cathode material batch stability and high quality requirements highlighted, can not be prepared to meet the needs of high-power tube development.

The purity of osmium powder is about 99.9% due to the scarcity of osmium raw materials and dependence on imports. In this paper, a high purity osmium powder was prepared from crude osmium by oxidative distillation and precipitation reduction processes. The osmium targets were prepared by cold isostatic pressing-hydrogen sintering (CIP-H_2) and hot-pressing sintering (HP) processes, respectively. Their properties such as density, purity and grain size were analyzed and characterized by drainage, GDMS and metallographic analysis in order to control the fine crystalline and high density process of the osmium targets. At the same time, it is difficult to detect the C element content of osmium powder or targets by conventional high-frequency combustion-IR absorption methods due to the chemical nature of Os, which is easily oxidized and volatile at low temperatures. The experiments are aimed at detecting the composition and distribution of C, O, N, S and other impurities in the osmium targets by means of XPS analysis, in order to provide guidance data for the applications of the osmium target coating and the performance of the cathode.

2. Experiment.

The purity of the raw osmium powder is approximately 99.9%. Osmium is separated from other impurities by oxidative distillation through self-made purification device, and is prepared by alkaline absorption and multi-stage hydrogen reduction, with a median particle size of 0.6 μm and a normal uniform distribution. The osmium target was prepared by powder metallurgy, and the effects of CIP-H_2 and HP on the density and grain size of osmium target were studied. The CIP-H_2 sintering in tungsten boat, sintering temperature 2100~2300°C, sintering time 6~12 h. HP sintering process using high purity graphite mold, sintering pressure 20~40 MPa, sintering temperature 1400~1700°C, sintering time 0.5~2 h.

The density of the osmium target was tested by using the Archimedes drainage method and the grain size of the target was tested by metallographic analysis. For the characterization of metallic impurities in osmium powder and targets by GDMS, a Nu Astrum-ES low pressure source DC glow discharge mass spectrometer with a dual focus mass analyzer consisting of an electrostatic field analyzer and a magnetic field mass analyzer was used. The compositional and structural analysis was carried out on the surface of the osmium target up to a certain depth from the surface by XPS. The XPS inspection equipment is a NEXSA X-ray photoelectron spectrometer system from Thermo Scientific, USA, with an incident light source of 1486.68 eV monochromate Al Kα rays and an energy resolution of $\Delta E \leq 0.5$ eV. The etching rate for Ta_2O_5 at this etching power was 0.56 nm/s. Since the etching rate is inversely proportional to the hardness (Mohs hardness: $Ta_2O_5 = 6$, $Os = 7$), the etching rate for the osmium target was calculated to be

0.48 nm/s. High purity osmium powder specimens and osmium targets were transferred to the sample chamber and vacuum desorbed for 2 hours when the vacuum level reached 10^{-7} Pa. The purpose was to use the high vacuum environment to remove the impurities adsorbed on the surface of the samples. Photoelectron spectra of the elements on the target surface and inside the target were collected to characterize the content and distribution of elements such as C and O in the target material.

3. Result and discussion

3.1 Purity

Os can produce volatile OsO_4 vapor under the action of strong oxidizing agents, and this feature can be used to separate Os from other metallic elements[8,9]. The type and content of impurity elements of osmium powder are closely related to the purification process. At present, only the standard 《YS/T681-2008 Osmium Powder》 specifies 99.90% and 99.95% impurity elements of the osmium powder, and there is no national or industry standards for higher purity osmium products.

In this experiment, the impurity content of the osmium was determined by GDMS method and the purity of the osmium targets was calculated using the subtraction method, involving impurity elements such as Au, Mg, Si, Fe, Pt, Ni, Al, Ir, Pd, Cu, Ag, Rh, Zn and Ru.

Table 1 shows the GDMS analysis results of crude osmium powder, and the osmium powder purified by oxidative distillation. It can be seen that the oxidative distillation process can effectively reduce the content of impurity elements such as Si, Fe, Ni and Zn, and the purity of osmium powder can be increased to 99.99% after purification by multiple oxidative distillation. The GDMS analysis results of hydrogen sintered osmium targets and hot press sintered osmium targets are shown in Table 2, and the purity of the targets is not less than 99.99%.

Table 1 Composition of crude osmium powder and purified osmium powder analyzed by GDMS

Project	*Impurity element content / 10^{-6}*													*Purity / %*
	Au	Mg	Si	Fe	Pt	Ni	Al	Ir	Pd	Cu	Ag	Rh	Zn	
Crude	<0.005	20.4	352	81.1	0.9	40.1	5.26	<10	0.035	6.48	1.2	0.017	11.4	99.90333
Purified 1	<0.005	0.64	31.9	8.34	0.14	2	0.63	<5	<0.005	2.53	0.03	<0.005	0.21	99.9954
Purified 2	<0.005	0.45	30.5	10.7	0.11	2.68	1.06	<5	<0.005	2.89	0.05	<0.005	0.16	99.9951

Table 2 Composition of CIP-H_2 sintering and HP sintering osmium target analyzed by GDMS

Project	*Impurity element content* / 10^{-6}													*Purity / %*
	Au	Mg	Si	Fe	Pt	Ni	Al	Ir	Pd	Cu	Ag	Rh	Zn	
CIP-H_2	<0.005	1.77	32.7	10.1	0.42	1.21	2.58	2.29	<0.005	1.93	<0.005	<0.005	5.59	99.9922
HP	<0.005	2.26	22.7	11.8	0.76	3.5	2.74	1.45	<0.005	2.04	<0.005	<0.005	3.20	99.9909

3.2 Effect of sintering process on target density and grain size

For powder metallurgy sintered block materials, their high density and fine grain size indicators are contradictory. This is because the powder metallurgy sintered block materials are usually not high density. To improve the density, it is necessary to increase the sintering temperature or extend the sintering time, which will lead to grain growth[10-12].

The CIP-H_2 sintering and the HP sintering processes have been investigated and optimized in order to obtain the finest possible grain size of the target with a high density. Fig.1 shows the variation of density with the osmium targets prepared by different sintering processes, and Fig.2 shows the metallographic of the osmium targets prepared by different sintering processes at different sintering temperatures. As can be seen from Fig.1 and 2, the density of the osmium targets prepared by both the CIP-H_2 sintering and the HP sintering processes gradually increases as the sintering temperature rises, and decreases slightly when the CIP-H_2 sintering temperature exceeds 2400°C and the HP sintering temperature exceeds 1700°C. This may be due to the fact that the crystal grains grow fast and the gas is not discharged in time as the sintering temperature increases. As a result closed pores are formed within the grains, leading to a reduction in density. Comparing the metallography of the osmium target grains prepared by the two processes with a similar density. It can be seen that at a target density of approximately 95%, the HP sintering process produces a target with a grain size of grade 12 and a grain size of ≤5.6 μm, while the CIP-H_2 sintering process produces a target with a grain size of grade 1.5 and a grain size of ≤213 μm. The grain size of the targets prepared by HP sintering is significantly smaller than that of the CIP-H_2 sintering process, which is due to the fact that the sintering temperature and sintering time of the HP sintering process are significantly lower than those of the CIP-H_2 sintering process under pressure assisted action. Admittedly HP is more conducive to promoting the densification of osmium sintered blanks and suppressing the grain size. After optimization of the HP sintering process, the densities of the prepared osmium targets are not less than 99%, while the grain size reaches grade 12.

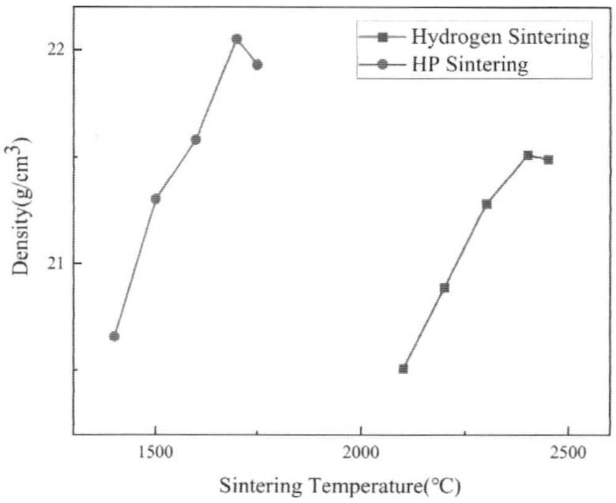

Fig.1 The influence of sintering process on the density of osmium target

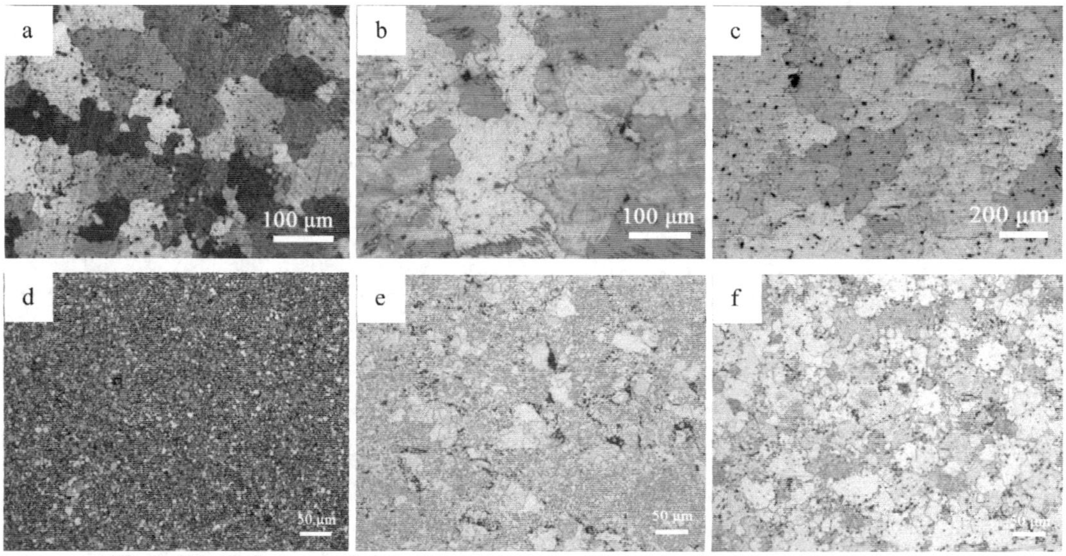

Fig.2 The influence of sintering process on the grain size of osmium target

(a) CIP-H_2 , 2300℃, 6 h, grain size number 4; (b) CIP-H_2 , 2400℃, 6 h, grain size number 3.5; (c) CIP-H_2 , 2450℃, 6 h, grain size number 1.5; (d) HP , 1600℃, 1.5 h, grain size number 12; (e) HP , 1700℃, 1.5 h, grain size number 8.5; (f) HP , 1750℃, 1.5 h, grain size number 6.5;

The fracture morphology analysis of the target with 99% density in the HP sintering, microstructure morphology as shown in Fig.3. The osmium target grains are uniformly small, with an average grain size of less than 10 μm, and the grain structure is closely connected. There is obviously no grain boundary melting phenomenon at the grain boundary, and only in the multi-grain contact position appears tiny holes.

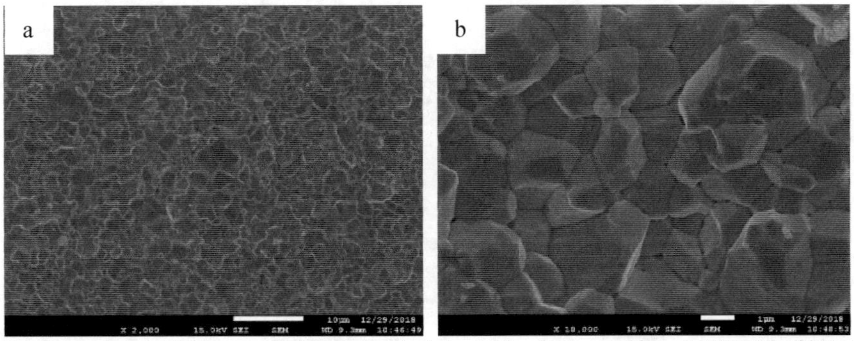

Fig.3 Microstructure of dense osmium target sintered by hot pressing (a)X2000 (b)X10000

3.3 XPS analysis of osmium target composition
The impurity elements such as carbon and oxygen in the target material have a great influence on the cathodic coating process and the activation of the ageing process. Too high a content will lead to a large amount of outgassing during the cathodic coating, which is not conducive to the adhesion of the osmium film to the porous tungsten substrate and the cathodic emission performance [13,14]. In this experiment, the GDMS method was used to detect the content of impurity elements in osmium, but due to the poor detection limits of nonmetallic element, the content of C and O in osmium targets could not be accurately evaluated. To solve this problem, the analytical technique of X-ray photoelectron spectroscopy was used to analyze the C and O elements in osmium targets.

Fig.4 shows the full XPS spectrum of an Ar^+ etched osmium target prepared by the hot press sintering, with etch depths ranging from 0 to 86.4 nm. The identification of the peaks shows that the presence of peaks for Os, C, O, N and S. The intensity of the spectral peaks can reflect the amount of elemental content. As can be seen in Fig.4, there are strong spectral bands in the region of binding energy 500-600 eV, 250-300 eV and 0-60 eV, corresponding to higher content of O, C and Os and lower content of N and S.

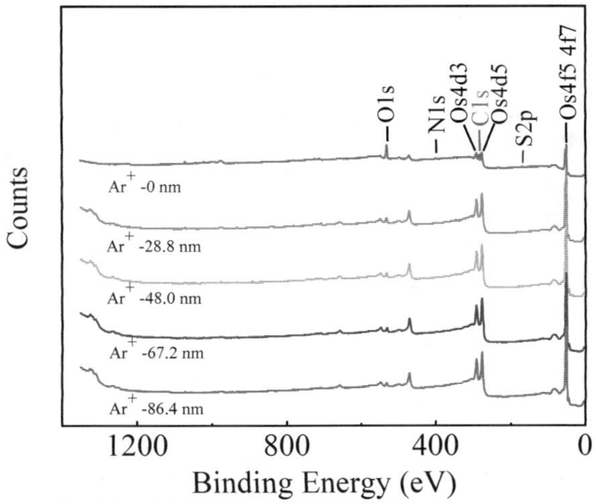

Fig.4 Full XPS spectra of osmium targets

The relative spectral intensities of the other elements were measured by the sensitivity factor method by using the spectral intensities of specific elements as a reference standard, and The curve of the osmium target element content and etching depth is shown in Figure 4. As can be seen from Table 3, the mass fractions of Os, C and O on the surface of the osmium target were 93.98%, 0.85% and 4.43%, respectively, when the target was not etched. This indicates that there is some carbon and oxygen present on the surface of the hot-press sintered osmium target. With the increase of etching depth, the mass fraction of C and O gradually decreases and the mass fraction of Os gradually increases.

Table 3 Percentage of element mass in osmium target at different etching depths / %

Etch deep（nm）	Os	C	N	O	S
0	93.9815	0.8507	0.3041	4.4341	0.4294
28.82	98.8894	0.1180	0.0598	0.8605	0.0721
48.0	99.1358	0.0609	0.0533	0.6955	0.0544
67.3	99.2103	0.0392	0.0442	0.6441	0.0619
86.4	99.2251	0.0379	0.0472	0.6256	0.0639

The existing M-type cathode preparation process usually etches the coating target before coating, and the etching depth is usually 5-6 μm[15]. Therefore, the etching process can completely remove the C and O impurities from the surface of the osmium target, improve the coating performance and effectively enhance the bonding force of the film[16]. The results of the XPS analysis of the osmium targets also indirectly show that the hot-press sintering process can fully meet the requirements of impurity control for M-type cathode coating.

4. Conclusion

1. Oxidation distillation purification of osmium can effectively reduce the content of Si, Fe, Ni, Zn and other impurity elements, after several times of oxidation distillation purification can obtain the purity of not less than 99.99% of osmium powder. Not only does it successfully solve the problem of lower purity and quality of imported osmium powder, but it also enables industrial scale-up and stable batch production.

2. CIP-H_2 and HP sintering can be used to obtain osmium targets with a density of not less than 95%. The HP sintering process is beneficial to obtain the osmium targets with a better combination of density and grain size. After optimization of the hot-press sintering process, 99.99% purity, 99.11% density, less than 12 grain size and less than 5.6 μm grain size were achieved.

3. The C and O content of the osmium target of HP gradually decreases with the increasing of the etch depth. When the etch depth of Ar^+ ion exceeds 67.3 nm, it tends to be stable, and the target mainly composed of Os. This depth is less than the etch depth of the cathodic coating process specification, and can be completely removed before the cathodic magnetron sputtering, ensuring a high purity, dense and strongly bonded osmium film.

4. After powder property repair and optimization of sintering process parameters, the osmium target reached high standards in terms of purity, density and grain size. We successfully realize the completely independent development and mass production of osmium targets for laminating cathodes, freeing itself from import dependence.

References

[1] Wang J, Wu G, Yang M H, Su X G. Development Status of High Power Millimeter Wave Traveling Wave Tube [J]. Vacuum Electronics, 2010, (03): 38-42.

[2] Shao W S, Li J, Yu Z Q, Cheng C, Zhang K, Wang H, Wang L S. Experimental study on cathode life of space traveling wave tube [J]. Chinese Space Science and Technology, 2017, 037(002):103-107.

[3] Green M C, Skinner H B, Tuck R A. Osmium-tungsten alloys and their relevance to improved M-type cathodes [J]. Applications of Surface Science, 1981, 8(1): 13-35.

[4] Fang C. Surface studies of Os/Re/W alloy - coated impregnated tungsten cathodes [J]. Journal of Vacuum Science & Technology A Vacuum Surfaces & Films, 1990, 8(3): 2329-2332.

[5] Chaharsoughi M S, Hadianfard M J, Shiezadeh M M. Study the Effect of Nanoemissive Materials on M-Type Cathode Performance [J]. Advanced Materials Research, 2014, 829: 772-777.

[6] Yin S Y, Zhang H L, Yang J X, Kui R X, Qian H J, Wang J X, Wang Y, Wang X X. Study on Surface Synchrotron Radiation Photoelectron Spectra of Cathode Coated with Osmium Film [J]. Journal of Electronics & Information Technology, 2011, (12): 250-255.

[7] Yin S, Yang J, Zhang H, Wang Y, Wang X. SPRES study on surface of barium tungsten cathodes coated with os film [J]. IEEE, 2010.

[8] Griffith W P. Osmium and its compounds [J]. Quarterly Reviews Chemical Society, 1965, 19(3): 254-273.

[9] Alexander I , Alan M . Method for the production of high purity osmium: US, US3536479 A[P]. 1970.

[10] Wang H, Xia M X, Li Y C, Liu X F, Cai X M, Bai R, Zhang X. Application and preparation technology of refractory metal sputtering target [J]. China Tungsten Industry, 2019(1):64-69.

[11] Wang S M, Zuo Y T, Du F Z, Zhang Z H, Wen M, Guo J M. Microstructure and sintering mechanism of high purity ruthenium target sintering process [J]. Precious Metals, 2019(S01): 1-5.

[12] Wang G D, Xiong N, Liu G H, Chen F G. Effect of powder properties on sintering density of pure rhenium [J]. Chinese Journal of Rare Metals, 2021, 45(04): 507-512.

[13] Yu Z Q, Gao Y J, Shao W S, Li J, Zhang K. Summary of poisoning characteristics of barium tungsten diffusion cathode[J]. Vacuum Electronics, 2017(01): 22-27.

[14] Shao W S, Zhang K, Li J, Yan S Q, Chen Q L, Gas poisoning investigations of scandate and M-type dispenser cathodes [J]. Applied Surface Science, 2003, 215(1-4): 54-58.

[15] Ma J, Wang Y C, Xie Q H, Lu W Y. Application of Cathode Etching and Coating [C]. Article series of the Eleventh Vacuum Academic Exchange Meeting of Jiangsu province. 2007.

[16] Wang P , Li J. A Novel Nanocomposite Scandia-Tungsten Coated Dispenser Cathode [J]. Vacuum Electronics, 2009, (03): 17-22.

Synthesis Methods and Influencing Factors of Metal Organic Framework Material MIL-53

Fangqin Li[1], Guanjun Wang[1*], Jianxing Ren[1] and Chenhao Sun[1]

[1]College of Energy and Mechanical Engineering, Shanghai University of Electric Power, Shanghai, 200090, China

*wzg2842299197@163.com

Abstract—Regarded as brand-new porous material, metal organic framework(MOF) has a certain ability to remove pollutants due to its high porosity, high specific surface area, adjustable micropore size and easy functionalization. Compared with other metal organic framework materials, MIL-53 has a relatively higher specific surface area and stronger hydrothermal stability, which leads to excellent performance and application prospects in the removal of pollutants. In this paper, synthesis methods of metal-organic framework material MIL-53 are introduced, and the influencing factors in the synthesis process using hydrothermal method are discussed, including the component ratio, type of organic solvent and pH value. Due to less dependence on residence time and organic solvent, mechanical method can reduce energy consumption and has latent value for future application. Among different influencing factors, reducing the pH of solution when synthesizing MIL-53 using hydrothermal method seems to save time and achieve rapid formation.

1. Introduction

Metal organic frameworks (MOFs) are porous materials with large specific surface area and adjustable pore size, which are formed by coordination of metal ions and multi-electron organic ligands [1,2]. The metal ions currently and mostly used in synthesis process mainly include Mg^{2+}, Cu^{2+}, Cr^{3+}, Al^{3+} and Fe^{3+} [3,4]. Férey et al. [5] carried out studies on the synthesis of MIL-n using trivalent metals, terephthalic acid(H_2BDC) and trimellitic acid(H_3BTC). MIL-53 is one of the typical representatives. Its synthesis process is generally carried out under hydrothermal conditions, forming a $MO_4(OH)_2$ ($M=Cr^{3+}$, Al^{3+}, Fe^{3+}) corner-sharing octahedron connected by terephthalate groups. Compared with other MOF materials, MIL-53 has a relatively higher specific surface area (up to 2000 m^2/g) and pore volume, while still maintaining good stability at high temperatures, thus showing excellent performance in pollutant removal research. However, the synthesis of MIL-53 usually takes plenty of time, and the yield is too low for industrial use, which affects future applications of the material, not to mention the cost and recycle ability. Therefore, improving the yield of product, and achieve rapid synthesis of MIL-53 are of great importance.

In this paper, synthesis methods of metal-organic framework material MIL-53 are introduced, and the influencing factors to hydrothermal method are discussed, including the component ratio, type of organic solvent, pH value and synthesis temperature.

2. Synthesis methods of MIL-53

2.1. hydrothermal method
The method is to put the ligand, metal salt and organic solvent in a container and react for a period of time under a certain pressure and temperature. The metal ions and organic ligands can complete the self-assembly process, thereby obtaining the required metal organic framework [6]. In specific, the raw materials for preparing MIL-53(Cr) by hydrothermal method are generally chromium nitrate [$Cr(NO_3)_3 \cdot 9H_2O$], terephthalic acid(H_2BDC) and water. The component ratio of $Cr(NO_3)_3 \cdot 9H_2O$, H2BDC and water is usually 1:1:280 or 1:1.5:280, and a small amount of hydrofluoric acid is added. After being stirred evenly, the mixture is transferred to the polytetrafluoroethylene reactor, which is put into an electric heating oven under airtight conditions The reaction is carried out in the container under a constant temperature which is usually 220°C, and the reaction time is up to 72 hours. After that, the reactor is cooled, the product is washed with deionized water, and the initially synthesized MIL-53 powder can be obtained. When it comes to MIL-53(Al), the metal salt is substituted for $AlCl_3 \cdot 6H_2O$.

The initial product of MIL-53 has crystalline terephthalic acid attached to the inner surface of the pores, so the product needs to be purified to dissolve and remove the excess terephthalic acid to achieve higher specific surface area. Due to the low solubility of terephthalic acid in water, solvents with greater solubility with terephthalic acid are necessary, such as ethanol, dimethylformamide (DMF), and ammonium fluoride.

The hydrothermal synthesis method is simple and safe, and has been widely used in the synthesis of metal organic frameworks. Nowadays most of the article reports still use this method to prepare MIL-53. However, since the time spent for heating in the reactor often reaches several days, the relatively longer synthesis time is an obvious disadvantage of the hydrothermal synthesis method.

2.2. microwave method
Microwave synthesis is a synthesis method that uses the function of microwave radiation heating to speed up the reaction process and achieve rapid crystallization. The energy transmission method of microwave synthesis is different from the that of the solvent method. The rapid crystallization is achieved by the direct interaction of electromagnetic waves and reactants. Microwave synthesis mainly depends on the contact of electromagnetic waves and movable charges to conduct energy. Solvent molecules and reactant ions are polarized under the electromagnetic field, and an electric current is formed in the solution. The electric resistance of the solution generates a lot of heat and accelerates the movement of reactant molecules.

Due to the direct interaction between radiation and the reactants, a better heating method is provided under microwave assistance, which can achieve a high heating rate and uniform heating effect for the entire sample, and crystals of smaller size can be synthesized. Gordon et al. [7] synthesized highly crystalline MIL-53(Fe) in 10 minutes from microwave irradiation and in 7 minutes from ultrasound irradiation. Therefore, this method has the advantages of shorter reaction time and uniform product distribution.

2.3. mechanical method
Mechanical synthesis is a method of converting mechanical energy into chemical energy to synthesize materials. Compared with hydrothermal method and microwave method, this method can obtain the target product through mechanical movement in a ball mill. Mao [8] mixed the chromium salt and H_2BDC uniformly. After the mixture was grinded for 30 minutes, the reactant was added to a polytetrafluoroethylene reactor. Under non-solvent conditions, MIL-100(Cr) and MIL-53(Cr) can be obtained under 220°C for 24h to 48h. The study found that when chromium chloride hexahydrate ($CrCl_3 \cdot 6H_2O$) was used as a chromium source for synthesis, the surface area (1310m^2/g) and pore volume (0.84cm^3/g) of MIL-53(Cr) prepared by this method were similar to those of the solvothermal method, and it had more excellent catalytic performance.

Since it requires less heating or organic solvents, mechanical synthesis can definitely reduce energy consumption and also help reduce environmental pollution during the synthesis process. To be more important, this method is suitable for the synthesis of large quantities of metal-organic framework materials, which provide latent value for future industrial use.

3. Influencing factors in synthesis of MIL-53 by hydrothermal method

3.1 the component ratio

When synthesizing MIL-53 by hydrothermal method, the ratio of metal salt, terephthalic acid and water is generally 1:1:280, and some ratios are 1:1.5:280. Different ratio leads to products with different status. Zhou et al. [9] tried to synthesize MIL-53(Cr) at a ratio of 1:1.5:280 and 1:2:280. By comparing the XRD results of the product, it was found that increasing the content of terephthalic acid in the solvent can make the product algin better with the standard spectrum, which may lead to better crystallinity. The specific surface area measured after the purification treatment by the reflux method was 1355.9m^2/g, which was close to many reported values. Combined with the SEM image, it can also be seen that several times of washing using DMF can remove the excess terephthalic acid inside and outside the pores.

The change of component ratio not only leads to different specific surface area, but also influences photocatalytic performance afterwards, since the ratio of metal and organic ligand affect the crystallization process to some extent. Liu et al. [15] prepared MIL-53(Fe) under different component ratio, and Rhodamine B(RhB) was used as model pollutant to test the photocatalytic activity of the products. After 180 min of visible light irradiation, the pseudo-first order reaction kinetics of RhB degradation over the MIL-53(Fe) at ratio of 1:2 was 0.0286, which was the highest among different component ratios including 1:1 and 1:1.5.

3.2 type of organic solvent

For hydrothermal method, pure water is sometimes selected as the solvent, and dimethylformamide (DMF) are often used as the solvent. These two methods have their own advantages and disadvantages. When pure DMF is used as the solvent, the yield of the sample can reach more than 80%, but with poor crystallinity. When pure water is used as the solvent, the low solubility of terephthalic acid ligand in water results in low yield (20%-30%), but the obtained sample has higher crystallinity. Although the solvent in the former method contains DMF, both methods still require purification steps to remove unreacted impurities. Cheng et al. [10] used a mixed solvent, which was a mixed solution of water and DMF to synthesize NH2-MIL-53(Al), trying to achieve high yield and high crystallinity at the same time. By comparing the morphology of the product with different proportions of water in the solvent, it was found that the crystal morphology, size and measured specific surface area of the product vary under different solvent ratios. Among them, the smallest grain size was 76nm, and the largest specific surface area was 1882 m^2/g, which was a big improvement compared with the value reported in the general articles.

3.3 pH

The reactants required to synthesize MIL-101 are the same as MIL-53, the main difference is the reaction time. The synthesis time of MIL-101 is generally 8-12h, while the residence time of MIL-53 is usually 72 hours, and there is a transition from MIL-101 to MIL-53 during the reaction. Zhao et al. [11] proposed that lowering the pH of the reaction solution was not helpful to the dissolution of H_2BDC and the formation of chromium trimers, while Khan et al. [12] proposed that the formation of chromium trimers during the synthesis process of MIL-101 is more important than the dissolution of H_2BDC, and appropriately lowering the pH of the solution may be helpful to the formation of chromium trimers.

Accordingly, Zhang et al. [13] used acetic acid instead of hydrofluoric acid as a mineralizer to synthesize MIL-101. It was found that after adding more than a certain amount of acetic acid, the product at a reaction time of 20h is no longer MIL-101, but MIL-53 instead. In addition, the pores of MIL-101 synthesized by adopting acetic acid as a mineralizer are cleaner, and the specific surface area was also increased, which showed that lowering the pH of the solution can accelerate the synthesis to a certain extent.

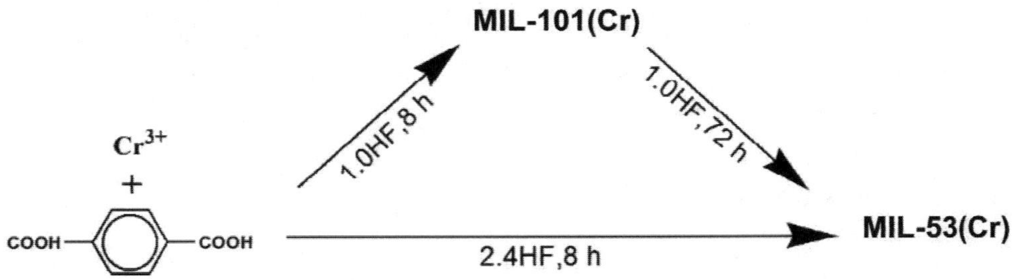

Fig.1 Different crystallization route for different content of HF added in the hydrothermal synthesis of MIL-53(Cr) [14]

There are also researches hoping to reduce the pH value by increasing the proportion of mineralizers. Li et al. [14] increased the proportion of mineralizer in the synthesis to 1.0Cr: 1.0BDC: 2.4HF: $284H_2O$, and observed the direct formation of MIL-53 during the 8h reaction, as is shown in Fig.1.

3.4 synthesis temperature

Another important factor in the synthesis of MIL-53 using hydrothermal method is the synthesis temperature. Different synthesis temperature may lead to different specific area value and photocatalytic activity. Liu et al. [15] prepared MIL-53(Fe) under different synthesis temperature and component ratio. It was found that the specific areas increased with the synthesis temperature rising from 130°C to 150°C. However, if the synthesis temperature was over 190°C and below 130°C, the crystal structure may be destroyed. Therefore, a suitable synthesis temperature may lead to better crystallization and more uniform morphology, which can be related to better photocatalytic performance.

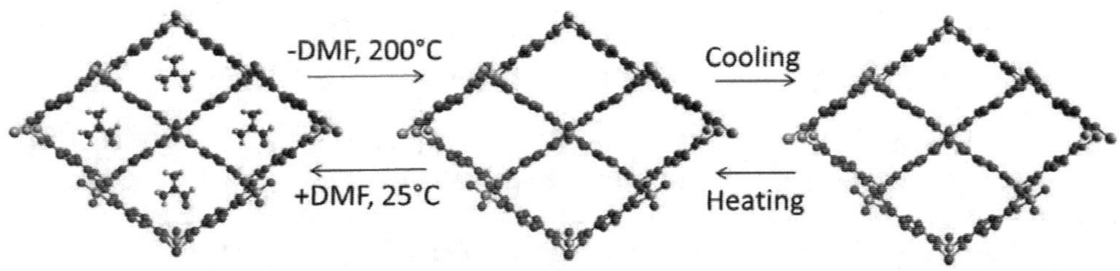

Fig.2 Stability of large-pore form of MIL-53(Al) after the removal of DMF and cooling to room temperature [16]

In addition, transition between large-pore and narrow-pore can be observed under typical synthesis temperature. William et al. [16] prepared MIL-53(Al) using DMF under different temperatures. It was found that the MIL-53(Al) synthesized in DMF at 120°C remained in the large-pore form under all conditions, as is shown in Fig.2, while the material synthesized at 220°C exhibited a gradual breathing transition to the large-pore form, allowing for increased CO_2 adsorption in the 1-20 bar region.

4. Conclusions

The metal organic framework material MIL-53 is widely used in the removal of environmental pollutants, such as antibiotics, organic dyes, aromatic compounds and heavy metals. However, the above results are still in the experimental research stage, while mass production and large-scale pollutant treatment are yet to be accomplished. There are still problems about the costs, actual adsorption and photocatalytic performance, long synthesis time, complex steps, low yield, and recycling performance. It is still difficult to carry out large-scale industrial applications. Reducing the pH of solution when synthesizing MIL-53 using hydrothermal method seems to save time and can achieve direct formation. Also, mechanical method can get rid of organic solvent, and save reaction time to some extent with less pollution derived from synthesis process. With deeper research on the high yield and rapid formation, MIL-53 as adsorbent and also catalytic material will definitely show great prospects in future applications.

References

[1] Kosal, M.E. and K.S. Suslick. (2000) Microporous porphyrin and metalloporphyrin materials. Journal of Solid State Chemistry., 152(1): 87-98.
[2] Ren, J., Dyosiba, X., Musyoka, N., et al. (2017) Review on the current practices and efforts towards pilot-scale production of metal-organic frameworks (MOFs). Coordination Chemistry Reviews., 352: 187-219.
[3] Eisenberg, D., Stroek, W., Geels, N.J., et al. (2016) A rational synthesis of hierarchically porous, N-doped carbon from Mg-based MOFs: Understanding the link between nitrogen content and oxygen reduction electrocatalysis. Physical Chemistry Chemical Physics., 18: 20778-20783.
[4] Teo, H.W.B, Chakraborry, A., Kayal, S. (2017) Evaluation of CH4 and CO2 adsorption on HKUST-1 and MIL-101 (Cr) MOFs employing monte carlo simulation and comparison with experimental data. Applied Thermal Engineering., 110: 891-900.
[5] Férey, G. (2008) Hybrid porous solids: past, present, future. Chem Soc Rev, 37(1): 191-214.
[6] Serre, C., Millange, F., Surble, S., et al. (2004) A route to the synthesis of trivalent transition-metal porous carboxylates with trimeric secondary building units. Angew Chem Int Ed., 43(46): 6286-6289.
[7] Gordon, J., Kazemian, H. and Rohani, G. (2012) Rapid and efficient crystallization of MIL-53(Fe) by ultrasound and microwave irradiation. Microporous and Mesoporous Materials., 162: 36-43.
[8] Mao, Y., Qi, H., Ye, G., et al. (2019) Green and time-saving synthesis of MIL-100(Cr) and its catalytic performance. Microporous and Mesoporous Materials., 274: 70-75.
[9] Zhou, M., Sun, L.B., Liu, X.Q. (2012) Preparation of Metal-Organic Frameworks MIL-53(Cr) and its Absorption Performance. Journal of Chemical Engineering of Chinese Universities., 26: 858-863.
[10] Cheng, X.Q., Zhang, A.F., Hou, K.K., et al. (2013) Size- and morphology-controlled NH2-MIL-53(Al) prepared in DMF-water mixed solvents. Dalton Transactions., 42: 13698-13705.
[11] Zhao, T., Jeremias, F., Boldog, I., et al. (2015) High-yield, fluoride-free and large-scale synthesis of MIL-101(Cr). Dalton Transactions., 44(38): 16791.
[12] Khan, N.A., Jun, J.W., Jhung, S.H. (2010) Effect of water concentration and acidity on the synthesis of porous chromium benzene dicarboxylates. European Journal of Inorganic Chemistry., 7: 1043-1048.
[13] Zhang, L.J., Li, F.Q., Ren, J.X., et al. (2019) Preparation of MIL-101(Cr) with acetic acid as mineralizer and its adsorption properties for CO2. Chemical Engineering., 47: 34-38+54.
[14] Li, X., Zhang, J., Shen, W.L., et al. (2016) Rapid synthesis of metal-organic frameworks MIL-53(Cr). Materials Letters., 255: 126519.
[15] Liu, N., Jing, C.W., Li, Z.M., et al. (2018) Effect of synthesis conditions on the photocatalytic degradation of Rhodamine B of MIL-53 (Fe). Materials Letters., 237: 92-95.
[16] Mounfield, W.P. and Walton, K.S. (2015) Effect of synthesis solvent on the breathing behavior of MIL-53(Al). Journal of Colloid and Interface Science., 447: 33-39.

Synthesis and Analysis of a Viscosity Modifying Admixture based on Acrylamide and 2-Acrylamide-2-methylpropanesulfonic acid

Tao Ni[1a*], Xin Song[1b*], Yongmei Wang[1c], Ke Feng[1d], Serina Ng[1e], Zhaoyang Liu[2f]

[1]Sichuan Tongdao Technology Co. LTD, Chengdu 610000, China

[2]Shijiazhuang Chang'an Yucai Building Materials Co., LTD, Shijiazhuang 051430, China

[a*]497739700@qq.com, [b*]songxin82@qq.com, [c]wangym0705@qq.com,

[d]1337098964@qq.com, [e]akfc@hotmail.sg, [f]85617489@qq.com

Abstract—Viscosity modifying admixture (VMA) can be employed to avoid the segregation of concrete. Here, we present a type of viscosity modifying agent (B2), prepared by free radical polymerization between acrylamide (AM) and 2-acrylamide-2-methylpropane sulfonic acid (AMPS). The structure and composition were characterized by FT-IR, NMR and light scattering. Hydroxy-propyl-methyl cellulose (HPMC) was used as a comparison to determine the suitability of B2 as a VMA. The results show that B2 is superior to HPMC, and can improve not only the homogeneity and bleeding resistancy, but also increase the compressive strengths of the concrete by 4 to 5 MPa across all testing periods, up to 28 days. A highlight of this admixture is the effective regeneration of nonuseable concrete to its non bleeding state with delayed addition. Finally, a mechanistic analysis of the viscosifying effect, which induced the increase in compressive strength is postulated in this article.

1. Introduction

The fresh concrete has to fulfill contradictory requirements of high flowability when it is being cast, and stability when it is at rest, in order to prevent segregation. High flowability is achieved using high performance of water reducer, while stability against segregation is achieved either by using a large quantity of fine materials or by using an appropriate viscosity agent (VMA) [1,2]. In addition, self-compacting concrete (SCC) has higher requirements on the fluidity and stability of concrete than traditional concrete, so VMA becomes a necessary admixture for the SCC [3]; VMA has been widely used to enhance the cohesion and stability of cement systems. The increase in VMA dosage is shown to exhibit a reduction in mini-slump and an increase in plate cohesion meter, Yield stress and plastic viscosity [4]. VMAS can be divided into two categories according to their chemical composition, one is inorganic chemicals and the other is organic chemicals. Inorganic VAMs are as following: natural zeolite [5], silica fume [6]; Organic VMAs have microbial polysaccharides, such as diutan gum, welan gum, cellulose derivatives (methyl cellulose, cellulose ether and cellulose filaments [7]), bacterial cell walls (BCW) [8] and rice husk ash (RHA) [9] and so on. Inorganic VAM has the following advantages: (1) Inorganic VMA cause an acceleration of hydration and higher compressive strength at the age of 1 day [10]; (2) Inorganic VMA causes the smallest change in the values of the air voids parameters. However, the dosage of inorganic VAM always exceeds 15% by weight of the cement, which leads to low cost-effective performance; Organic VMA has the following advantages :(1) the dosage is much lower than the former, which is not more than 0.05% of the cement; (2) VMA was found to be very effective in reducing both crack area and crack width, and the crack reductions were correlated with surface tension, bleeding and weight changes [11]. The mechanism of action in each case is different and related to its molecular structure. The mechanism of action in each case is different. In the case of welan gum and cellulose, the long chain polymer molecules adhere to the periphery of water molecules, thus imbibing and fixing a part of the mixing water. Some other VMAs adsorb on cement particles and increase viscosity by promoting inter-particle attraction [12].

At present, organic VMA has been extensively and deeply studied by scholars. For a given dosage of VMA, diutan gum showed a high apparent viscosity than welan. For any given dosage of the super plasticizer, the diutan gum grouts demonstrate a greater yield value and apparent and plastic viscosity than welan gum [13,14]. Cellulose ether adsorption seems to slow down the nucleation of calcium silicates at the surface of the cement grains in the tens of minutes following mixing, and bridge the cement grains by adsorbed ether molecule[15]. Cellulose ethers have the most high water retention capability. Increasing cellulose ethers addition will decrease slip-resistance[16]. According to the literature, the disadvantages of organic VMA also were found: (1) The organic VMA show almost no influence on early cement and the development of compressive strength [17]. The initial setting time is shown to be delayed by the incorporation of high range water reducer and VMA with the latter additive exhibiting a greater influence on retardation of setting [18] (2) organic VMA causes an happens in the content of the air pores;

2. Research Significance

As we knew that organic VAM has low dosage, good cost-efficiency, but which leads to reduce the concrete compressive strength[19]. In the article, a new organic VMA was studied based on the molecular design by introducing sulfonic group in order to avoid retarding effect on the concrete, as the sulfonic group was not chelated by calcium ions, and thus will not affect the compressive strength of the concrete; The introduction of amino group to the molecular structure would improve the hydrophilicity of molecules and early strength of concrete; The linear soft polymer was designed to block water molecules in mortar.

3. Experimental Investigation

3.1 Raw materials and instruments

Acrylamide (AM), industrial grade, purity> 99.0%; 2-acrylamide-2-methylpropyl sulfonic acid (AMPS), industrial grade, purity > 98.0%; Acetone, analytically pure; Azo diisobutamidine hydrochloride initiator, industrial grade; Deionized water; Nitrogen, 99.9999% purity.

There were two types of chemical admixture used in the concrete experiment, including ether based polycarboxylate superplasticizer (GK-3000) and VMA. The GK-3000 behaves a water reduction of 30% and a solid content of 21%. Two VMAs chosen for this study are hydroxypropyl methylcellulose (HPMC) with a relative molecular mass of $1.5*10^6$g/mol and a new type of VMA by self-made. Both the VMAs are water solution with mass fraction of 2%.

Table 1 Chemical and physical properties of cement and coal ash

	SiO_2(%)	Al_2O_3(%)	Fe_2O_3(%)	CaO(%)	SO_3(%)	MgO(%)	K_2O(%)	Na_2O(%)	LoI(%)	Specific surface area(cm^2/g)
Cement	22.12	4.75	3.14	64.3	2.12	1.91	1.02	0.24	1.6	3410
Fly ash	54.1	27.4	6.35	2.18	0.14	0.78	1.61	0.35	92.91	3600

Table 2 Grading of artificial sand

Sieve diameter (mm)	sieve passing (g)	sieve percentage passing (%)	cumulative percentage passing(%)	fineness modulus
4.75	6.50	1.3	1.3	
2.36	98.00	19.6	20.9	
1.18	93.50	18.7	39.6	
0.600	89.00	17.8	57.4	2.70
0.300	71.00	14.2	71.6	
0.150	52.00	10.4	82.0	
0.075	48.50	9.7	91.7	
Sieve bottom	41.50	8.3	100.0	

3.2. Synthesis of AM-AMPS copolymer
At first, 120 mL of deionized water was placed in a 500 mL three-neck round bottom flask equipped with a stirrer, mercury thermometer. After the deionized water in the flask was heated to 60°C, purged N_2 into the water for 25min. Next, 15.6 g of AM and 5.3 g AMPS were added in the flask. Then, 0.03 g azo disobutamidine hydrochloride (initiator) was dissolved in the other 30 mL of deionized water, and was fed continuously into the reaction vessel over a period of 2 h at 60 °C using a peristaltic pump. When addition of the solution was completed, the mixture was curing for another 2 h to finalize the reaction. After that, another 200 mL deionized water and 3 g 30wt% NaOH solution were fed into the vessel, and the stirrer was accelerated in order to dissolve the production. At last, the solution was pure into 725 mL of deionized water, which was stirred for 20min, then a new VMA was obtained and named as B_2.

3.3 Performance test method

3.3.1 FT-IR measurement
Perkin Elmer Spectrum 100 Fourier transform infrared spectrometer is used to perform the test by using the thin film method with the resolution of 1cm^{-1} and the scanning times of 32 times.

3.3.2 NMR measurement
The hydrogen spectra of the refined samples were measured by INOVA 400 NMR apparatus with D_2O as the solvent and methyl silane (TMS) as internal standard.

3.3.3 Light scattering measurement
Static and dynamic light scattering were measured by Brookhaven light scatterer. The wavelength of the laser used for light scattering is 532nm, and the angle of the static light scattering is from 15°to 155°. The scattering Angle selected for dynamic light scattering is 90°.

3.3.4 Rheological measurement
HAAKE RS150 torque rheometer was used for steady-state rate scanning and dynamic frequency scanning experiments. All tests were carried out at 25°C. The diameter and cone Angle of the lamina were 40mm and 4°, respectively, and the plate spacing was 150 μm.

3.3.5 Concrete test
In order to compare the effect of B2 and HPMC on properties of concrete, concrete have been mixed with a water-cement of 0.46, 0.29kg PCE (made by Shijiazhuang Chang'an Yucai Building Material Co., LTD, China) is added to 1m^3 concrete to make the concrete slightly bleeding, and the slump flow is 600mm. The dosage of B2 is 0.01%, 0.015%, 0.03%, 0.05% of concrete, and the dosage of

HPMC(2wt%) solution is 0.01%,0.03%,0.05%, 0.07% of concrete. The mixture proportions are given in Table 3. Cementitious materials and some of the water were firstly added in mixer and stirred for 30 s. Water containing dissolved VMA and SP was then poured and mixed for 60 s. Finally, gravels were added and mixed for another 60 s. The prepared fresh concrete mixture was measured by the slump flow test, bleeding test, flow time of mixture from the inverted slump cone, air content test, in accordance with Chinese Standard GB/T 50080-2016. Furthermore, compressive strength on 100×100×100 square (at 3, 7, 28 days) were assessed. In order to study the effect of B2 on regeneration of nonuseable concrete to its non-bleeding state with delayed addition. Cementitious materials and some of the water were firstly added in mixer and stirred for 30 s. Water containing SP which dosage is 1.5% of concrete to make the concrete bleeding was then poured and mixed for 60 s. Then, gravels were added and mixed for another 60 s. Finally, B2 which dosage is 0.2% of concrete was add and mixed for 30s. The prepared fresh concrete mixture are also measured by the slump flow test, bleeding test, flow time of mixture from the inverted slump cone, air content test.

Table 3. Mixture proportion of concrete

W/C	Unit weight（kg/m^3）				
	C	F	S	G	W
0.46	290	70	805	1025	165

4. Experimental Results and Discussion

4.1 Composition and structure of copolymer

4.1.1 Infrared and NMR spectra

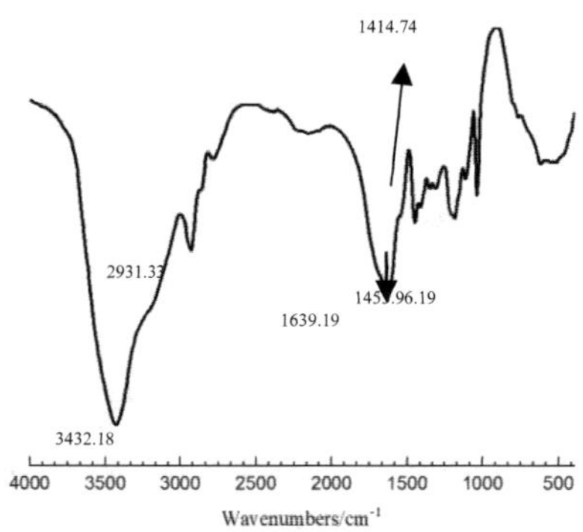

Fig.1 FT-IR spectroscopy of the copolymer

As shown in Fig.1, there is a strong peak at 3432.18 cm^{-1}, which is the stretching vibration absorption peak of N-H on the amide group. The stretching vibration absorption peak of methylene at 2931.33 cm^{-1} is a moderate intensity peak. The bending vibration absorption peaks of $-CH_3$ at 1414.74 cm^{-1} and 1453.96 cm^{-1} are two weak peaks, and the stretching vibration absorption peak of carbonyl group at 1639.19 cm^{-1} is strong. Because the sample had been extracted with acetone as the solvent for 72 hours before the infrared test, the monomer and low molecular homopolymers had been removed, so the infrared results preliminarily indicated that AM and AMPS had formed a copolymer.

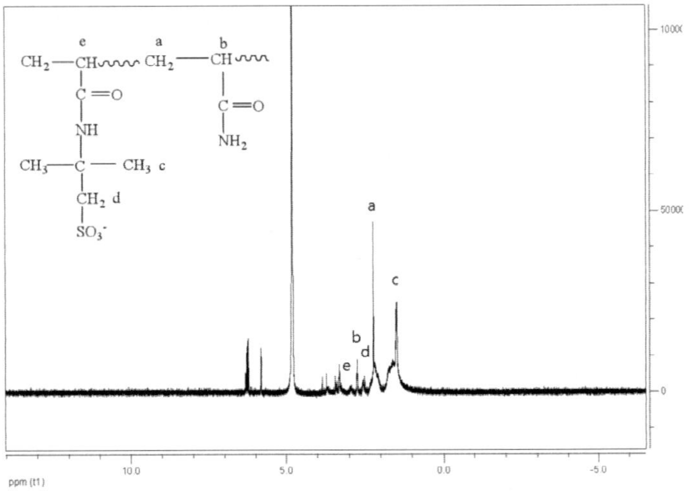

Fig.2 NMR spectroscopy of the copolymer

In Fig.2, a wide peak appears at 1.47ppm of chemical shift, which should be the result of the superposition of two $-CH_3$ (c) peaks in AMPS unit. Due to the influence of electron-withdrawing group $SO3^-$, the chemical shift of $-CH_2$ increases. The chemical shifts of $-CH_3$ were found at 2.2ppm and 2.5ppm, and the peak strength at 2.2ppm was the largest (except the solvent peak), as could be attributed to the methylene peak (a) of the acrylamide unit contained the most hydrogen. The peak at 2.5ppm was assigned to the methylene peak (d) in the AMPS unit, whose chemical shift was greater than that of the -CH peak (e) in the AMPS unit, which was also affected by the anion of sulfonic acid group. However, the chemical shift of methylene peak (b) at 2.7ppm, which was attributed to acrylamide. Based on the above analysis, it is further proved that the AMPS-AM copolymer has been successfully synthesized.

4.2 Light scattering characterization

The zimm diagram of the copolymer in the pore fluid of the cement slurry determined by static light scattering is shown in Fig.3. The average molecular weight (M_W), the mean square root mean square radius of rotation (Rg) and the second dimension coefficient (A_2) of the B_2 sample were obtained, which are shown in Table 4.

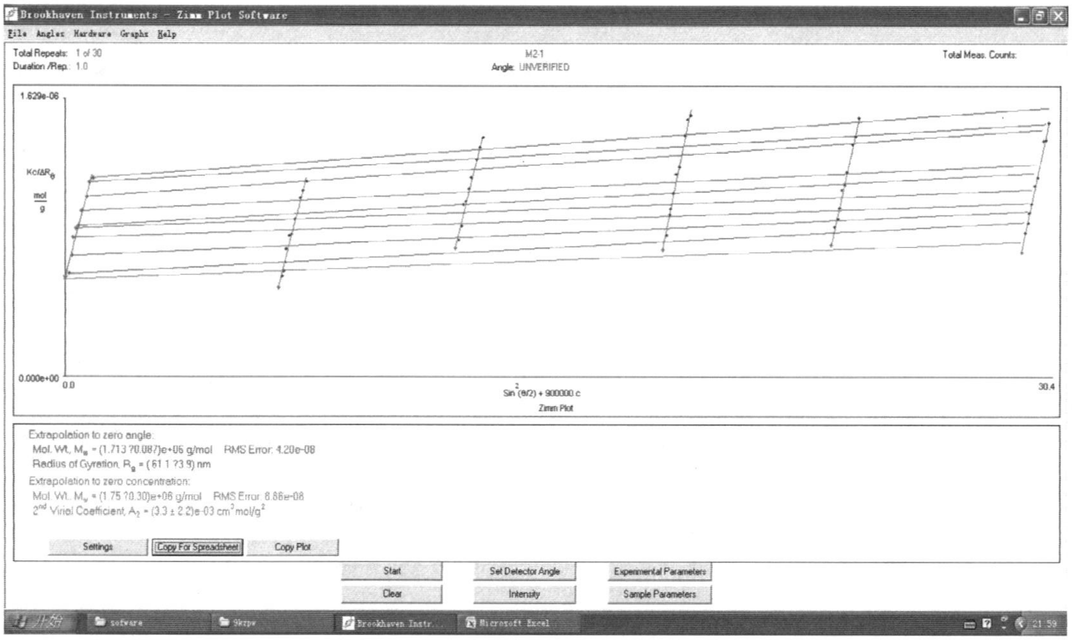

Fig.3 Zimm plot for the B2 sample

Table. 4 Parameters of static light scattering for the B_2 sample

Name	Mw(10^6)/g/mol	Rg(nm)	A_2(10^{-3}, cm^3mol/g^2)
B_2	1.73	61.1	3.3

The testing results of static light scattering (Figure 3 and Table 4) show that the Mw of the B_2 reached 1.73 x 10^6 g/mol, whose root mean square radius of gyration is 61.1 nm. The second virial coefficient between B_2 and the pore fluid is 3.3 x 10^{-3} cm^3mol/g^2. The above results indicate that the copolymer contained in B_2 has a high molecular weight and a strong interaction with water molecules in the pore solution, and the linear molecules are speculated that the conformation of the polymer is relax in the solution.

Further dynamic light scattering test was performed on the pore solution of the copolymer, and the concentration of the polymer solution was 150mg/L. The test results were shown in Fig.4. It can be seen from the figure that the distribution of copolymer solution particle size (D) is multi-dispersive, which is distributed in the range of 2.0-2.8nm, 26.3-36.2nm and 94.5-210.2nm, respectively, and the number of particles in the range of 94.5-210.2nm is the largest. The hydraulics radius of the polymer in solution, Rh, is equal to D/2. Rh can be considered as the radius of the equivalent hard sphere with the same translational diffusion coefficient as the polymer. If the polymer chain is an extended Gaussian chain, the solvent molecules in the space occupied by the polymer chain do not move with the polymer chain when the polymer chain is diffused. At this point, Rh is much less than Rg, indicating that the polymer molecule is rigid. But when the polymer chain forms a curled ball, the solvent molecules move along with the polymer. At this condition, Rh is greater than Rg. Therefore, Rg/Rh of scattered particles is used to investigate the microscopic conformation of scattered particles in solution [7]. The average hydraulics radius Rh is 61.1nm (123.1/2), and the ratio of Rg/Rh is 1.0. The difference between them is not significant, which indicates that the copolymer presents a flexible curl shape and can move together with the solvent molecules.

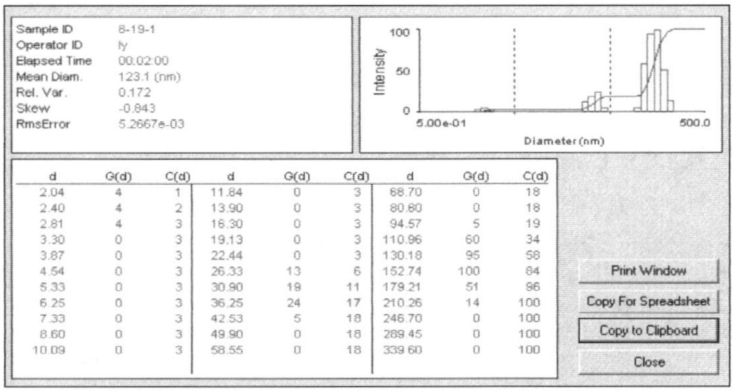

d	G(d)	C(d)	d	G(d)	C(d)	d	G(d)	C(d)
2.04	4	1	11.84	0	3	68.70	0	18
2.40	4	2	13.90	0	3	80.80	0	18
2.81	4	3	16.30	0	3	94.57	5	19
3.30	0	3	19.13	0	3	110.96	60	34
3.87	0	3	22.44	0	3	130.18	95	58
4.54	0	3	26.33	13	6	152.74	100	84
5.33	0	3	30.90	19	11	179.21	51	96
6.25	0	3	36.25	24	17	210.26	14	100
7.33	0	3	42.53	5	18	246.70	0	100
8.60	0	3	49.90	0	18	289.45	0	100
10.09	0	3	58.55	0	18	339.60	0	100

Fig.4 Dynamic light scattering for the copolymer

Based on the above analysis, it is concluded that amide units introduced in the copolymer molecular chain greatly enhances interaction between water molecules and the polymer, which leads to more water molecules controlled by the polymer; Rigidity and ion repulsion of sulfonic acid in the molecular structure in the pore fluid can increase of Rh and are beneficial to improve the water retention performance.

4.3 Rheological properties

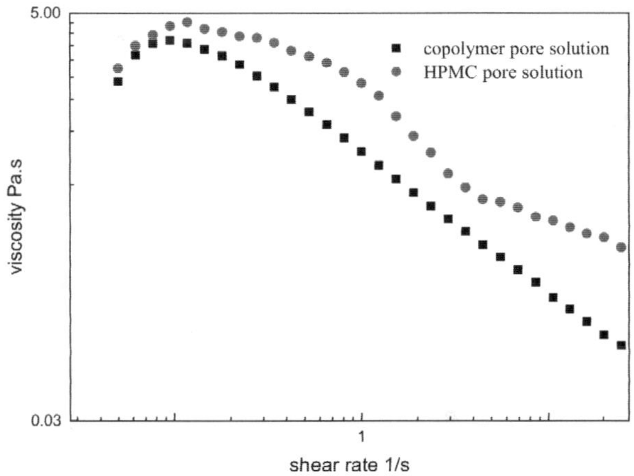

Fig.5 Shear-viscosity curve of the copolymer water solution

The steady-state rheological diagram of the copolymer in the pore fluid was shown in Fig.5, and the concentration of the pore solution is 1500mg/L. It can be seen from the figure that the viscosity of the synthesized copolymer in the pore solution can reach more than 3 Pa.s at the low shear rate, while the viscosity of the HPMC in pore solution is above 4 Pa.s at the low shear rate, which indicates that both of them have good viscosity retention rate, namely good water retention when they are static. On the rheological curve of shearing and viscosity, the viscosity of copolymer and HPMC in the pore fluid increased slowly with the increase of shear rate at beginning, which showed a tendency of shear-thickening. Then a plateau appeared and began to decrease with the increase of shear rate. At high shear rate, the apparent viscosity of the copolymer in pore solution decreases linearly with the increase of shear rate, which the characteristic property of shear-thinning was appeared. The viscosity of HPMC decreases less than that of copolymer at high shear rate. The property of the shear-thickening of the two VMAs at low shear rates can be explained by the orientation and entanglement effects of the molecular chains. In the pore solution, the molecular chains of the VMA are entangled or hydrogen-bonded to form a series of physical crosslinking points. Due to the thermal movement of the molecules, these physical crosslinking points are constantly disintegrated and formed, and their positions are constantly changing. In the low shear rate region, the physical crosslinking points destroyed by shearing can be reconstructed in time, and the molecular orientation increases the viscosity of the copolymer solution. When the shear rate increases to a certain extent, the destructed speed of the physical crosslinking points is greater than its reconstructed speed, so that the friction resistance decreases and the solution appears shear-thinning behavior. In addition, due to the stronger intermolecular interaction of HPMC, the physical cross-linking network formed in solution has stronger shear resistance than linear copolymer molecules.

4.4 Properties of concrete

In order to compare the effect of B2 and HPMC on properties of concrete, concrete have been mixed with a water-cement of 0.46, 0.29kg PCE (made by Shijiazhuang Chang'an Yucai Building Material Co., LTD, China) is added to $1m^3$ concrete to make the concrete slightly bleeding, and the slump flow is 600mm.

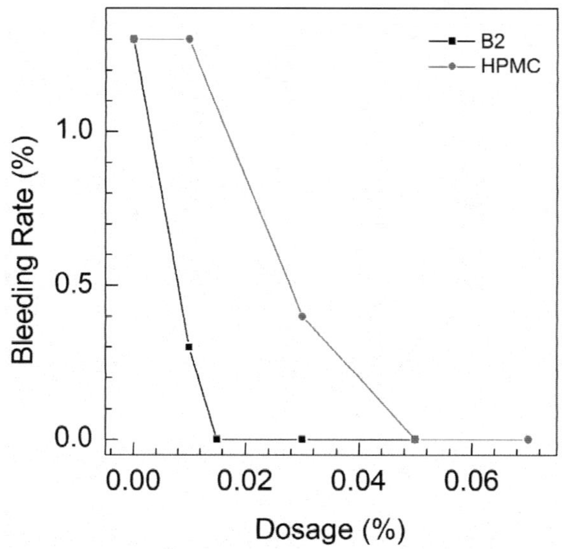

Fig.6 Curves of bleeding rate versus dosage for concrete added with B2 and HPMC

Fig.6 gives the Curves of bleeding rate versus dosage for concrete added with B2and HPMC. It can be seen that the bleeding rate of concrete decreased significantly after treated with B2, and decrease to 0% when the dosage of B2 is 0.015%. However, the bleeding rate of concrete is not changed when the dosage of HPMC is low, and it dropped to 0 while the dosage of HPMC is 0.05%, which is higer than B2. It can be explained, that hydrophilic groups of VMA molecules (such as the hydroxypropyl in HPMC, the acylamino in B2) can adsorb and associate water molecules via hydrogen bond. Then Van der Waals attractive forces produced by adjacent B2 chains can retard the free motion of water therein and even block water molecules as the chains are intertwined, thus presenting a network at low concentration. Whereas HPMC molecules had shorter and semi-rigid main chain composed of glucosyl rings, the chains were difficult to intertwind. Therefore, the free motion of water is retarded only when the dosage of HPMC is high, the bleeding rate of concrete is also decreased.

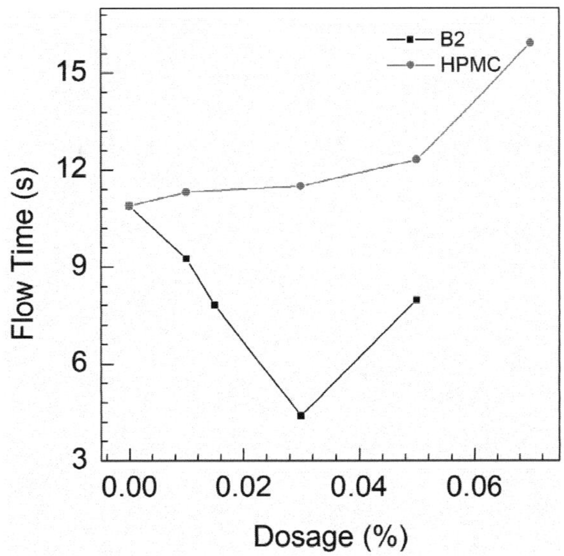

Fig.7 Curves of flow time versus dosage for concrete added with B2 and HPMC

The curves of flow time versus dosage for concrete added with B2 and HPMC are presented in Fig.7. It can be seen that the flow time of concrete is decreased and then increased with the increase of B2 dosage, and the flow time attain the minimum when the dosage of B2 is 0.03%. But the flow time of concrete ascend with the increase of HPMC dosage. It is all known that the flow time of concrete has a great relationship with the plastic viscosity and workability of concrete. As 2.2.1 said, B2 chains are entangled or hydrogen-bonded to form a series of physical crosslinking points when concrete does not flow, consequently the workability of concrete is excellent. When concrete flows, mortar added with B2 appears shear thinning behavior, thus the increase of plastic viscosity is not obvious. Therefore, with the increase of B2 dosage, the flow time does not increase. Whereas HPMC molecules had shorter and semi-rigid main chain, and the free motion of water is retarded only when the dosage of HPMC is high, otherwise, mortar added with HPMC does not appear shear thinning behavior significantly, thus the plastic viscosity of concrete increases significantly as HPMC dosage is high. So that the flow time of concrete increase with the increase of HPMC dosage.

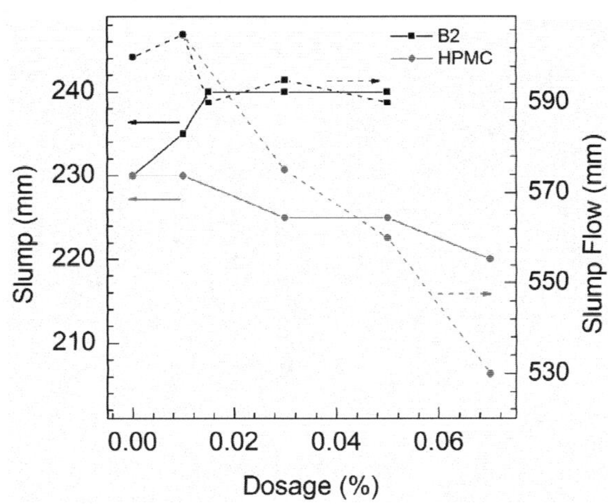

Fig.8. Curves of slump and slump flow versus dosage for concrete added with B2 and HPMC

As it can be seen in Fig.8, with the increase of B2 dosage, the slump and slump flow of concrete do not change much, but with the increase of HPMC, the slump and slump flow of concrete decreased. Due to the molecular structure, there is little effect of B2 on the slump and slump flow of concrete. However, on one hand, the plastic viscosity of concrete increases significantly as HPMC dosage is high, on the other hand, due to the introduction of hydrophilic group (hydroxypropyl) and hydrophobic group (methoxy) in its molecular structure, when its dosage is high, it will introduce bubbles into concrete, further increasing the viscosity of concrete. Therefore, the concrete slump and slump flow decrease with the increase of HPMC content.

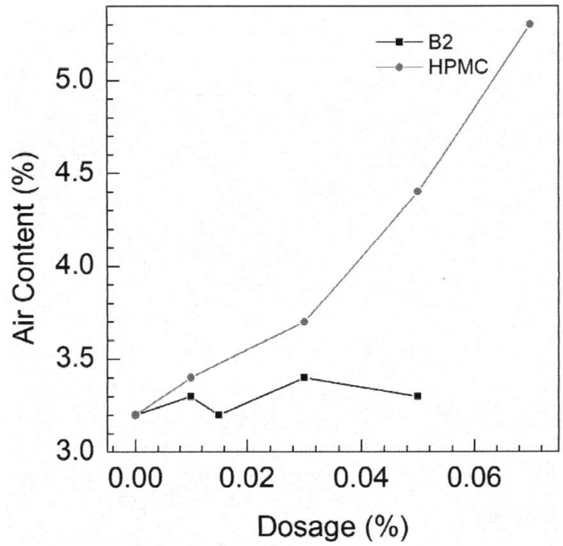

Fig.9. Curve of concrete air content versus dosage for concrete added with B2 and HPMC

Fig.9 shows the variation curve of concrete air content with B2 and HPMC content. With the increase of B2 content, the air content of concrete has no obvious change, but with the increase of HPMC content, the air content of concrete keeps increasing. This also proves that the introduction of hydrophilic (hydroxypropyl) and hydrophobic (methoxy) groups into the molecular structure of HPMC will introduce bubbles into the concrete. The increase of concrete air content will inevitably lead to the decrease of concrete strength.

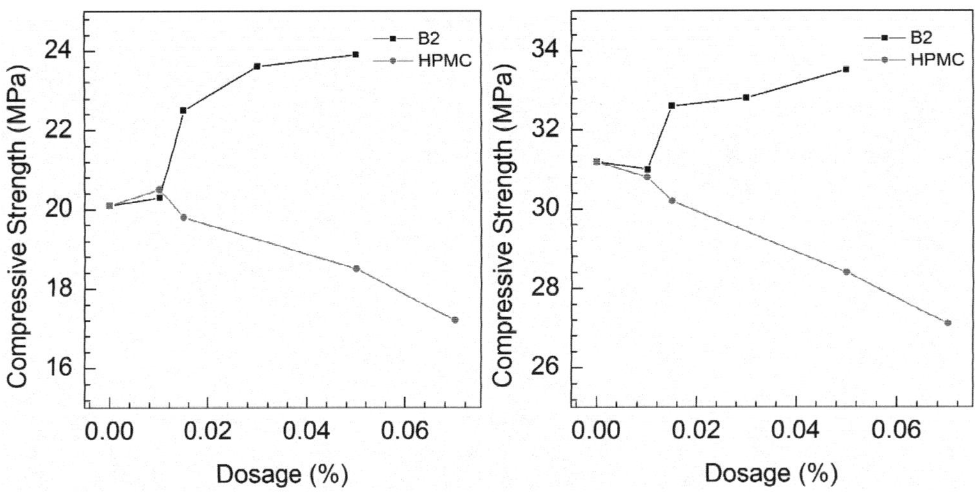

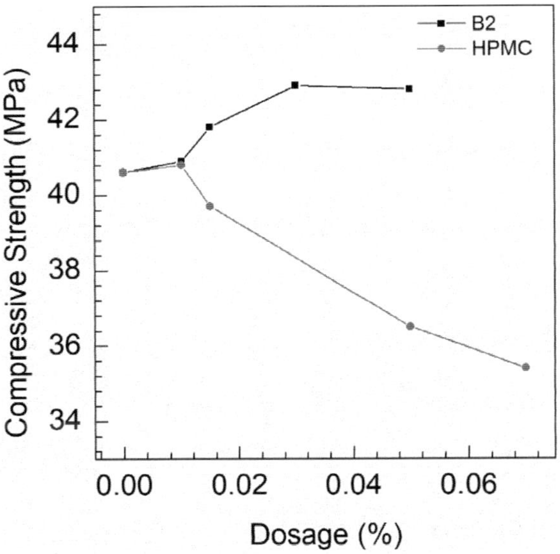

Fig. 10. Curve of concrete strength versus dosage for concrete added with B2 and HPMC
(a: 3d, b: 7d, c: 28d)

As can be seen from Figure 10, B2 can increase the strength of concrete, especially the 3d strength. HPMC will reduce the strength of concrete, and with the increase of HPMC content, concrete 3d, 7d and 28d strength continues to decrease, this is mainly because with the increase of HPMC content, concrete air content continues to increase, resulting in a continuous decline in strength. The air content of the concrete mixed with B2 does not change significantly, and B2 contains amino, the N atom in the amino has a pair of unshared electrons, it is easy to form covalent bonds with metal ions, complex, and to form a relatively stable complex with metal ions. These complexes form a number of soluble zones in the solution, thus increase the diffusion rate of the hydration products and improve the early strength of concrete.

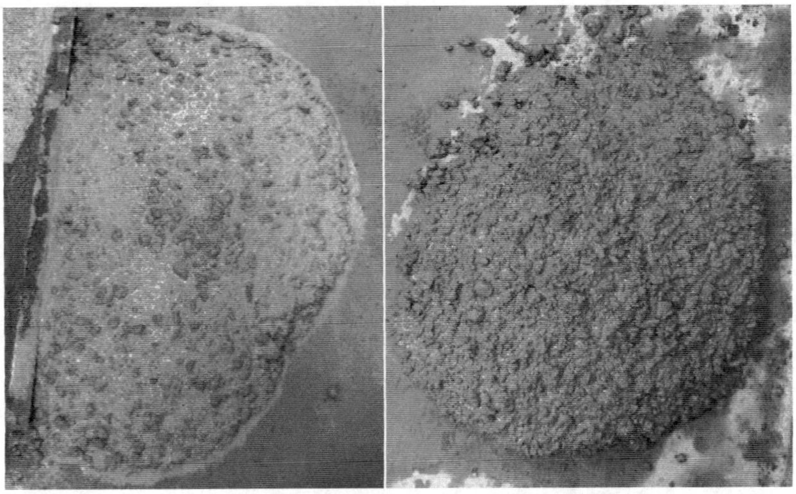

Fig. 11. State of bleeding concrete before and after addition of B2
(A: before addition, B: after addition)

Table 5 The performance of fresh concrete

Sample	Bleeding rate (%)	Flow time (s)	Slump (mm)	Slump flow (mm)
Before	1.6	11.2	240	620
After	0	6.7	240	545

As can be seen from Fig. 11 and the table 5, after the addition of B2 into the bleeding concrete, the concrete aggregate and slurry are no longer separated, the flow time of concrete reduces, and the concrete shows good workability.

5. Conclusion

A new type of viscosity modifying agent (B2) was prepared by free radical polymerization between acrylamide (AM) and 2-acrylamide-2-methylpropane sulfonic acid (AMPS) in the article. The structure and composition were characterized by FT-IR, NMR and light scattering. Hydroxy-propyl-methyl cellulose (HPMC) was used as a comparison to determine the suitability of B2 as a VMA. The results show that B2 is superior to HPMC, and can improve not only the homogeneity and bleeding resistancy, but also increase the compressive strengths of the concrete by 4 to 5 MPa across all testing periods, up to 28 days. A highlight of this admixture is the effective regeneration of nonuseable concrete to its non bleeding state with delayed addition. Finally, a mechanistic analysis of the viscosifying effect, which induced the increase in compressive strength is postulated in this article.

References

[1] Konsta-Gdoutos M.S.(2006) A study of the interaction between viscosity modifying agent and high range water reducer in self compacting concrete. Measuring, Monitoring and Modeling Concrete Properties. Springer, Netherlands, pp.449–454.

[2] Beata Ł.(2013) The type of air-entraining and viscosity modifying admixtures and porosity and frost durability of high performance self-compacting concrete. Construction and Building Materials, 40:659–671.

[3] Hamid R.S.(2017) An investigation of the effects of limestone powder and Viscosity Modifying Agent in durability related parameters of self-consolidating concrete (SCC). Construction and Building Materials, 156: 152–160.

[4] Mohammed S.(2019) Effect of mix proportions on rheology and permeability of cement grouts containing viscosity modifying admixture. Construction and Building Materials, 212: 687–697.

[5] Şahmaran M., Özkan N.(2008) Evaluation of natural zeolite as a viscosity-modifying agent for cement-based grouts. Cement and Concrete Research, 38: 930–937.

[6] Mouhcine B., Xavier R.(2015) Influence of silica fume and viscosity modifying agent on the mechanical and rheological behavior of self compacting concrete. Construction and Building Materials, 84: 103–110.

[7] Kamal H. K.(1998) Viscosity-Enhancing Admixtures for Cement-Based Materials - An Overview. Cement and Concrete Composites, 20: 171-188.

[8] Ruoting P., Jun L.(2015) Use of bacterial cell walls as a viscosity-modifying admixture of concrete. Cement & Concrete Composites, 55: 186–195

[9] Shazim A.M., Muhammad A. S.(2011) Utilization of Rice Husk Ash as viscosity modifying agent in Self Compacting Concrete. Construction and Building Materials, 25: 1044–1048.

[10] Andreas L., Frank W.(2007) The effect of viscosity modifying agents on mortar and concrete", Cement & Concrete Composites, 29: 341–34.

[11] Shih-Tang L.(2010) Effect of viscosity modifying agent on plastic shrinkage cracking of cementitious composites", Materials and Structures, 43: 651–664.

[12] Coralie B.(2014) Cellulose ethers and yield stress of cement pastes", Cement and Concrete Research, 55: 14–21

[13] Mohammed S.(2006) Rheological properties of grouts with viscosity modifying agents as diutan gum and welan gum incorporating pulverised fly ash. Cement and Concrete Research, 36: 1609–1618
[14] Lachemi M., Hossain K. M. A.(2004) Self-consolidating concrete incorporating new viscosity modifying admixtures. Cement and Concrete Research, 34: 917 – 926.
[15] Marco C., Anne D.(2013) Influence of organic thickening admixtures on the rheological properties of mortars: Relationship with water-retention. Construction and Building Materials, 38: 950–961
[16] Alyaa M. and Nihad T. K. A.(2020) Ultra-High Early Strength Cementitious Grout Suitable for Additive Manufacturing Applications Fabricated by Using Graphene Oxide and Viscosity Modifying Agents. Polymers, 12: 2900
[17] Khayat K. H. and Yahiat A.(1997) Effect of welan gum-high-range water reducer combinations on rheology of cement grout", ACI Materials Journal, September-October, pp.365-372.
[18] Junyi Z., Xiaojian G.(2020) Improvement of viscosity-modifying agents on air-void system of vibrated concrete. Construction and Building Materials, 239: 117843.
[19] Barbara L., Frank W., Corinne A., Effect of temperature on the pore solution, microstructure and hydration products of Portland cement paste. Cement and Concrete Research. V.37, 2007, pp.483–491
[20] Feng Y., Billon L, Grassl B, et al(2002) Hydrophobically associating polyacrylamides and their partially hydrolyzed derivatives prepared by postmodification. Synthesis and characterization". Journal of Polymer, Britain, 43: 2055-2064.
[21] Joseph S., Mohammed S.(2020) The effect of a polyacrylic acid viscosity modifying agent on the mechanical, thermal and transport properties of hemp and rapeseed straw concrete. Construction and Building Materials, 235: 117536

Study on Organic Modification of Lanthanum Oxide

Yunhan SUN[1], Xiaoning Wang[1], Qingxiu JIA[1*], Jian YANG[2]

[1] School of Materials Design and Engineering, Beijing Institute of Fashion Technology, Beijing, China 100029;

[2] Beijing Beihua New Rubber Special Material Technology Co., Ltd, Beijing, China 100029

*Correspongding author's e-mail: jiaqingxiu@163.com,

Abstract—In this paper, lanthanum oxide was modified by silane coupling agent (KH560), titanate coupler NDZ-201 and dopamine hydrochloride (DP). The microstructure and chemical structure of modified La_2O_3 were characterized by scanning electron microscopy (SEM) and X-ray photoelectron spectroscopy (XPS), and the dispersion of modified lanthanum oxide in polypropylene (PP) was investigated. The results show that the three modifiers can improve the organic properties of lanthanum oxide, and the interfacial compatibility between dopamine modified La_2O_3 and PP is better than the other two modifiers.

1. Introduction

Rare earth has a special 4f electron layer structure, which can absorb energy particles from 38.9kev to 63kev, and can effectively compensate the lead in the weak energy absorption region. It is a good radiation protection material. A variety of rare earth/polymer radiation protection materials have been developed, such as rare earth/NR composite[1-2], rare earth / PP composite fibers[3-5], and rare earth doped barium tellurate glass[6].

As inorganic materials, the interface between rare earth and organic polymers is very poor, resulting in poor mechanical properties of the composites. Therefore, organic modification of rare earth is needed to improve the compatibility, interfacial strength and dispersion between rare earth and polymer matrix. Acrylic acid, methacrylic acid and silane coupling agent are generally used for organic modification of rare earth [1-6]. However, the structure and modification methods of these modifiers are relatively simple, and the effects of modifier types on the structure and properties of rare earth and its dispersion in polymers have not been studied in detail.

In this paper, silane coupling agent KH560, titanate coupler NDZ-201 and dopamine were used as modifiers to investigate the effects of different modifiers on the structure and properties of rare earth lanthanum oxide and its dispersion in polypropylene matrix. The results can provide a reference for the preparation of rare earth / polymer composites.

2. EXPERIMENTAL

2.1. *Materials*

Polypropylene (PP), spinning grade, density 0.9g/cm3, melt index 25g/min(235℃,2.16kg), Sinopec Group; Lanthanum oxide (nano grade, ≥99.99%), China Beijing Nonferrous Metals Research Institute; dopamine hydrochloride (analytical purity, ≥ 99.9%), titanate coupler NDZ-201 (analytical purity,

≥99%), and KH560 (analytical purity, ≥98%), Beijing Chemical Reagent Company. All other reagents used are commercially available.

2.2. Organic modification of La_2O_3

First, the modifier and the organic solvent were mixed in a certain ratio to obtain a solution, the pH value was adjusted, and then lanthanum oxide was added to the above solution, and ultrasonic treatment was performed for a certain period of time to carry out sufficient organic modification. Then, the modified lanthanum oxide was put into a blast drying oven and dried at 40°C to remove the solvent, and finally lanthanum oxides modified with different modifiers were obtained. The solution ratio and experimental conditions used in three modification processes are shown in Table 1.

Table 1 Condition of organic modification on La_2O_3 with three modifiers

Modifider	Solution	Weight ratio of modifier to La_2O_3	pH value	Time of ultrasonic processing	Blast drying conditions
silane coupling agent (KH560)	absolute ethanol: deionized water : KH560: acetone=70:28:1:1 (v:v)	1:25	3~5 adjusted with acetic acid	0.5 h	40 ℃, 24h
dopamine hydrochloride (DP)	absolute ethanol: deionized water: dopamine=50:49:1 (v:v)	1:25	8 adjusted with trimethylol aminomethane	0.5 h	40 ℃, 24h
titanate coupler NDZ-201 (NDZ201)	ethanol:NDZ201=99:1 (w:w)	1:25	7	0.5 h	40 ℃, 24h

2.3. Preparation of modified La_2O_3 blends with PP

The modified lanthanum oxide and PP were added to the twin-screw Haake extruder at a mass ratio of 30:100, blended for 10 minutes under the conditions of 210 °C and 100 r/min, and extruded to obtain a modified lanthanum oxide/PP composite.

2.4. Characterization

Scanning electron microscopy (SEM) Surface observation of rare earth particles and composite fibers before and after modification was carried out using a Japan Electron JSM-7500F type field emission scanning electron microscope, and the surface gold spraying method was used to prepare the observation samples with a thickness of about 10 nm.

X-ray photoelectron spectroscopy (XPS) The relative content of La2O3 powder elements was measured using a Model 250 X-ray photoelectron spectrometer manufactured by Excalab.

3. Results and Discussions

3.1. Morphology of La_2O_3 particles

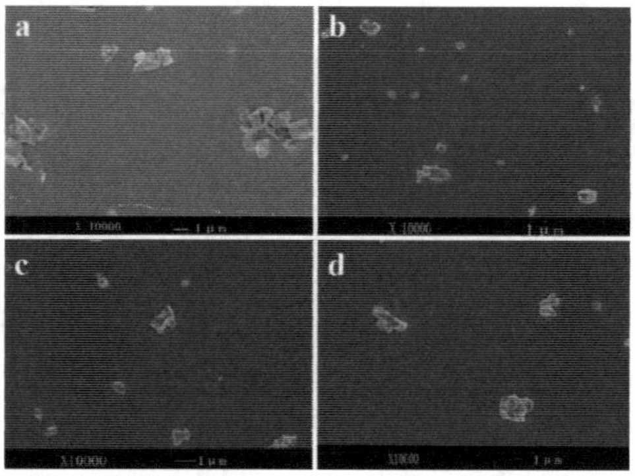

Fig.1 SEM micrographs of (a) unmodified La_2O_3; (b) KH560 modified La_2O_3; (c) DP modified La_2O_3; (d) NDZ201 modified La_2O_3

The micro morphology of La_2O_3 before and after modification is shown in Figure 1. It can be seen that the shape of unmodified La_2O_3 particles is irregular, the particle size is about 1 ~ 2 μ m, and some particles agglomerate. After modification with KH560 and DP, the particle size of La_2O_3 is less than 1 μ m, and there is no agglomeration between particles. This may be due to the grafting of modifier molecules on the surface of La_2O_3, which increases the molecular spacing of modified particles and reduces particle agglomeration. Although the particle size of NDZ201 modified Lanthanum oxide decreased, some particles still agglomerated, but the particle size of the agglomerated lanthanum oxide was smaller than that of unmodified La_2O_3.

3.2. XPS analysis

Figure 2 shows the total XPS spectrum of modified La_2O_3 obtained using different modifiers, and Table 2 shows the XPS elemental content of modified La_2O_3. From Figure 2 and Table 2, it can be seen that C, O and La elements were detected in the unmodified La_2O_3, with the highest content of O elements, followed by the content of C elements and the lowest content of La elements. The presence of carbon elements may be due to the presence of $La_2O_2(CO_3)$ in the system. The KH560 modified La_2O_3, the DP modified La_2O_3 and the NDZ201 modified La_2O_3 have more Si elements, N elements, and P elements compared with the unmodified La_2O_3, indicating that the modifier KH560, dopamine, and NDZ201 have been successfully grafted to La_2O_3 surface.

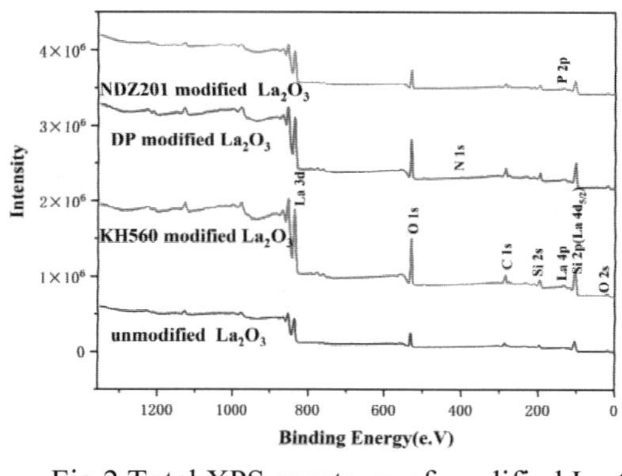

Fig.2 Total XPS spectrum of modified La_2O_3

Table 2 XPS elemental content of modified La_2O_3

	C 1s(%)	O 1s(%)	La 3d(%)	Si 2p(%)	N 1s(%)	P 2p(%)
unmodified La_2O_3	26.67	36.02	10.65	—	—	—
KH560 modified La_2O_3	12..44	21.31	5.61	60.64	—	—
DP modified La_2O_3	35.42	48.9	12.82	—	2.86	—
NDZ201 modified La_2O_3	29.18	44.76	13.63	—	—	12.34

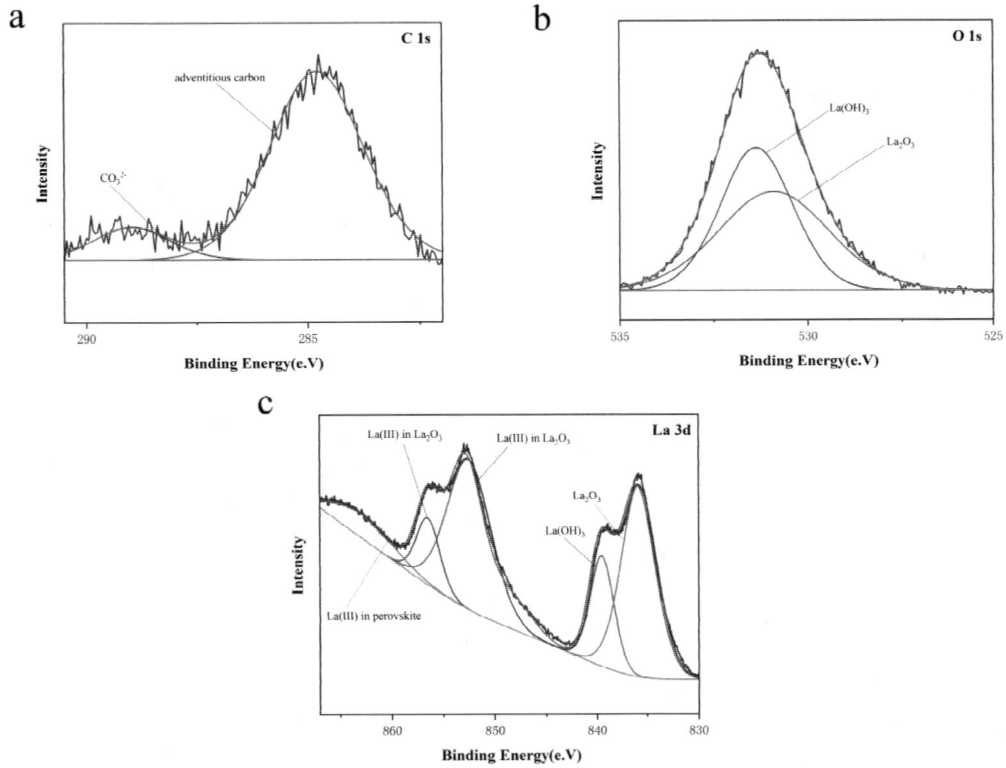

Fig.3 XPS spectra of unmodified La_2O_3 (a) C 1s high-resolution XPS spectrum; (b) O 1s high-resolution XPS spectrum; (c) La 3d high-resolution XPS spectrum

The elemental high fraction XPS spectra of unmodified La_2O_3, KH560 modified La_2O_3, DP modified La_2O_3 and NDZ201 modified La_2O_3 are shown in Fig.3, 4, 5, and 6, respectively. In the C 1s spectrum of La_2O_3 shown in Fig. 3(a), the 284.8 eV peak originates from the indeterminate carbon in chalcocite, while 289.7 eV corresponds to $La_2O_2(CO_3)$ in La_2O_3; in the O 1s spectrum of La_2O_3 shown in Fig. 3(b), the 530.9 eV and 531.3 eV correspond to La_2O_3 and $La(OH)_3$, respectively; in La 3d spectra of La_2O_3 shown in Fig.3(C), 835.9eV, 839.5eV, and 852.7eV correspond to La_2O_3, $La(OH)_3$, and La(III) in La_2O_3, respectively.

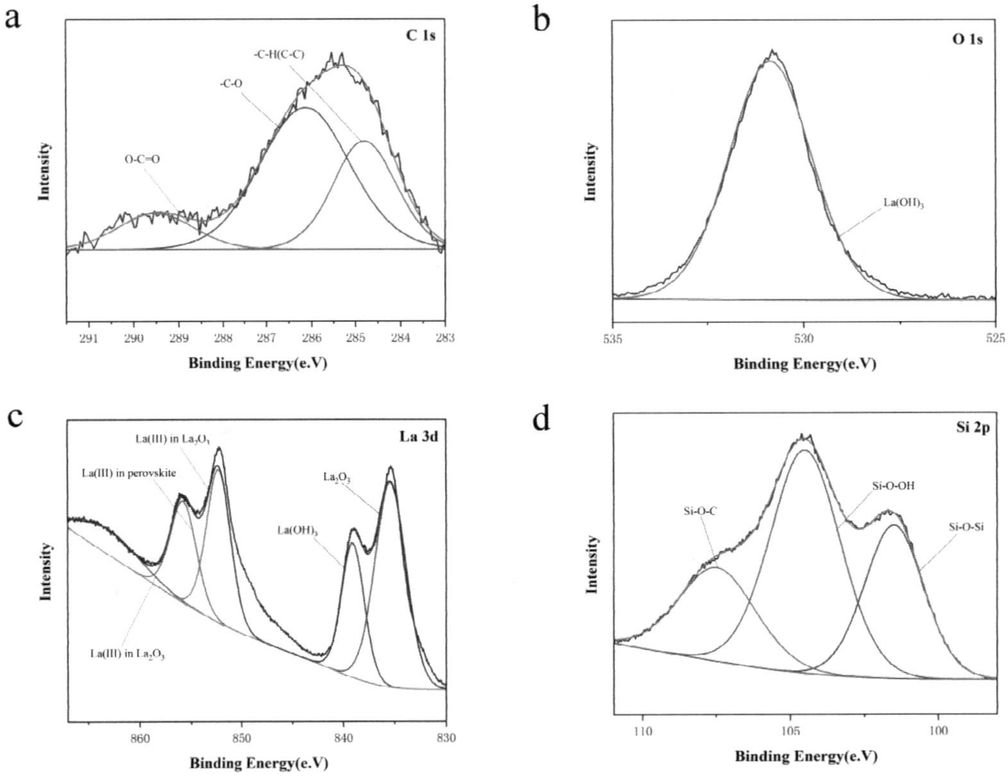

Fig.4 XPS spectra of KH560 modified La_2O_3 (a) C 1s high-resolution XPS spectrum; (b) O 1s high-resolution XPS spectrum; (c) La 3d high-resolution XPS spectrum;

The O 1s spectrum of KH560 modified La_2O_3 was fitted to the split peaks and the results are shown in Fig.4(b). From Fig.4(b), it can be seen that the fitted O 1s spectrum has only one peak, which is located at the binding energy of 530.8 eV. It is known that this place is O in the structure of $La(OH)_3$. A split-peak fit to the Si 2p spectrum of KH560 modified La_2O_3 is shown in Fig.4(d). From Fig. 4(d), it can be seen that the fitted Si 2p spectrum has three peaks located at binding energies of 101.3 eV, 104.5 eV, and 107.6 eV, respectively. Among them, the peak at 101.3eV is Si with Si-O-Si structure; the peak at 104.5eV is Si with Si-O-OH structure; and the peak at 107.6eV is Si with Si-O-C structure. From the structure of KH-560, there is only one chemical state of element Si, i.e., Si-O-C structure; the appearance of Si-O-Si structure, further surface KH 560 has chemically reacted with the surface of La_2O_3 to generate new chemical bonds.

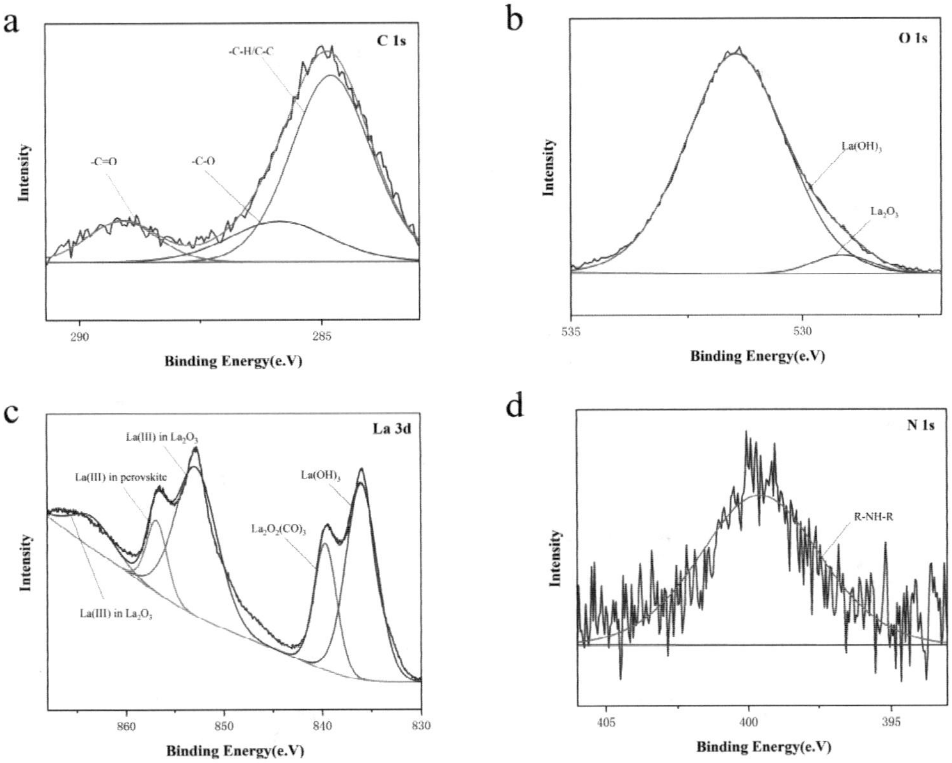

Fig.5 XPS spectra of DP modified La_2O_3 (a) C 1s high-resolution XPS spectrum; (b) O 1s high-resolution XPS spectrum; (c) La 3d high-resolution XPS spectrum; (4) N 1s high-resolution XPS spectrum

The C 1s spectrum of DP modified La_2O_3 was fitted to the split peaks and the results are shown in Fig.5(a). From Fig. 5(a), it can be seen that the fitted C 1s spectrum has three peaks located at binding energies of 284.8 eV, 286.1 eV, and 289.6 eV, respectively. Among them, the peak at 284.8eV is C in the C-H/C-O structure; the peak at 285.9eV is C in the -C-O structure; and the peak at 289.1eV is C in the -C=O structure. The N 1s spectrum of DP modified La_2O_3 was fitted to the split peaks, and the results are shown in Fig.5(d). From Fig.5(d), it can be seen that the fitted N 1s spectrum has one peak located at the binding energy of 399.6 eV. It is known that this location is the N in the R-NH2-R structure. The appearance of the N 1s peak spectrum further proves that dopamine chemically reacted with the surface of La_2O_3 and successfully grafted on the surface of La_2O_3.

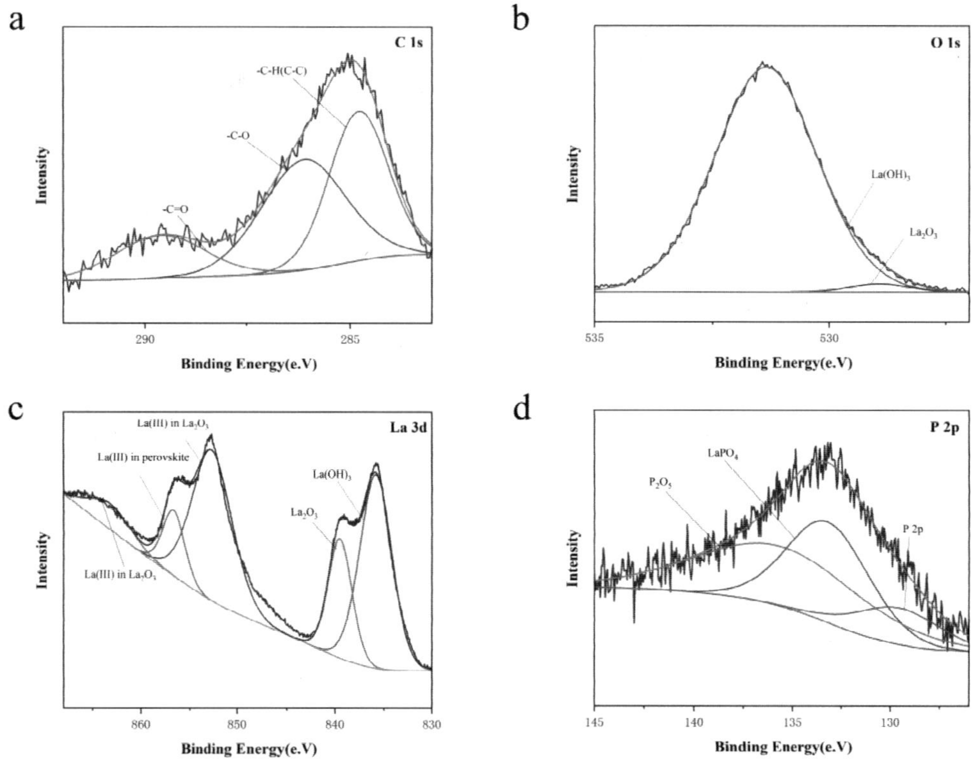

Fig.6 XPS spectra of NDZ201 modified La_2O_3 (a) C 1s high-resolution XPS spectrum; (b) O 1s high-resolution XPS spectrum; (c) La 3d high-resolution XPS spectrum; (d) P 2p high-resolution XPS spectrum

The C 1s spectrum of NDZ201 modified La_2O_3 was fitted to the split peaks and the results are shown in Fig.6(a). From Fig.6(a), it can be seen that the fitted C 1s spectrum has three peaks located at binding energies of 284.8 eV, 286.1 eV, and 289.6 eV, respectively. Among them, the peak at 284.8 eV is C in the C-H/C-O structure; the peak at 286.1 eV is C in the -C-O structure; and the peak at 289.6 eV is C in the -C=O structure. The P 2p spectrum of KR- La_2O_3 was fitted to the split peaks, and the results are shown in Fig.6(d). From Fig.6(d), it can be seen that the fitted P 2p spectrum has three peaks located at the binding energy of 129.4eV, 133.4eV, and 136.6eV, respectively. Among them, the peak at 129.4eV is the P in the P 2p orbital; the peak at 133.4eV is the P in the $LaPO_4$ structure; and the peak at 136.6eV is the P in P_2O_5. The comparative analysis with the elemental spectrum of La_2O_3 further proves that the phthalate coupling agent reacted chemically with the surface of La_2O_3 and successfully grafted on the surface of La_2O_3.

3.3. Morphology and modified La_2O_3/PP composites

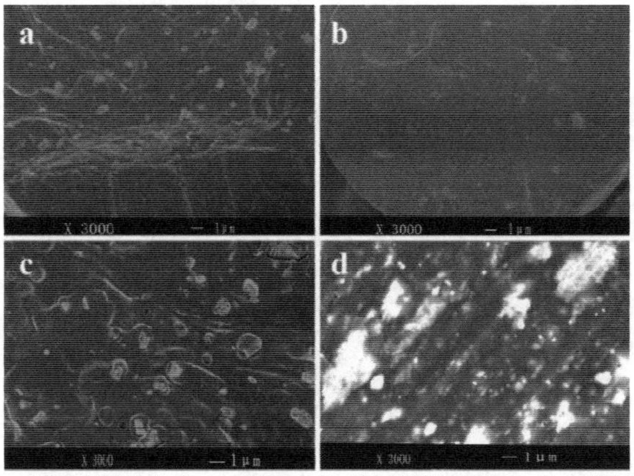

Fig.7 SEM micrographs of (a) KH560 modified La_2O_3/PP composite; (b) DP modified La_2O_3/PP composite; (c) NDZ201 modified La_2O_3/PP composite; (d) unmodified La_2O_3/PP composite

Figure 7 shows the SEM micrographs of La_2O_3/PP composites modified by different modifiers. Unmodified La_2O_3 agglomerates seriously in polypropylene matrix (Fig. 7 (d)), and many particles precipitate from the matrix. The interfacial bonding and filler dispersion of three modified La_2O_3/PP composites are obviously better than those of unmodified La_2O_3/PP composite. Compared with three kinds of modified Lanthanum oxide, NDZ201 modified La_2O_3 still has agglomeration, and the dispersed particle size is relatively large. KH560 modified La_2O_3 is well blended with PP matrix, and the dispersed particle size decreases significantly. In DP modified lanthanum La_2O_3/PP composites, rare earth is evenly dispersed without aggregation, and the cross-section of the composite is very smooth which means a good interface bonding between La_2O_3 and matrix.

4. Conclusion

Modification of La_2O_3 by silane coupling agent (KH560), titanate coupler NDZ-201 and dopamine hydrochloride (DP) and the dispersion of modified rare earth in PP matrix were investigated. The three modifiers can improve the organic properties of lanthanum oxide and the interfacial bonding between lanthanum oxide and PP. Among them, DP has the best modification effect on lanthanum oxide. The DP modified La_2O_3 can disperse evenly in PP matrix with good interface bonding. KH560 also can improve the dispersion of La_2O_3 in PP. However, the dispersion of NDZ201 modified La_2O_3 in the matrix is still poor. In general, dopamine and KH560 can effectively improve the interfacial bonding between lanthanum oxide and polymer, and improve its dispersion in polymer matrix, which is conducive to high-performance lanthanum oxide/polymer composites or fiber materials.

Acknowledgments
The authors sincerely acknowledge the financial support from the Project of Top Young Talents of Beijing Excellent Talents Training Program, and the Innovation Teambuilding Program of Beijing Institute of Fashion Technology (BIFTTD201904). The authors declare that they have no conflict of interest.

References

[1] SAMBHUDEVANA S, SANKARA B, AA S, et al. Synthesis and characterization of organically modified GD_2O_3/natural rubber composites as effective X-ray shielding materials[C]. National conference on innovations in Chemical sciences and technology. 2015
[2] YUAN XL. Studies on the preparation and properties of X-ray shielding thermoplastic polyurethane composites [D]. Beijing University of Chemical Technology, 2010

[3] GU CY, SUN K, ZHANG YF, JIA QX. Preparation and performance study of X-ray resistant rare earth/polypropylene composite fibers[J]. Rare Metal Materials and Engineering,2017,46(09):2633-2638.
[4] LOU PF, JIA QX, WANG FD, XIANG CL. Preparation and performance study of radiation-proof laminated composites[J]. Journal of Beijing Institute of Fashion (Natural Science Edition),2019,39(02):10-15.
[5] LOU P, TENG X, JIA Q, et al. Preparation and structure of rare earth/thermoplastic polyurethane fiber for X - ray shielding [J]. Journal of Applied Polymer Science, 2019, 136(17): 47435.
[6] VANI P, VINITHA G, SAYYED M, et al. Effect of rare earth dopants on the radiation shielding properties of barium tellurite glasses [J]. Nuclear Engineering and Technology, 2021, 53(12). DOI: 10.1016/J.NET.2021.06.009

Journal of Physics: Conference Series **2194** (2022) 012033 doi:10.1088/1742-6596/2194/1/012033

Study on Synthesizing Polycarbonate with Triphosgene

Yongsheng E[1a*], Ping Lv[1b], Fang Yang[1c]

[1]School of Biological and Chemical Engineering, Liaoning Institute of Science and Technology, Benxi, Liaoning 117004, China

[a*]eyongsheng@lnist.edu.cn, [b]eys73@126.com, [c]377619497@qq.com

Abstract—Polycarbonate is a widely used thermoplastic engineering plastic with excellent properties. Using triphosgene and Bisphenol A as raw materials, polycarbonate is synthesized by interfacial polycondensation. The effects of catalyst dosage, molar ratio of raw material, reaction temperature and reaction time on the molecular weight of the product are investigated, and the optimal reaction conditions are obtained.The structure of the product is verified by infrared absorption spectroscopy. The rotary viscometer measures the intrinsic viscosity of the sample and calculates the viscosity average molecular weight of the sample to be 50000. The glass transition temperature of the product is detected to be 148.5℃ by dilatometer. The polycarbonate synthesized by the new method meets the quality requirements.

1. Introduction

Polycarbonate(PC), mainly bisphenol A(BPA) type polycarbonate, is a kind of excellent thermoplastic engineering plastics containing carbonate group in the molecular chain, with outstanding impact resistance and creep resistance, and good heat resistance and cold resistance [1]. It is widely used in automobile, electrical and electronic, building materials, communication equipment, optical disc and daily necessities and other fields[2].

There are two main reports on the synthesis methods of polycarbonate. One is the phosgene interface method, which uses phosgene and bisphenol A as raw materials to prepare polycarbonate through interfacial polycondensation; the other is the melt transesterification method, which uses diphenyl carbonate and bisphenol A as raw materials to prepare polycarbonate through melt polycondensation. The phosgene interface method uses highly toxic phosgene, which brings risks to the experiment or production process. The melt transesterification method needs to be carried out under high temperature and high vacuum conditions, especially the absolute pressure of the latter stage of the reaction is required to be less than 1mmHg [3]. The operating conditions are extremely harsh, and high temperature easily affects the color of the product. Triphosgene（BTC）, also called solid phosgene, is a stable solid compound. Because it is relatively safe in transportation and storage, it has been used more and more widely instead of phosgene in industrial production. Using the triphosgene to synthesize polycarbonate, which not only avoids the highly toxic phosgene, but also enables the reaction to proceed at near normal temperature and pressure. It is a good new method for polycarbonate production. In this paper, we choose triphosgene instead of phosgene and bisphenol A to synthesize polycarbonate by interfacial polycondensation,, and its structure and thermal properties are characterized, which provids technical basis for industrial production.

2. Experiment

2.1. Reagents and instruments
Triphosgene (99%), bisphenol A (99%), triethylamine (AR), dichloromethane (AR), sodium hydroxide (AR). The experimental instruments include WQF-510A FTIR infrared spectrometer, DHJ-8s rotary viscometer, and 30cm dilatometer.

2.2. Synthesis of polycarbonate
A condensing tube, a dropping funnel and a stirrer are installed on the 250ml three-necked flask. Put the flask in the water bath. Add 10.27g bisphenol A and sodium hydroxide solution into the flask, heat up to 40℃, stir until bisphenol A is completely dissolved, then add 0.6g trethylamine, and slowly add the pre-prepared triphosgene dichloromethane solution(5.64g triphosgene is dissolved in 100mL dichloromethane), and then continue the reaction for 60min, then separate the dichloromethane solution of the lower layer, wash with deionized water to neutrality, recover the solvent dichloromethane in the rotary evaporator. The remaining white fibrous solid is washed with hot acetone to obtain the product.

The polymerization equation is shown in Figure 1.

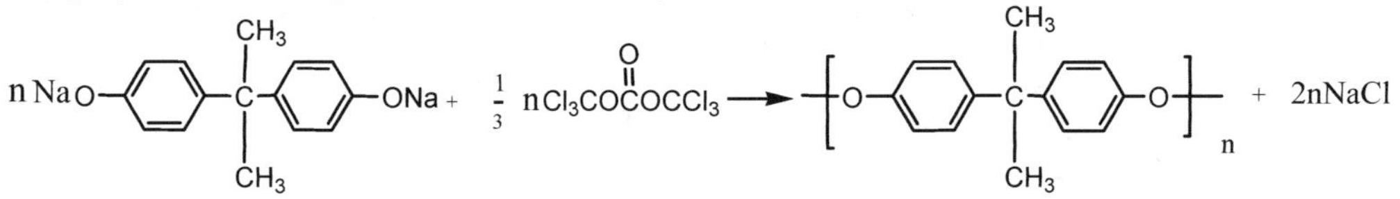

Fig.1 The polymerization equation of polycarbonate

2.3. Testing and characterization
The structure of the product is characterized by the infrared spectrometer, the potassium bromide tablet is pressed, and the scanning range is set to 400~4400 cm-1. The dilatometer is used to detect the glass transition temperature of the product. The rotary viscometer is used to measure the intrinsic viscosity of the sample at 20°C, and then the viscosity average molecular weight of the sample is calculated by the Mark-Houwink formula [4].

Mark-Houwink formula: $[\eta] = kM^{\alpha} \Rightarrow M = ([\eta]/k)^{1/\alpha}$

Where, [η] is the intrinsic viscosity, M is the viscosity average molecular weight, k is the proportional constant, k =1.11×10-2, α is the expansion factor, α=0.82[5].

3. Results and Discussions

3.1. The effect of catalyst dosage on product molecular weight
The molecular weight reflects the degree of polymerization of the polymer and has a direct impact on its performance. Therefore, this experiment uses the molecular weight of the product as the basis for selecting the best reaction conditions.

In this experiment, triethylamine is selected as the reaction catalyst [6]. Take bisphenol A 10.27g (45mmol) and triphosgene 5.64g. Keep the reaction temperature at 40°C and the reaction time for 60 minutes. Investigate the effect of the amount of catalyst triethylamine on the molecular weight of the product.

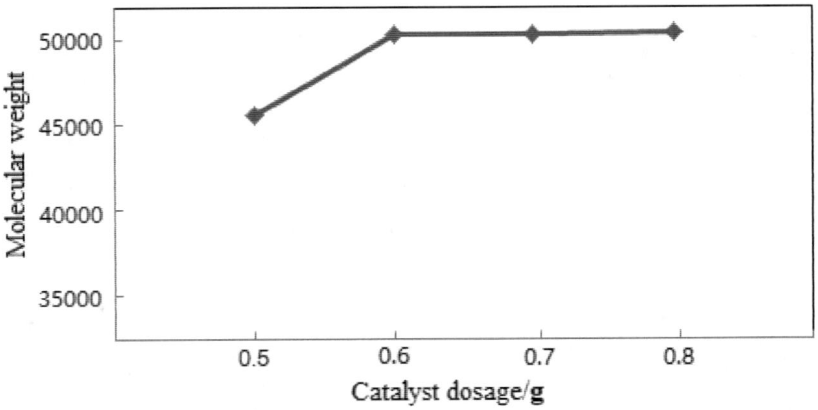

Fig.2 The effect of catalyst dosage on product molecular weight

It can be showed from Figure 2 , when the amount of triethylamine is low, the molecular weight of the product increases with the increase of the amount of triethylamine. When the amount of triethylamine is 0.6g, the catalytic effect is basically achieved, and it is meaningless to continue to increase the amount of catalyst. From the perspective of saving, the optimal amount of catalyst is 0.6g, that is, 0.0584g/g (BPA). At this time, the maximum viscosity average molecular weight of the product polycarbonate is 50000.

3.2. The effect of raw material molar ratio on product molecular weight

Take bisphenol A 10.27g (45mmol) and catalyst 0.6g. Keep the reaction temperature at 40°C and the reaction time for 60 minutes. Investigate the effect of raw material molar ratio on the molecular weight of the product.

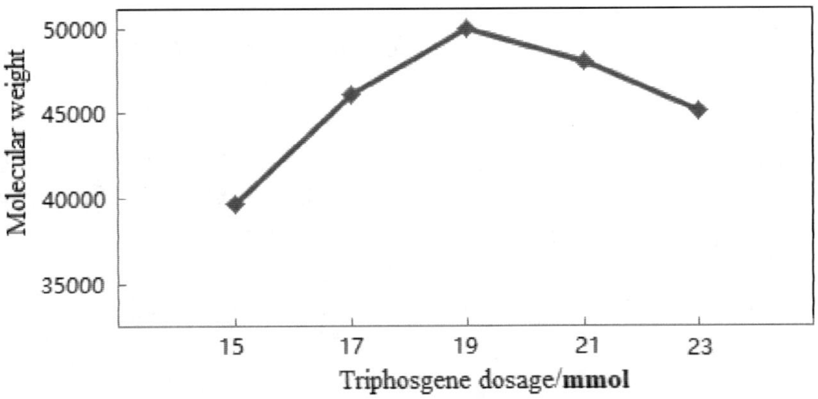

Fig.3 The effect of raw material molar ratio on product molecular weight

According to the polymerization equation, the theoretical mole ratio of raw material bisphenol A and triphosgene is 3:1. Due to the hydrolysis reaction of triphosgene in the alkaline solution of sodium hydroxide, a part of triphosgene is wasted, so triphosgene should be appropriately excessive. When bisphenol A is 45mmol, triphosgene should be greater than 15mmol. It can be seen from Figure 3 that the molecular weight of the product gradually increases with the increase of the amount of triphosgene, but when the amount of triphosgene is too much, the excess triphosgene will act as a blocking agent, which restricts the polymerization reaction and reduce the molecular weight. According to Figure 3, when the amount of triphosgene is 19mmol (5.64g), the product has the largest molecular weight. At this time, the molar ratio of bisphenol A to triphosgene is 3:1.27.

3.3. The effect of the reaction temperature on product molecular weight

Take bisphenol A 10.27g, catalyst 0.6g and triphosgene 5.64g. Keep the reaction time for 60 minutes. Since the boiling point of dichloromethane is 40°C, when investigating the reaction temperature of 45°C and 50°C, we replace the solvent with chloroform, which has similar properties to dichloromethane. Investigate the effect of the reaction temperature on product molecular weight.

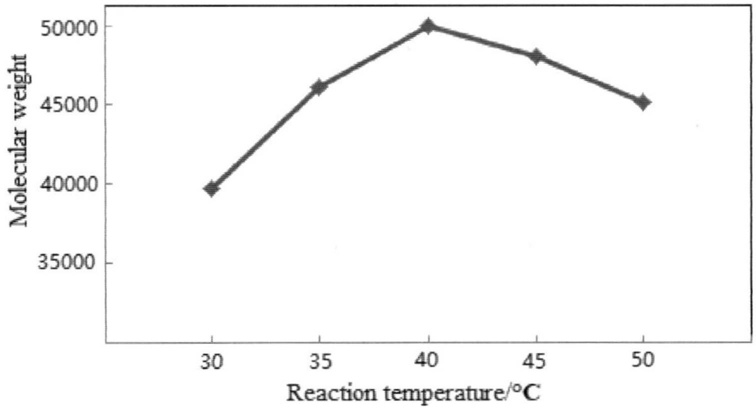

Fig.4 The effect of the reaction temperature on product molecular weight

It can be seen from Figure 4 that the molecular weight of the product first increases and then decreases with the reaction temperature increases. When the reaction temperature is low, the reaction rate is slow and the reaction system is viscous, which is not conducive to the progress of the reaction. With the reaction temperature increases, the reaction rate, the degree of polymerization and molecular weight all increase. When the reaction temperature reaches 40°C (the boiling point of dichloromethane), the molecular weight of the product is the largest. Because it is a reflux reaction, the reaction temperature at this time is also very well controlled. We replace the solvent with chloroform and continue to increase the reaction temperature, the side reactions such as hydrolysis and oxidation intensify, and the molecular weight decreases. Therefore, the optimal reaction temperature in this experiment is 40°C.

3.4. The effect of reaction time on product molecular weight

Take bisphenol A 10.27g, catalyst 0.6g and triphosgene 5.64g. Keep the reaction temperature at 40°C. Investigate the effect of reaction time on product molecular weight.

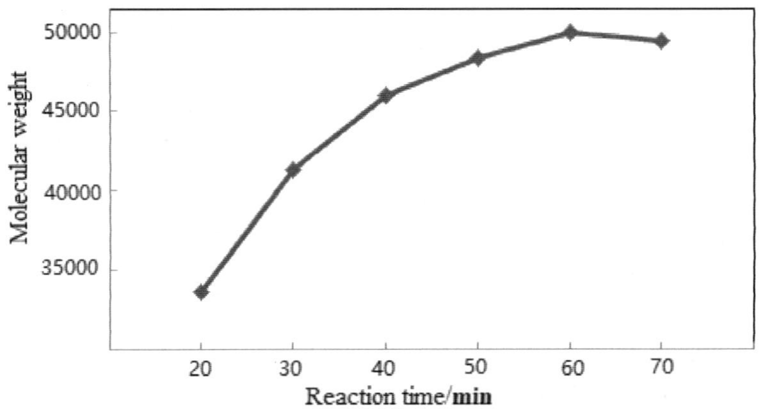

Fig.5 The effect of reaction time on product molecular weight

It can be seen from Figure 5 that the degree of reaction polymerization and molecular weight gradually increases with the reaction time increases. When the reaction time reaches 60min, the molecular weight of the polycondensation reaction is the highest. When the reaction time is prolonged, the side reaction increase and the molecular weight of the product decrease slightly. The optimal reaction time is 60min.

3.5. *Product characterization*

3.5.1*Product structure analysis*

Figure 6 shows the infrared absorption spectrum of the product. In the figure, 1774cm-1 is the characteristic absorption peak of C=O stretching vibration, indicating that there is C=O in the sample; the absorption peak at 1502cm-1 is caused by the stretching vibration of the benzene ring skeleton, indicating that the sample contains benzene ring; 1230cm-1, 1191cm-1 and 1160cm-1 are caused by the stretching vibration of C-O, indicating that there is C-O in the sample[7]; 1016cm-1 and 825cm-1 correspond to the fingerprint peaks of the para aromatic ring; 2969 cm-1 is saturated with C-H stretching vibration peak, indicating that there is -CH3 in the sample [8]. It can be seen from the infrared absorption spectrum that the product is polycarbonate.

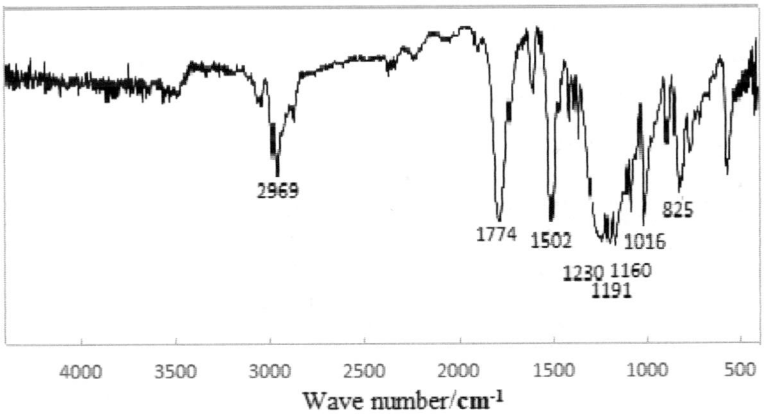

Fig.6 FT-IR spectrum of polycarbonate

3.5.2*Product glass transition temperature determination*

Figure 7 is the relationship graph of capillary liquid column height and temperature.

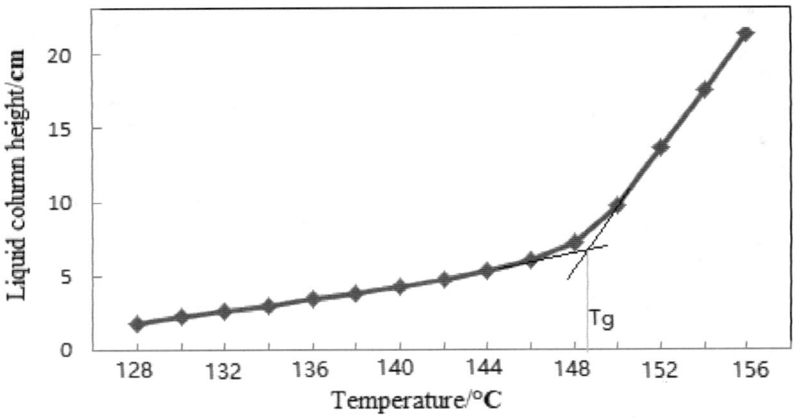

Fig.7 Relationship graph of capillary liquid column height and temperature

The temperature corresponding to the turning point of the slope of the two straight lines in the figure is the glass transition temperature of the product, which is 148.5°C, which is consistent with the polycarbonate prepared by the phosgene interface method and the melt transesterification method.

4. Conclusion

Based on the results and discussions presented above, the conclusions are obtained as below:

(1) Using bisphenol A and triphosgene as raw materials, triethylamine as catalyst, polycarbonate is synthesized by interface polycondensation method. The optimal reaction conditions are: the molar ratio of raw materials n (BPA): n (BTC) is 3: 1.27, catalyst dosage is 0.0584g/g (BPA), the reaction temperature is 40℃, and the reaction time is 60min.

(2) The product is proved to be bisphenol A type polycarbonate by infrared absorption spectroscopy. The maximum viscosity average molecular weight is calculated to be 50000. The glass transition temperature of the product is 148.5℃. The polycarbonate produced by the new method is consistent in quality with the existing polycarbonate.

(3) The new method not only avoids the highly toxic phosgene, but also enables the reaction to proceed at near normal temperature and pressure, which provides important guiding significance for industrial production. Further optimization of reaction conditions to improve the molecular weight of products will be the direction of future research.

Acknowledgments

This work was financially supported by t the Scientific Research Funds of the Educational Department of Liaoning Province in 2021 (General Project)（LJKZ1057）.

References

[1] Yi Luo, Qinsha Xiao. (1997)Engineering Plastics Polycarbonate and Its Developing Trends. Plastics, vol.26, no.5, pp.12-15.
[2] Guihua Zhang, Jianzhou Niu, Xingtian Guo. (2018)Analysis of Polycarbonate Production and Market Application Prospect in China. Chemical Industry, vol.36, no.5, pp.45-56.
[3] Yuhong Chen, Lipeng Du, Kai Wang, et al. (2016)Studies on Synthesis and Thermal Properties of Polycarbonate Derived from 9，9-Bis(4-hydroxyphenyl) Fluorene. Acta Polymerica Sinica, no.9, pp. 1273-1280.
[4] Riguang Jin, Youqing Hua. (2000)Polymer Physics, 2rd ed., Beijing: Chemical Industry Press, pp.92.
[5] Xiaoyu Wang, Meijun Chen, Yang Cui. (2013)Research on the relationship between molecular weight and viscosity of polycarbonate. Gansu Science and Technology, vol.29, no.12, pp.18-19.
[6] Xiaoting Chen, Kaijun Li, Xudong Tang. (2007)Synthesis of the High Molecular Weight Polycarbonate by Interfacial Polycondensation of Bisphenol A and Triphosgene. Journal of Tianjin University of Science and Technology, vol.22, no.3, pp.48-50.
[7] Sweileh B.A., Al-Hiari Y.M., Kailani M.H. (2010)Synthesis and Characterization of Polycarbonates by Melt Phase Interchange Reactions of Alkylene and Arylene Diacetates with Alkylene and Arylene Diphenyl Dicarbonates. Molecules, vol.15, no.5, pp.3661-3682.
[8] Yongcheng Ning. (2000)Structure Identification of Organic Compounds and Organic Spectroscopy, 2rd ed., Beijing: Science Press, pp.332-340.

The effect of magnetic field on freezing point of water

YanMin Li[1a*], HaiFei Lin[1b*], ZhenDong Yang[1c*]

[1]College of Civil Engineering and Architecture, Shandong University of Science and Technology, Qingdao 266590, China

[a*]1728756632@qq.com, [b*]2571275982@qq.com, [c*]y344282892@163.com

Abstract: Cold storage technology can realize energy storage and release according to demand and balance the power grid by "shifting peak and filling valley". Ice cold storage system has the outstanding advantages of large cold storage density and simple system, and is the frontier direction of current cold storage technology research and application. However, there are some problems such as low phase transition temperature of cold storage medium and high energy consumption under cold storage condition. The research shows that magnetic field can increase the freezing point of water and aqueous solution by about 3.0-5.0K, which is beneficial to further increase the economic benefit and save energy in refrigeration condition based on the refrigeration demand.

1. Introduction

The physical and chemical properties of water or aqueous solution will change after magnetic field treatment. Magnetic treatment technology is widely used in practical applications. At present, the cooling energy of air conditioning is getting higher and higher, even reaching more than 40% in some cities [1]. Air conditioning power consumption increases peak load in summer, resulting in insufficient power consumption in peak period and surplus power consumption in trough period, and the gap is gradually increasing, exacerbating the contradiction of unbalanced peak-valley load in China's power grid [2]. In order to solve this trend, ice storage air conditioning system has been widely used as an important means to balance the load of power grid and to shift peak load and fill valley. The application rate of ice storage air conditioning system exceeds 70% in China. Although the ice storage air conditioning system has many advantages compared with conventional air conditioning, the operation efficiency and cooling capacity of the unit are greatly attenuated due to the low phase change temperature of ice, and there are still areas to be improved [3]. In order to make more effective use of ice storage technology, it is necessary to further increase the phase transition temperature of the storage medium, so as to reduce the energy consumption of refrigeration. Studies have shown that magnetized water has a very different freezing temperature from ordinary water. Inaba et al. [4] used a high-resolution differential scanning calorimeter to measure that the freezing point of H_2O and D_2O under the condition of 6T magnetic field intensity was higher than that without magnetic field, and the phase transition temperature increased with the increase of magnetic field intensity. Magnetic field can affect phase equilibrium conditions and dynamic behavior of water freezing. Magnetized water or aqueous solution can not only be used for frozen preservation of food and frozen biological products, but also helps to prevent scale prevention and corrosion inhibition [5-8]. Based on the deficiency of the cold storage medium in the ice storage system, this paper evaluates the influence of magnetic field on the energy consumption under the cold storage condition by introducing the change of the phase change temperature of the water after magnetic field treatment, hoping to promote the application and promotion of the ice storage air conditioning system in China.

2. Influence of magnetic field on freezing point of water

2.1 Study on the effect of magnetic field on water and aqueous solution

Scholars at home and abroad have done a lot of research on the influence of magnetic field on water, and analyzed that the main reasons for the influence of magnetic field on water are the nuclear spin and electron orbital motion of atoms in water molecules, molecular dipole characteristics and intermolecular hydrogen bond characteristics. The structure of water molecule and the diagram of electric dipole under static magnetic field were shown in Fig.1[9]. In general, the magnetic field affects the hydrogen bond state, magnetic torque and activation energy in aqueous solution through polarization and Lorentz force, thus causing the change in the physical and chemical properties of water.

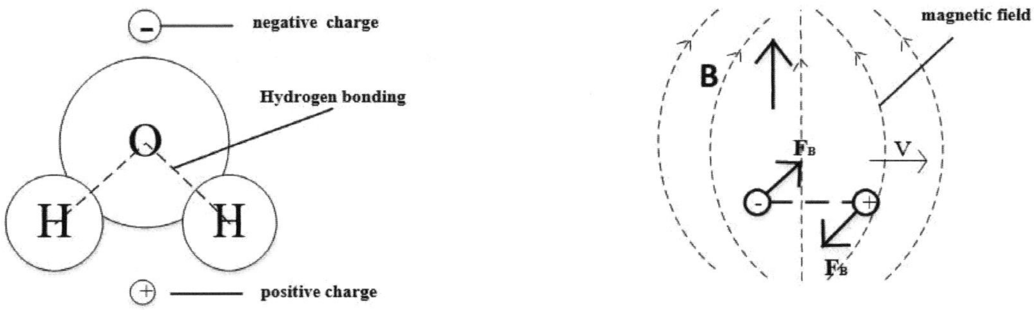

Fig. 1 The structure of the water molecule and diagram of the electric dipole in static magnetic field [9]

The influence of magnetic field on water and solution is mainly reflected in the change of some physical and chemical properties of water, such as the change of refractive index, surface tension and viscosity. Some studies have studied the effect of magnetic field on water surface tension by using molecular dynamics simulation method, and the results show that the surface tension coefficient of water decreases. Analysis shows that magnetic field increases the average distance between water molecules, breaks partial hydrogen bonds, and weakens the interaction between water molecules [10]. With the increase of magnetic field intensity and magnetization time, the solution viscosity decreased significantly, which was also related to lorentz force and hydrogen bond fracture [11]. Some studies have also found that magnetic field can significantly increase the evaporation rate of water by treating liquid with magnetic field, and its influence intensity is related to magnetic field intensity gradient [12]. At the same time, some studies have shown that the magnetic field with a strength of 0.16-0.53t can reduce the friction coefficient of liquid, and it is analyzed that the magnetic field changes the bonding tightness of hydrogen bond [13]. Haruki Hosoda [14] et al. measured the refractive index of water and electrolyte. Under the magnetic field intensity of 10T, the refractive index of pure water and electrolyte solution with low concentration increased by about 0.1%, which was also attributed to the magnetic field enhancing the stability of hydrogen bond. Toledo et al. [15] believed that magnetic field weakened the hydrogen bonds in the clusters of large water molecules, and the clusters of large molecules were split into small clusters with strong hydrogen bonds.

2.2 Effect of magnetic field on freezing point of water

Najibiet al. [16] proposed a simple equation by correlating the freezing point to hydrate equilibrium temperature in salt and/or organic inhibitor, as equation (1). The change of freezing point of aqueous solution can be calculated as long as the phase equilibrium of hydrate is known. Therefore, in this paper, the influence of magnetic field on freezing point of water and NaCl solution is quantitatively analyzed by using the previously measured data of hydrate phase balance [17、18]. Table 1 show the change of freezing points of water and NaCl solutions with different concentrations in the presence and absence of magnetic field.

$$T_s = T_w - 0.683\Delta T \quad (1)$$

T_s and T_w are the hydrate equilibrium temperatures in bulk water and salt inhibitors with or without magnetic field, respectively. ΔT is the change of freezing point.

Tab. 1 The change of freezing point of water and NaCl solutions with different concentrations in the presence and absence of magnetic field

	CO_2 Hydrate (Magnetic field) [17]		CO_2 Hydrate (Without magnetic field) [18]		
	p/MPa	T/K	p/MPa	T/K	ΔT/K
Pure water	2.4-3.5	281.7-283.9	2.4-3.5	278.9-281.6	3.6
0.029NaCl	2.4-3.5	280.9-283.0	2.4-3.5	277.3-280.4	4.4
0.056NaCl	2.4-3.5	279.5-282.6	2.4-3.5	276.1-279.5	4.7
0.110NaCl	2.4-3.5	277.8-281.1	2.4-3.5	273.9-277.3	5.5

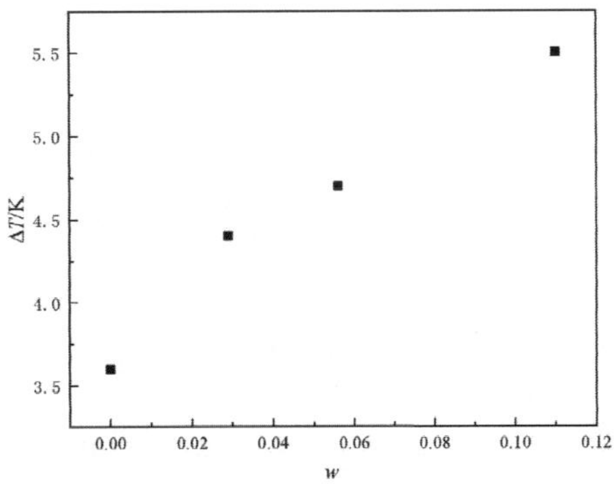

Fig.2 The change of freezing point of pure water and NaCl solution with different concentrations in the presence and absence of magnetic field

As can be seen from Tab.1, compared with the absence of magnetic field, magnetic field can increase the freezing point of pure water by about 3.6K, and in (0.029, 0.056 and 0.110) NaCl solutions by about 4.4, 4.7 and 5.5K, respectively. As can be seen from Figure 2, with the increase of NaCl solution concentration, the magnetic field has a greater influence on the freezing point of the solution. As is known to all, when the ice storage system is running under the condition of cold storage, the lowest temperature of ice storage is about -6.5°C due to the relatively low phase change temperature of ice. The temperature of coil and encapsulated ice storage equipment is also about -5.6°C, so that the evaporation temperature of the refrigerator is 8-10°C lower than that of the conventional non-cold storage system. As a result, the energy efficiency of cold storage air conditioning system is usually lower than that of conventional air conditioning system under the same load. Studies have shown that when evaporation temperature increases by 1°C, COP increases by 3%-4% [19]. It can be concluded that if the water treated by magnetic field is used in the cold storage system, the evaporation temperature of the refrigerant can be increased by about 3.6°C, which will greatly improve the refrigeration capacity and efficiency of the cold storage system, and achieve the purpose of reducing the refrigeration energy consumption and operation cost.

3. Conclusion

By studying the magnetic field can improve water and aqueous phase transition temperature of about 3 to 5 k, then applying magnetized water in the cold storage condition can decrease caused by low phase transition temperature of the cold storage conditions of energy consumption, for the cold storage system is more secure, the use of economy provides the train of thought, so as to promote the development of cold storage air conditioning system in our country.

Reference

[1] Wang, Y. G., Pei, X. Y., & Zhao, L. N. (2008). Economy analysis and comparison of ice-storage air-conditioning system for an office building. Power Demand Side Management.

[2] Technology. (2007). Economic analysis of an ice-storage air-conditioning system. Refrigeration and Air-Conditioning.

[3] Zhimei, L. U., Junqiang, Y. U., Pei, L., & Zhang, L. (2010). New research on phase change materials for air conditioning cool storage system. Materials Review.

[4] Inaba, H., Saitou, T., Tozaki, K. I., & Hayashi, H. (2004). Effect of the magnetic field on the melting transition of H_2O and D_2O measured by a high resolution and supersensitive differential scanning calorimeter. Journal of Applied Physics, 96(11), 6127-6132.

[5] Hossein, Kiani, and, Da-Wen, & Sun. (2011). Water crystallization and its importance to freezing of foods: a review. Trends in Food Science & Technology.

[6] Song, J., Liu, B., Guan, W., & Zhang, D. (2016). Effect of DC magnetic field on freezing process of onion cells. Journal of Refrigeration.

[7] Kobe, S., G Dražić, Cefalas, A. C., Sarantopoulou, E., & J Stražišar. (2008). Nucleation and crystallization of CaCO3 in applied magnetic fields. Crystal Engineering, 5(3), 243-253.

[8] Yanagiya, S. I., Sazaki, G., Durbin, S. D., Miyashita, S., Nakajima, K., & Komatsu, H., et al. (2000). Effects of a magnetic field on the growth rate of tetragonal lysozyme crystals - sciencedirect. Journal of Crystal Growth, 208(1), 645-650.

[9] A, X. H., A, Y. P., & B, Z. M. (2016). Effect of magnetic field on optical features of water and kcl solutions. Optik, 127(16), 6371-6376.

[10] Zhang, J. (2002). Molecular dynamics simulation of the surface tension of water under a magnetic field. Journal of Qingdao University (Natvral Science).

[11] Niu, X., Kai, D., & Fu, X. (2011). Experimental study on the effect of magnetic field on the heat conductivity and viscosity of ammonia–water. Energy & Buildings, 43(5), 1164-1168.

[12] Nakagawa, J, Hirota, N., Kitazawa, K., & Shoda, M. (1999). Magnetic field enhancement of water vaporization. Journal of Applied Physics, 86(5), 2923-2925.

[13] Wang, Y., Zhang, B., Gong, Z., Gao, K., Ou, Y., & Zhang, J. (2013). The effect of a static magnetic field on the hydrogen bonding in water using frictional experiments. Journal of Molecular Structure, 1052, 102-104.

[14] Hosoda, H., Mori, H., Sogoshi, N., Nagasawa, A., & Nakabayashi, S. (2004). Refractive indices of water and aqueous electrolyte solutions under high magnetic fields. Journal of Physical Chemistry A, 108(9), 1461-1464.

[15] Toledo, E., Ramalho, T. C., & Magriotis, Z. M. (2008). Influence of magnetic field on physical–chemical properties of the liquid water: insights from experimental and theoretical models. Journal of Molecular Structure, 888(1-3), 409-415.

[16] Najibi, H., Mohammadi, A. H., & Tohidi, B. (2006). Estimating the hydrate safety margin in the presence of salt and/or organic inhibitor using freezing point depression data of aqueous solutions. Industrial & Engineering Chemistry Research, 45(12), 189-222.

[17] Zhao JR. (2020). Effect of static magnetic field on phase transition characteristics of carbon dioxide hydrate [D]. Shandong University of Science and Technology.

[18] Sun, S. C., Liu, C. L., & Ye, Y. G. (2013). Phase equilibrium condition of marine carbon dioxide hydrate. Journal of Chemical Thermodynamics, 57(none).

[19] Ji, L., Wei, X., Yu, Z., Sun, Z., Zhang, R., & Qiao, B. (2018). Energy efficiency limits of cool storage air conditioning system. Heating Ventilating & Air Conditioning.

Preparation and application of low molecular weight polyether polycarboxylate superplasticizer

Bobo Liu* Kuliang Ke* Yinggang Yang* Donghui Miao* Chong He*

KZJ New Materials Group Co., Ltd. Xian 710500, Shanxi, China

*Corresponding author, E-mail: vikerliu@163.com

Abstract. In this paper, a low molecular weight polycarboxylate superplasticizer was designed and synthesized by using low molecular weight polyether macromonomer with molecular weight of 800 as the side chain of polycarboxylate superplasticizer; The ratio of oxidant and reductant in inorganic salt initiator system is 2:1, and that in organic water solvent initiator system is 4:1. The amount of initiator in the two systems accounts for 5% and 2% of the total amount of polyether macromonomer, respectively; GPC and Fourier transform infrared spectroscopy tests showed that the number average molecular weight of the water reducer was about 24000, the conversion rate of polyether macromonomer was more than 90%, and the molecular structure of the polymerization product met the design requirements. The influence of the polymer on the fluidity and bleeding rate of manufactured sand concrete is also discussed.

1. Introduction

Polycarboxylate superplasticizer, as the third generation of concrete superplasticizer, has a unique comb structure and steric effect. Compared with the first and second generation of superplasticizers, it improves the performance of concrete mixture more significantly. In recent years, it has been widely used at home and abroad. Polycarboxylic acid water reducing agent is polymerized based on organic polymer materials. It has strong designability of molecular structure and large space for performance adjustment. There are many studies on its performance and molecular structure in China. Zhou Keli, Liu Jin, Li Zhen and others believe that the new polycarboxylic acid water reducer with short side group and low molecular weight has good dispersion, good fluidity retention of cement paste and no bleeding segregation [1]. He Yan, Kong Yaning, Wang Xiaofu and others found that the ratio of acid to ether, the molar ratio of chain transfer agent to polyether have a great influence on the molecular weight of polycarboxylic acid water reducer, and the molecular conformation curls with the increase of polyether molecular weight [3]. Therefore, the molecular weight and conformation can be designed by adjusting the ratio of acid to ether and the molar ratio of chain transfer agent to polyether. Wu yingzhe and Huang Furen pointed out that the relative molecular weight increase of polycarboxylate water reducer affects the fluidity of cement paste and the slump of concrete [2].

At present, when domestic polycarboxylic acid water reducing agent encounters some poor sand and gravel materials, the fluidity of concrete will become poor and easy to bleed. For example, when the fineness modulus is greater than 3.0, the concrete will leak stone, have poor fluidity and serious bleeding; In view of the above problems, it is difficult to solve the water reducing agent produced by conventional materials and processes. In this paper, organic and inorganic redox systems are used as polymerization initiators, polyether macromonomer with molecular weight of 800 is used for room temperature polymerization, and a polyether water reducing agent with number average molecular weight of 24000 is obtained. The water reducing agent in this paper has good adaptability in coarse machine-made sand and low bleeding rate of mixed concrete. The water reducing agent has the advantages of simple production operation, no special equipment, no pollution in production, and is convenient for market promotion and application. At the same time, the water reducing agent solves the problem of the decline of concrete construction performance caused by the deterioration of the quality of sand and gravel materials.

2. Test

2.1. Raw material

Raw materials for raw material test synthesis are shown in Table 1 below. Raw materials for concrete test are as follows: Cement: conch P.O 42.5; Fly ash: Datang Shenglong class II fly ash; Sand: machine-made sand, fineness modulus MX = 3.2; Stone: Double graded basalt (small stone 5 ~ 16mm, large stone 5 ~ 31.5mm); Additive pce0: polycarboxylic acid water reducer is synthesized in this test, with solid content of 50% and water reduction rate of 32%; Admixture pce1: commercially available polycarboxylic acid water reducer, with solid content of 49.8% and water reduction rate of 32%.

Table 1. Chemical raw materials

Name	Essential information
Isopentenyl polyoxyethylene ether	Industrial products, Produced in Guangdong, molecular weight 800
Methacrylic acid	Analytical pure, Produced by Sinopharm group
Acrylic acid	Analytical pure, Produced by Sinopharm group
Potassium hypophosphite	Analytical pure, Produced by Sinopharm group
Sodium bisulfite	Analytical pure, Produced by Sinopharm group
Hydrogen peroxide	Analytical pure, Produced by Sinopharm group
Organic reducing agent (f)	Industrial products, Clariant Chemical (Huizhou) Ltd.

2.2. Synthetic test method

In this experiment, the amount of fixed modified isopentenyl polyoxyethylene ether is m, and the amount of other raw materials is calculated according to the mass ratio of m, including 13% m of acrylic acid, 3% m of methacrylic acid, 0.4% m of reducing agent F, 1.7% m of sodium bisulfite, 1.6% m of hydrogen peroxide and 3.3% m of potassium hypophosphite. The admixture preparation operation in the synthesis stage of this test is as follows:

(1) Prepare liquid A and liquid B. Liquid A is composed of acrylic acid, methacrylic acid and some deionized water; Liquid B consists of reducing agent F, sodium bisulfite and deionized water;

(2) Add a certain amount of modified isopentenyl polyoxyethylene ether macromonomer, hydrogen peroxide, potassium hypophosphite and some deionized water to the four mouth flask equipped with thermometer and stirrer, and stir to the state of aqueous solution;

(3) Add liquid A and B dropwise at the same time. The dropping of liquid A is completed in 3H and the dropping of liquid B is completed in 3.5H;

(4) Hold for half an hour, add sodium hydroxide and adjust the pH value to about 5 ~ 7;

(5) Make up water until the solid content is 50% to obtain polycarboxylic acid water reducer mother liquor pce0.

2.3. Test methods

The test method is as follows:

(1) Gel permeation chromatography (GPC) was characterized by Waters1515 Isocratic HPLP pump/Waters2414 differential detector, and the molecular weight and conversion rate of the sample PCE0 and commercial sample PCE1 were obtained by gel chromatography;

(2) The synthetic samples were analyzed by PE spectrum 100 infrared chromatograph (FT-IR) produced by American PE company. The solution is coated on the KBr wafer to form a thin liquid

film, dried under the infrared lamp for determination, and its main molecular structure is analyzed to judge whether the sample meets the expected goal of molecular design;

(3) The fluidity of cement paste shall be tested in accordance with GB / T 8077-2008 test method for homogeneity of concrete admixtures. It mainly tests the performance of synthetic water reducing agent pce0 and commercial water reducing agent pce1 in the fluidity test of cement paste at the initial and 1H and 2H, and the comparison of the state of cement paste;

(4) The concrete test shall be carried out in accordance with the relevant provisions of GB 8076-2008 concrete admixtures. It mainly tests the difference of fluidity, bleeding rate and other performance indexes between synthetic water reducing agent pce0 and commercial water reducing agent pce1 in mixing C30 concrete. See Table 2 concrete raw materials and mix proportion for concrete mix proportion.

Table 2. Concrete raw materials and mix proportion

Material variety	Cement	Fly ash	Machine-made sand (Mx=3.2)	Crushed stone (5～16mm)	Crushed stone (5～31.5mm)	Tap water
mix proportion (kg/m3)	280	80	870	300	680	155

3. Test results and discussion

3.1. Gel chromatographic test results

In this experiment, the PCE0 and commercial samples PCE1 were tested by gel permeation chromatography (GPC). The test results were shown in Figure 1 Gel chromatogram, and the data analysis was shown in Table 3 GPC test results.

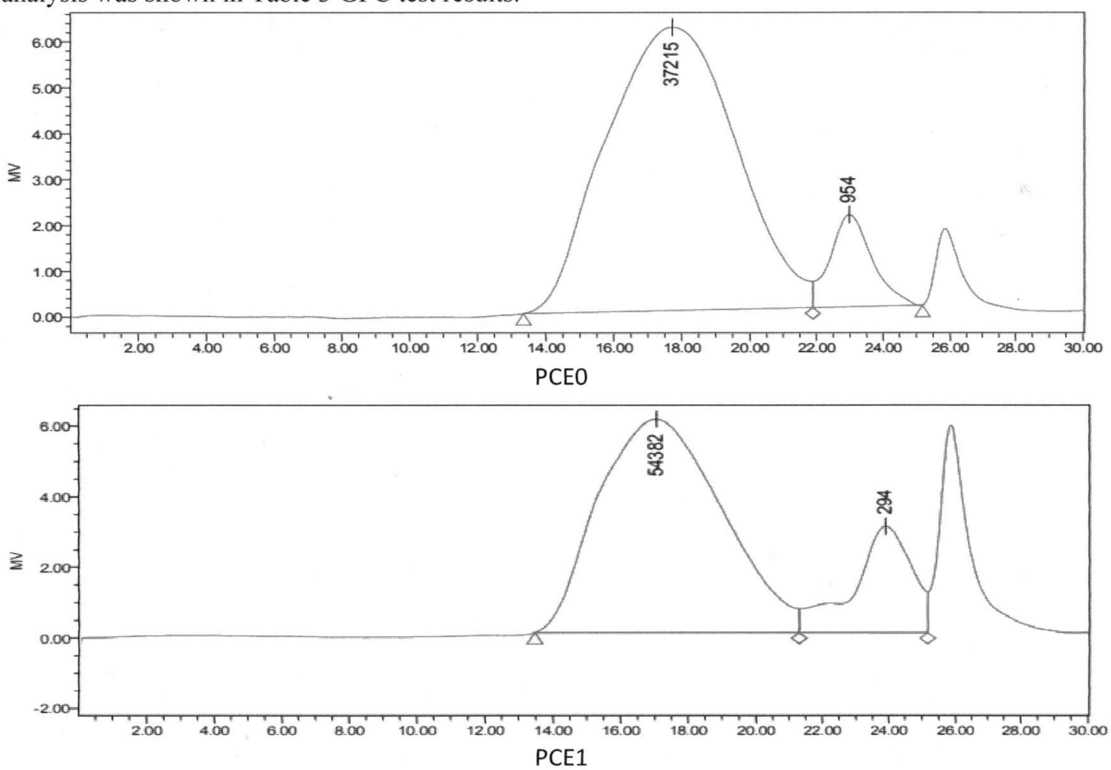

Fig. 1 Gel chromatogram

Table 3. Concrete raw materials and mix ratio

Sample	Mn	Mw	MP	Mz	Polydispersity	Total area	The measure of area /%	Retention time (min)	Conversion rate /%
PCE0	24663	65906	37215	178264	2.672272	1820880	90.35	17.729	90.35
PCE1	32579	83800	54382	204474	2.572222	1909249	80.67	17.058	80.67

According to the data in Figure 1 and table 3, the number average molecular weight of polycarboxylic acid water reducer sample pce0 in this test is more than 24000, the weight average molecular weight is more than 65000, and the viscosity average molecular weight is more than 37000. According to the calculated area, the conversion rate is about 90%; The number average molecular weight of the commercial sample pce1 is more than 32000, the weight average molecular weight is more than 83000, and the viscosity average molecular weight is more than 54000. The calculated area shows that the conversion rate is about 80%; In this experiment, the modified isopentenyl polyoxyethylene ether monomer with molecular weight of 800 was selected, and its weight average, number average and viscosity average molecular weight were lower than those of conventional polycarboxylic acid water reducer; The polycarboxylic acid water reducer pce0 polymerized in the organic and inorganic double initiator system has higher polymerization conversion than the conventional polycarboxylic acid water reducer on the market; The polymerization conversion rate of this test is about 10% higher than that of conventional products, which makes this test product have higher economic performance price ratio, improve the utilization rate of raw materials and be more environmentally friendly; The number average molecular weight of the test product is more than 8000 lower than that of the conventional product, which provides more space for the performance improvement and functional particularity of the product.

3.2. Fourier infrared spectrum test results

The mother liquor pce0 was tested by Fourier infrared spectrometer, and the obtained infrared spectrum is shown in Fig. 2.

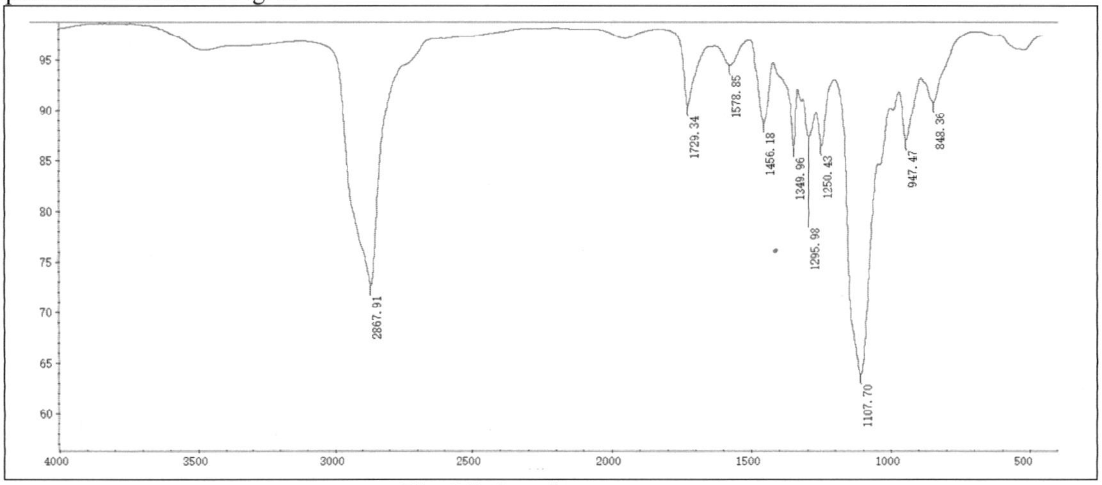

Fig. 2 Infrared spectrogram

According to the infrared spectrum analysis in Figure 2, 2869.91cm-1 is the stretching vibration peak of saturated carbon-C-H, 1725.71cm-1 is the stretching vibration peak of - C = O in carboxyl functional group, 1579.52cm-1 is the C = C skeleton vibration peak, 1454.64cm-1 is the bending absorption peak of saturated carbon-C-H, 1350.08cm-1 is the in-plane bending vibration absorption peak of saturated carbon-C-H, and 1250.26cm-1 is the stretching vibration peak of carbon oxygen-C-O in alcohol ether, 1107.61 cm-1 is the characteristic absorption peak of ether bond-O - and 847.43 cm-1 is the trans vibration peak of alkyne. The infrared spectrum of pce0 and the above analysis show that the mother liquor pce0 reaches the expected functional group structure, and there is no unreacted C = C unsaturated double bond.

3.3. Fourier infrared spectrum test results

The cement paste test results of pce0 and pce1 in this test are shown in Table 4 cement paste test results below.

Table 4. Test results of cement paste

Water reducing agent model	Dosage of water reducing agent (%)	Fluidity of cement paste (mm)			State
		0h	1h	2h	
PCE0	0.2	220	240	230	The homogeneity is good after standing
PCE1	0.2	225	240	200	Slight bleeding at the edge after standing

It can be seen from the test data of cement paste fluidity in Table 4 that under the same dosage of pce0 and pce1, the initial paste fluidity and the paste fluidity after 1h are close, but the pce1 paste fluidity obviously loses faster after 2h, indicating that the water reducer pce0 synthesized in this test has better slump retention performance than pce1; Through the static state of cement slurry mixed with two water reducing agents, it can be seen that pce1 has slightly poor water locking performance, while pce0 has good water locking performance. Therefore, it can be inferred that pce0 water reducing agent has better uniformity in concrete.

3.4. Concrete application experiment

In the experiment, the low molecular weight polyether polycarboxylic acid water reducer pce0 synthesized in this paper and the commercial water reducer pce1 are tested for concrete performance under the same slump. The results are shown in Table 5 concrete test results.

Table 5. Concrete test results

Water reducing agent model	Dosage of water reducing agent (%)	Expansion / Slump (mm)		Gas content (A/%)	Strength (MPa)		Bleeding rate ratio (B/%)
		0h	1h		7d	28d	
PCE0	0.15	590/230	570/230	1.6	37.5	44.6	1.0
PCE1	0.15	590/220	560/230	1.4	34.0	42.8	4.5

It can be seen from the experimental data of concrete in Table 5 that the water reduction rate of low molecular weight polyether water reducer pce0 synthesized in this test is equivalent to that of commercial water reducer pce1. Under the same slump, the fluidity of concrete is close, and the slump loss and fluidity loss in 1H are equivalent; Compared with pce1, the air content of pce0 mixed concrete is slightly higher, so the concrete slurry is fuller; The 7d and 28d strength of concrete, pce0 mixed concrete is significantly higher than pce1; Finally, the bleeding rate of pce0 mixed concrete is much lower than that of pce1 mixed concrete, which effectively solves the problem of easy bleeding of coarse machine-made sand concrete. According to the results of gel permeation chromatography (Fig. 1), the polycarboxylate superplasticizer synthesized by polyether macromonomer with 800 molecular weight is shorter. The average molecular weight of the superplasticizer is lower because of its short side chain structure. Therefore, the steric hindrance effect of the superplasticizer is not obvious, thus reducing the sensitivity of the superplasticizer. At the same time, due to the reduction of the molecular weight of the side chain, the overall molecular weight of the water reducer is reduced, which is more conducive to the rapid adsorption of polycarboxylic acid water reducer in the cement slurry, the concrete has good slurry starting effect, the bleeding rate of the mixed coarse machine-made sand concrete is lower than that of the conventional water reducer, and the adaptability is better.

4. Conclusion

In conclusion, the following conclusions can be drawn:

(1) Due to the short side chain, the average molecular weight of polycarboxylic acid water reducer synthesized from low molecular weight polyether macromonomer is lower than that of conventional polycarboxylic acid water reducer, and the number average molecular weight is only more than 24000; Under the double initiator system of organic aqueous solution and inorganic salt, the conversion of polymerization is 10% higher than that of conventional water reducer, and the polymerization efficiency is high;

(2) Through the Fourier infrared spectrum test, it can be known that the molecular main structure and functional groups of the low molecular weight polyether water reducer pce0 in this test are in line with the molecular structure of ether polycarboxylic acid water reducer, and the reaction of main monomers such as acrylic acid and methacrylic acid in the product is complete, which meets the design goal;

(3) Due to its low average molecular weight, low molecular weight polyether water reducer has more flexible and efficient adsorption on cement slurry, and has good slurry starting effect in coarse mechanism concrete. On the premise of maintaining the workability of concrete such as high fluidity and low slump loss, compared with conventional water reducer, concrete strength is higher and bleeding rate is lower.

References:

[1] Synthesis and action mechanism of low molecular weight polycarboxylate water reducer containing short side group [J]. Wang Zhou Keli, Liu Jin, Li Zhen. Journal of materials science and engineering. 2011,29 (02): 295-300.

[2] Structural analysis and performance study of polycarboxylic acid water reducer with different molecular weight [J]. Wu yingzhe, Huang Furen, Luo Hui, Shao Qiang. Guangzhou architecture. 2019,47 (02): 13-16.

[3] Effect of polycarboxylic acid water reducer with different carboxyl density on the properties of cement paste [J]. He Yan, Kong Yaning, Wang Xiaofu, Shui Liangliang. Journal of building materials. 2018 (02): 185-188.

[4] Plank J, Pöllmann K, Zouaoui N. Synthesis and performance of methacrylic ester based polycarboxylate superplasticizers possessing hydroxy terminated poly (ethylene glycol) sidechains [J]. Cement and Concrete Research, 2008, (38): 1210-1216.

[5] Principle and Application of Concrete Admixtures [M]. China Planning Press, Edited by Chen Jiankui, 1997.

[6] Development course and research and development direction of polycarboxylic acid water reducing agent [J]. Wang Ling, Gao Ruijun. Concrete World. 2012 (11).

Effect of shear thickening property of polymer on relative permeability

Wenlong Zhang[1,2,3], Jian Hou[1,2,3*], Yongge Liu[1,2,3], Qingjun Du[1,2,3]

[1]Key Laboratory of Unconventional Oil & Gas Development, China University of Petroleum (East China), Ministry of Education, Qingdao, 266580, PR China

[2]Laboratory for Marine Mineral Resources, Qingdao National Laboratory for Marine Science and Technology, Qingdao, 266237, PR China

[3]School of Petroleum Engineering, China University of Petroleum (East China), Qingdao, 266580, PR China

*Corresponding author: houjian@upc.edu.cn (J. Hou)

Abstract: The relative permeability at a high Darcy velocity was estimated by combining a numerical simulation model that considers the rheology of polymer in shear thickening region. After verifying the robustness of the method, the effect of shear thickening property of polymer on relative permeability was studied. The results show that when Darcy velocity exceeds 17.61 m/d, the apparent relative viscosity of polymer solution begins to increase, and the rheology of polymer changes from shear thinning to shear thickening behavior, which reduces water relative permeability. For water relative permeability model, as Darcy velocity increases, water relative permeability at residual oil decreases in a power function and exponent of water relative permeability model increases linearly.

1. Introduction

In polymer flooding, the polymer dissolved in water will change the relative flow between oil and water by increasing the aqueous-phase viscosity, which will affect relative permeability [1]. In addition to directly changing polymer viscosity, the change of polymer rheology caused by the change of Darcy velocity in shear thinning region will also affect the relative permeability [2]. Darcy velocity varies greatly for multiphase flow in porous media (e.g., radial flow from near-wellbore to far-field), which causes polymer to exhibit different rheological properties (e.g., shear thinning and shear thickening) [3]. However, the effect of shear thickening properties of polymer on relative permeability has not been studied.

The existing methods for calculating relative permeability are mainly steady-state method and unsteady-state method [4]. The steady-state method, which is difficult to achieve oil-water stable flow, is not suitable for high Darcy velocity in shear thickening region. Because of the limitation of mathematical model, the unsteady-state method fails to directly obtain relative permeability at high Darcy velocity in shear thickening region. Relative permeability at a high Darcy velocity was therefore estimated by combining a numerical simulation model that considers the rheology of polymer in shear thickening region. With that, the effect of shear thickening property of polymer on relative permeability was studied.

2. Estimation Method of Relative Permeability

The relative permeability was estimated by automatically matching actual and predicted production data, in which the actual production data was obtained by coreflooding experiment, the predicted data was obtained by numerical simulation model, and the automatic history matching was performed through an optimization algorithm.

2.1. Characterization of polymer properties in numerical model

Based on the black oil model, a numerical simulation model of oil-water two-phase polymer flooding was established. The mass conservation equation [5] of polymer component is as follows,

$$\nabla\cdot\left(\frac{kk_{rw}c_p}{\mu_p B_w r_k}(\nabla p_w-\rho_w g\nabla D)\right)+\nabla\cdot\left[d_p\varphi\left(1-F_p\right)S_w\nabla c_p\right]+q_w c_p=$$
$$\frac{\partial}{\partial t}\left[\frac{\varphi\left(1-F_p\right)S_w c_p}{B_w}\right]+\frac{\partial\left[\left(1-F_p\right)\left(1-\varphi\right)\rho_b\hat{c}_p\right]}{\partial t} \tag{1}$$

where, k and k_{rw} refer to absolute and relative permeability respectively; μ_p is apparent viscosity of polymer; r_k is reduction factor of permeability; d_p is polymer diffusion coefficient; F_p is inaccessible pore volume; $\hat{c}_p$ is adsorbed mass concentration of polymer; c_p is mass concentration of polymer; φ is porosity.

As a non-Newtonian fluid, the apparent viscosity of polymer is usually characterized by a power function [6]. Polymer solution tends to show shear-thinning behavior at low and medium shear rates and shear-thickening behavior at high shear rate, which are characterized by different power functions. A representation model of apparent viscosity of polymer in shear-thickening region was established by Galindo-Rosales et al. [7] in log-log plot. As follows,

$$\log\left(\mu_{\mathrm{p}}\right)=\log\left(\mu_{\max}\right)+\frac{\log\left(\mu_{\min}\right)-\log\left(\mu_{\max}\right)}{1+\left(K\cdot\log\left(\gamma\right)\right)^m} \tag{2}$$

where, $\mu_{\max}$ is apparent viscosity at maximum shear rate in shear-thickening region; $\mu_{\min}$ is apparent viscosity at minimum shear rate in shear-thickening region; γ is shear rate; K and m are constants.

A formula is introduced to calculate the shear rate at different Darcy velocity in polymer-flooding numerical simulation model [8].

$$\gamma=a\cdot\left(\frac{3b+1}{4b}\right)^{\frac{b}{b-1}}\cdot\frac{v_{\mathrm{D}}}{\sqrt{kk_{\mathrm{rw}}\varphi s_{\mathrm{w}}}} \tag{3}$$

where, a is a constant with default value of 6; b is a power exponent and its default value is 0.5.

In addition, other physical parameters of polymer, including inaccessible pore volume, residual resistance factor, diffusion and adsorption, are also considered in the numerical model.

2.2. Automatic matching of production data

The optimization algorithm updates relative permeability by changing the parameters of relative permeability model, which makes polymer-flooding numerical model generate new predicted production data. In the iterations, when the latest predicted production data reaches the best match with the production data obtained by coreflooding experiment, the automatic matching of production data stops, and the latest relative permeability in numerical model is the optimal estimated relative permeability.

The relative permeability model is characterized by a power law function [9] as shown in Eq. (4) and (5).

$$k_{rw}(s_w)=k_{rw}(s_{or})\cdot\left(\frac{s_w-s_{wi}}{1-s_{or}-s_{wi}}\right)^{n_w} \tag{4}$$

where, s_w refers to water saturation; s_{wi} refers to irreducible water saturation; s_{or} refers to residual oil saturation; $k_{rw}(s_{or})$ refers to water relative permeability at residual oil; n_w refers to exponent of power-law model of water relative permeability.

$$k_{ro}(s_w) = \left(\frac{1 - s_w - s_{or}}{1 - s_{or} - s_{wi}} \right)^{n_o} \tag{5}$$

where, k_{ro} refers to oil relative permeability; n_o refers to exponent of power-law model of oil relative permeability.

By observing Eq. (4) and (5), it is found that water relative permeability model is controlled by water relative permeability at residual oil and exponent, and oil relative permeability model only limited by exponent. The relative permeability can be changed by adjusting these model parameters. In order to realize the optimal matching between actual and predicted production data faster and more accurately, Levenberg-Marquardt method is applied to perform automatic matching. Since coreflood is carried out at a constant rate, only cumulative oil production along with injection pressure are regarded as actual production data.

2.3. Verification of estimation method of relative permeability

A numerical simulation model of polymer flooding with predefined relative permeability replaces coreflooding experiment to provide actual production data, and the predefined relative permeability is therefore regarded as actual relative permeability. The numerical model providing predicted production data is input into a set of random relative permeability curve (grey curve in Fig. 2) at the beginning of automatic matching. As long as the relative permeability estimated by the proposed method matches well with the predefined one, it means that the method is robust. It is noted that the square section with equal area is used to represent the circular section of core in numerical model. In addition, the core and fluid parameters of the numerical model are all measured by experiments (see Table 1 and Fig.1).

Table 1 Parameters of numerical simulation model

Parameter	Value	Parameter	Value
Polymer concentration, kg/m^3	1.8	μ_{max}, mPa·s	21.8
Injection rate, ml/min	6.5	K	0.277
TDS of brine, mg/L	10500	m	53.2
Brine permeability, $\times 10^{-3}\mu m^2$	750	d_p, $\times 10^{-8} m^2/s$	7.3
Viscosity of brine, mPa·s	0.78	F_p	0.21
Viscosity of oil, mPa·s	42.5	φ, %	19.55
Irreducible water saturation	0.215	r_k	3.05

The average error of relative permeability curve (η) is defined to represent the error between the estimated and actual relative permeability curve, which is written as,

$$\eta = \frac{\sum_{i=1}^{n} \left| k_{rl,i} - k_{arl,i} \right|}{\sum_{i=1}^{n} k_{arl,i}} \tag{6}$$

where, k_{arl} refers to actual relative permeability; i refers to point i on relative permeability curve.

Looking at Fig. 2, we can find that the random relative permeability curves at the beginning of automatic matching are significantly away from the actual ones, but the estimated relative permeability curves match the actual ones quite well. The average error of oil relative permeability curve is 3.5%, and the average error of water relative permeability curve is 4.1%, which means the robustness of the estimation method of relative permeability of polymer flooding is good.

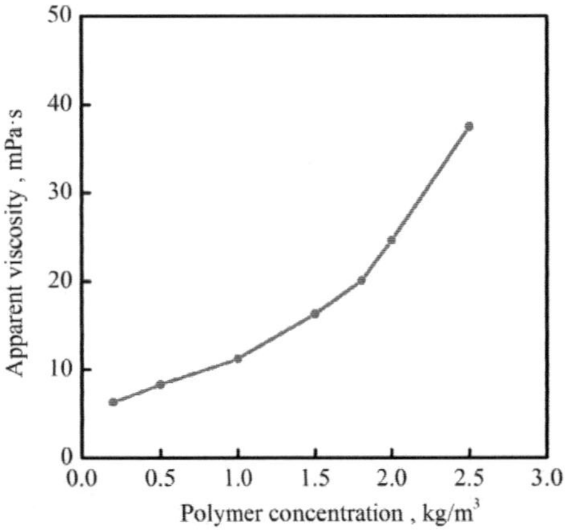

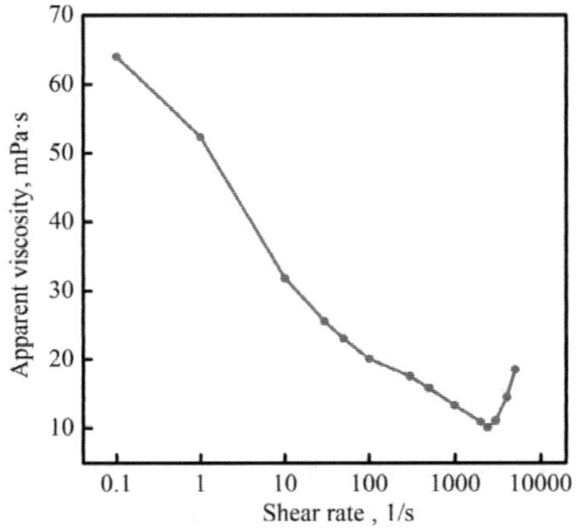

(a) Apparent viscosity-polymer concentration curve (b) Apparent viscosity-shear rate curve
Fig. 1 Apparent viscosity of polymer

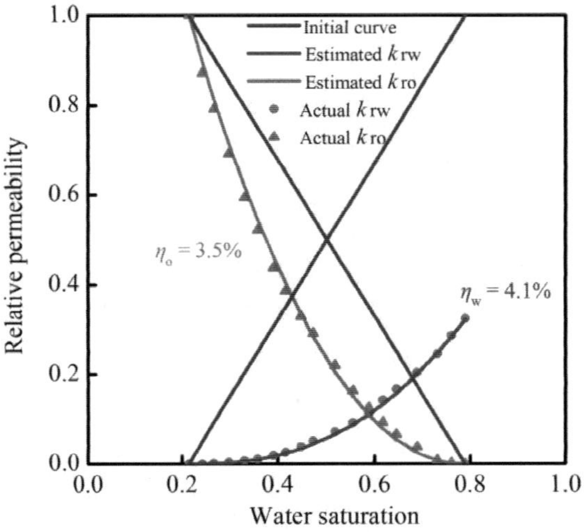

Fig. 2 Comparison of actual and estimated relative permeability

3. Effect of shear thickening property of polymer on relative permeability

3.1. Determination of shear thickening region

Apparent relative viscosity (μ_{rp}) was introduced to describe the rheology of polymers flowing in porous media [10], and its expression is as follows,

$$\mu_{rp} = \frac{F_r}{F_{rr}} \tag{7}$$

where, F_r refers to resistance factor; F_{rr} refers to residual resistance factor.

The polymer solution with concentration of 1.8 kg/m^3 flowed through the core at different Darcy velocities (ratio of injection rate to core cross-sectional area), and the apparent relative viscosity was measured according to Eq. (7). As shown in Figure 3, the critical Darcy velocity of shear thinning region and shear thickening region is 17.61 m/d, and the corresponding critical injection rate is 6 ml/min.

3.2. Relative permeability under the influence of shear thickening

6 coreflooding experiments were completed (see Table 2), of which case 1 for comparison was water flooding experiment, and the other 5 experiments were polymer flooding experiments in the shear thickening region (i.e., injection rate from 6 ml/min to 10 ml/min). The relative permeability curves at different injection rates were then estimated by the proposed method.

As shown in Fig. 4, the Darcy velocity in shear thickening region mainly affects water relative permeability, and all water relative permeability curves of polymer flooding are lower than that of water flooding. Further observation shows that water relative permeability decreases with the increase of Darcy velocity. This is because apparent relative viscosity of polymer increases as Darcy velocity increases in shear thickening region (see Fig. 3). By studying water relative permeability models at different Darcy velocities in shear thickening region, it can be found water relative permeability at residual oil endpoint decreases in a power function as Darcy velocity increases (see Fig. 5(a)), and exponent of water relative permeability model increases in a linear function as Darcy velocity increases (see Fig. 5(b)).

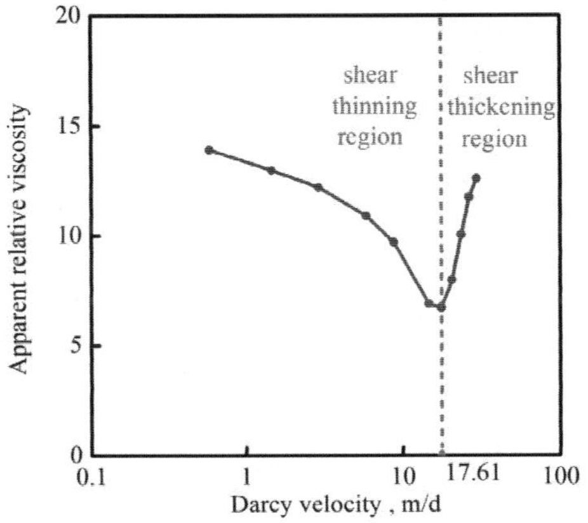

Fig. 3 Rheology of polymer flowing in core

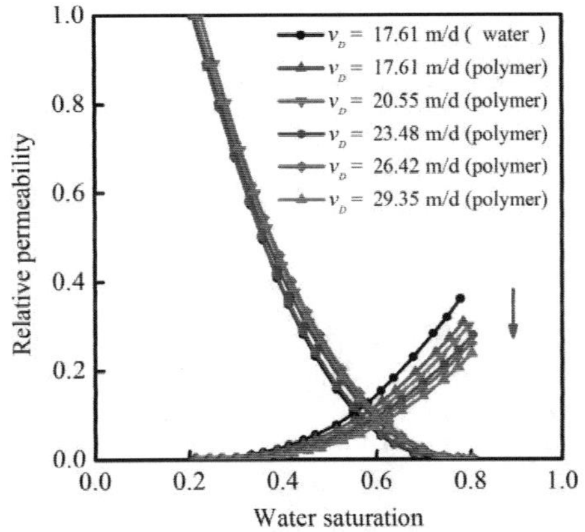

Fig. 4 Relative permeability curve

Table 2 Parameters of coreflooding experiment

Case	k, 10^{-3} μm^2	φ, %	Injection rate, ml/min	Darcy velocity, m/d	Residual resistance factor	s_{wi}	s_{or}
1	721	19.23	6	17.61	-	0.216	0.221
2	686	18.65	6	17.61	3.05	0.211	0.215
3	755	19.77	7	20.55	3.01	0.226	0.205
4	673	18.96	8	23.48	3.19	0.209	0.195
5	666	19.33	9	26.42	3.63	0.213	0.198
6	712	19.11	10	29.35	3.96	0.216	0.197

4. Conclusion

Based on the study of varieties of relative permeability curves under different Darcy velocities in shear thickening region, the key conclusions are as below,

(1) The robustness of the proposed relative permeability estimation method is proved by good matching between actual and estimated relative permeability.

(2) The critical Darcy velocity of shear thinning region and shear thickening region is 17.61 m/d in this work.

(3) For water relative permeability model, as Darcy velocity increases, water relative permeability at residual oil decreases in a power function and exponent of water relative permeability model increases linearly.

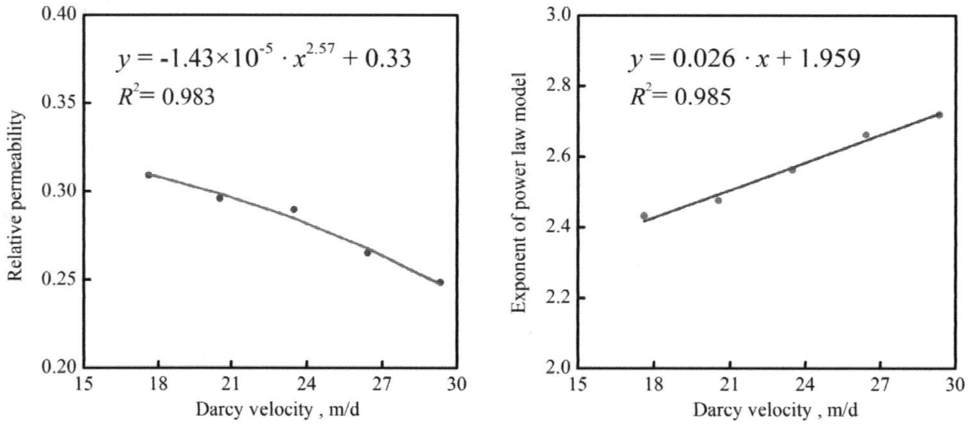

(a) Water relative permeability at residual oil endpoint (b) Exponent of power-law model
Fig. 5 Effect of Darcy velocity on water relative permeability model

Acknowledgments
This work was financially supported by the National Natural Science Foundation of China (Grant No. 51804324) and the Important National Science and Technology Specific Projects of China (Grant No. 2016ZX05011-003).

References
[1] Koh, H., Lee, V. B., Pope, G. A. (2018) Experimental investigation of the effect of polymers on residual oil saturation. SPE J., 23(1): 1-17.
[2] Zhang, W., Hou, J., Liu, Y., Du, Q., Cao, W., Zhou, K. (2021) Study on the effect of polymer viscosity and Darcy velocity on relative permeability curves in polymer flooding. J. Petrol. Sci. Eng., 200: 108393.
[3] Seright, R. S., Fan, T., Wavrik, K., de Carvalho Balaban, R. (2011) New insights into polymer rheology in porous media. SPE J., 16(1): 35-42.
[4] Honarpour, M., Koederitz, L., Harvey, A. H. (2018) Relative permeability of petroleum reservoirs. CRC press.
[5] Liu, Y., Hou, J., Liu, L., Zhou, K., Zhang, Y., Dai, T., Cao, W. (2018) An inversion method of relative permeability curves in polymer flooding considering physical properties of polymer. SPE J., 23(5): 1929-1943.
[6] Manica, R., De Bortoli, A. L. (2004) Simulation of sudden expansion flows for power-law fluids. J. Nonnewton Fluid Mech., 121(1): 35-40.
[7] Galindo-Rosales, F. J., Rubio-Hernández, F. J., Sevilla, A. (2011) An apparent viscosity function for shear thickening fluids. J. Nonnewton Fluid Mech., 166(5-6): 321-325.
[8] Cannella, W. J., Huh, C., Seright, R. S. (1988) Prediction of xanthan rheology in porous media. SPE Annual Technical Conference and Exhibition, Houston City, USA, 2-5 October, SPE-18089-MS.
[9] Chen, S., Li, G., Peres, A. M., Reynolds, A. C. (2008) A well test for in-situ determination of relative permeability curves. SPE Reserv. Eval. Eng., 11(1): 95-107.
[10] Dupas, A., Henaut, I., Rousseau, D., Poulain, P., Tabary, R., Argillier, J.-F., Aubry, T. (2013) Impact of polymer mechanical degradation on shear and extensional viscosities: toward better injectivity forecasts in polymer flooding operations. SPE International Symposium on Oilfield Chemistry, Woodlands City, USA, 8-10 April, SPE-164083-MS.

The effects of temperature on surfactant solution: A molecules dynamics simulation

Pang Feng [1a*], **Linlin Wang** [1b*]

[1]College of Safety and Ocean Engineering, China University of Petroleum, Beijing 102249, China

[a*]1152582467@qq.com, [b*]linlin.wang@cup.edu.cn

Abstract—We use molecules dynamics simulation to investigate the effects of temperature on surfactant solution, both the microcosmic and macroscopic performance of the surfactant solution. Results show that the increase of temperature has bidirectional influence on the capability of the surfactant solution. From the microscopic point of view, the thermophoresis of the free water molecules has arisen in the forward of the temperature, which causes the hydrogen bonding between molecules fracture and reconnection. And the micelles keep close to and detached from each other for these. From a macro perspective, viscosity will decrease with temperature. But when the temperature is about 392.15K, the viscosity-temperature curve will have a peak value. Combining the micro results and macro results of the molecules dynamics simulation, we can clearly draw a conclusion that the effects of temperature on the surfactant solution never are monotonic.

1. Introduction

Since oilfield engineers of Mitchell Energy & Development Corporation first applied a polymer-free water-based fracturing fluid to the shale gas fracturing operation of Barnett and achieved good results in 1997 [1-3], the viscoelastic surfactant based fracturing fluid (VES) has become one possible replacement of guar and polymer in the fracturing fluid systems for the advantages of gel to be broken easily, simple operation, no need for crosslinker and biocides [4-9]. The VES frequently consist of viscoelastic surfactants [10-11]. Therefore, the high temperature has a negative impact on the performance of the VES [12], but the reservoirs are increasingly deep. Researchers have done much experimental work to synthetize various dilute cationic, cationic, zwitterionic, non-ionic and mixed surfactants or other auxiliary agents to improve the viscoelasticity of the solution at high temperature [13-16], which shows us that the temperature surely play a great role in the measure of the performance of the surfactant solution. However, despite the large amount of experimental data on surfactant solution, the understanding of the link among forming microstructures, their macroscopic rheological properties and the external temperature is still lacking.

As the computers developing faster and faster, the molecules dynamics simulation has become an efficient tool for the investigation of the performance of surfactant solution [17]. As a mesoscale simulation algorithm, the dynamic mean field density functional method (MesoDyn) is proposed and has successfully been used to investigate the density distribution of various surfactant solutions [18–22]. MU et al. put this simulation method into studying the self-assembled morphologies of polyelectrolytes [23]. Elistratova et al. investigate the effect of the external temperature in aqueous solutions of the triblock copolymers [24]. However, they cannot link the self-assembled morphologies and the rheological properties of the surfactant solution to the external temperature. In this work, we apply the MesoDyn method to simulate the self-assembled morphologies of the surfactant solution in different external temperature, apply the shear flow method to simulate the rheological properties of the surfactant solution in different external temperature, and associate with each other.

In this paper, Section 2 describes two simulation methods, field density functional method and shear flow method, to simulate the self-assembled morphologies and calculate shear viscosity, respectively. The computational details of the simulation are given in Section 3. Section 4 first presents and discusses the results of the simulation, and then the performance of the surfactant solution can be achieved in different external temperature. In Section 5, we draw conclusions from this work.

2. Method

2.1. Theory

2.1.1. Theory in MesoDyn

The Mesodyn method is based on the dynamic mean field density functional method, which is a coarse-grained method for simulating the dynamics and equilibrium of complex fluids at long time and large space scales. And it is on account of the idea that the free energy F of an inhomogeneous liquid is a function of the local density function ρ. From the free energy, all thermodynamic functions can be derived [25-26].

The model used in the MesoDyn project consists of beads of various types D, O… with interactions described by harmonicoscillator potentials for the intramolecules interactions (Gaussian chain) and a mean field potential for all other interactions [27]. Each bead is of a certain component type representing covalently bonded groups of atoms such as those given by one or a few structural units of a polymer chain. The dynamics of the system is described by a set of functional Langevin equations. They represent diffusion equations in the component densities, which take account of the noise in the system [28].

2.1.2. Shear flow method

The shear flow methods are as similar as real experiments: the cause is an appropriate field or gradient, which is imposed on the system, then the ensemble average of the effect, the resulting flux, is measured and the ratio of flux and field gives the viscosity [29]. In the shear flow simulations, we have to tackle two problems: for one thing, we must mechanically impose the temperature. To generate a fix shear flow, we have to use a moving periodic boundary condition, which is called Lees–Edwards boundary condition [30]. For another, the temperature is enforced by constantly pumping energy into the system. Hence, we must get rid of the heat by applying an appropriate thermostat.

2.2. Module

2.2.1. MesoDyn module

In MesoDyn simulation, the surfactant molecules structures were simplified. A mass model was proposed for real-time elastic deformable modeling that each particle represented a particular group of molecules structure, and different particles were connected with each other. Considering the special molecules of the surfactants, five kinds of MesoDyn particles were defined. Groups of '–$CH_2CH_2CH_2CH_2$–', '–CH_2NHCO–', '–CH_2CH_2NOH–', '–CH_2OH', and 'H_2O' were represented by letters D, S, O, T and W in sequence. These letters correspond to the colors deep blue, red, green, light blue, and purple, respectively. For example, simplified MesoDyn topology of the surfactant is D4S1O1T2 (see Fig.1).

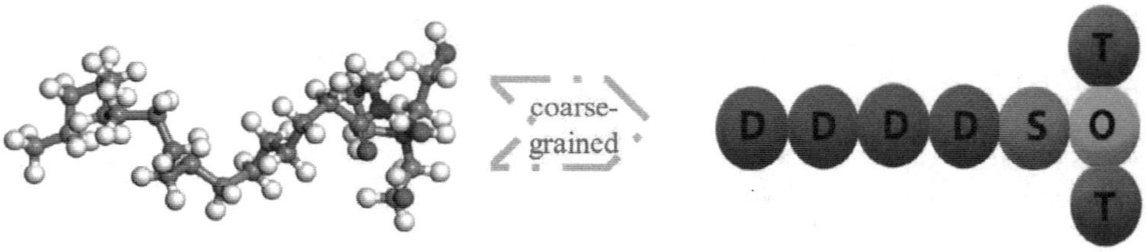

Fig.1 the coarse-grained cationic surfactant

2.2.2.Rheological properties simulation module

In molecules dynamics (MD) simulation, we put the surfactant and water molecules in a cube to simulate rheological properties of the surfactant solution. Fig.2 are the cubic structures of the surfactant in aqueous solution using molecule model. And we mark the distribution of surfactant molecules in the system by red symbols.

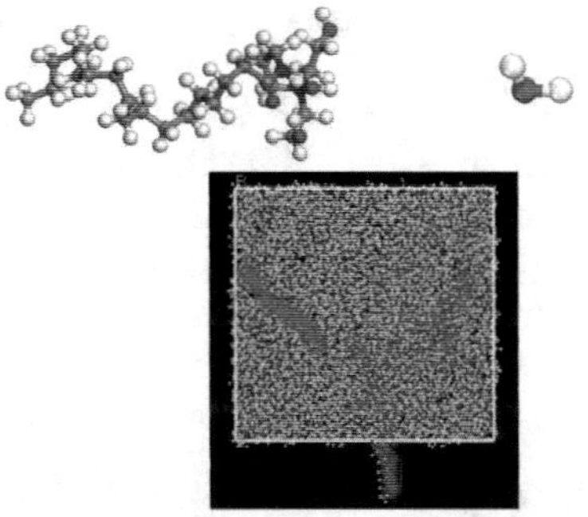

Fig.2 the cubic structures of the surfactant solution

2.3.Simulation parameters

2.3.1.MesoDyn simulation parameters

In the simulation of the surfactant solution, interactions between the surfactant molecules and water were accurately determined by Flory–Huggins interaction parameters. Flory–Huggins interaction parameters coincide with the relationship as follows:

$$\chi_{ij} = \frac{V_{mol}(\delta_i - \delta_j)^2}{RT} \quad (1)$$

where χ_{ij} is Flory–Huggins parameter, δ_i and δ_j are solubility parameters, V_{mol} is the molar volume, R is the molar gas constant and T is the temperature. Repulsion parameters in the surfactant solution are listed in Table 1.

Table 1 Flory–Huggins interaction parameters

	D	O	S	T	W
D	0.0	6.0	3.5	2.7	18.8
O	7.8	0.0	0.5	1.2	2.2
S	5.3	0.1	0.0	0.1	5.3
T	2.2	1.2	0.5	0.0	7.8
W	18.8	2.7	3.5	6.0	0.0

The dimensions of the simulation lattice are 32 × 32 × 32. The bond length is set to 1.1543 to ensure isotropy of all grid-restricted operators [31], to ensure a stable numerical algorithm, as an approximation, all bead diffusion coefficients of types D, S, O, T and W were $1.0 \times 10^{-7} cm^2, s^{-1}$. The volumes of the beads were 0.05, 0.03, 0.03, 0.03 and 0.01 nm^2, respectively. The shear rate in simulation was 0.001

1/ns. The simulation temperatures were from 262.15 K to 402.15K, and the time step was 50.0 ns. The noise-scaling parameter was 100, and the compressibility parameter was fixed at 10.0 [32-33].

2.3.2.Rheological properties simulation parameters

The dimensions of the simulation lattice are 67.92 × 67.92 × 67.92. The forcefield is compass, to ensure a stable numerical algorithm for the fact that there are both molecules and ions. The simulation temperatures were from 262.15 K to 402.15K, and the time step was 50.0 ns. The noise-scaling parameter was 100, the pressure was 18MPa, the ensemble was NPT, and the Initial velocities were random.

3. Results and discussions

3.1.The self-assembly of the surfactant in different temperature

We simulated the surfactant solution with 3% surfactant under different temperatures from 272.15K to 412.15K for 5,000 steps. The snapshots after 5,000 steps are given in Fig.3. We can observe that every surfactant molecule self-assembles into different micelles from 272.15K to 412.15K. These micelles are randomly distributed in the simulation box and entwined with each other, and they are worm-like, sphere-like and stick-like. When the temperatures are 272.15K and 282.15K, the outermost shell of the micelles are all O beads which are hydrophilic groups. However, the W beads occupy the outermost shell of the micelles after 292.15K (e.g. (c) (d) (e) (f) (g) (h) (i) (j) (k) (l) (m) (n) (o)). In addition, in Fig.3, there is an intriguing pattern that the beads are scattered as layers like the KFC chicken rolls or chocolate cream. Finally, it is obviously that the micelles have been changing from 272.15K to 412.15K.

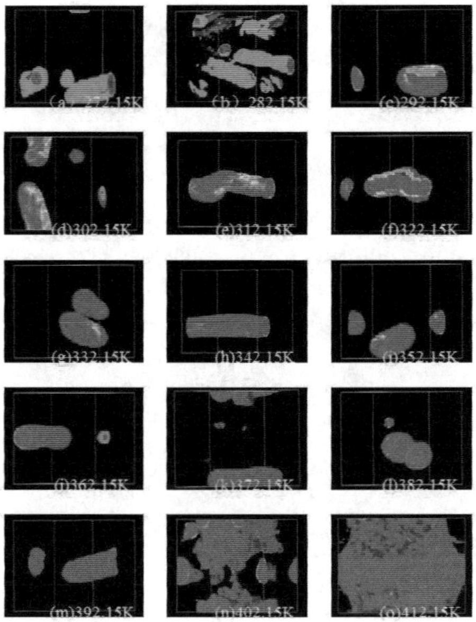

Fig.3 Snapshots of system with 3% surfactant fraction under temperature of 272.15K (a), 282.15K (b), 292.15K(c) and et al. after 5,000 simulation steps.

3.2.The models of the micelles

It is clearly that the beads distribute layer by layer like the coiling block in Fig.3, in a storyboard (see in Fig.4) for example. The W beads occupy the outermost shell of the micelles, near them are the hydrophilic groups, including O beads and T beads, and the intermediate group, S beads, link the hydrophilic groups and the hydrophobic groups, and the hydrophobic groups called D beads are enveloped in the innermost layer of the micelles.

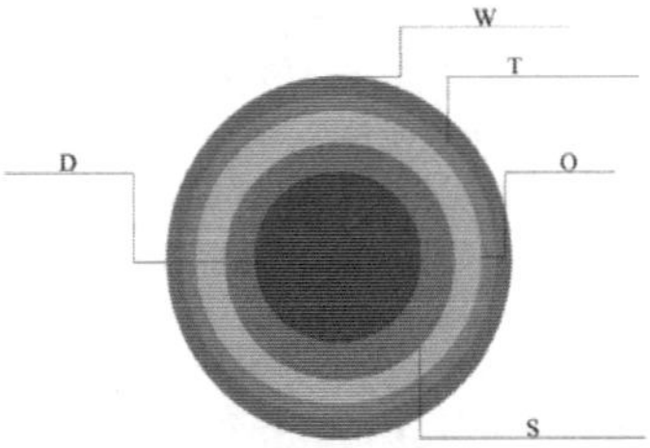

Fig.4 the storyboard of the micelles

Although the models of the micelles are shown as the coiling block like the KFC chicken rolls or chocolate cream because they consist of a great quantity of beads, they should be the layers like the cakes. We can see it in the Fig.5 that the upper and the lower layer should be the water molecules, and the interface layer should be the surfactant molecules.

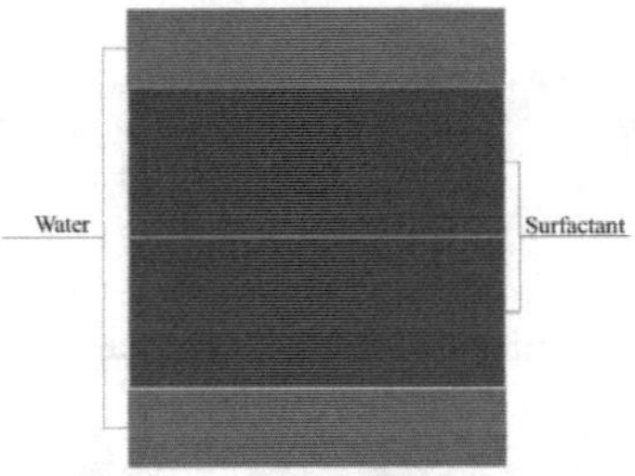

Fig.5 the model of the layers

And if we put the water and surfactant molecules into the system, we can observe the phenomenon shown in the Fig.6.

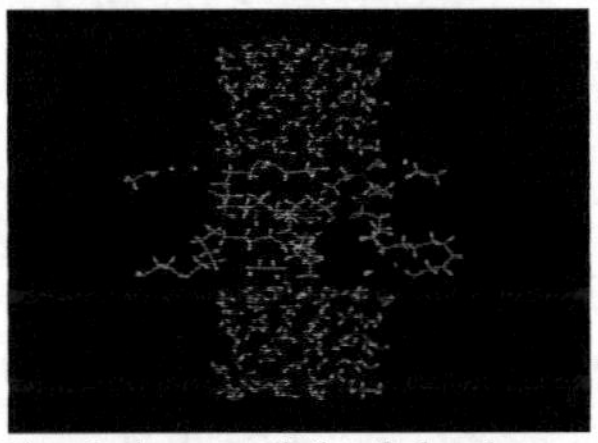

Fig.6 the model of the layers

3.3. Thermophoresis of the free water molecules

Obviously, we can survey in Fig.3 that the 'green robes' of the micelles become the 'purple robes' suddenly when the temperature rise from 272.15K to 292.15K. But there is not the changing of the beads' color, instead, it is the movement of the molecules (see in Fig.7), which is known as 'Thermophoresis'.

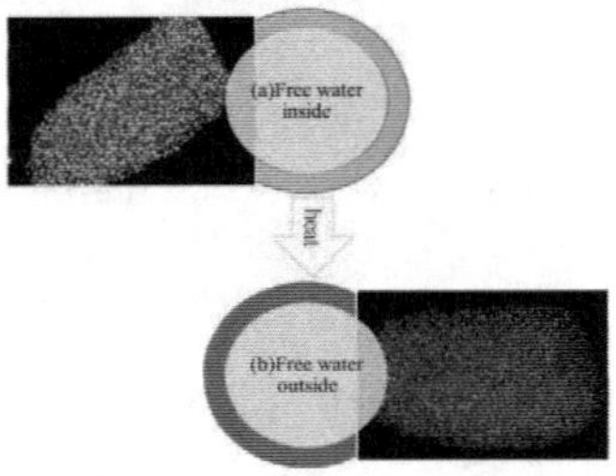

Fig.7 the movement of the free waters after heated

Thermophoresis is the time-averaged movement caused by the Brownian motion of particles within a fluid subjected to a steady temperature gradient [34]. The free water molecules impulses inside of the micelles, on the average, drive the free water particles toward the outside. Hence, the 'green robe' changes the 'purple one'. In other words, the network structures of the micelles are destroyed in the process that the free water particles go toward the outside.

3.4.The hydrogen bonding between the surfactant and water

The aggregation of the surfactant molecules essentially dominated by the interactions between their polar groupheads in aqueous phase. The nitrogen-bearing and oxygen-bearing surfactant molecules may form Hydrogen bondings with water or theirselves. And HBs are one of the key mechanisms behind the formation of various supramoleculer structures from small moieties [35]. In this simulation, the HBs play a great role in the self-assembly of the micelles.

Firstly, at the low external temperature, the micelles are close together, which can be seen in the Fig.8. There is no doubt that the hydrogen bonding has played a great role in the process that a lot of molecules gather together, because the hydrogen bondings between the two surfactant molecules connect them together (see in Fig.9).

272.15K 282.15K

Fig.8 the micelles at the low external temperature

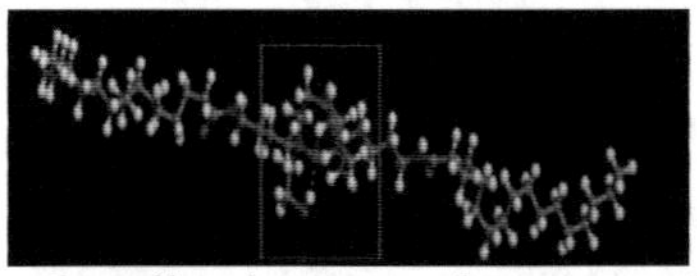

Fig.9 the hydrogen bonding between the two surfactant molecules

Secondly, resulting from the thermophoresis of the free water molecules, the hydrogen bonding between the two surfactant molecules has to break down, and the hydrogen bonding between the surfactant molecule and the water molecule (see in Fig.10) instead of it. Yet, the number of the water molecule is not larger. Thus, the hydrogen bonding between the two water molecules (see in Fig.11) cannot be built that there are coexistence of green and purple shell for micelles (see in Fig.12). All these clarify that the micelles give up entanglement for each other.

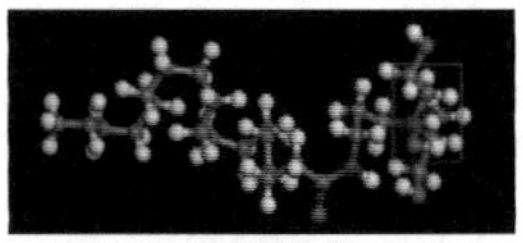

Fig.10 the hydrogen bonding between the surfactant molecule and the water molecule

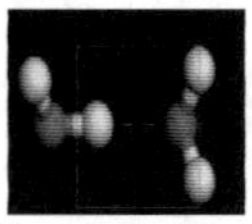

Fig.11 the hydrogen bonding between the two water molecules

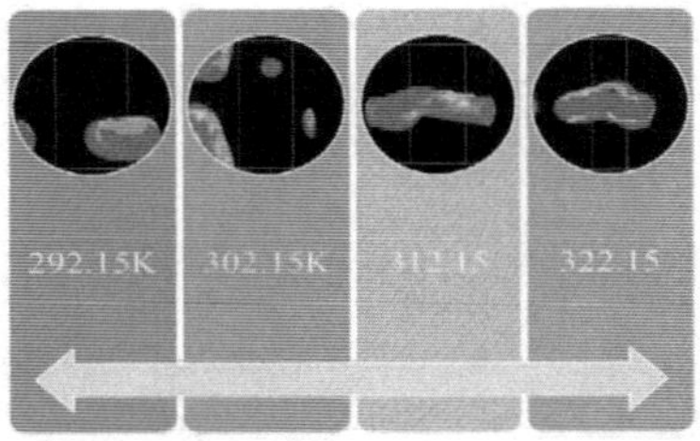

Fig.12 the coexistence of green and purple shell

Thirdly, with an increasing number of free water molecules rushing out of network structure, the hydrogen bonding between the two water molecules gradually gets to be established. Even more, the hydrogen bonding links the surfactant molecules and the water molecules all together like the Fig.13. The distance between micelles is narrowed for it.

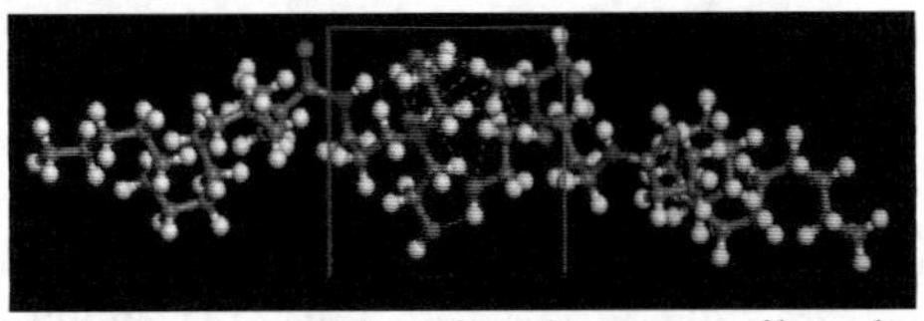

Fig.13 the hydrogen bonding between all molecules

Finally, as the external temperature continues to rise, the structure of the micelles is going to suffer devastating damage, which is shown in Fig.14.

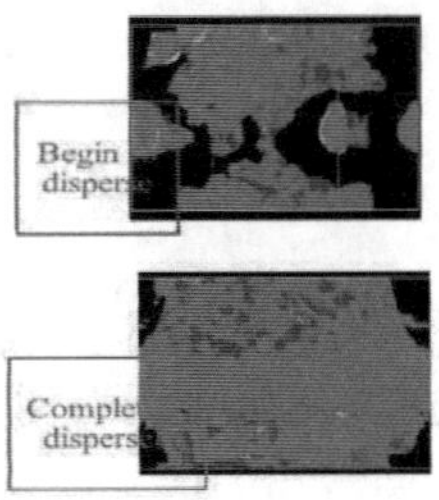

Fig.14 the dispersal of the micelles

3.5. The rheological properties of the surfactant solution in different temperature

In order to acquire the rheological properties of the surfactant solution under macroscopic conditions, almost 10000 molecules and ions were aggregated to a box whose volume is $67.92 \times 67.92 \times 67.92$ Å^3 (see in the Fig.15), which was used to calculate the rheological properties of the surfactant solution in different temperature. The shear viscosity of the system should be one of the foremost parameters to measure the rheological properties of the surfactant solution, which need to be attached great importance during the whole process of reservoir reconstruction.

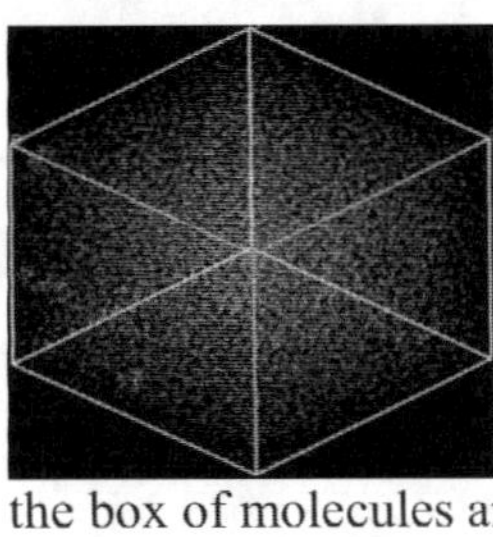

Fig.15 the box of molecules and ions

Accordingly, we simulated and calculated the shear viscosities of the surfactant solution from 272.15K to 412.15K, and we recorded every 10K. Consequently, we got a set of data of the shear viscosities of the surfactant solution, and they will be shown in the Fig.16.

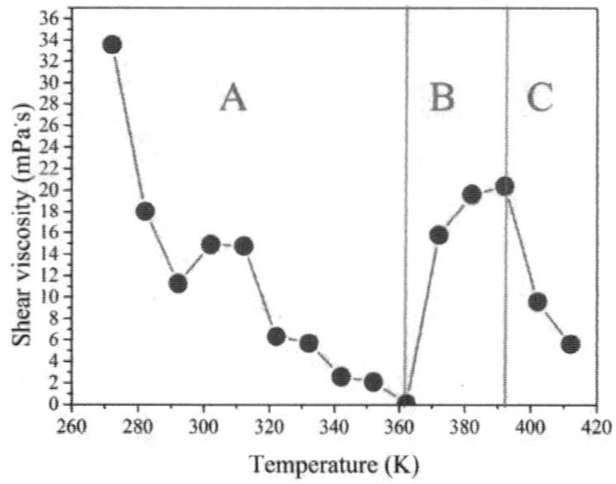

Fig.16 Viscosity of systems as a function of temperature

From the Fig.16, you can find that we divided the whole picture into three sections, called Section A, Section B and Section C. In Section A, the viscosities of the surfactant solution continue to reduce, but they increase abruptly in the Section B, and they go down too in Section C. Combining the phenomenon with the theory about the thermophoresis and hydrogen bonding in the previous article, it is not hard to find that the free water molecules get rid of the network structure of the surfactant molecules for the increasing temperature, which break off the hydrogen bonding between the surfactant molecules, and cut down the entanglement of micelles. So the viscosities in the Section A have been decreasing steadily. In the Section B, the number of the free water molecules is larger and larger that the water molecules and the surfactant molecules would like to bond together to become the structure shown in Fig.13, which bringing about the micelles bound more tightly and the viscosities upper. However, the temperature continues to go up, which resulting to the structural collapse of all the molecules. Unfortunately, the viscosities must decline.

4. Conclusion

The effect of the temperature on the surfactant solution should be bidirectional in fact. In pace with increasing of the temperature from 272.15K to 412.15K in the simulation, the viscosities showed markedly changes. That is to say, the rheological properties of the surfactant solution ought not to be thought to change simply from wellhead to the target reservoir for the increasing of the formation temperature.

Thermophoresis of the free water molecules is the most directive expression of the rise in the temperature. The free water molecules extricate themselves from the network structure of the surfactant molecules, which destroys the structure between the surfactant molecules from different micelles, and isolates the micelles.

The changes of hydrogen bonding between different molecules are the most important reason that the shear viscosities go up and down. Hydrogen bonding plays an important role in drawing the micelles closer and far them away.

The micelles keep close and detach from each other is the simplest reason that the viscosities go up and down. When we improve micelles' entanglement, it could increase the viscosity, vice versa.

With the exception of temperature, other factors such as the types and the values of sludge addition of counterion, molecule topological structure, the types and the value of sludge addition of surfactant, shear rate and so on also affect the rheology of surfactant solutions. We will put these influencing factors into our work for some time to come.

References

[1] N. W. Ray, J. L. Hunter, A. C. Brake, P. A. Fagin, N. Steinsberger. *ADM; revista de la Asociación Dental Mexicana*. 1998, 23 (1), 73-88
[2] J. Wang, S. L. Wang, W. Lin, Z. H. Kang, Q. You. *Journal of Moleculer Liquids*, 2017, 241,563-569
[3] M. Samuel, R. J. Card, E. B. Nelson, J. E. Brown, P. S. Vinod, H. L. Temple, Q. Qu, D. K. Fu. 1997, SPE 38622
[4] J. C. Mao, H. Zhang, W. L. Zhang, J. M. Fan, C. Zhang, J. Z. Zhao. *Journal of Industrial and Engineering Chemistry*, 2018, 60, 133-142
[5] K. D. Eisenbraun, T. M. Kwong, *Electrically operated telephone cradle*. US6035036, 2000.
[6] S. M. Centurion, M. Rengifo, M. N. Cortona, F. Padilla, A. L. Hurtado. 2006, SPE 103879
[7] J. B. Crews, T. Huang, W. R. Wood. *SPE Drill*. 2008, 23 (1), 41
[8] C. N. Fredd, T. N. Olsen, G. Brenize, B. W. Quintero, T. Bui, S. Glenn, C. L. Boney. 2004, SPE 91433
[9] L. Whitlock, A. Sanders, P. Knox. 2015, SPE 173791
[10] B Lin, T Liu, Z Zhao, J Xiang. *Drilling Fluid & Completion Fluid*. 2011, 28 (4), 64-66
[11] Z. Z. Yang, G. G. Zhou, Y. J. Lu, C. Y. Zhu. *Drilling Fluid & Completion Fluid*. 2005, 22 (1), 48-50
[12] J. C. Mao, X. J. Yang, Y. N. Chen, Z. Y. Zhang, C. Zhang, B. Yang, J. Z. Zhao. *Journal of Petroleum Science and Engineering*. 2018, 164, 189-195
[13] A. Baruah, A. K. Pathak, K. Ojha. *Chemical Engineering Science*. 2015, 131, 146-154
[14] X. P. Ma, Z. X. Zhu, L. Y. Dai, L. Y. Liu, W. Shi. *Russian Journal of Applied Chemistry*, 2016, 89, (12), 2016-2026
[15] M. Yu, Y. Mu, G. Q. Wang, H. A. Nasr-El-Din. 2012, SPE 142264
[16] A. S. Elhag, C. Da, Y. S. Chen, N. Mukherjee, J. A. Noguera, S. Alzobaidi, P. P. Reddy, A. M. AlSumaiti, G. J. Hirasaki, S. L. Biswal, Q. P. Nguyen, K. P. Johnston. *Journal of Colloid and Interface Science*. 2018, 522, 151-162
[17] S. S. Dai, J. H. Zhang, T. L. Zhang, Z. Y. Huang, H. P. Quan, H. S. Lu, X. W. Zhao. *New J. Chem.* 2016, 40, 5143
[18] X. Zhao, Z. W. Wang, S. L. Yuan, J. Lu, Z. N. Wang. *RSC Adv.* 2017, 7, 20293
[19] J. H. Cho. *Macromoleculer Research*. 2018, 26 (4), 380-387
[20] D. Mu, J. Q. Li, S. Wang. *Colloid and Polymer Science*. 2015, 293 (10), 2831–2844
[21] Q. H. Hao, B. Miao, Q. G. Song, X. H. Niu, T. J. Liu. *Polymer*. 2014, 55 (16), 4281-4288
[22] N. Rakkapao, V. Vao-soongnern. *Journal of Polymer Research*. 2015, 21 (12), 1-1
[23] D. Mu, J. Q. Li, S. Y. Feng. *Phys.Chem.Chem.Phys.* 2017, 19, 31011
[24] J. Elistratova, A. Mustafina, A. Litvinov, V. Burilov, A. Khisametdinova, V. Morozov, R. Amirov, Y. Burilova, D. Tatarinov, M. Kadirov, V. Mironov, A. Konovalov. *Colloids and Surfaces A: Physicochemical and Engineering Aspects*. 2013, 422, 126-135
[25] J. Fraaije, B. Van Vlimmeren, N. Maurits, M. Postma, O. Evers, C. Hoffmann, P. Altevogt, G. Goldbeck-Wood. *J Chem Phys.* 1997, 106, 4260
[26] Y. M. Li, G. Y. Xu, A. M. Chen, S. L. Yuan, X. R. Cao. *J. Phys. Chem. B*. 2005, 109 (47): 22290–22295
[27] M. Doi, S. F. Edwards, *Oxford Science Publications* 1986, Oxford, UK.
[28] Y. M. Li, G. Y. Xu, Y. Y. Zhu, Y. J. Wang, H. J. Gong. *Colloids and Surfaces A: Physicochemical and Engineering Aspects*. 2009, 334, 124–130
[29] S. W. Meng, J. Zhang, Y. Wang, X. T. Li, C. Wu, T. F. Hou, L. J. Xiao, G. W. Lu. *Molecular Simulation*. 2015, 41 (9), 772–778
[30] A. W. Lees, S. F. Edwards. *J Phys C*. 1972, 5, 1921–1928.
[31] Y. Y. Li, T. J. Hou, S. L. Guo, K. X. Wang, X. Xu. *J. Phys. Chem. Chem. Phys.* 2000, 2, 2749.
[32] N. M. Maurits, B. A. C. V. Vlimmeren. *Phys. ReV. E*. 1997, 56,816.
[33] B. A. C. V. Vlimmeren, J. G. E. M. Fraaije, *Comput. Phys.Commun.* 1996, 99, 21.
[34] E. E. Michaelides. *International Journal of Heat and Mass Transfer*. 2015, 81, 179-187

[35] R. Kothari, H. H. Winter, J. J. Watkins. *Macromolecules.* 2014, 47 (22), 8048-8055

Experimental study on hygrothermal accelerated aging effects of transparent fire resistive wood structure coatings

Tianhao Zhang[1a, *], Hao Huang[1b], Xinguo Ge[1c], Zejiang Zhang[1d], Pingli Li[1e]

[1] Sichuan Fire Research Institute of MEM, Chengdu, Sichuan Province, 610036, China

[a*]zhangtianhao@scfri.cn, [b]huanghao@scfri.c[n]@scfri.cn, [c]gexinguo@scfri.cn, [d]zhangzejiang@scfri.cn, [e]lipingli@scfri.cn

Abstract—In this paper, the durability of fire performance of a transparent fire resistive coating for wood structure was studied using hygrothermal accelerated aging. The aging effects was analysed using SEM/EDS and fire test. The results show that after 21 days of aging, the flame-retardant components gradually precipitate to the surface. The coating basically loses its expansion ability and the char layer shows a low-strength lamellar structure. The flame-retardant time of the coating decreased by 43.48% from 23 min to 13 min.

1. Introduction

Organic cultural relics, especially wood-structured buildings, are highly flammable, which will cause huge economic losses and irreparable damage to historical and cultural heritage after fire. In recent years, fires of cultural relics have occurred frequently, such as the fire of the National Museum of Brazil in 2018 and the fire of Notre Dame Cathedral Paris in 2019, which have caused extremely painful losses. Transparent fire resistive wood coatings can effectively reduce the flammability of the coated materials to delay the fire event of the substrate and prevent the rapid spread of the fire, or to delay the destruction time of the substrate structure by improving the thermal insulation[1]. At present, environmentally friendly halogen-free intumescent fire resistive coatings for wood structures are mainly achieved by physically or chemically mixing intumescent flame-retardants in polymer resin substrates to achieve the purpose of fire prevention. [2-6] Sol-Gel, Plasma and other surface treatment methods have also been developed and a series of transparent fire resistive coatings have been prepared. [7-10].

However, most of the chemical components in fire resistive coatings are organic, it is reasonable to infer that they react with the exposed environment and that the basic performance especially fire protection function of fire resistive coatings deteriorates over time [11]. At present, there are very few reported researches on the aging performance of fire resistive wood coating, among which most of the scholars focus on the adhesion loss [12, 13] after aging rather than the fire performance.

It should be noted that most of the durability and fire performance indicators of fire resistive coatings are based on independent evaluation systems. Few studies can effectively correlate the durability and weather resistance of coatings with fire performance. However, the occurrence of fire is uncertain, and it is necessary to ensure that the fire resistive coating can retain fire performance after a period of use. Therefore, it is very meaningful to study the durability of fire performance for transparent wood structure fire resistive coatings.

In this paper, experimental study on hygrothermal accelerated aging effects of transparent fire resistive wood structure coatings was carried out. Fire performance, surface morphology and distribution of element during the aging progress were characterized. The aging mechanism of transparent fire resistive coating for wood structure was tentatively proposed by analyzing the changing of fire properties and element distribution during the aging process.

2. Experimental

2.1 Materials

Transparent coating: two-component, water-borne fire resistive coating for wood structure, self-made.

Specimen: three-layer plywood with white beech veneer, commercial products

2.2 Specimen preparation

Transparent fire resistive wood structure coating was coated on three-layer plywood with white beech veneer with the dimension of 300*150*5 mm. In order to prevent side intrusion or other factors from affecting the aging behavior, a mixture of paraffin wax and rosin with the ratio of 1:1 was used to seal the edge of each specimen with a depth of at least 5 mm. After construction, all samples should be cured at a constant environment of 23±2 °C and 50±5 RH% for at least 14 days.

2.3 Test method

Hydrothermal aging was carried out by Sartec SS-7123X (SARTEC TESTING INSTRUMENTS Co., Ltd). All samples are arranged vertically at a certain distance, as shown in Figure. 1. Total aging time were set as 21 days and the sampling frequency was set as every 3 days. The aging method is shown as below.

-4 h (23±3) ℃, (80±5) %RH
-16 h (40±3) ℃, (50±5) %RH
-4 h (5±3) ℃, (50±5) %RH

Figure. 1 Hydrothermal aging test instrument and sample placement

Scanning Electron Microscope (SEM)/ Energy Dispersive Spectroscopy (EDS) was carried out using FEI Inspect F50(FSEM, USA) and Super Octane (EDAX, USA). Fire test was carried out by a self-made test furnace and the heating curve was set as ISO 834 (T=345lg(8t+1) +20) as shown in **Figure. 2**. Two K-type thermocouples were used to monitor the backside temperature and the flame-retardant time was defined as the time that the average backside temperature rise reaches 220 ℃.

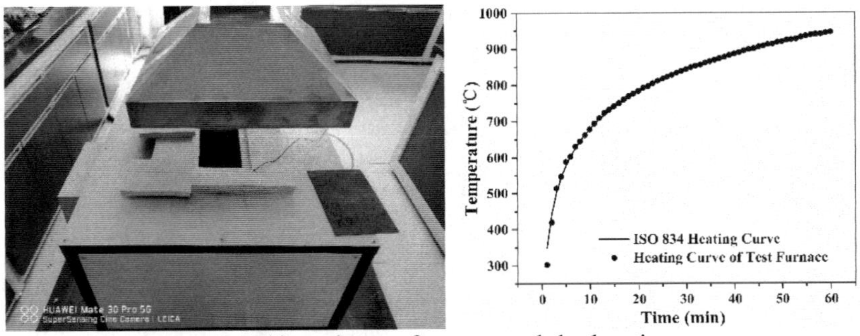

Figure. 2 Self-made test furnace and the heating curve.

3. Results and Discussions

3.1 Surface appearance

Digital pictures of transparent fire resistive coating after 0-21days of accelerated aging are shown in Figure. 3. It can be seen that the transparency of the coating changed significantly after 3 days of aging and then deteriorated gradually. As the aging progresses, obvious cracks appeared on the surface of the coating and components precipitation occurred on the coating surface.

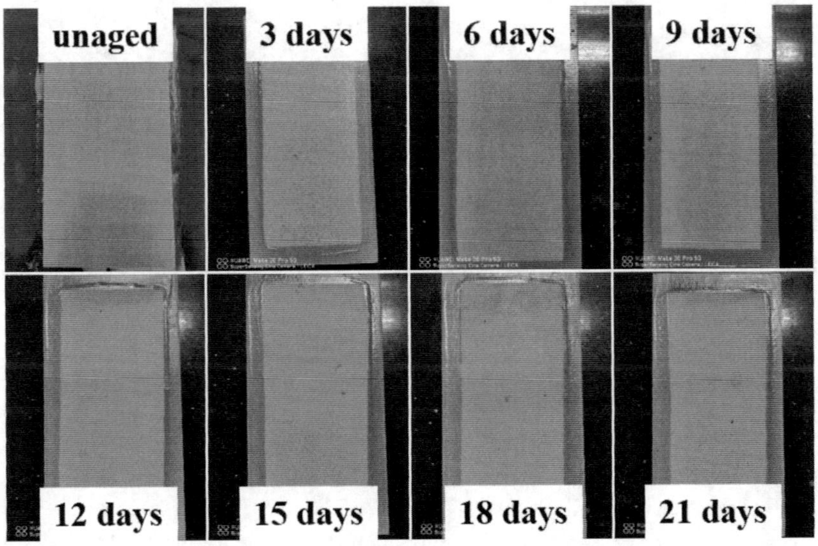

Figure. 3 Digital pictures of transparent fire resistive coating during aging methods

3.2 Fire performance of the coatings

Fire resistant time of coatings after 21 days of accelerated aging methods are shown in Figure.4 (a) and each specimen's backside temperature curves are shown in Figure.4 (b). It can be seen that the fire resistant time of the coating decreased rapidly during aging and finally attenuated by 43.48 % after 21 days of aging from 23 min to 13 min.

Digital pictures of char layers after fire test are shown in **Figure. 5**. It can be seen that with the aging, the expansion performance of the coating decreases gradually, and the integrity of the char layer formed in the fire test is gradually destroyed. After 21 days of aging, the coating basically loses its expansion ability and the char layer shows a low-strength lamellar structure.

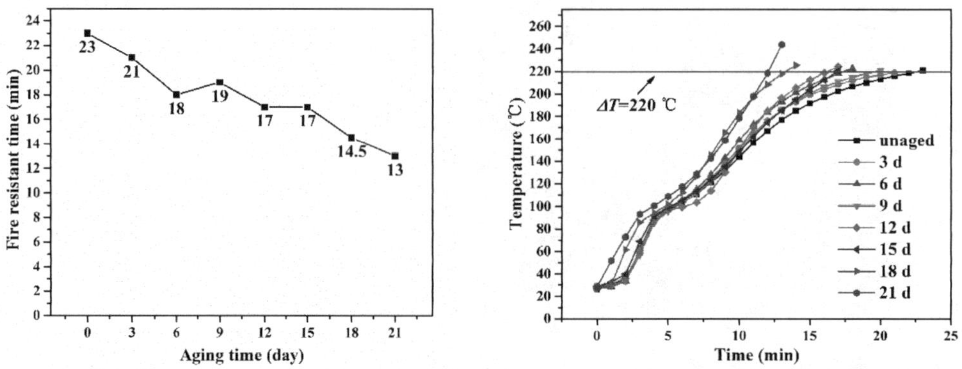

Figure.4 (a). Fire resistance times of coatings during aging progress; **(b).** backside temperature

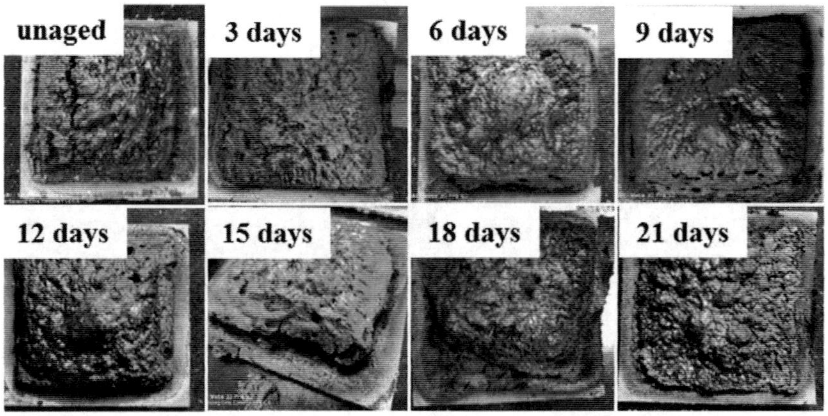

Figure 5. Digital pictures of char layers after fire test during 0-21 days of accelerated aging

3.3 Elemental analysis of the coating

SEM pictures of transparent coating after 0-21 days of accelerated aging are shown in Figure. 6. Obvious precipitation of components can be found on the surface and the surface gradually becomes unsmooth with aging. Mass fractions of main flame-retardant elements from EDS are shown in Table. 1. We can see that the mass fraction of P on the surface increased which should be caused by the precipitation of flame retardants. The mass fraction of C and O first increase with the gradually degradation of base resin. Based on these results, we can preliminarily judge that the aging mechanism is the precipitation of flame retardants which contains P such as APP. However, further extensive analysis should be done to comprehensive explain the relationship of changes in P element and the decrease of fire performance.

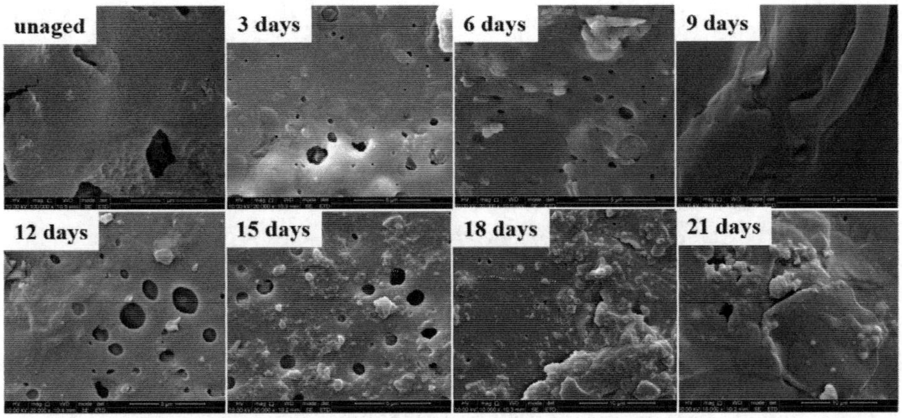

Figure. 6 SEM pictures of transparent fire resistive coating during aging methods

Table. 1 Mass fraction of surface elements during aging methods

Elements (wt. %) \ Aging Time	0 d	3 d	6 d	9 d	12 d	15 d	18 d	21 d
C	34.62	35.76	43.49	43.95	42.83	39.62	39.17	46.78
N	21.67	19.98	8.68	4.49	16.37	17.21	13.09	13.43
O	35.25	37.66	41.39	42.19	23.05	22.45	25.93	23.65
Si	0.76	0.62	0.52	0.80	1.32	1.69	1.43	1.00
P	6.66	4.39	4.95	6.76	11.95	13.74	16.36	12.13
S	1.04	1.58	0.97	1.82	4.47	5.29	4.01	3.01

4. Conclusion

In this paper, the durability of fire performance of a transparent fire resistive coating for wood structure was studied using hygrothermal accelerated aging. The aging effects was analysed using SEM/EDS and fire test. The following conclusions can be drawn:

1. After 21 days of aging, the expansion capacity of the fire resistive coating is basically lost, and the char layer shows a low-strength lamellar structure. The flame-retardant time of the coating decreased by 43.48% from 23 min to 13 min.
2. The increasing mass fraction of P element proves that flame-retardant components gradually precipitate to the surface with aging. Based on this, a preliminarily aging mechanism of flame retardants precipitation is proposed.

Acknowledgments
This work was financially supported by National Natural Scientific Foundation of China (No. U19A2045).

References

[1] Shan, Y. (2010). Exploitation of ultra-thin intumescent fireproof coating for wood structure. New Chemical Materials, 28(7): 112-115
[2] Gu, J. W.; Zhang G. C.; Dong S. L.; Zhang, Q. Y.; Kong, J. (2007) Study on preparation and fire-retardant mechanism analysis of intumescent flame-retardant coatings. Surface & Coatings Technology, 201(18): 7835-7841.
[3] Shi, Y.; Wang, G. (2016) The novel silicon-containing epoxy/PEPA phosphate flame retardant for transparent intumescent fire resistant coating. Applied Surface Science, 385, 453-463.
[4] Shanshan, C.; Xiang, L.; Yang, L.; Junqi, S. (2015) Intumescent flame-retardant and self-healing superhydrophobic coatings on cotton fabric. ACS Nano, 9 (4), 4070-4076.

[5] Chan, S. Y.; Si, L.; Lee, K. I.; Ng, P. F.; Lei, C.; Yu, B.; Yuan, H.; Yuen, R. K. K.; Xin, J. H.; Fei, B. (2017) A novel boron–nitrogen intumescent flame retardant coating on cotton with improved washing durability. Cellulose, 25 (1), 843-857.

[6] Radmard, B.; Poznysz, E. R.; Sargeant, J. A.; Waryold, K.; Giroux, D. J. Fire resistant coatings for wood veneer panels. U.S. Patent 9,975,314[P]. 2018-5-22.

[7] Cheng, X. W.; Liang, C. X.; Guan, J. P.; Yang, X. H.; Tang, R. C. (2018) Flame retardant and hydrophobic properties of novel sol-gel derived phytic acid/silica hybrid organic-inorganic coatings for silk fabric. Applied Surface Science, 427, 69-80.

[8] Grancarić, A.; Colleoni, C.; Guido, E.; Botteri, L.; Rosace, G. (2017) Thermal behaviour and flame retardancy of monoethanolamine-doped sol-gel coatings of cotton fabric. Progress in Organic Coatings, 103, 174-181.

[9] Farag, Z. R.; Krüger, S.; Hidde, G.; Schimanski, A.; Jäger, C.; Friedrich, J. (2013) Deposition of thick polymer or inorganic layers with flame-retardant properties by combination of plasma and spray processes. Surface & Coatings Technology, 228 (9), 266-274.

[10] Alongi, J.; Colleoni, C.; Malucelli, G.; Rosaceb, G. (2012) Hybrid phosphorus-doped silica architectures derived from a multistep sol–gel process for improving thermal stability and flame retardancy of cotton fabrics. Polymer Degradation & Stability, 97 (8), 1334-1344

[11] Wang, L. L.; Wang, Y. C.; Li, G. Q. (2013). Experimental study of hydrothermal aging effects on insulative properties of intumescent coating for steel elements. Fire Safety Journal, 55(JAN.), 168-181.

[12] Vlad-Cristea; Riedl; Blanchet; Jimenez-Pique. (2012). Nano characterization techniques for investigating the durability of wood coatings. EUR POLYM J, 48(3) (-), 441-453.

[13] Williams, R. S. (2005). Don't get (sun) burned: exposing exterior wood to the weather prior to painting contributes to finish failure. Journal of architectural coatings. July: 56-60.

Experimental study on bond strength failure of GFRP lap joints

Xiao Xiao[1a*],Gui Xiong[1b*], Hao Dang[1c]

[1]School of Civil Engineering, University of South China, Hengyang 421001, China.

[a*]2005000112@usc.edu.cn, [b*]1044487414@qq.com, [c]dangh2019@126.com

Abstract—Bonding has many advantages such as high adhesive strength, peel resistance, good toughness, shrinkage and corrosion-resistance, compared with bolting, welding and other connection patterns, which is the basic connection of the composite materials. The mechanical properties of the bond depend on the type and geometric parameters of the bond. In this paper, experiments were conducted to study the bond strength failure of GFRP single and double lap joints and explore related influencing factors. The variable factors of test specimens include bond type, length and thickness. The quasi-static tensile test was carried out at room temperature to analyze multiple influencing factors of lap joints and characteristic of GFRP bond failure strength. The results showed that the failure of the bond usually occurred at the interface of bond, and bond types, length and thickness have significant effects on the failure of bond. Failure strength increased firstly and then decreased with the increase of adhesive thickness. With the increase of bonding length, the failure mode of the double-lap joint become complex failure mode, which is accompanied by GFRP delamination near the surface of fiber of the bonding.

1. Introduction

Glass fiber reinforced polymer (GFRP) composite materials have been recognized and gradually promoted in aerospace, aviation, architecture and other fields due to the specific durability, specific modulus and geometry[1]. In order to maintain the integrity of the structure, bonding has many advantages such as high adhesive strength, peel resistance, good toughness, shrinkage and corrosion-resistance, compared with bolting, welding and other connection patterns. Bonding is the most essential connection of the composite materials though there are many conbination partterns of composites[2-3].

Over the years, many scholars had propose many theories to explore the property of the bonding such as mechanical interlocking theory, bonding adsorption theory, chemical bond theory, diffusion theory and so on[4]. In fact, the complex mechanical phenomena in the bond structure system could not be explained by only one theory. It was found that for the double lap joints with ladder-like slot lap plate end, sub-maximum shear stress and peel stress were significantly decreased at overlapping end, joint static strength and fatigue strength were improved significantly compared with the traditional double lap joints[5]. Based on the existing main failure criteria of adhesive, the maximum stress or maximum strain criterion was applied to the composite failure criterion. But for composite material, the maximum stress or maximum strain criterion is unsuital actually because the failure of molecular cohesive force at the interface should be considered to be conbined with the failure of adhesive force at the interface.

As for the double-lap bonded components, it was found that the material properties, geometric parameters of the adhesive and the bonding form have a great influence on the stress of the bonded structure, as well as the strength performance of the bonded structure. Obtaining structural stress distribution accurately is one of the preconditions for predicting bond strength. In this paper, the bond strength performance of GFRP lap joints under different adhesives type, lapping form, bond length and bond thickness were studied. Meanwhile, the stress-strain relationship and failure mode of bonding were analyzed. The aim of this study is to provide a more suitable method for failure characteristics of ductile adhesive.

2. Experimental program

2.1.Materials

The single and double lap joints were included in GFRP lap joints as the specimens in this test. All specimens were made of pultruded GFRP composite material. Araldite epoxy resin adhesive (AE) and Hysol E-20HP adhesive (HE) were used respectively. The mechanical properties of GFRP materials and adhesive were summarized in Table 1 according to the manufacturer provided. These specimens were divided into two groups, namely, 8 GFRP single lap joints and 32 GFRP double lap joints. The details of some specimens are showed in Table 2.

Table 1 Mechanical properties of adhesive

Material name	Properties	Value
GFRP	Young's modulus E_G (GPa)	32.2
	Poisson's ratio μ_G	0.32
	Tensile Strength f_{u1} (MPa)	307
	Shear strength f_{s1} (MPa)	26.7
Araldite epoxy resin adhesive	Young's modulus E_{AB} (GPa)	1850
	Poisson's ratio μ_{AB}	0.33
	Shear strength f_{s2} (MPa)	560
Hysol E-20HP	Tensile Strength f_{u2} (MPa)	5700
	Poisson's ratio μ_{HP}	0.35
	Ductility (%)	8

2.2.Specimens

The specimens of the lap joint for tensile test are shown in Fig.1. The thickness of the lap plate is equal to that of the GFRP laminates bonded. According to the literature[6], sandpaper was used to polish the bonding surface and anhydrous alcohol was used to clean the impurities on the bonding surface to ensure the cleanliness of the bonding surface. The gasket was used to control the thickness of the adhesive, whose thickness is 0.1mm. The adhesive was evenly distributed in the overlapping area of GFRP laminates, and cured at room temperature for more than 12 hours, so that the bond strength of the samples reached the test conditions, which indicated that the specimens were made and formed[7].

Table 2 Details of some specimens

Specimen number	Adhesive	Size(mm)	Effective bond length (mm)	Effective bond thickness (mm)
S-1-40	Epoxy resin AB Adhesive	200×20×5	40	0.5
S-2-50		200×20×5	50	0.5
S-3-60		200×20×5	60	0.5
S-4-70		200×20×5	70	0.5
S-5-0.5	Hysol loctite E-20 HP	200×20×5	60	0.5

S-6-0.6	Adhesive	200×20×5	60	0.6
S-7-0.7		200×20×5	60	0.7
D-AE-1	Epoxy resin AB Adhesive	200×20×5	40	0.5
D-AE-2		200×20×5	50	0.5
D-AE-4		200×20×5	70	0.5
D-AE-5		200×20×5	40	0.6
D-HE-5	Hysol loctite E-20 HP Adhesive	200×20×5	40	0.6
D-HE-6		200×20×5	50	0.6
D-HE-7		200×20×5	60	0.6
D-HE-8		200×20×5	70	0.6

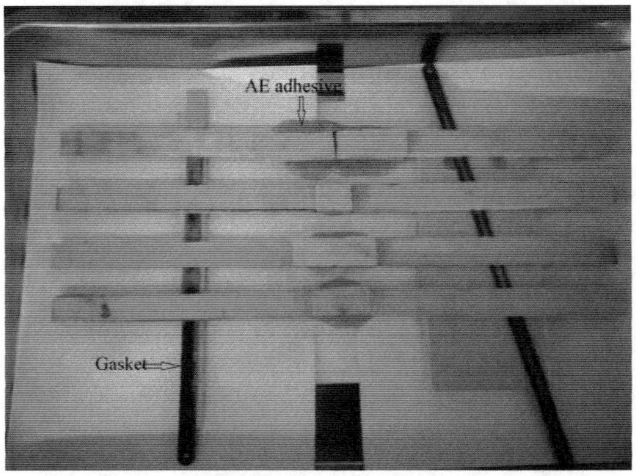

Fig. 1 specimens

Fig. 2 Loading experiment

2.3. Test equipment and loading

In order to test the tensile strength of specimens and the energy dissipation performance of adhesive in bonding, electrohy-draulic servo universal testing machine WA-FY600C controlled by microcomputer was used to test the tensile properties according to ISO 527-1. Firstly, the upper end of the test specimen was clamped and fixed completely. Then, the lower end was clamped as well. A uniformly stretching along the longitudinal direction at a constant speed of 0.5mm/min until failured was conducted[8-9], as shown in Fig.2. The failure strength of the bond joints was obtained by dividing the failure load F by S of the bond area.The failure modes of the joints and the curves of the tensile force changing with the displacement were also obtained, as shown in Fig.3 and 4.

Three sets of experiments, which variable factors composed of adhesive, bond length and bond thickness, were conducted respectively. The single-variable controll method was adopted and each variable was repeated three times during the experiment.

3. Results and Discussion

3.1. Failure mode

Tensile failure modes for GFRP lap joints are different because of using different adhesive and geometry of laps. The failure failure of specimens with Araldite epoxy resin adhesive (AE) is mainly caused by cohesion failure and adhesion failure. When the thickness of the adhesive layer is maintained at 0.5mm, adhesion failure occurred in the peeling way, as shown in Fig.3 (a) and (b). However, the failure failure of the specimens with Hysol loctite E-20 HP Adhesive(HE) mainly occurred in the form of cohesion failure, adhesion failure, substrate stratification failure and mixed failure, as shown in Fig.3 (c) and (d). The reason of brittle failure is that the clamping force was too large, which made the bond stress mutation and affected the stress distribution of specimens.

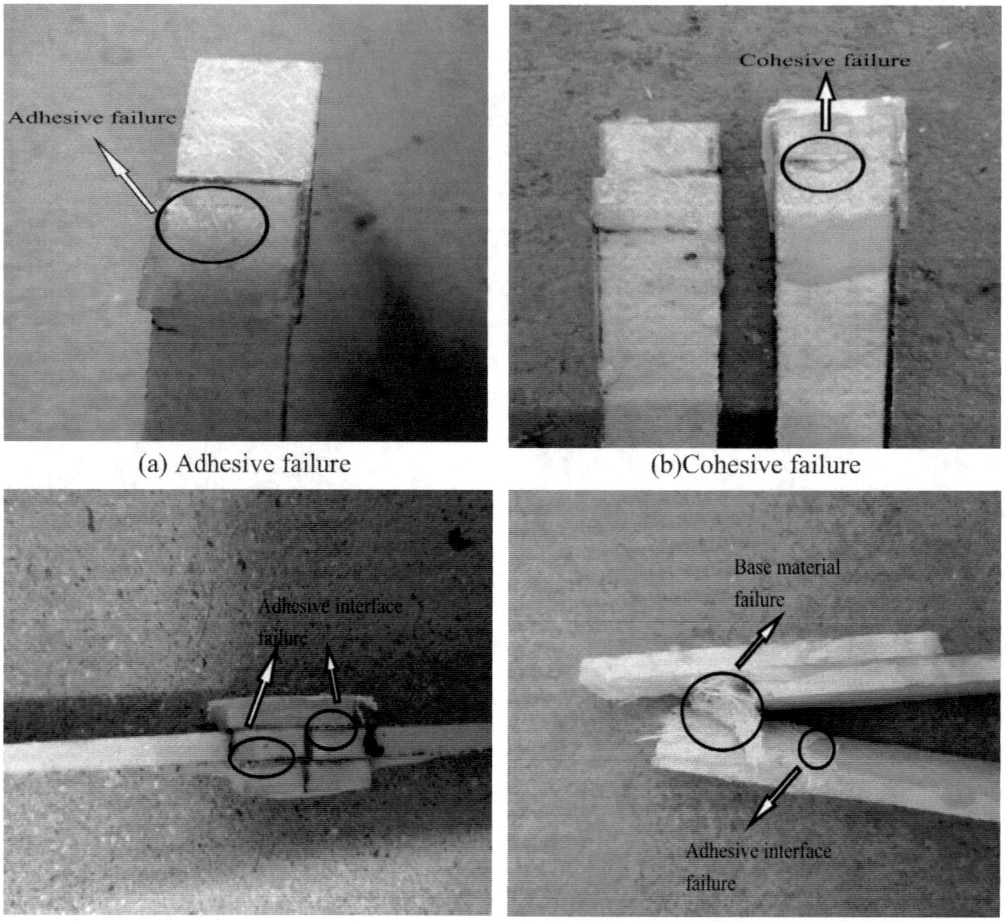

(a) Adhesive failure (b)Cohesive failure

(c) Adhesive interface failure (d) Mix failure

Fig. 3 Failure mode of specimens

3.2.Load-displacement curves

The force-displacement curves of the single-lap joints were obtained, as shown in Fig.4 (a). The results showed that the force increases with the increase of bonding length. The force increases by 1.08kN, 0.66kN, 0.48kN, respectively. For specimen S-4-70, capacity force increased by 2.27kN, which was more than half of that for specimen S-1-40. When the bonding thickness increases by 0.1mm, the applied load increases by 0.336kN on average. Kept the loading rate of specimens of S-5-0.5, S-6-0.6, S-7-0.7 and adhesive layer length of 60mm the same, there can be found a force drop appeared in the middle. The reason that is the internal bond changed with a slight sound in the process of the experiment, resulted to stress redistribution. Furthermore, when thickness is between 0.6mm and 0.7mm, the failure load increased almost 0.794kN, experimental results are shown in Fig.4(b).

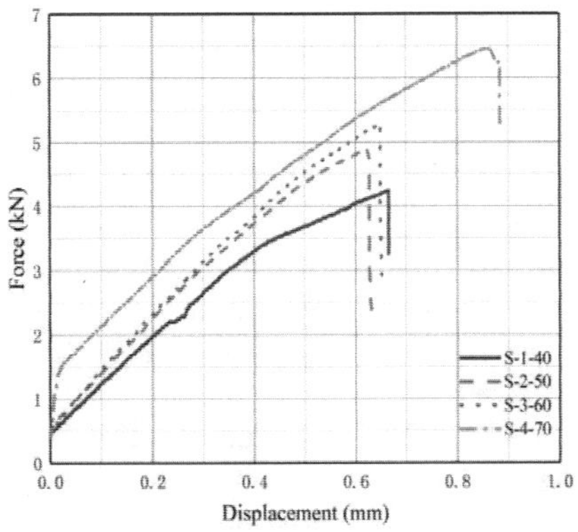

(a) AE adhesive

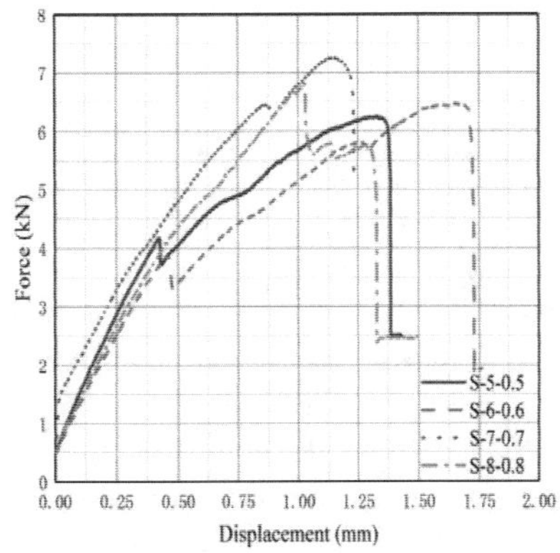

(b) HE adhesive

Fig.4 Force-displacement curves for single lap

3.3.Stress-strain analysis

Influence of bond thickness of double-lap bond strength failure were conducted by testing D-AE-1, D-AE-2, D-AE-4 and D-AE-5 with AE adhesive. The stress-train curves of tensile test are shown as Fig.5(a). Compared D-AE-1 to D-AE-5, it is found that the slope of D-AE-5 curve is higher at the beginning of loading to the failure. The final failure load is also higher and increasing by about 25%. Compared with D-AE-2 and D-AE-4, it is found that the change rate of D-AE-4 curve decreased at the beginning of loading and increased until failure, compared with D-AE-2. The failure load increased by about 12.5%.

In order to test effect of bond length of double lap on the failure, the specimens of D-HE-5, D-HE-6, D-HE-7 and D-HE-8 were selected, the experiment results are showed in Fig.5(b). when bonding length kept at 70mm, the maximum stress value was 101.97MPa and the minimum stress value was 77.63MPa. The reason why the curve is different is that the center line of the bonding member for D-HE-5 was offset. It can be seen that the initial slope (elastic modulus) of four specimens is very close, indicated that bond performance with HE adhesive is reliable.

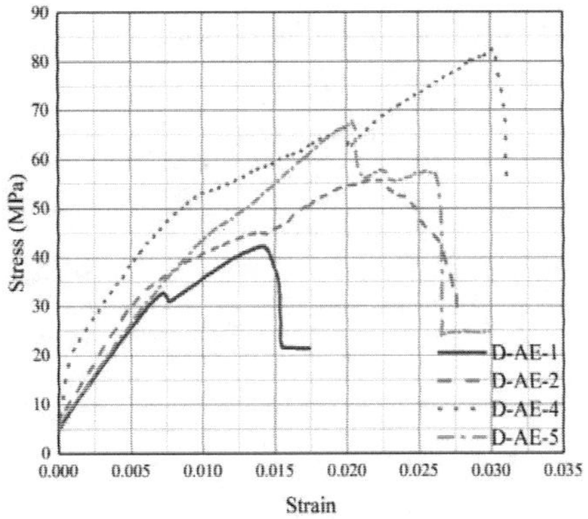

(a) AE adhesive

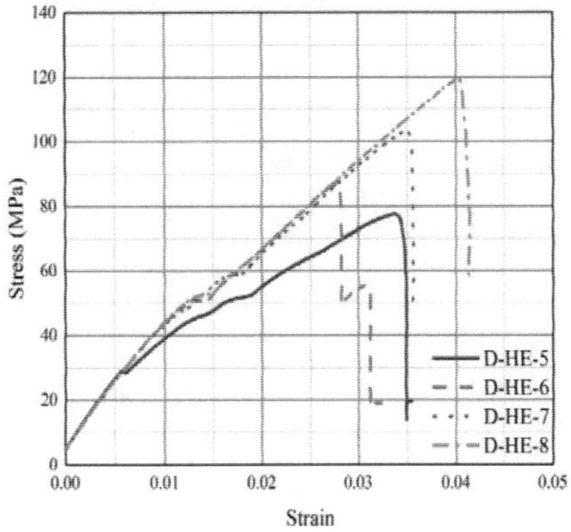

(b) HE adhesive

Fig. 5 Stress-strain curves for double lap

4. Conclusions

In this paper, tensile experiments were conducted to analyze several characteristics of bond failure and feasibility, including GFRP single and double lap bonding. The results showed that the double-lap bonding was effective for GFRP, which verified the effectiveness of the proposed method. The following conclusions were drawn:

(1) According to the comparative analysis of the experiments, the failure modes of specimens mainly occurred at the bond interface, failure mode was mainly adhesion failure, cohesion failure and the superposition mode of the two failure.

(2) Through the comparison of GFRP single and double lap bonding performance it was found that bond strength using double lap for GFRP members was larger than that using single lap method.

(3) By comparison of AE and HE adhesive, it was found that HE adhesive had better bond performance and pseudo ductility under the same experimental conditions due to the fiber on the bonding surface of GFRP plate was delamination, and the tensile failure mode was occurred.

(4) Bond performance of GFRP double lap was closely related to bond thickness. The optimal bonding thickness was 0.6mm. With increasing of bond thickness, the bond strength increased, then decreased slightly. The failure strength also increased linearly with increasing of bond length.

Acknowledgements

The authors gratefully acknowledge the financial support provided by key scientific research project of Education Department of Hunan Province (grant No: 20A426) and Natural Science Foundation of Hunan Provincial (grant No: 2021JJ50015). The experimental work was conducted in the Department of Civil Engineering at University of South China, and the technical support from Mr. Zhenfu Chen and Mr. Zhiheng Zhang (Structural Laboratory of University of South China) is acknowledged.

References

[1] Ku H and Wang H, 2011, A review on the tensile properties of natural fiber reinforced polymer composites, Compos. Pt. B-Eng. 42(4):856-873.
[2] Petrie E M. Handbook of adhesives and sealants[M]. McGraw-Hill Education, 2007.
[3] Yanhua Zhang, Zhu Libin, Tan Haiyan, et al. Adhesives and bonding technology [M]. Beijing: Chemical Industry Press, 2018.
[4] D5573-99 Standard practice for classifying failure modes in (FRP) joints.
[5] Chuang W Y, Tsai J L. Investigating the performances of stepwise patched double lap joint[J]. International Journal of Adhesion & Adhesives,2013, 42:44-50.
[6] Guan-xia Yang. Study on bonding repair and reinforcement technology and mechanical properties of composite materials [D]. Tianjin Polytechnic University,2019.
[7] D3164-03 Standard test method for strength properties of adhesively bonded plastic lap-shear sandwich joint in shear by tension loading.
[8] ISO 527–1 Determination of tensile properties.
[9] Pascoe J A, Zavatta N, Troiani E, et al. The effect of bond-line thickness on fatigue crack growth rate in adhesively bonded joints[J]. Engineering Fracture Mechanics, 2020, 229: 106959.

Preparation of amphiphilic block copolymers by Reversible Addition Fragmentation Chain Transfer polymerization

Na Hu[123*]

[1] The Key Laboratory of Food Colloids and Biotechnology, Ministry of Education, School of Chemical and Material Engineering, Jiangnan University, Wuxi 214122, Jiangsu Province, P.R, China;

[2] Enviromental and Material Engineering, Jiangyin Polytechnic College, Jiangsu Jiangyin, 214405, China;

[3] Amcor Flexibles Jiangyin Propack Packing Co.ltd.

E-mail: eleanor_hu@126.com

Abstract: Polystyrene macromolecular chain transfer agent (PS-CTA) was firstly synthesized by RAFT of styrene (St) in THF using benzyl dithiobenzoate (BDTB) as chain transfer agent and azobisisobutyronitrile (AIBN) as initiator. It was found that the molar ratio of monomers to chain transfer agents and reaction time had obvious effects on the molecular weight of PS-CTA, and the distribution of molecular weight (M_w/M_n) decreases with the increase of the ratio of chain transfer agent to initiator by using gel permeation chromatography (GPC). The amphiphilic block copolymers polystyrene-*b*-poly(4-vinylpyridine) (PS-*b*-P4VP) were synthesized using PS-CTA by RAFT. The structure, glass transition temperature (T_g) and molecular weight of the diblock copolymer were characterized by nuclear magnetic resonance (^{1}H-NMR), differential scanning calorimetry (DSC) and gel permeation chromatography (GPC). The results showed that the structure of PS-*b*-P4VP was the target product. The segment ratio of PS-*b*-P4VP could be effectively controlled by changing the polymerization conditions in a large range.

1. Instruction

Controllable/active free radical polymerization method can passivate a large number of reactive radicals to make them dormant, establish a rapid dynamic balance between a small amount of growth radicals and a large number of dormant radicals, greatly reduce the concentration of reactive radicals, and reduce the possibility of double radical termination and chain transfer. Therefore, controllable/active free radical polymerization method provides a good way to control the structure and properties of polymers[1-2]. In 1998, Rizzardo found an active radical polymerization regulated by disulfide ester to obtain polymers with narrow molecular weight distribution, they called this polymerization method reversible addition fragmentation chain transfer (RAFT) polymerization[3]. RAFT polymerization provides a more effective and cheaper way to prepare active polymers and design polymer structures[4]. In addition, there are a wide range of monomers that can carry out RAFT polymerization, including almost all monomers suitable for other active polymerization and some monomers that can't carry out active polymerization at present. Moreover, RAFT polymerization adopts by selecting chain transfer agents with different structures, it is convenient to introduce end group functional groups into polymer materials; in addition, block and gRAFT copolymerization that can't or is difficult to obtain by other synthesis technologies can be synthesized by using macromolecular chain transfer agent.

In this paper, polystyrene macromolecular chain transfer agent (PS-CTA) was firstly synthesized by RAFT using benzyl dithiobenzoate (BDTB) as chain transfer agent. PS-CTA with controllable molecular weight and the distribution of molecular weight (M_w/M_n) was prepared. According to the determination results of GPC, the effects of the amount of chain transfer agent and initiator, reaction time on the conversion, molecular weight and molecular weight distribution were investigated. The synthesized PS-CTA was used to regulate the monomer 4VP for solution RAFT reaction to synthesize amphiphilic block copolymers PS-*b*-P4VP.

2. Experimental

2.1. Materials
4-Vinylpyridine (4VP, Acros Organics) was dried over CaH2 and distilled at 55 oC before use. Benzyl dithiobenzoate (BDTB) was synthesized and purified according to the reported methods[5]. All the other solvents and materials were purchased from Shanghai Chemical Reagent Station without further purification.

2.2. Characterization methods
Nuclear magnetic resonance (1H NMR, Bruker AVANCEIII, 400 MHz) spectrum was used to identify the structures of the polymers using chloroform-d (CDCl3) as solvent and tetramethylsilane (TMS) as the internal standard. The molecular weight and distribution of the polymers were determined by gel-permeation chromatography (GPC, Agilent-1100, USA) using DMF as eluent and polystyrene as standard with a flow rate of 1 mL/min at 35 oC. The thermal properties were analyzed by differential scanning calorimetry (DSC 822e, Mettler Toledo), DSC analysis was carried out by scanning the samples at a heating rate of 10oC/min from 20oC to 180oC.

2.3. Synthesis of PS-CTA
PS-CTA was synthesized by RAFT of St using BDTB as chain transfer agent and AIBN as initiator in THF. The detailed procedure was as follows: AIBN, BDTB and 4 g St (the typical feed molar ratio was n[St]:n[BDTB]:n[AIBN]= 50/100/150/200:2:1) were added into the reaction flask at room temperature, and then 4 mL solvent THF was added. Stirring or ultrasonic oscillation for a period of time to mix the reactants. The sample was degassed under nitrogen purge with stirring for 30min. Then the polymerization was performed at 70 oC for scheduled time, precipitated from methanol and then filtered under vacuum. Such process was repeated twice, and the obtained powder dried in a vacuum oven overnight at 30 oC.

2.4. Preparation of amphiphilic block copolymers PS-b-P4VP
Amphiphilic block copolymers PS-b-P4VP was synthesized by RAFT of 4VP using PS-CTA and AIBN as initiator. The detailed procedure was as follows: PS-CTA, AIBN and 4 g 4VP (the typical feed molar ratio was n[4VP]:n[PS-CTA]: n[AIBN] = 1500:10:1) were added into the reaction flask, and then 4 mL solvent dimethylformamide (DMF) was added. The sample was degassed under nitrogen purge with stirring for 30min. Then the polymerization was performed at 70oC for scheduled time, the amphiphilic block copolymer was obtained after precipitated from ether, and filtered under vacuum. Such process was repeated twice, and the obtained powder dried in a vacuum oven overnight at 30oC.

3. Results and discussion

3.1. Effect of molar ratio of monomer and chain transfer agent on PS-CTA

Tab.1 The effect of different PS-CTA ratio on the M_n ahd PDI

n[St]:n[BDTB]: n[AIBN] (mol/mol)	Conversion %	$M_{n(th)}$ $\times 10^{-3}$	$M_{n\,(GPC)}$ $\times 10^{-3}$	M_w/M_n
50:2:1	23.1	1.98	6.87	1.08
100:2:1	30.1	3.91	10.1	1.23
150:2:1	20.0	3.90	10.8	1.23
200:2:1	25.5	6.47	13.9	1.31

Polymerization Conditions: 70°C, reaction time 5 h

The lower conversion of St at 5 h might be due to the slow polymerization and induction effect of chain transfer agent, the RAFT polymerization rate of St was lower than that of conventional free radical polymerization. The retarding effect might due to the fact that most of the growth radicals formed after the initiation reaction of the RAFT process were transformed into intermediate radicals, which accumulated until the addition cleavage equilibrium was established without directly initiating the growth, resulting in the retarding phenomenon. It depended on the concentration of the initial RAFT reagent, and the polymerization rate of the system decreased with the increase of RAFT reagent. The induction period also prolonged with the increase of chain transfer agent concentration, which might due to the slow reinitialization rate of leaving groups in the early stage of polymerization. It could be seen from Tab.1 the molecular weight of PS-CTA could be controlled by adjusting the ratio of initiator to monomer. Thus, it was easier to control the segment ratio of block copolymer. At the same time, it was found that the Mw/Mn with the increase of the feed ratio of monomer to chain transfer agent, so it was necessary to control the appropriate feed ratio to maintain narrow molecular weight distribution.

3.2. Effect of reaction time on PS-CTA

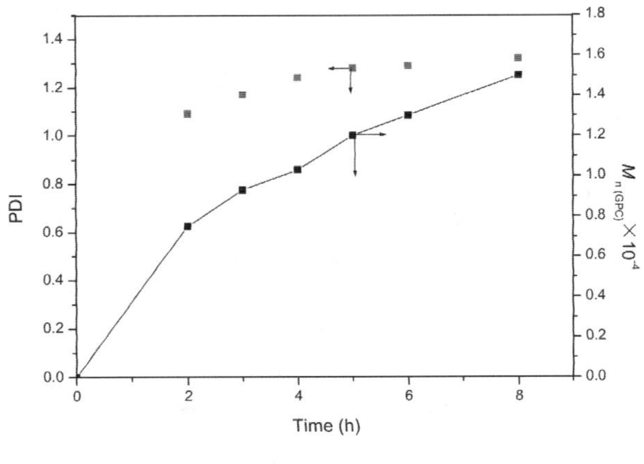

Fig.1 Dependence of M_n on Reaction time
(Conditions: 70 °C, n[St]:n[BDTB]: n[AIBN]=200:2:1)

The M_n and M_W/M_n of the synthesized PS-CTA were determined by GPC. The results showed that M_n increased steadily with the increase of reaction time, as shown in Fig.1. It shows that the M_W/M_n of the PS-CTA obtained was relatively stable under different polymerization reaction times, which accorded with the characteristics of active polymerization.

3.3. ^{1}H-NMR spectrum of PS-CTA and PS-b-P4VP

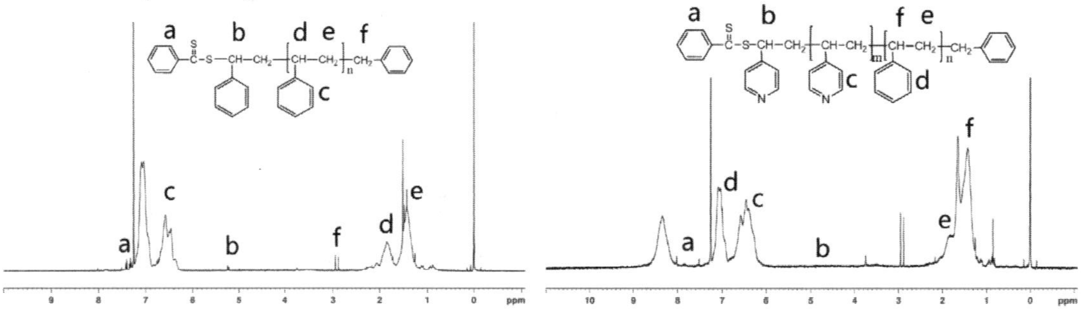

(a) (b)

Fig.2 ^{1}H-NMR spectrum of PS-CTA(a) and PS-*b*-P4VP(b)

Fig.2(a) showed ^{1}H-NMR spectrum of PS-CTA, it showed that the chemical shift δ4.7 and δ7.9 correspond to the methylene adjacent to sulfur on the end group and two H on the benzene ring adjacent to sulfur, which indicated that the chain transfer agent had been connected to the macromolecule. The peaks c, d and e represent the peaks of polystyrene, which proved that PS-CTA with clear structure. Fig.2(b) showed ^{1}H-NMR spectrum of PS-*b*-P4VP, the two absorption peaks at δ6.4 and δ8.3 belonged to the absorption peaks of n-ortho and h-meso on the pyridine ring of P4VP chain segment, there were protons on methyl and methylene at peaks e and f. a and b were methylene adjacent to sulfur on the end group and two H on the benzene ring adjacent to sulfur respectively. The above analysis showed that the amphiphilic block copolymer PS-*b*-P4VP was successfully synthesized by RAFT.

3.4. DSC analysis of PS-CTA and PS-b-P4VP

The DCS curves of PS-CTA and PS-*b*-P4VP were shown in Fig.3. Two glass transition temperatures (T_g) appearred on the DCS curves of PS-*b*-P4VP in Fig.3, which belong to the T_g of PS segment and P4VP segment of PS-*b*-P4VP respectively. It could be preliminarily judged from the DSC curves that PS-*b*-P4VP was synthesized successfully.

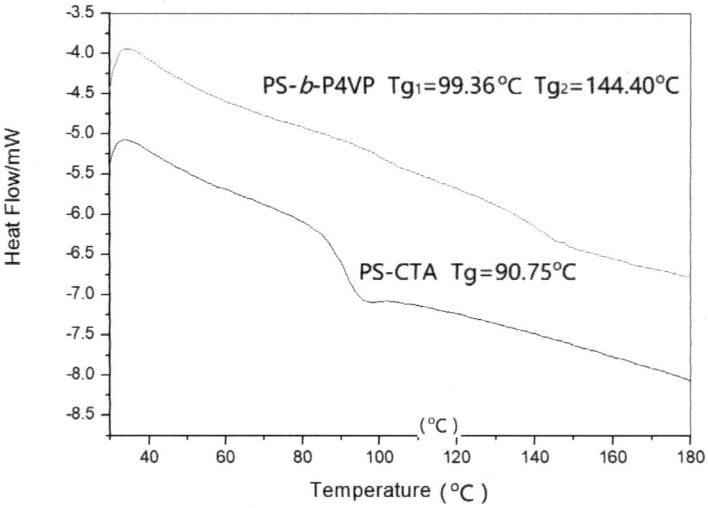

Fig.3 The DSC curves of PS-CTA and PS-*b*-P4VP

4. Conclusions

PS-CTA was prepared via RAFT. With the increase of the molar ratio of monomer, chain transfer agent and initiator in the reaction system, or reaction time, Mn of PS-CTA gradually increased. The MW/Mn of the PS-CTA obtained was relatively stable under different polymerization reaction times, which accorded with the characteristics of active polymerization. The polymerization rate in the initial reaction stage was very slow, which was significantly lower than that in the conventional free radical polymerization. The amphiphilic block copolymers PS-b-P4VP were successfully synthesized using PS-CTA by RAFT. It was expected to be used in the fields of catalysis, heavy metal loading, medical treatment and so on.

Acknowledgments
The authors would like to thank Qinglan project of Jiangsu Province of China, Postdoctoral research funding program of Jiangsu Province 2019K007 and Soft science research project of Wuxi Association for science and technology KX-21-C165 for financially supporting this research.

References
[1] Webster, O. W. (1991) Living polymerization methods. Science, 251: 887-894.
[2] Matyjaszewski, K. (1996) Cationic polymerizations: mechanisms, synthesis, andapplications. New York: Marcel Dekker,.
[3] Hawthorne, D. G., Moad, G., Rizzardo, E., Thang, S. H. (1999) Living radical polymerization with reversible addition fragmentation chain transfer (RAFT): Direct ESR observation of intermediate radicals. Macromolecules, 32: 5457-5459.
[4] Chiefari, J., Mayadunne, R. T. A., Moad, G., Rizzardo, E., Thang S. H. (1999) PCT Int. Appl. WO 9931144 A1 990624; Chem.Abstr, 131, 45250.
[5] Le, T. P., Moad, G., Rizzardo, E., Thang, S. H. (1998) PCT Int. Appl. WO 9801478 A1 980115; Chem Abstr, 128, 115390.

Syntheses of Coordination Polymer Glass Membranes and Their Gas Separation Performance Research

Jiaming Wang[1]

[1]School of Chemistry and Chemical Engineering, Beijing Institute of Technology, Beijing 100081, China

[a]wangjm960310@163.com

Abstract—Purity requirements for gas in daily life and industrial production have been increasing. Coordination polymer glass membranes can solve the problem of brittle and fragile inorganic membranes and retain some porosity. We obtained the novel CP material, to obtain self-supporting glass membranes by means of sheeting-melting-cooling, the selectivity of the a_gCd-P-dmbIm membrane in H_2/CO_2, H_2/N_2 and H_2/CH_4 are 68.8, 33.8 and 49.5 respectively. Its hydrogen permeance is 235 GPU, and its permeability is 73621 barrer. The results show that the diffusion coefficient of H_2 is much greater than that of CO_2. Relevant mechanism proved that the gas molecular size sieving properties of coordination polymers is the key factor to achieve efficient gas permeability and selectivity. a_gM-P-dmbIm is hydrogen purification membrane. Coordination polymer glass membranes have great potential in gas separation research.

1. Introduction

In the past two decades, Humans began to use gas separation methods based on physical and chemical force to separate and purify single-component gases. However, these traditional separation methods have disadvantages such as large energy consumption, large investment, and low recovery rate. As a kind of separation technology with high purification efficiency, membrane separation technology has the advantages of low energy consumption, easy operation, and small footprint, and is expected to be widely used in the future. Coordination polymer glass membranes can solve the problem of brittle and fragile inorganic membranes and retain some porosity, but the system has been less studied at present, the glass membranes aperture of 0.62 Å affects its screening of gases with smaller molecules.

For most materials, the melt usually forms crystal nuclei and grows into a crystalline substance near the freezing point temperature, but glassy materials will form a solid state with glass-like characteristics. The glassy state is isotropic and gradually softens when heated[1-3]. In 2012, Kitagawa[4] combined into a new coordination polymer material $[Zn(HPO_4)\text{-}(H_2PO_4)_2](ImH_2)_2$, which improved the crystal specific surface area and ion conductivity at 160 °C. In 2013, the research team of Umeyama[5] reported the crystal-glass transition behavior of a series of coordination polymers formed by metallic zinc ions, phosphoric acid and azole compounds. In 2016, Bennet[6] report the glass behavior of the molten cold extraction of the metal-organic framework material of the scientific research team. Compared with the existing glass types (inorganic non-metal, organic non-metal and organic metal) at that time, the glass retains the connection mode of metal and ligand. And has good mechanical properties. In 2018, the research team of Bennet and Coudrt[7] jointly discussed the thermal behavior of ZIF-4 material and discussed its liquid porous structure. ZIF-4 changed from a crystalline structure to an amorphous structure during the melting behavior. ZIF-8 has explored and found that its

T_m is higher than ZIF-4, which is caused by its coordination bond strength and viscosity. In 2020, Li, Pan, and Jiang[8] studied an important work. They developed the ZIF-62 coordination polymer glass membrane and used its gas separation characteristics to test the gas sieving ability. H_2/CH_4, CO_2/N_2, CO_2/CH_4 are 50.7, 34.5, and 36.6.

Aiming at the problems of high processing temperature, low selectivity, and less research. In this work, coordination polymer glass membrane with high flux and high gas selectivity was designed through a reasonable material structure, and its processability and gas separation performance were solved at the same time.

2. Methods and Materials

It is designed to synthesize coordination polymer materials by replacing the metal connecting unit and the type of imidazole or replacing the phosphoric acid with a longer chain organic phosphine. M-P-dmbIm (metal-ligand-ligand): $M(OAc)_2 \cdot 2H_2O$ (1 mmol), 5,6-dimethylbenzimidazole (292.3 mg, 2 mmol), phosphoric acid (210 μL, 3 mmol) and ethanol (500 μL) were manually pulverized in a mortar for 30 min to obtain a white powder. The obtained powder is thoroughly washed with dichloromethane three times, and each time it is stirred for 3 h. Finally, it was dried in an oven at 70 °C for 10 h to obtain the final coordination polymer product. (25~30 °C, 30~40 RH%)

M-P-dmbIm powder into a powder in a mortar, take 100 mg of the powder and put it into the tablet press abrasive tool to fully flatten it. The continuous force is less than 30 s under 1.5 MPa. The diameter of the compressed tablet is 2 cm, thickness 200~300 μm, calcined in a tube furnace for 0.5 h under vacuum conditions, after cooling to room temperature, a self-supporting glassy polymer membranes a_gM-H-dmbIm (the abbreviation of glass material after glass state conversion) is automatically formed.

3. Test Results and Discussions

3.1. Basic structure characterization of coordination polymers

XRD is no peak from metal salt and ligand in the diffraction peak, which is a new diffraction peak, which proves the synthesis of a new coordination polymer. in Fig.1a. Between 2500-3150 cm^{-1}, the N-H stretching vibration broadband of dmbIm almost disappears, indicating that dmbIm has been deprotonated. The P-O stretching peak of phosphoric acid at 1011 cm^{-1} has a significant shift from the original 1018 cm^{-1}. This is due to the deprotonation of H_3PO_4. These features indicate that H_3PO_4 and dmbIm should be involved in the coordination. in Fig.1b.

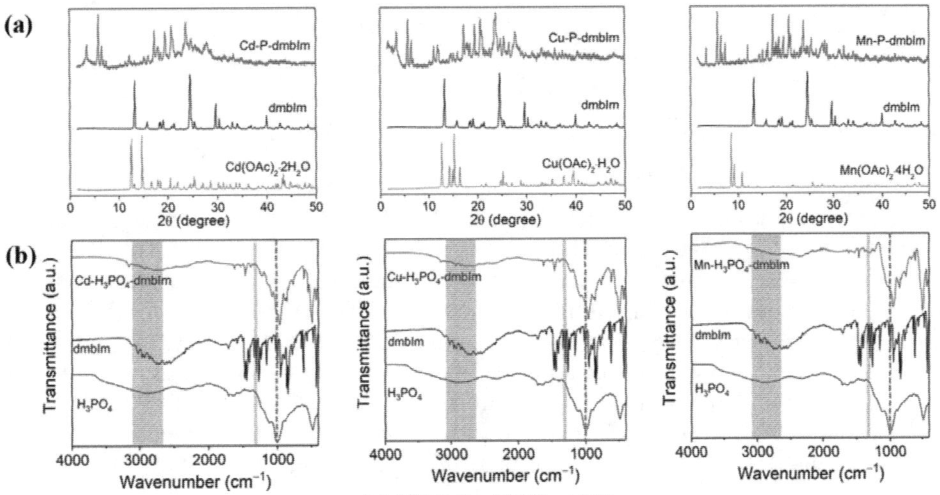

(a) XRD(b) FTIR-ATR

Fig.1 XRD and FTIR-ATR for M-P-dmbIm

Journal of Physics: Conference Series **2194** (2022) 012041 doi:10.1088/1742-6596/2194/1/012041

TGA and DSC measurements were performed under nitrogen as a protective atmosphere to study its melting behavior. In the TG test, the three samples all showed weight loss behavior at about 200 °C. in Fig.2a. Combined with the DSC test results, the weight loss temperature corresponds to an absorption/exothermic peak, which is the decomposition temperature (T_d) of the sample. in Fig.2b. All three samples exhibited endothermic characteristics before thermal decomposition, which corresponds to their melting temperature (Tm), representing their transition from solid to liquid. It is worth noting that the Tm values of the three samples are in the range of 162 172 °C, which is much lower than the ZIF glass transition temperature (>400 °C), proving that they can be melt-processed at a lower temperature. After the melting temperature was obtained in the first round of the DSC test cycle, the second round of the DSC test scan was performed immediately to obtain the glass transition temperature (T_g). Cd–P–dmbIm exhibited an endothermic behavior during the second scan, which corresponds to its T_g. In the same way, the T_g of Cu-P-dmbIm and Mn-P-dmbIm can be obtained. The ratio of the obtained T_g to Tm is the glass formation ability (GFA) of the material. GFA generally defines the difficulty of glass recrystallization. The larger the value, the more difficult it is to recrystallize the material in the glass state. These three materials all have ultra-high GFA, which may be due to the large steric hindrance of the ligand hindering the recrystallization process of the coordination polymer in the liquid state.

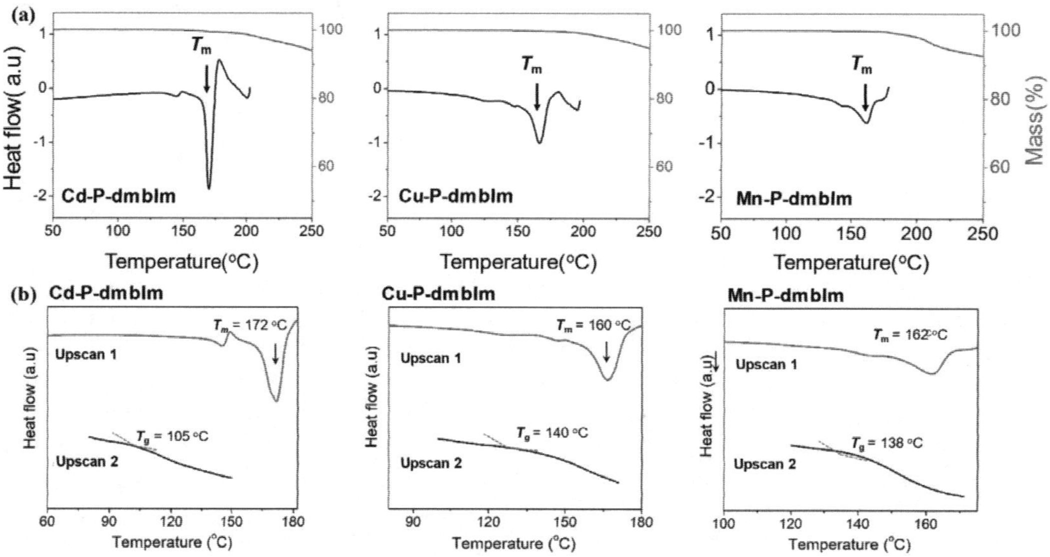

(a) TGA(b) DSC

Fig.2 TGA and DSC for M-P-dmbIm

3.2. Basic structure characterization of coordination polymers glass

It can be seen that the coordination polymer glass has no diffraction peak after the glass state transition in XRD, which proves that the material has changed from a crystalline ordered structure to a disordered structure. in Fig.3a. Similar to the results obtained from crystalline materials, it proves that the ligand still retains coordination in the glass structure in FTIR-ATR. in Fig.3b. Both M-P-dmbIm and a_gM-P-dmbIm have characteristic peaks at 1.5 ppm in the ^{31}PNMR spectra. in Fig.1c. Due to the single coordination $H_2PO_4^-$, the peak positions of the crystalline state and the glass state can be preliminarily judged that they have similar coordination. Way. The obtained glass materials were subjected to elemental analysis and ICP characterization, and the element compositions and ratios of the obtained three glasses were basically similar to the element contents of the previously obtained crystalline materials.

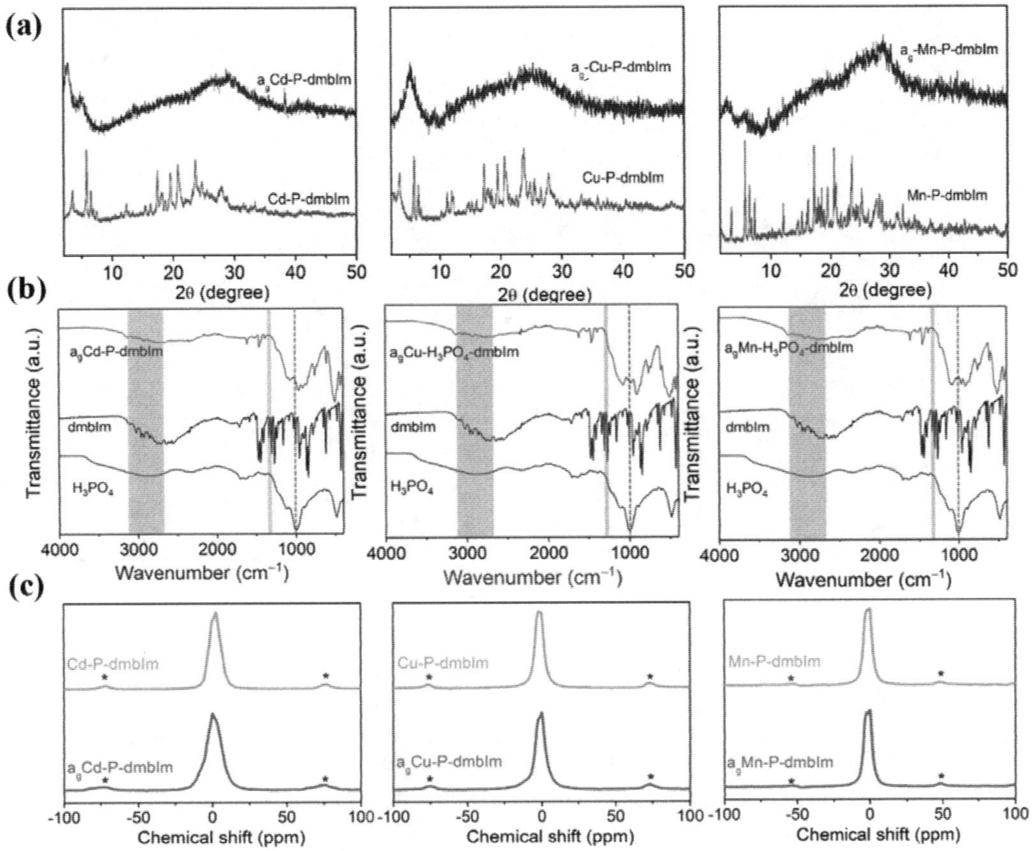

(a) XRD (b) FTIR-ATR (c) ^{31}P NMR
Fig.3 XRD; FTIR-ATR and ^{31}P NMR for a_gM-P-dmbIm

Through the tableting-melting-cooling method, a complete and defect-free glass membranes material can be obtained. The microscopic morphology of coordination polymers and glassy coordination polymers can be studied by SEM. Due to the crystal characteristics of inorganic materials, The plane SEM image of the M-P-dmbIm tablet shows the discrete distribution of crystal grains with large gaps, and the particle size is in the order of micrometers in diameter. On the contrary, the glass membranes material has no obvious cracks to prove that it is defect-free, and a_gM-P-dmbIm has the characteristics of uniformity and compactness, indicating that a_gM-P-dmbIm has successfully transformed from a crystal to a glass state. In addition, the cross-sectional SEM image 3.6 shows that a_gM-P-dmbIm exhibits uniform thickness, and the a_gCd-P-dmbIm membranes, a_gCu-P-dmbIm membranes, and a_gMn-P-dmbIm membranes have thicknesses of 313, 425, and 297 μm. in Fig.4.

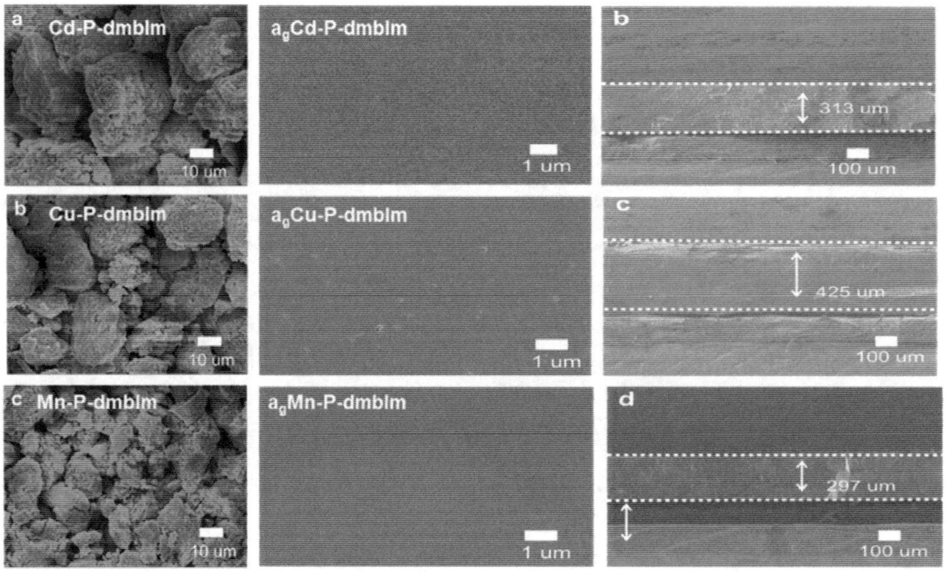

Fig.4 SEM for M-P-dmbIm and a_gM-P-dmbIm

Next, test the gas separation performance of the glass membrane at a constant inlet pressure of 1 bar. First, the single air permeability of the glass membrane is tested, and there is an obvious cut-off effect. The illustration shows the selective comparison and found the actual choice. in Fig.5a.The performance is close to the ideal and the selectivity is far greater than the Knudsen diffusion selectivity, which proves to be a good gas separation membrane material. In addition, the two-component mixed gas (1/1, v/v) membrane separation test was performed on the three materials. All three glasses showed gas separation performance similar to the ideal selectivity, a_gCd–P–dmbIm selectivity for H_2/CO_2, H_2/N_2 and H_2/CH_4 is the highest, respectively, 64.1, 34.4 and 54.4. in Fig.5b. The hydrogen flow rate is as high as 74950 barrer, which proves that these three glass materials are ideal hydrogen separation membrane materials.

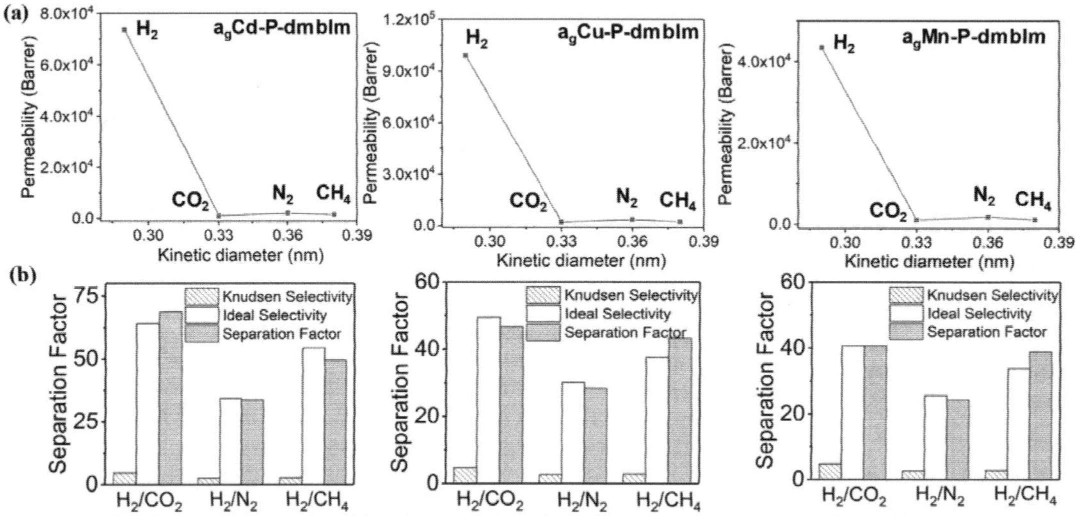

(a) Single gas test results of a_gM-P-dmbIm (b) Mixed gas test results of a_gM-P-dmbIm

Fig.5 Gas test results

The glass material shows good gas selectivity much higher than its corresponding Knudsen diffusion coefficient. Putting the glass material into the Robeson upper limit graph, it can be seen that the separation performance of the three gas pairs far exceeds the corresponding upper limit, breaking the gas permeability-selectivity game effect, and further judgement may be gas molecular sieving. in Fig.6.

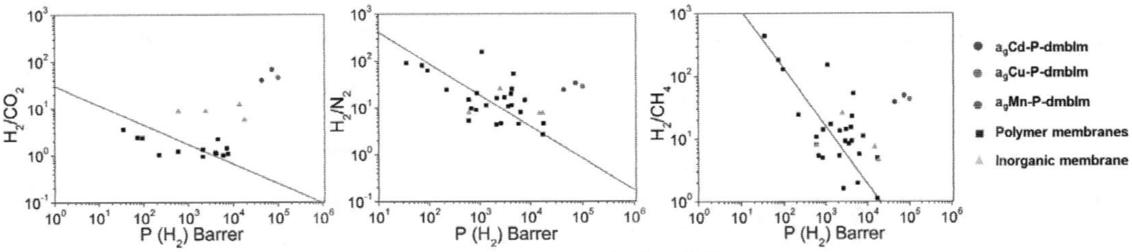

Fig.6 Robeson upper bound graph for different gas groups

The gas transport mechanism of the a_gM-P-dmbIm membranes is further studied. In the dissolution diffusion model, both the diffusion coefficient (D) and the solubility (S) have a crucial effect on the permeability (P). In terms of the mechanism of dissolution and diffusion, the diffusion coefficient of a gas is related to its permeability and solubility. The permeability is obtained through testing. The graph shows the adsorption curve of hydrogen and carbon dioxide gas on the glass membrane. The adsorption capacity of the material can be understood through the adsorption curve, and the solubility of the gas can be calculated. By calculation, the hydrogen diffusion coefficients of the three glass membranes are 547, 2650, and 622, which are much larger than the diffusion coefficients of carbon dioxide. Proved that hydrogen is easier to diffuse in the material than carbon dioxide. The gas solubility is calculated from the saturated adsorption capacity at 298 K. Through the ratio of permeability to solubility, the diffusion coefficient of the material to different gases can be calculated. The results are listed in Table 3.6. The hydrogen permeability of a_gCd-P-dmbIm, a_gCu-P-dmbIm, and a_gMn-P-dmbIm are 74.939, 100.70, and 41.985 ($\times 10^{-9}$ cm^2 $cmHg^{-1}$), respectively, and a_gCd-P-dmbIm, a_gCu-P-dmbIm and a_gMn-P-dmbIm are 1.17, 2.03, and 1.03, respectively ($\times 10^{-9}$ cm^2 $cmHg^{-1}$). From the results, it can be concluded that the H_2 diffusion coefficient of the three glasses is much higher than that of CO_2, which proves that H_2 is easier to diffuse in the material. in Fig.7.

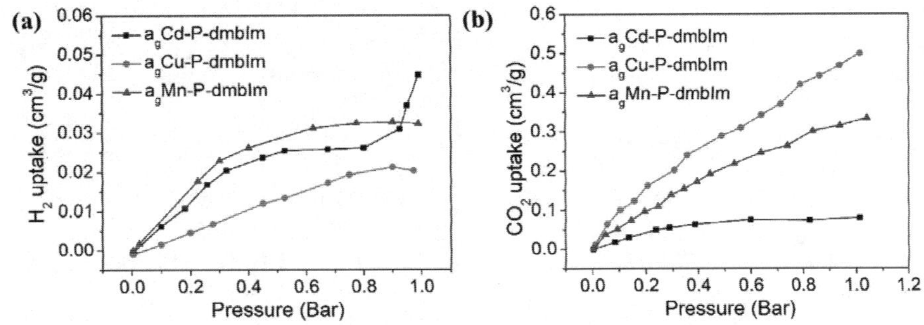

(a) Adsorption curve of H_2 (b) Adsorption curve of CO_2

Fig.7 Adsorption curve of H_2 and CO_2

4. Conclusion

Based on the results and discussions presented above, the conclusions are obtained as below:

(1) In view of the selection of coordination polymer materials, new materials were synthesized by changing the metal connection unit and ligand types. Through XRD pattern, FTIR-ATR spectrum and

melting point test M-P-dmbIm, confirm that it has been coordinated and connected and combined with thermogravimetric and DSC test results to analyze its melting point between 162 172 °C

(2) Three kinds of glass membranes were successfully prepared by pressing and cooling at a high temperature. The material changed from a crystalline structure to an amorphous structure before and after the glass state, and the proportion of the element group remained unchanged. The gas selectivity test was carried out by the Wicke-Kallenbach method, and the results showed that the three glass membranes all showed good separation performance, which exceeded the Robeson upper limit.

(3) The gas separation mechanism of a_gM-P-dmbIm was studied. The adsorption capacity of carbon dioxide and hydrogen was tested at 298 K, and the diffusion coefficients of the two materials in the glass material were calculated. The diffusion coefficients of hydrogen were both far greater than the diffusion of carbon dioxide. As a result of the coefficient, it is believed that this is the reason why the a_gM-P-dmbIm membrane material has good hydrogen gas selectivity.

In order to better study the performance of a_gM-P-dmbIm film and meet the needs of practical applications, future research is looking forward to two aspects: (1) Exploring the self-healing properties of the material; (2) Effectively controlling the thickness by adding support materials to meet the requirements of industrialization .

Acknowledgments

This work was financially supported by the National Natural Science Foundation of China (Grant No. 21922502, 21674012,21625102, 21971017), National Key Research and Development Program of China (2020YFB1506300), Beijing Municipal Science and Technology Project (Z201100007520005), Beijing Institute of Technology Research Fund Program and Analysis and Testing Center of Beijing Institute of Technology, Ceshigo Research Service.

References

[1] Gaillac R, Pullumbi P, Beyer K A. (2017) Liquid metal-organic frameworks. Nature Materials. 16(11): 1149-1154.

[2] Kitagawa S, Kitaura R, Noro S. (2004) Functional porous coordination polymers. Angewandte Chemie International Edition. 43(18): 2334-2375.

[3] S. Horike, D. Umeyama, M. Inukai, T. Itakura, S. Kitagawa. (2012) Synthesis and Transport Properties of Novel MOF/PIM-1/MOF Sandwich Membranes for Gas Separation. Membranes. 21(13): 2580-2582.

[4] Fuoco A, Khdhayyer M R, Attfield M P. (2017) Coordination Network Based Ionic Plastic Crystal for Anhydrous Proton Conductivity. J. Am. Chem. Soc. 134(18): 7612-7615.

[5] Umeyama D, Horike S, Inukai M. (2013) Integration of Intrinsic Proton Conduction and Guest-Accessible Nanospace into a Coordination Polymer. J. Am. Chem. Soc. 135(30): 11345-11350.

[6] Bennett T D, Yue Y Z, Li P. (2016) Melt-Quenched Glasses of Metal-Organic Frameworks. J. Am. Chem. Soc. 138(10): 3484-3492.

[7] Bennett T D, Horike S. Liquid. (2018) glass and amorphous solid states of coordination polymers and metal organic frameworks. Nature Reviews Materials. 3(11): 431-440.

[8] Wang Y, Jin H, Ma Q. (2020) A MOF Glass Membrane for Gas Separation. Angewandte Chemie. 132(11): 4395-4399.

Preparation and Catalysis of Pd Loaded PS/Fe_3O_4@PPy/Pb Magnetic Composite materials

Xue Fang[1a], Guiming Su[1b,*], Haijian Jiang[1c], Mingyue Chen[1d], Tianzhi Li[1e], Meihui Song[1f], Xiaochen Zhang[1g]

[1]Institute of Advanced Technology, Heilongjiang Academy of Sciences, Harbin, Heilongjiang, China

[a]email: haerbinxiaoxue@163.com, [b]email: fangxue@g68.onexmail.com, [c]email: jhj-992@163.com, [d]email: fengtian391@163.com, [e]email: liverpool_1985@163.com, [f]email:smh1980@126.com, [g]email:xiaochen966@sohu.com

Abstract: Fe_3O_4 particles were prepared by chemical co-precipitation method. Then, the functionalized PS nanoparticles were mixed with the prepared Fe_3O_4 particles, the PS/Fe_3O_4 magnetic nanocomposite were obtained by electrostatic interaction; the pyrrole monomer was added, and the pyrrole monomer undergoes oxidative polymerization on the surface of the composite. The PS/Fe_3O_4@PPy composite coated with polypyrrole was obtained. Finally, the PS/Fe_3O_4@PPy/Pb magnetic composite was prepared by the deposition-precipitation method. The morphology and structure of the composite particles were analyzed by scanning electron microscope, X-ray diffractometer, Fourier transform infrared spectrometer, and thermogravimetric analyzer. The concentration of organic dye molecules was determined by ultraviolet-visible spectrophotometer. The results showed that the dispersion of the PS/ Fe_3O_4@PPy/Pb magnetic composite was excellent; the PS/Fe_3O_4@PPy/Pb magnetic composite showed excellent catalytic performance in the reduction of methylene blue with sodium borohydride as reducing agent. The composite could be recycled and reused many times.

1. Introduction

With the rapid development of the current industry, the problem of water pollution has attracted attention from all walks of life[1]. Especially with the vigorous development of textiles, papermaking, cosmetics and other industries[2]. It is a pollution problem that cannot be ignored of the dye wastewater generated. The water quality of dye wastewater is complex, deep in color, and large in discharge[3]. It is a kind of organic matter that is highly toxic, difficult to biodegrade, and easy to accumulate. It is very harmful to the human body. Direct discharge will cause great damage to the ecological environment and drinking water. harm. Therefore, its modification or functionalization to remove pollutants in water has become a research hotspot in recent years[4,5].

For sewage treatment, the traditional method is to use adsorbents. Traditional adsorbent materials mainly include montmorillonite, activated carbon, clay, silica, zeolite, etc. Their disadvantage is that they are easy to saturate and difficult to recycle[6,7]. Nanomaterials have always received widespread attention due to their excellent performance. Magnetic nanoparticles in nanomaterials have become a good separation and catalyst due to their large specific surface area, easy functionalization, and good separation performance. Used to treat wastewater and remove heavy metals. In recent decades, Fe_3O_4 magnetic nanocomposite materials have attracted attention in the field of environmental remediation as adsorbents[8,9]. Its advantage is that it can achieve the purpose of solid-liquid separation only under the action of an external magnetic field, which provides a faster and more convenient way for the repeated use of adsorbent materials. Conditions, can save a lot of money in production. Therefore, Fe_3O_4 magnetic nanocomposite materials adsorb dyes in water as a new development direction that year[10].

Fe_3O_4 nanoparticles are easy to aggregate due to the dipole force between the particles, which reduces the intrinsic magnetic properties, resulting in weak magnetic response and reduced specific surface area[11-13]. Coating Fe_3O_4 nanoparticles in a dense shell layer to form magnetic composite microspheres with a core-shell structure is one of the effective methods to solve this problem. In this study, Fe_3O_4 particles were prepared by the chemical co-precipitation method, and PS/Fe_3O_4 nanocomposite particles were used as seeds[14-16]. The PPy nanocomposite was successfully deposited on the surface of the seed particles by the reaction of pyrrole polymerization. Finally, the deposition-precipitation method was used to obtain a multifunctional PS/Fe_3O_4@PPy/Pb magnetic composite. More importantly, due to the catalytic effect of PdNPs and the enrichment effect of PPy in the PS/Fe_3O_4@PPy/Pd nanocomposite, it exhibits excellent performance in the catalytic degradation of methylene blue (MB) and can be recycled and reused.

2. Materials and Methods

2.1 materials

Ferric chloride hexahydrate ($FeCl_3$•$6H_2O$), ferrous chloride tetrahydrate ($FeCl_2$•$4H_2O$), polystyrene microspheres (PS), pyrrole monomer, palladium chloride, absolute ethanol. Methylene blue(MB): analytically pure, Sinopharm Chemical Reagent Co., Ltd.; all water used is ultrapure water.

2.2 Characterizations

HD 2015W electric stirrer; KQ 5200E ultrasonic instrument; Fourier Transform Infrared Spectrometer: Model Cary630; The structures of nanoparticles and nanocomposites were investigated by Lambda7600 -FT-IR. A JEOL JSM-6700F scanning electron microscope (SEM) with primary electron energy of 3 keV employed to examine the surface morphologies of products. Inorganic content of the composite spheres was measured by DSC-404 F3 and TG-209 F3. X-ray diffraction (XRD) data was collected on a X'Pert PRO X-ray diffractometer . New century general T6 ultraviolet and visible (UV–Vis) spectrophotometer was employed for analysis of MB dye reduction.

2.3 Methods

2.3.1 Preparation of Fe_3O_4 nanoparticles

At room temperature, 12.2 g $FeCl_3$•$6H_2O$ and 5.9 g $FeCl_2$•$4H_2O$ were added to a 100 mL round bottom flask, and 50 mL water was added in the flask. Under mechanical stirring, 25 mL ammonia water was added in and maintained the reaction for 30 mins. After repeated washing and magnet decanting, the resulting sample was dried in an oven at 60 ℃ and stored as a solid for later use.

2.3.2 Preparation of polystyrene/ferroferric oxide nanocomposite particles(PS/Fe_3O_4)

A certain mass of PS particles and Fe_3O_4 nanoparticles were weighed in a 100 mL round bottom flask, the mass ratio of PS particles and Fe_3O_4 nanoparticles were 1:1.25, a certain volume of ultrapure water was added. The system was maintained 30 mL and the pH was adjusted to 3. After mechanical stirring for 5 hours, the supernatant was poured by magnet decanting and washed with ultrapure water repeatedly until the supernatant was clear. The sample was dried in an oven at 60 ℃ for later use.

2.3.3 Preparation of polystyrene/ferroferric oxide@polypyrrole nanocomposite particles (PS/Fe_3O_4@PPy)

the pyrrole monomer was polymerized on the surface of the composite to form PS/Fe_3O_4@PPy nanocomposites.

2.3.4 Preparation of polystyrene/ferroferric oxide/palladium@polypyrrole nanocomposite particles(PS/ Fe_3O_4 @PPy/Pd)

50 mg PS/Fe_3O_4@PPy nanocomposites were dispersed in 30 ml of water under 60℃ water bath conditions, 20 mg of $PdCl_2$ was added in the system with mechanical stirring for 24 h. The precipitate was centrifuged and washed with water for three times. Then the precipitate was dried in oven under 60℃ for 12 h.

3. Results and discussion

The whole procedure to prepare PS/Fe_3O_4@PPy /Pd composites is illustrated in Fig. 1.

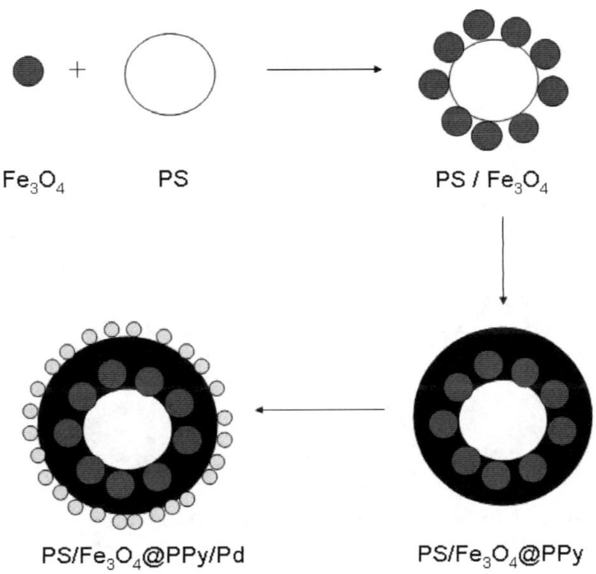

Fig.1 Synthesis route to PS/Fe_3O_4@PPy/Pd composites

3.1 Analysis of the surface morphology of the composites

It can be seen from the picture of SEM in Fig. 2(a), the micrographs of PS nanoparticles that the surface of the PS nanoparticles is smooth and uniformly spherical.

Fig.2 SEM of PS nanoparticles(a); PS/Fe_3O_4 nanocomposites(b)

Compared with Fig.2(a), Fig.2(b) shows the PS/Fe_3O_4 nanocomposite particles loaded with Fe_3O_4 nanoparticles. The surface of the PS nanoparticles becomes rough and has obvious particles. This is because the surface of the PS nanoparticles is negatively charged, which provides conditions for the positive and negative combination of Fe_3O_4 nanoparticles. Fe_3O_4 nanoparticles synthesized by co-precipitation method are deposited on the surface of PS nanoparticles by electrostatic self-assembly at pH=3. Then the pyrrole monomer undergoes oxidative polymerization on the surface of the PS.

3.2 XRD analysis of magnetic nanocomposites

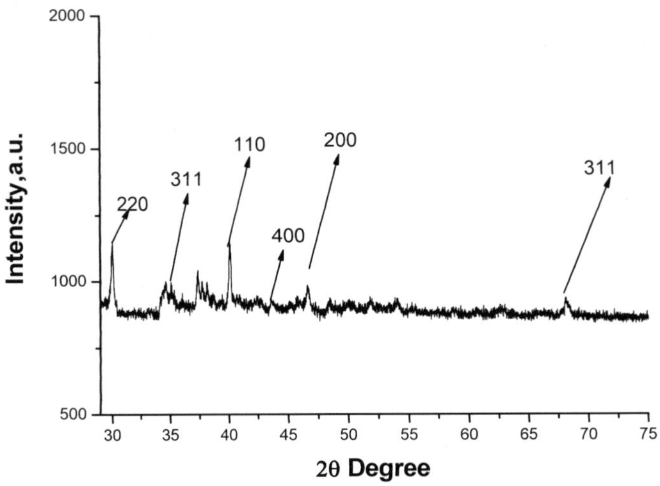

Fig.3 XRD patterns of PS/Fe_3O_4@PPy/Pd composites

In order to verify the prepared PS/Fe_3O_4@PPy/Pd nanocomposite particles, x-ray diffraction (XRD) was used to analyze them. XRD was an important method to characterize the prepared materials, the phase structure of nano-materials can be obtained by XRD. As shown in Fig. 3 is the XRD pattern of Fe_3O_4 nanoparticles. The (220) , (311) and (400) crystal faces of Fe_3O_4 nanoparticles in 30°,35.5° and 43.3° can be seen, respectively, showing the face-centered cubic structure of the prepared Fe_3O_4 nanoparticles. At the same time, there are new characteristic diffraction peaks, which belong to the (110), (200) and (311) crystal planes of PD nanoparticles in 39.8°, 46.4° and 67.6°, which also indicate the formation of PD nanoparticles in the process of PPy formation.

3.3 Analysis of membrane structure

The temperature stability of the PS/Fe_3O_4@PPy/Pd nanocomposites in the system can be analyzed by infrared spectroscopy. Fig. 4 shows the infrared spectra of magnetic composite materials with different heating temperatures. As shown in Fig. 4(a) is the characteristic curve of PS. The characteristic peaks appear at 2980, 2830, 1495 and 750cm^{-1}. The Fig. 4(b) is the PS/Fe_3O_4 nanocomposite particle at peaks appear at the same position, indicating that PS is contained therein, and the structure of PS is not damaged by loading Fe_3O_4 nanoparticles. The characteristic peaks of Fe_3O_4 are not obvious, mainly because: 1) The detection of inorganic substances by FTIR is not as good as that of organic functional groups; 2) The absorption peak of Fe_3O_4 nanoparticles is covered by the absorption peak of PS nanoparticles. In Fig. 4(c), the absorption peaks at 1568 and 1450 cm^{-1} are attributed to the C-C asymmetric and symmetric stretching vibration peaks in the pyrrole ring, respectively. The absorption peak at 1360cm^{-1} is the C-N stretching vibration peak. The characteristic peak at 1050 cm^{-1} is the C-H bending vibration peak on the pyrrole ring. These vibration peaks can indicate that PPy is formed on the PS/Fe_3O_4 nanocomposite particles.

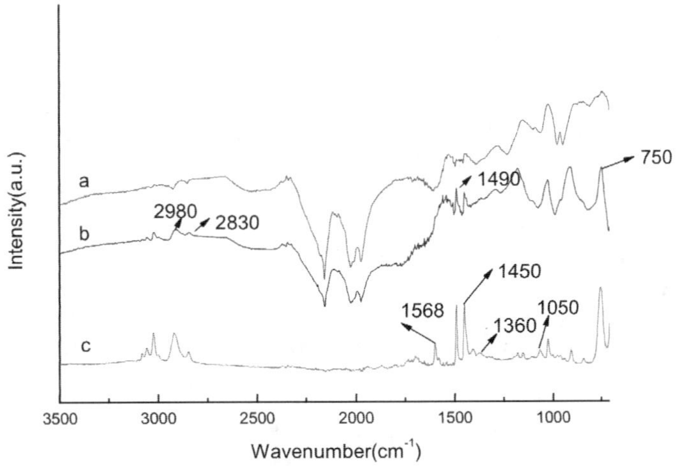

Fig. 4 FT-IR spectra of PS nanoparticles(a); PS/Fe_3O_4 nanocomposites(b); PS/Fe_3O_4@PPy composites(c)

3.4 Catalytic analysis of magnetic materials

MB is a typical industrial pollutant. Because of its stability in water, it is more complicated to handle. Therefore, MB is selected as the characteristic dye, and $NaBH_4$ is used as a reducing agent to catalyze the degradation of MB to explore the PS/Fe_3O_4@PPy/Pd nanocomposites. The catalytic performance of the particles as a catalyst. The degradation process of MB can be monitored by its absorbance change at the maximum wavelength of 665 nm. Fig. 5 shows the UV spectra of (a) the initial MB solution, (b) the MB solution after adding the reducing agent $NaBH_4$ and reacting for 30 minutes, and (c) the UV spectrum of the MB solution after adding the reducing agent $NaBH_4$ and the magnetic catalyst for 30 minutes.

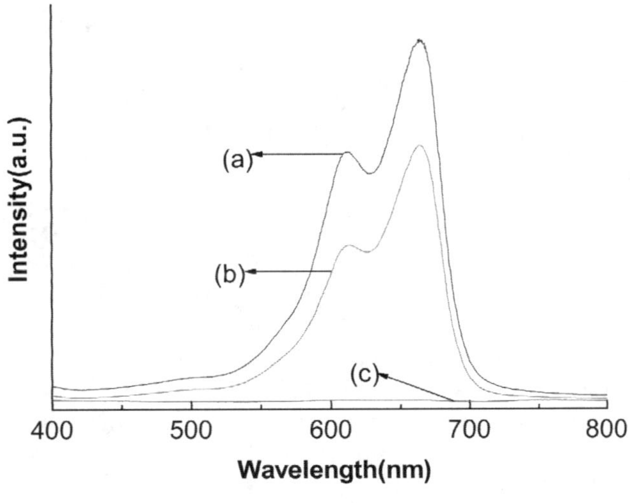

Fig.5 Vis spectra of he initial MB solution(a); the MB solution after adding the reducing agent $NaBH_4$ after reaction for 30 mins(b); the UV spectrum of the MB solution after adding the reducing agent $NaBH_4$ and the magnetic catalyst for 30 mins(c)

We prepared (0.2 mg/mL) aqueous solution of MB 3.5 mL. The reaction process was monitored by Vis spectra, as illustrated in Fig. 5a. The original aqueous MB dye solution is dark blue in color and the kmax appears at 665 nm (curve a). 2 mL $NaBH_4$ (10 mg/mL)was added to the system. Upon the addition of $NaBH_4$ without PS/Fe_3O_4@PPy/Pd catalysts, the absorbance intensity at kmax of MB decreases, but does not completely vanish after reaction for 30 minutes (curve b). The solution color only becomes light blue. After 10.0 mg of the PS/Fe_3O_4@PPy/Pd composites were added, the absorbance at kmax completely disappears within 30 minutes (curve c). The solution color turns colorless as seen. It could be concluded that the PS/Fe_3O_4@PPy/Pd composites had excellent catalytic performance in reduction of MB dye with $NaBH_4$ as a reducing agent.

4. Conclusions

(1) Using PS/Fe_3O_4 nanocomposite particles as the carrier, based on the principle of "swelling-diffusion-interfacial polymerization", based on the redox effect between palladium chloride and pyrrole monomer, PPy coating on the surface of PS/Fe_3O_4 nanocomposite particles is formed, and loaded with palladium on the surface.

(2) Considering the harm caused by methylene blue dye to the water system, using the enrichment effect of PPy and the catalytic performance of PdNPs, we carried out catalytic degradation of methylene blue dye. The results showed that PS/Fe_3O_4@PPy/Pd nanoparticles with different shell thicknesses can be effective degrade methylene blue.

(3) This method not only simplifies the process conditions of the conventional method, but also solves the problem of poor adhesion between metal and polymer. The biggest advantage of this method is that the doped nano-palladium can be distributed in the polymer system with the original particles, so as to best reflect or maintain the unique properties of the nano-palladium. Therefore it is anticipated that this kind of catalysts will have great potential for further practical applications.

References

[1] Yao, T.J., Cui, T.Y., Fang, X., Yu, J., Cui, F., Wu, J. (2013) Preparation of yolk/shell Fe_3O_4@polypyrrole composites and their applications as catalyst supports. J. Chemical Engineering Journal, 225: 230-236.
[2] Xue F, Guiming S, Meihui S, et al. (2021) Preparation of Polyimide@Polypyrrole/ Palladium hollow composites with applications in catalysis. J. Materials Science, 27:192-196.
[3] Qiu, J. (2018) Preparation of magnetic multifunctional polymer nanocomposite particles and the research of properties. D. Jiangnan University.
[4] Juang D.F, Chen P.C. (2021) Treatment of polluted river water by a new constructed wetland. J. Int J Environ Sci Tech., 4(4): 481-488.
[5] Tesh S J, Scott T B. (2014) Nano-composites for water remediation: a review. J. Adv Mater., 26(35): 6056-6068.
[6] Ghasemi E, Heydari A, Sillanpää M. J. (2017) Superparamagnetic Fe_3O_4@EDTA nanoparticles as an efficient adsorbent for simultaneous removal of Ag(I), Hg(II), Mn(II), Zn(II), Pb(II) and Cd(II) from water and soil environmental samples. J. Microchem J., 131: 51-56.
[7] Dong C, Lu J, Qiu B, et al. (2018) Developing stretchable and graphene-oxide-based hydrogel for the removal of organic pollutants and metal ions. J. Appl Catal, B., 222: 146-156.
[8] An H, Liu J, Li X, et al. (2018) The fate of cyanuric acid in biological wastewater treatment system and its impact on biological nutrient removal. J. Environ Manage, 206: 901-909.
[9] Muhammad Ekramul Mahmud H N, Huq A K O, Yahya R. (2016) The removal of heavy metal ions from wastewater/aqueous solution using polypyrrole-based adsorbents: a review. J. RSC Adv., 6(18): 14778-14791.
[10] Zare E N, Lakouraj M M, Ramezani A. (2015) Effective adsorption of heavy metal cations by superparamagnetic poly(aniline-co-m-phenylenediamine)@Fe_3O_4 nanocomposite. J. Adv Poly Tech., 34(3): 1-11.
[11] Zhou Q, Wang Y, Xiao J, et al. (2017) Fabrication and characterisation of magnetic graphene

oxide incorporated Fe_3O_4@polyaniline for the removal of bisphenol A, t-octyl-phenol, and alpha-naphthol from water. J. Sci Rep., 7(1): 11316-11327.
[12] Wang J, Bi L, Ji Y, et al. (2014) Removal of humic acid from aqueous solution by magnetically separable polyaniline: Adsorption behavior and mechanism. J. Colloid Interface Sci., 430(430): 140-146.
[13] Bin M, Aiqin W. (2015) One-Pot fabrication of multifunctional superparamagnetic Attapulgite/Fe_3O_4/polyaniline nanocomposites served as an adsorbent and catalyst support. J. Mater Chem A., 3(1): 281-289.
[14] Zheng Y, Wang W B, Zhu G, et al. (2013) Enhanced selectivity for heavy metals using polyaniline-modified hydrogel. J. Ind Eng Chem Res., 52(13): 4957-4961.
[15] Fang X H, Xu X C, Wang S S, et al. (2013) Adsorption kinetics and equilibrium of Cu(II) from aqueous solution by polyaniline/coconut shell-activated carbon composites. J. Environ Eng., 139: 1279-1284.
[16] Sheng D, Zhang G S, Xi W, et al. (2015) Preparation and performance of polyacrylonitrile fiber functionalized with iminodiacetic acid under microwave irradiation for adsorption of Cu(II) and Hg(II) . J. Chem Eng J., 276: 349-357.

Journal of Physics: Conference Series **2194** (2022) 012043 doi:10.1088/1742-6596/2194/1/012043

Preparation of nickel-coated graphene and evaluation of infrared interference performance

Kai Li[1a], Xuanyu Wang[1b*], Yanqing Gao[2a], Wenjie Dong[1c], Huiying Li[1d]

[1]Institute of NBC Defense, Beijing 102205, China

[2]State Key Laboratory of Infrared Physics, Shanghai Institute of Technical Physics, Chinese Academy of Sciences, Shanghai 200083, China

[1a]lkyanhuo@163.com, [1b*]wxycrs@aliyun.com, [2a]gyq _1982@mail.sitp.ac.cn

Abstract—In this paper, reduced graphene oxide (RGO) was prepared by oxidation-reduction method, and then RGO/Ni composite was prepared by chemical plating method. The RGO and RGO/Ni samples were characterized by the methods of SEM, TEM, XPS, EDS, XRD, etc. The complex refractive index of the samples in 2~14μm infrared wavebands was measured by ellipsometry method, and the infrared absorption characteristic was calculated by DDA method. Then, the infrared interference performance was tested by the method of Fourier transform spectrometer connected with smaller smoke box, and the dispersivity was also comprehensively evaluated by Carr index method. It is found that electroless nickel plating not only can ease the secondary agglomeration between the RGO layers, improve its dispersivity, but also change the electromagnetic properties of graphene surface and significantly improve its infrared absorption. From the infrared interference performance test, it can be seen that the average extinction coefficients of RGO smoke to 3~5 μm and 8~14 μm infrared is $2.56m^2/g$ and $1.8m^2/g$, respectively. While the average infrared extinction coefficients of RGO/Ni in the corresponding wavebands is $2.9m^2/g$ and $2.55m^2/g$, respectively. It further indicates that the surface modification of electroless nickel plating can greatly promote the infrared interference performance of RGO. In particular, the far-infrared extinction performance has been obviously improved.

1. Introduction

Smoke screen has been used in military for thousands of years. The earliest "smoke" mainly appeared in ancient wars as a means of concealment and communication. It was not until the explosion of World War I that the use of smoke screen was officially transformed into combat support operations. With the rapid development of photoelectric detection technology and the practical application of precision guided weapons, the requirements of battlefield survival for smoke protection ability are becoming higher and higher. Because infrared guidance has the advantages of high precision, strong anti-interference ability and good concealment, it has become one of the main guidance methods of many long-range precision strike weapons. At this time, the conventional smoke screen that can only resist visible light viewing equipment can no longer meet the needs of combat. At present, the development of new infrared interference smoke screen materials has become an important topic in the field of passive interference.

Graphene is a two-dimensional nano carbon material with single atom thickness. It has a special microstructure. Compared with traditional smoke screen materials, graphene not only has lower density, larger specific surface area, excellent conductivity and suspension performance, but also has the unique properties of micron and nano materials. It has great application value in infrared interference. Theoretically, the extinction ability of smoke to infrared mainly depends on its attenuation performance and suspension performance. According to a few literature reports, the infrared extinction ability of graphene depends on its extinction mechanism dominated by absorption attenuation. Although graphene powder shows good near-infrared interference performance, its far-infrared interference performance needs to be further improved. Besides, the problem of poor dispersion caused by the easy agglomeration of graphene power has not been effectively solved [1-3]. However, graphene's unique two-dimensional structure and huge specific surface area provide unlimited possibilities for the preparation of wide band composite infrared interference materials. Graphene is used as the base material to load electromagnetic wave loss components such as metal, ferrite and polymer. Each component can form complementary advantages, which can reduce the density of composites, maintain good suspension performance, expand the electromagnetic wave loss mechanism of new materials and improve the dispersion performance of materials. It is expected to obtain better infrared interference effect.

In this paper, graphene was prepared by redox method. Aiming at the problems of limited far-infrared interference performance and easy agglomeration, the surface of graphene was modified by electroless nickel plating, and the material was characterized. The infrared absorption characteristics of the material were calculated based on DDA method, and the infrared interference performance and dispersion performance of graphene before and after modification were experimentally studied. This study can provide more data basis and method reference for the follow-up application research of graphene in infrared interference.

2. Experiment

2.1. Chemical reagents and instruments

The chemical reagents used for preparation of RGO mainly include flake graphite, concentrated sulfuric acid, potassium permanganate, 30% hydrogen peroxide, absolute ethanol, etc. The chemical reagents used for preparation of RGO/Ni mainly include stannous chloride, palladium chloride, concentrated hydrochloric acid, nickel sulfate, sodium hypophosphite, sodium citrate, ammonium chloride, ammonia, etc. The chemical reagents used in the experiments are analytical pure, and the water used is deionized water.

The instruments used in the experiments mainly include model Sl-234 electronic balance, model Hh-2 electric heating constant temperature water bath pot, model SHB-III circulating water multipurpose vacuum pump, model KQ-250DB numerical control ultrasonic cleaner, model LC-LX-L60D desktop centrifuge, model 101-2BS electrothermal blowing, model FD-1A-80 freeze dryer, model SK-B08123K vacuum/inert atmosphere tube furnace.

2.2. Experimental methods

Graphene (RGO) was prepared by improved Hummers method and thermal reduction method. The surface of RGO was modified by electroless nickel plating method. Electroless nickel plating process mainly includes sensitization/activation, degumming, electroless plating. The sensitization/activation process adopts colloidal palladium activation process, which can obtain sufficient Pd^+ catalytic active center on the surface of RGO. Degumming can remove the protective layer of alkaline stannide on the surface of RGO and maintain high catalytic activity. Colloidal palladium solution and conditions are shown in Tab.1, and the degumming solution and conditions are shown in Tab.2. The sodium hypophosphite alkaline system is used for electroless plating experiment. The loading capacity of RGO powder is 2/L. the composition of electroless nickel plating solution and conditions are shown in Tab.3.

Tab.1 Compositions of the colloid-palladium solution and operating conditions

$SnCl_2{\cdot}2H_2O$(g/L)	$PdCl_2$(g/L)	HCl(mL/L)	T(℃)	Time(min)
50	0.5	300	45	20

Tab.2 Compositions of the degumming solution and operating conditions

HCl(mL/L)	*T(℃)*	*Time(min)*
100	40	10

Tab.3 Compositions of the plating solution and operating conditions

$NiSO_4{\cdot}6H_2O$ (g/L)	$NaH_2PO_2{\cdot}H_2O$ (g/L)	$C_6H_5Na_3O_7{\cdot}2H_2O$ (g/L)	NH_4Cl (g/L)	*pH*	*T* (℃)
20	20	15	30	8~9	65

2.3. Characterization of samples

The characterization instruments mainly include SEM (*model QUANTA FEG450, FEI*), TEM (*model Talos F200X, Thermo Scientific*), AFM (*model Multimode V, Bruker*), XPS (*model ESCALAB 250Xi, Thermo Scientific*), Lab Ram ARAMIS (*model Thermo Fisher DXR, Thermo Scientific*), XRD (*model D8, Bruker*), EDS (*model X-Max, Oxford*) and Laser particle size analyzer (*model BT-9300H, Bettersize*).The SEM-AFM instruments were used for characterizing the micromorphology of samples, the XPS-XRD instruments were used for characterizing the structural characteristics of samples, the EDS instrument was used for characterizing the element composition of samples and the Laser particle size analyzer was used for characterizing the particle size distribution of samples.

2.4. Measurement of complex refractive index

The real part and imaginary part of the complex refractive index in the mid-far infrared bands of RGO and RGO/Ni samples were measured by ellipsometry [4]. The instrument used was IR-VASE Mark II variable temperature infrared elliptical polarization spectrometer (*J. A. Wallam company*).

Since the imaginary part of the complex refractive index of one material is closely related to its absorption, the relationship between the imaginary part and the absorption coefficient is shown as Eq. (1)[5]. According to Eq. (1), the absorption coefficients of RGO and RGO/Ni samples in infrared bands can be calculated. Concurrently, the infrared extinction characteristics of RGO and RGO/Ni particles can be calculated by DDSCAT 7.3 program based on DDA method [4,6-8]. The values of particles size used in calculations were taken from the characterization results of each sample.

$$A(\lambda) = 4\pi k / \lambda \tag{1}$$

where k is the imaginary part of the complex refractive index, $A(\lambda)$ is the absorption coefficient.

2.5. Infrared interference performance test

Fourier transform spectrometer (*model WQF-530, Rayleigh*) and smaller smoke box (*length × width × height is 16 cm × 13 cm × 14 cm, volume is about 0.0015 m^3 and optical path is 0.16 m*) were used to test the infrared interference of RGO and RGO/Ni samples. The layout images of the test system are shown in Fig.1.

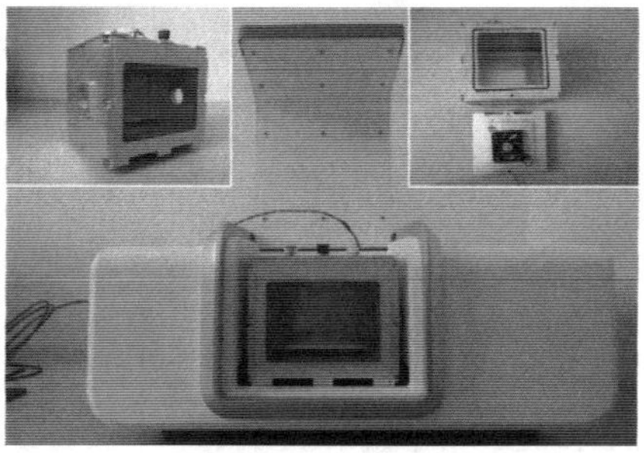

Fig.1 The physical images of smoke box and infrared interference test system

Test methods: (1) use Fourier spectrometer to quickly scan and collect the background transmission of blank smoke box. (2) Weigh a certain quality of sample into the smaller smoke box, turn on the fan for mixing, use the regulator to adjust the speed to make the sample disperse quickly in the smoke box, and start to scan the transmission of the sample quickly after the smoke screen is stable for 30s. After each test, store the data and empty the smoke screen in the smoke box. (3) Calculate the extinction coefficient of smoke screen to infrared according to Lambert Beer law [9], and the calculation formula is shown in Eq. (2).

$$T_s = \exp(-\alpha \cdot C \cdot L) \tag{2}$$

where T_s is the infrared transmission (%), α is the infrared extinction coefficient (m^2/g), C is the mass concentration of smoke screen (g/m^3), L is the optical path of smoke box (m). Due to the smaller size of the smoke box, it is difficult to obtain the real-time concentration of the smoke screen. Therefore, we use the average mass concentration of the smoke screen instead of the real-time concentration in the calculation. The average mass concentration is equal to the ratio of the mass of the added test sample to the volume of the smoke box. In order to reduce the errors of extinction coefficient calculated by single concentration, the average extinction coefficient of smoke screen is calculated by testing different mass concentrations and corresponding transmissions. The calculation formula is shown in Eq. (3), which can be transformed from Eq. (2).

$$-\ln\overline{T} = \alpha \cdot CL \tag{3}$$

As shown in Eq. (3), where $\ln\overline{T}$ is the natural logarithm of the average infrared transmission, CL is the area density of the smoke screen (g/m^2). By carrying out a linear quasi cooperative diagram between $-\ln\overline{T}$ and CL, we can obtain the slope, which is the average infrared extinction coefficient of the smoke screen.

2.6. Dispersion performance test

The evaluation parameters measured by powder comprehensive characteristic instrument (*model BT-1000, Bettersize*) include angle of repose (θ_r), collapse angle (θ_f), difference angle (θ_d), plate angle (θ_s), compression (C_p), dispersion (D_s) and homogeneity (M_f), etc. Then, the Carr index method [10] was used to find out the corresponding index to each parameter, and the dispersibility of the samples was comprehensively evaluated through the flowability index (F_w)and floodability index (F_d). The calculation formulas of F_w and F_d is shown in Eq. (4) and Eq. (5).

$$F_w = \theta_{r(index)} + C_{p(index)} + \theta_{s(index)} + M_{f(index)} \tag{4}$$

$$F_d = F_{w(index)} + \theta_{f(index)} + \theta_{d(index)} + D_{s(index)} \tag{5}$$

3. Results and discussion

3.1. Characterization analysis

3.1.1. Characterization results of RGO

The micromorphology characteristics of RGO is shown in Fig.2, the structural characteristics of RGO is shown in Fig.3, the particle size distribution of RGO is shown in Fig.4

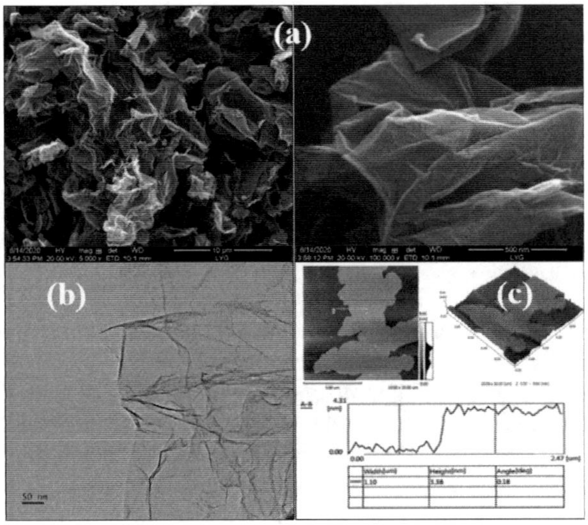

Fig.2 SEM(a), TEM(b), AFM(c) of RGO

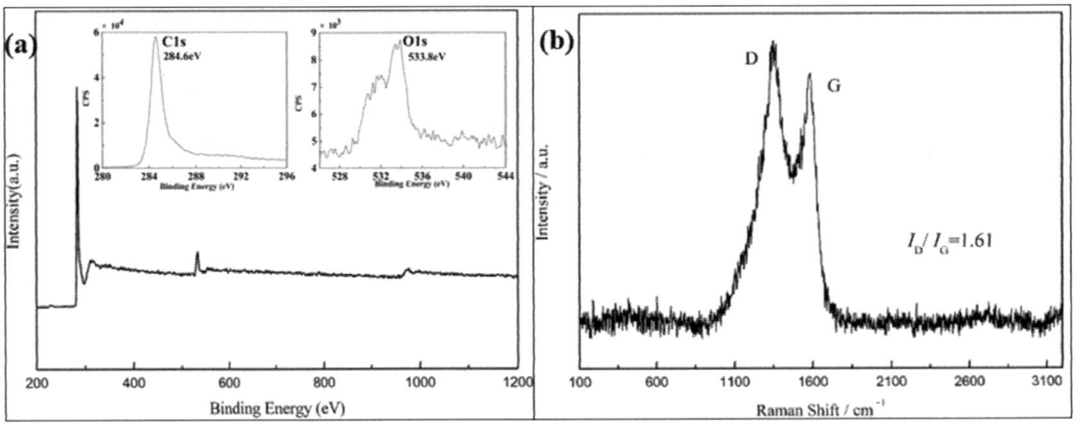

Fig.3 XPS(a), Raman (b) of RGO

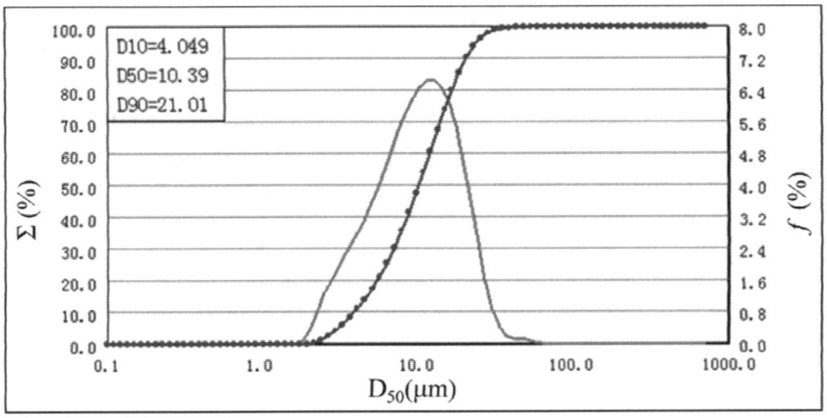

Fig.4 DLS of RGO

As shown in Fig.2, there is stacking and agglomeration between RGO particle lamellae. The lamellar size is micron, ranging from several microns to more than ten microns; The particle lamellae are folded and have different degrees of bending, which is due to the large specific surface area and high surface energy of RGO. The particle lamellae will maintain its own thermodynamic stability through bending; RGO particles have a thin layer structure and the edge of the lamella is tilted. Combined with the height curve and phase image, it is analyzed that the thickness of most particle lamellae in RGO samples is about 5nm.

As shown in Fig.3, the XPS full spectrum of RGO contains obvious C1s peak (284.6 EV) and O1s peak (533.8 EV). The C1s peak comes from C-C (284.6 EV), C-O (286.4 EV) and C = O (288.5 EV). In the C1s high-resolution spectrum, the peak shapes of C-O and C = O are not obvious, and the key position of C is mainly C-C bond, indicating that the oxygen content in the sample is not high, After calculation, the contents of C and o are 93.82% and 6.18% respectively, in which a small amount of oxygen is introduced during the reaction, indicating that the reduction degree of graphene sample is sufficient, but not thorough enough. There are obvious G and D peaks in the Raman spectrum of the sample. The G peak near 1590 cm^{-1} is caused by the E_{2G} stretching vibration of C-C, and the D peak near 1350 cm^{-1} is the embodiment of the disorder degree of graphene. The high intensity of D peak indicates that there are many structural defects due to the oxidation of graphite in the reaction process. After fitting, it is calculated that the intensity ratio (I_D/I_G) of D peak to G peak is 1.61, indicating that the thermal reduction degree of graphene is not good enough, and there are still some oxygen-containing functional groups that have not been reduced.

As shown in Fig.4, the equivalent volume particle size distribution of RGO sample shows typical lognormal distribution characteristics, indicating that the sample is well dispersed in ethanol medium, in which the agglomeration is not obvious. The D_{50} of RGO sample is about 10.4 μm and the D_{90} is about 21 μm, indicating that there are more particles with small size in the samples.

3.1.2. Characterization results of RGO/Ni

The micromorphology characteristics of RGO/Ni is shown in Fig.5, the structural characteristics and element composition of RGO/Ni is shown in Fig.6, the particle size distribution of RGO/Ni is shown in Fig.7

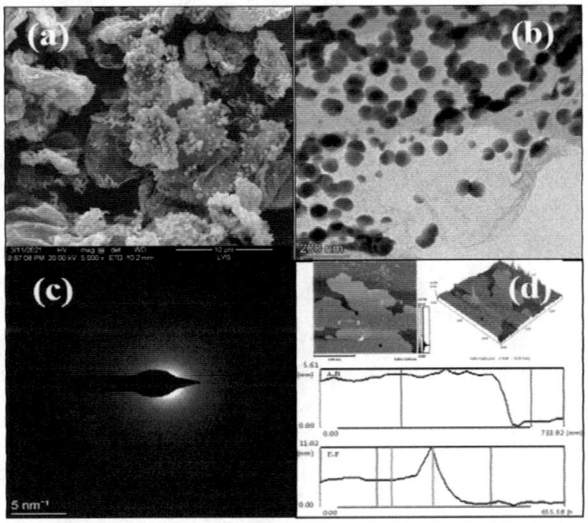

Fig.5 SEM(a), TEM(b), SEAD(c), AFM(d) of RGO/Ni

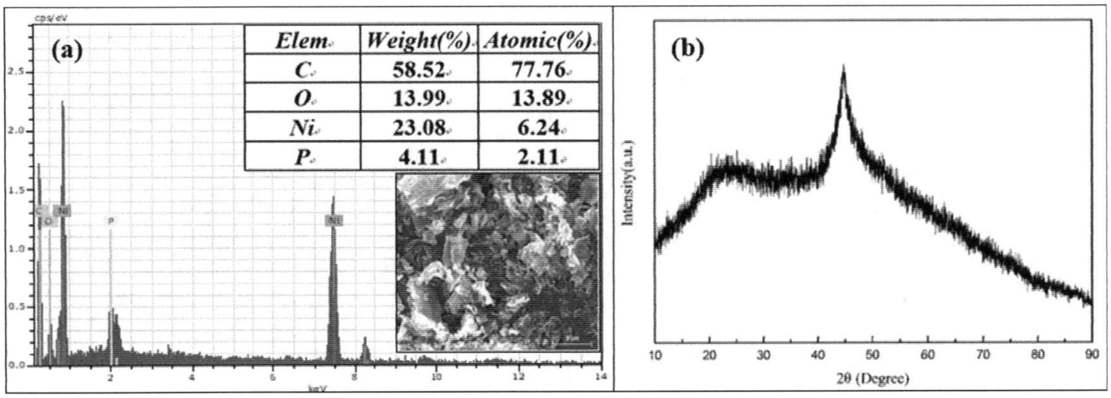

Fig.6 EDS(a), XRD(b) of RGO/Ni

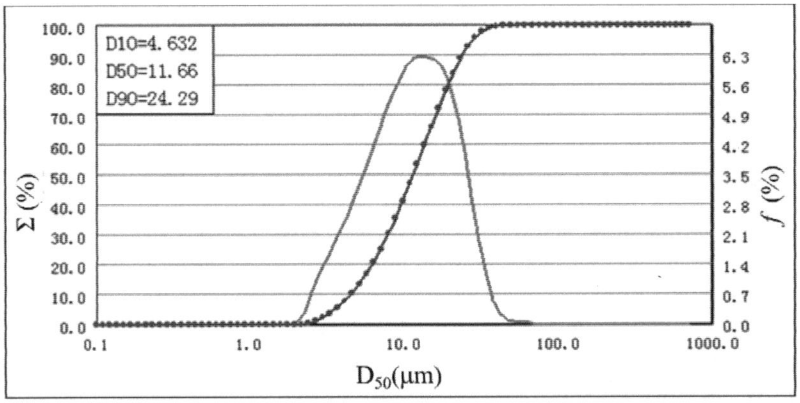

Fig.7 DLS of RGO/Ni

As shown in Fig.5, there are many nickel nanoparticles coated on the upper and lower surfaces of RGO. Most particles exist in a single form, with a little local aggregation. The agglomeration phenomenon between RGO/Ni particles is reduced and the dispersion is improved. This is because the large surface area, bending and surface folds of graphene effectively prevent the aggregation of nickel nanoparticles, at the same time, the loading of nickel nanoparticles also prevents the secondary agglomeration of RGO layers. The distribution and size of nickel nanoparticles coated on the surface of RGO thin layers are relatively uniform. The shape of nickel nanoparticles is approximately flat sphere, and its average diameter is about 50 nm; The electron diffraction patterns of RGO/Ni samples show a wide and bright halo ring and multiple weak halo rings, indicating that the nickel nanoparticles loaded on the surface of RGO have a certain microcrystalline structure; In the AFM diagram of the sample, the thickness of RGO layer within the regions of A to B and C to D is about 5.6 nm and 3.5 nm respectively, and the thickness of metal nickel coating within the region of E to F is about 11 nm. It can be calculated that the average thickness of RGO/Ni layer is about 16 nm.

As shown in Fig.6, the main elements contained in RGO/Ni sample are carbon and nickel, in addition to a small amount of oxygen and a small amount of phosphorus, it further shows that metal nickel has been successfully deposited on the surface of RGO. The existence of oxygen indicates that the oxygen-containing groups have not been completely removed during the preparation of RGO, while the existence of phosphorus may be due to the free electrons produced by the dehydrogenation side reaction of hypophosphite ion. The nickel coating is composed of metal nickel and a small amount of phosphorus. In the XRD spectra of the sample, the peak near 25°(2θ) is the amorphous diffraction peak of RGO. In addition, a broadened diffraction peak appears at 44.5°(2θ), which corresponds to the (111) crystal plane of nickel. The broadening of the peak shape indicates that the crystallinity of metallic nickel in the coating is low and tends to transition to amorphous state, which is basically consistent with the analysis results of electron diffraction pattern above.

As shown in Fig.7, the equivalent volume particle size distribution of RGO/Ni sample also shows the characteristics of typical lognormal distribution. The D_{50} is about 11.7 μm, the D_{90} is about 24.3 μm. It shows that the deposition of nano-nickel particles increases the thickness of the particle layers, while the equivalent volume of RGO/Ni particles also increases.

3.2.Calculation and analysis of infrared absorption characteristics

The complex refractive index curves, the absorption coefficient curves, infrared extinction curves of RGO and RGO/Ni are shown in Fig.8 to Fig.10 respectively.

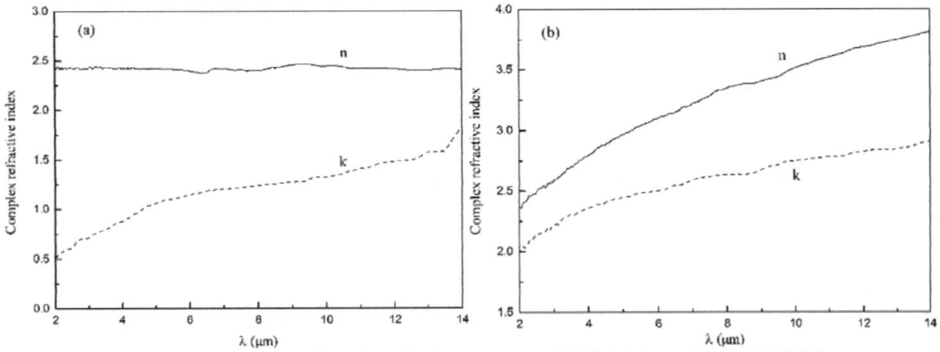

Fig.8 Complex refractive index curve of RGO(a) and RGO/Ni(b)

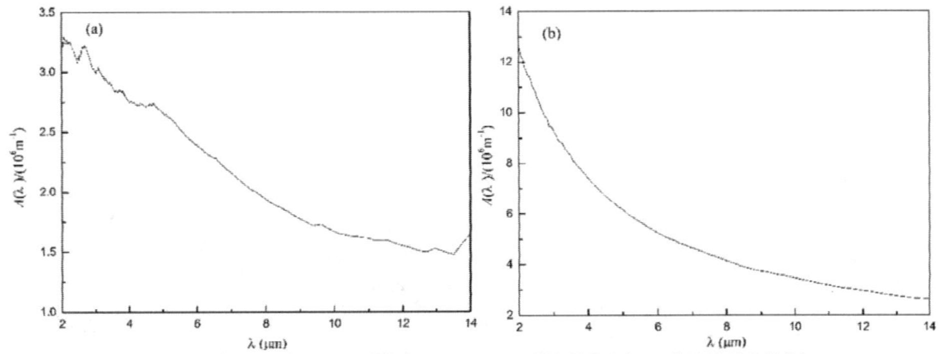

Fig.9 Absorption coefficient curve of RGO(a) and RGO/Ni(b)

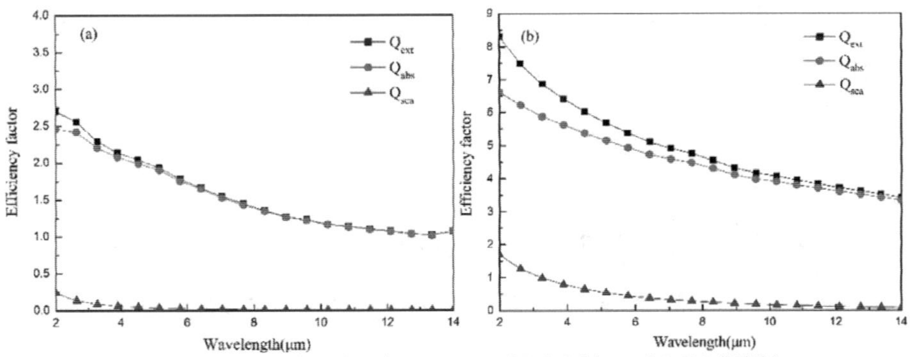

Fig.10 Infrared extinction curve of RGO(a) and RGO/Ni(b)

As shown in Fig.8, the real part(n) of RGO does not change significantly with the wavelength, and its value is basically stable near 2.5. The imaginary part (k) of RGO, the real part (n) and imaginary part (k) of RGO/Ni gradually increase with wavelength, with the values in the range of 0.5-1.8, 2.4-3.8, 2-2.9, respectively.

As shown in Fig.9, the absorption coefficient of RGO sample basically decreases with the wavelength, and the absorption coefficient is the largest in the near-infrared band, which indicates that the absorption of RGO in the near-infrared band is strong, while its absorption in the mid-far infrared band is relatively weak. In the same infrared band, the absorption coefficient of RGO/Ni is greatly improved compared with RGO, indicating that the modification of electroless nickel plating effectively improves its infrared absorption capacity, especially the increase of the far-infrared absorption capacity is obvious.

The infrared extinction curves of RGO circular sheet particles with diameter of 5μm, thickness of 5 nm and RGO/Ni circular sheet particles with diameter of 5μm, thickness of 16 nm is shown in Fig.10. From the comparison, it can be seen that the extinction, absorption and scattering efficiency factors (Q_{ext}, Q_{abs}, Q_{sca}) of RGO/Ni particles are greater than those of RGO particles, indicating that the modification of electroless nickel plating can greatly improve the infrared extinction of RGO, especially the increase of the infrared absorption capacity is obvious.

3.3.Results of infrared interference performance test

The transmission and average extinction coefficient of RGO smoke in 3~5 μm , 8~14 μm infrared bands are shown in Fig.11 and Fig.12. The results of RGO/Ni smoke in the same bands are shown in Fig.13 and Fig.14.

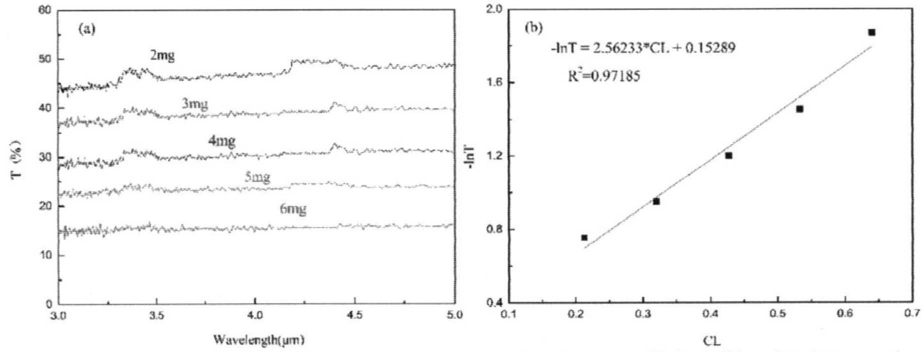

Fig.11 Infrared transmission(a) and average extinction coefficient(b) of RGO smoke in 3~5 μm wavebands

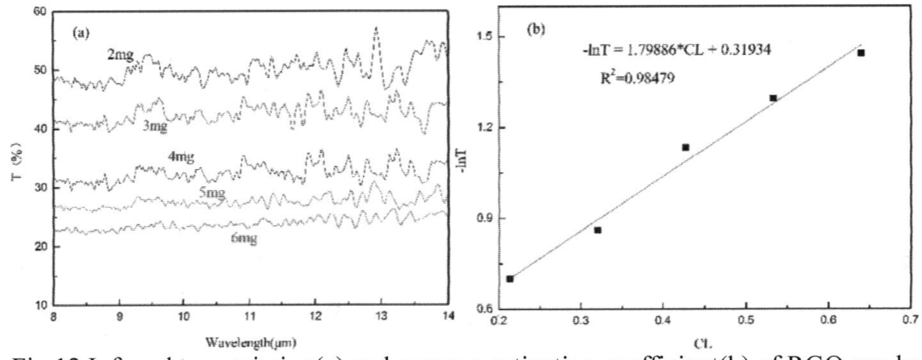

Fig.12 Infrared transmission(a) and average extinction coefficient(b) of RGO smoke in 8~14 μm wavebands

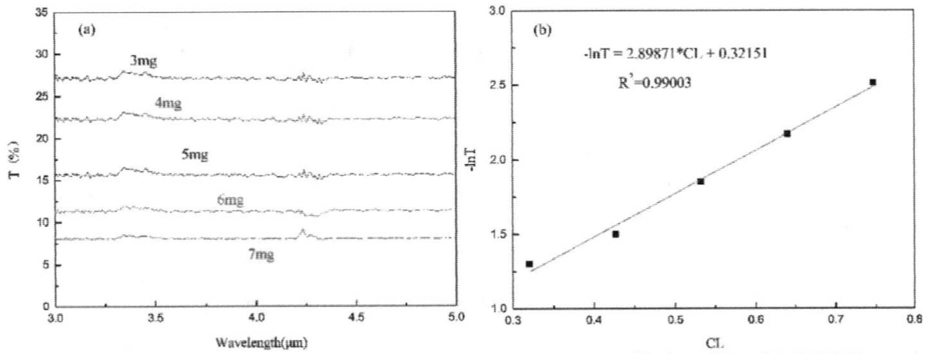

Fig.13 Infrared transmission(a) and average extinction coefficient(b) of RGO/Ni smoke in 3~5 μm wavebands

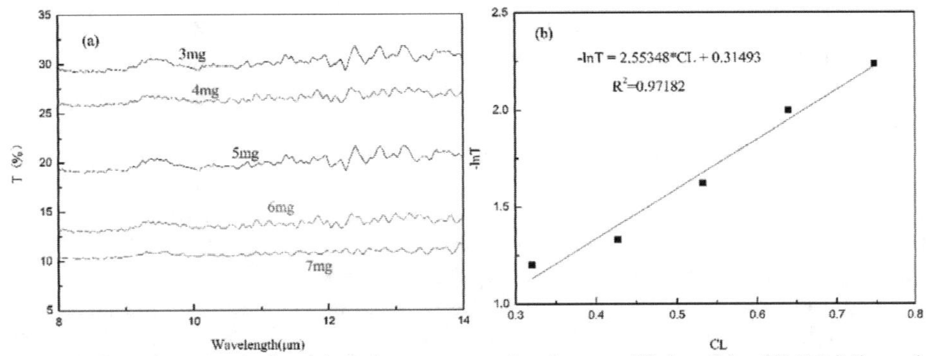

Fig.14 Infrared transmission(a) and average extinction coefficient(b) of RGO/Ni smoke in 8~14 μm wavebands

The extinction coefficient of smoke particles is an important parameter to characterize its shielding ability. The larger the extinction coefficient, the better the extinction effect and the stronger the interference ability. As shown in Fig.11 and Fig.12, it is calculated that the average extinction coefficient of RGO smoke screen to 3~5μm and 8~14 μm infrared is 2.56m^2/g and 1.8m^2/g respectively, both of which are higher than the extinction coefficient of traditional infrared interference materials , such as copper powder and graphite [1], indicating that RGO smoke has good near-mid infrared interference performance, and its far-infrared interference performance needs to be further improved.

As shown in Fig.13 and Fig.14, the average extinction coefficient of RGO/Ni smoke screen to 3~5 μm and 8~14 μm infrared is 2.9m^2/g and 2.55m^2/g respectively. The results show that nickel plated graphene has excellent infrared extinction performance, and its infrared interference performance is greatly improved than that before modification, especially the far-infrared interference. increases obviously.

3.4. Results of dispersion performance test

The result of dispersibility between RGO and RGO/Ni samples was shown in Tab.4. The dispersibility of one material is related to its floodability. As shown in Tab.4, it can be seen that the dispersion of RGO sample is far less than 50%, indicating that the splash tendency of RGO powders is very general, and its dispersion and floatability maybe also not good enough. Meanwhile, it can be seen that F_w and F_d of RGO powders is 57 and 52 respectively. Although the flowability of RGO is better than its floodability, but its dispersibility is still far from the relevant national military standards of flake graphite infrared interference smoke agent. After electroless nickel plating, the dispersion of RGO/Ni powders is 33.6%, and its F_w and F_d is 61 and 62 respectively. The evaluation results show that electroless nickel plating can improve the dispersion of RGO powders, prolong the floating time of smoke particles, and increase the effective shielding time of smoke screen to infrared.

Tab.4 Comparison of dispersibility between RGO and RGO/Ni powders

Evaluation parameters	RGO		RGO/Ni	
	measured value	index	measured value	index
θ_r	53.3°	12	45.3°	15
C_p	31.8%	10	33.5%	7
θ_s	68.2°	12	51.6°	16
M_f	2.74	23	2.67	23
F_w		57		61
F_w	57	24	61	25
θ_f	48.5°	12	41.3°	15
θ_d	4.8°	4	4°	3
D_s	17.6%	12	33.6%	19
F_d		52		62

4. Conclusion

Based on the results and discussions presented above, the conclusions are obtained as below:

(1) The characterization results show that the distribution and size of nano-nickel particles on RGO surface are uniform, which prevents the secondary stacking of RGO layers. The shape of nano-nickel particles is approximately flat sphere, and its average diameter is about 50 nm. The nickel nanoparticles deposited on the surface of RGO have a certain microcrystalline structure, and the content of phosphorus in the nickel coating is very low.

(2) The analysis results of infrared absorption characteristics show that the modification of electroless nickel plating changes the electromagnetic characteristics of the surface of RGO and effectively improves the infrared absorption ability, especially the far-infrared absorption ability.

(3) The results obtained from the tests show that electroless nickel plating on the surface of RGO can improve its dispersibility and far-infrared interference properties, which can be used as a reference for the expansion of modification methods of RGO.

References

[1] Liu Qinghai, Liu Haifeng, Dai Xiaodong, et al. Infrared interfering performance of graphene smoke screen[J]. Infrared Technology, 2019, 41(11): 1071-1076. (in Chinese)

[2] CHEN Hao, GAO Xinbao, XU Xingchun, et al. Middle and Far Infrared Interference Properties of CNT/graphene/carbon Composites Smoke Screen[J]. Chinese Journal of Energetic Materials, 2019, 27(3): 249-254. (in Chinese)

[3] Zheng Enshuang, Lei Chaoshuai, Li Jian, et al. The preparation of super lightweight magnetic Fe3O4/graphene/carbon aerogels and their use in electromagnetic interference shielding[J]. New Carbon Materials,2020,35(6):707-715. (in Chinese)

[4] Li Kai, Wang Xuanyu, Gao Yanqing, et al. Complex refractive index and extinction performance of graphene in infrared bands[J]. Infrared and Laser Engineering, 2021, 50(4): 20200246. (in Chinese)

[5] Fan Jieping, Liu Huimin, Tian Qiang. The imaginary part of dielectric function and the absorption coefficient[J]. College Physics, 2009,28(3):24-25. (in Chinese)

[6] Draine B T., Flatau P J. Discrete-dipole approximation for periodic targets: theory and tests[J]. J. Opt. Soc. Am. A, 2008, 25: 2593-2703.

[7] Flatau P J, Draine B T. Fast near field calculations in the discrete dipole approximation for regular rectilinear grids[J]. Optics Express, 2012,20:1247-1252.

[8] Draine B T, Flatau P J. DDSCAT 7.3(CP)[EB/OL]. http://code.google.com/p/ddscat/.

[9] Wang Xuanyu. Technical basis of pyrotechnics [M]. Beijing: Tsinghua University Press,2017. (in Chinese)

[10] Cui Ling, Sasabe Shuji, Shimizu Kenji, et al. Development and application for measuring method of powder flowability and floodability[J]. China Powder Science and Technology,2012, (1):72-77. (in Chinese)

Study on performance stability of Polycarboxylate Superplasticizer prepared at different temperatures

Lina Zhong[a], Zhaopeng Wang[b], Zeyu Lin[c], Yunhui Fang[d*]

KZJ New Materials Group Co., Ltd. Xiamen 361101, Fujian, China

[a]18959212113@163.com, [b]514204325@qq.com, [c]610785719@qq.com,
[d*] owen@xmabr-kzj.com

Abstract—The effects of initial dropping temperature, amount of ferrous sulfate and dropping time on the performance of Polycarboxylate Superplasticizer were studied, and the measures to maintain the performance stability of Polycarboxylate Superplasticizer from 15℃ to 40 ℃ were discussed. The results show that the stability of the process is better when the initial dropping temperature is 20℃-30℃. When the temperature is higher than 30℃, the stability of the process can be achieved by prolonging the dropping time or reducing the amount of $FeSO_4$; When the temperature is lower than 15℃, the performance can be improved by increasing the amount of $FeSO_4$, so as to achieve the stability of the process.

1. Introduction

Polycarboxylate Superplasticizer has the advantages of designable molecular structure, good dispersion and dispersion retention, green environmental protection and so on. It meets the needs of modern concrete engineering and is gradually becoming the preferred admixture for preparing high-performance concrete [1-4]. In recent years, the function of polycarboxylate Superplasticizer has been continuously developed, and the preparation technology has been continuously improved. The preparation process has gradually changed from early heating process to normal temperature process [5-6]. In different seasons, the normal temperature ranges from 15℃ to 40℃. When polycarboxylate Superplasticizer is prepared by room temperature process, different reaction temperature will directly affect the activity of free radical and comonomer, so as to affect the performance of water reducer[7-9]. Therefore, the study of reaction temperature and factors affecting process temperature plays an important role in maintaining the stability of polycarboxylic acid high-performance water reducer.

In this project, the effects of initial dropping temperature, FeSO4 dosage and dropping time on the performance of water reducer were studied, and the measures to maintain the performance stability from 15℃ to 40 ℃ were discussed.

2. Experimental

2.1. Raw material

Raw materials include Methyl allyl polyethylene glycol ether (HPEG, Mn=2400), Acrylic acid (AA), Hydrogen peroxide solution (H_2O_2), Vinyl acetate, mercaptoacetic acid, ascorbic acid, $FeSO_4$, Sodium hydroxide solution (NaOH), 30 wt%. All raw materials are industrial grade.

2.2. Test raw materials

Cement (C): The physical performance of cement is shown in Tab 1.

Tab 1. The physical performance of cement

Varieties.	Specific surface area / (g/cm^3)	Standard consistency water consumption/%	Setting time/min		Flexural strength/MPa		Compressive strength/MPa	
			Initial	Initial	3d	28d	3d	28d
Hongshi P.O 42.5	368	27.8	230	355	5.5	8.7	29.3	52.4

Sand (S): Natural sand, fineness modulus is 2.7, mud content is 0.8%.

Stone.(G): Crushed stone G1 with a nominal diameter of 5mm-20 mm, and G2 gravel with a nominal diameter of 16mm-31.5 mm.

Water.(W): Tap water.

2.3. Synthetic route

Firstly, a certain amount of water, HPEG, $FeSO_4$ and H_2O_2 are added to the reactor, and the aqueous solution of VC, aqueous solution of mercaptoacetic acid and mixed aqueous solution of acrylic acid and vinyl acetate are added to the reaction vessel by dropping for room temperature reaction. The dropping time is 2h, and then the constant temperature is 1h. The obtained product is neutralized with Sodium hydroxide solution to pH 6 ~ 7, Finally, polycarboxylate Superplasticizer PCE with solid content of 50% was prepared..

2.4. Concrete test

The slump, expansion and strength of concrete shall be tested according to the methods specified in GB 8076 "concrete admixtures". The concrete mix ratio (kg/m3) is m(C) : m(S) : m(G1) : m(G2) : m(water) = 360:770:742: 318:175, and the amount of PCE is 0.2% of the amount of cementitious material.

3. Results and discussion

3.1. Effect of initial dropping temperature on properties of PCE

The initial reaction temperature will affect the activation energy and free radical decomposition rate of the polymerization system[10]. The higher the temperature, the lower the activation energy required by the system, and the faster the free radical decomposition rate, which will affect the performance of the prepared water reducer. In this project, the effects of different initial temperatures of 15 ℃, 20 ℃, 25 ℃, 30 ℃, 35 ℃, 40 ℃ and 45 ℃ on the performance of water reducer are studied. The test results are shown in Table 2.

Tab 2. Effect of initial dropping temperature on properties of PCE

Temperature/℃	(Slump and Slump expansion)/mm				Compressive strength/MPa		
	0h		**1h**		**3d**	**7d**	**28d**
15	210	570	195	370	26.9	30	39.4
20	210	590	205	410	26.8	31.1	39.2
25	210	600	210	390	27.7	31.5	40.6
30	215	590	180	390	26.7	30.8	39.7
35	220	585	205	340	27.7	31.7	39.4
40	215	560	180	300	27.6	31.3	39.8
45	195	510	165		27.2	31.3	39.9

The results show that when the process temperature is lower than 20℃ and higher than 35℃, the 0h slump expansion of concrete will become smaller, that is, the dispersion effect of water reducer will be reduced. When the temperature is lower than 20℃ and higher than 30℃, the 1h slump expansion of concrete will become smaller, that is, the dispersion retention of the water reducer decreases. Considering comprehensively, the process has good stability from 20℃ to 30℃.

3.2. Effect of $FeSO_4$ dosage on the performance of PCE

Some studies have shown that [9-10], $FeSO_4$ will coordinate with the carboxyl group of acrylic acid, shift the common electron cloud of carbon carbon double bond in the monomer to carbonyl carbon, enhance the activity of double bond, and improve the conversion rate of monomer. $FeSO_4$ can accelerate the decomposition of hydrogen peroxide and the production of active free radicals. $FeSO_4$ will reduce the energy barrier of the reaction and promote the progress of the reaction, resulting in more concentrated heat release, which is not conducive to the control of reaction temperature under high temperature conditions. Therefore, the effect of ferrous sulfate content on the performance of water reducer was investigated at 40 ℃. The test results are shown in Table 3.

Tab 3. Effect of FeSO4 dosage on the performance of PCE

$FeSO_4$/**wt%**	(Slump and Slump expansion)/mm				Compressive strength/MPa		
	0h		**1h**		**3d**	**7d**	**28d**
0	210	585	205	385	27.5	30.7	39.5
0.0015	210	590	205	380	27.2	30.3	39.8
0.003	215	570	200	345	27.1	31.4	39.6
0.0045	215	555	185	310	28.0	31.8	41.0
0.006	205	520	165		27.0	31.1	40.1
0.0075	220	480	140		28.0	32.0	39.8

The results show that under the condition of 40 ℃, with the increase of the amount of $FeSO_4$, the dispersion and dispersion retention of PCE decrease, indicating that the amount of $FeSO_4$ should be reduced or not used under the condition of high temperature.

3.3. Effect of dropping time on the performance of PCE

The length of dropping time will affect the exothermic reaction of polymerization, and then affect the performance of water reducer [11]. Under the condition of 40 ℃, when the dosage of ferrous sulfate is 0.0045%, the effect of dropping temperature on performance is investigated. The test results are shown in Table 4.

Tab 4. Effect of dropping time on the performance of PCE

time/h	(Slump and Slump expansion)/mm				Compressive strength/MPa		
	0h		**1h**		**3d**	**7d**	**28d**
2	220	550	205	310	26.2	31.5	41.2
2.5	205	560	205	360	26.2	31	40.7
3	220	580	200	385	26.2	32	40.6
3.5	210	575	195	380	26.4	30.5	39.6

The results show that the dispersion and dispersion retention of the samples are improved with the extension of dropping time, which shows that prolonging the dropping time is conducive to the stability of the polymerization reaction at higher temperature, and the performance of dropping time of 3H is the best.

4. Conclusiong

Based on the results and discussions presented above, the conclusions are obtained as below:

(1) The initial dropping temperature of the process is between 20 ℃ and 30 ℃, and the stability of the process is good.

(2) When the temperature is higher than 30 ℃, the stability of the process can be achieved by prolonging the dropping time or reducing the amount of ferrous sulfate.

(3) When the temperature is lower than 15 ℃, the performance can be improved by increasing the amount of ferrous sulfate, so as to achieve the stability of the process.

References

[1] Shanshan Qian et al. Synthesis, characterization and working mechanism of a novel polycarboxylate superplasticizer for concrete possessing reduced viscosity[J]. Construction and Building Materials, 2018, 169 : 452-461.

[2] Lina Zhong. Synthesis and properties of mud resistant polycarboxylate superplasticize [J]. New building materials, 2018,45 (05): 41-44 + 48.

[3] Yunhui Fang. Preparation and characterization of viscosity reducing polycarboxylate superplasticize with different molecular structures [J]. New building materials, 2017,44 (06): 104-108 + 121.

[4] Yunhui Fang. Application of polycarboxylate superplasticize molecular design in prestressed high strength concrete pipe pile [J]. New building materials, 2012,39 (08): 32-35 + 67.

[5] Hao Wang, Jianjun Yun, Ranran Ye, et al. Theoretical study on the process of synthesizing polycarboxylate superplasticizer at room temperature [J]. Commercial concrete, 2014 (5): 27-29.

[6] Shunguan Lin. Study on process and performance of polycarboxylate superplasticizer synthesized at room temperature [J]. Concrete, 2020 (12): 77-79 + 84.

[7] Shaomin Zhang. Preparation and properties of polycarboxylate superplasticizer from new polyether macromonomer at low temperature [J]. Silicate bulletin, 2020,39 (09): 2844-2848.

[8] Xiaolu Chen. Study on the method and performance of synthesizing polycarboxylate superplasticizer at room temperature [J]. New building materials, 2014,41 (12): 80-83.

[9] Qiumin Xiong, Xiujuan Guo, Wei Dang, Xiaobin Chen, Jican Guan, Guangzhou Li, Han Xu. Effect of synthesis temperature on the performance of polycarboxylate superplasticizer [J]. Guangdong building materials, 2020,36 (07): 2-4.

[10] Jianjun Yun, Jianshu Tang, Meiqing Xu, Shumeng Min , Jingyao Wang. Temperature rise influencing factors and control measures for the preparation of polycarboxylate superplasticizer at room temperature [J]. Concrete world, 2021 (10): 61-64.

[11] Yu Zeng, You Sun, Ao Fan, Long Xiao, Wang Yuanfu, Fang Shichang. Synthesis and dispersion of new polycarboxylate superplasticizer [J]. Shandong chemical industry, 2021,50 (19): 12-13 + 16.

An Easy Preparation of $BaTiO_3$/PAM organic/inorganic hybrid material Using Surface-Initiated Process

X. F. Wei, *1,3 **W. Z. Li,**2,3 **H.F.Chen,**3 **Mengge Lv,** 3 **H. W. Wang,** 1*1,2,3 **and X.W. Qi***1,2,3

[1] School of Materials Science and Engineering, Northeastern University, No. 3-11 Wenhua Road, Shenyang, Liaoning 110819, P. R. China

[2] School of Materials Science and Engineering, Beijing Institute of Technology, Beijing 100081, China

[3] School of Resources and Materials, Northeastern University at Qinhuangdao, 143 Taishan Road, Qinhuangdao, Hebei 066004, P. R. China

*To whom correspondence should be addressed. Tel:18630332906

E-mail address: whwdbdx@126.com (H. W. Wang)

Abstract—We developed an easy surface-initiated polymerization process to prepare BaTiO3/PAM organic/inorganic hybrid hydrogel material, which would have potential applications in intelligent soft substances. As various organic–inorganic hybrid nanoparticles can be synthesized by this property, it has broad applications in fields such as nano hybrid material, nanocomposites and medicine.

1. Introduction

Comprehending and utilizing the unique functionalities of nanoparticles has attracted considerable research attention.[1,2] To realize the full advantages of these excellent properties on nanoscale,[3] the nanoparticles should be combined with other materials that possess the desirable properties, to form hybrid composites.[4,5] For example, in the field of polymer chemistry the development of organic/inorganic hybrid composites has opened up exciting new horizons, making the preparation of composite multifunctional materials possible.[6,7] In addition, the advent of organic/inorganic mixture composite materials has boosted significant advancements in fields such as biomaterials [8] and supramolecular chemistry.[9]

Here we imparted the initiation property to $BaTiO_3$ by peroxidizing it in a mixed solution of H_2O_2 and NaOH, and then the surface-modified $BaTiO_3$ nanoparticles were added in the mixture solution of reducing agent $FeCl_2$ and AM under N_2 conditions. After 30 min, the mixture changed into a brownish-yellow sticky gel, which confirmed the formation of the hybrid organic (PAM)/inorganic ($BaTiO_3$) composite ($BaTiO_3$@PAM). The chemical process associated with the formation of this composite is shown in Fig. 1.

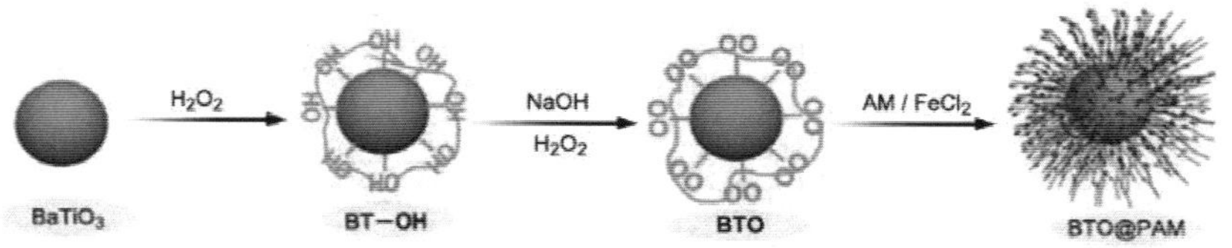

Figure 1. Chemical process underlying the formation of PAM/ $BaTiO_3$ hybrid material.

2. Experimental

2.1. Materials

Acrylic acid (AA, from Beijing Xingjin Chemical Factory), Acrylamide (AM, Shantou Xilong Chemical Factory Guangdong), Potassium persulfate (KPS, Beijing Hengye Zhongyuan Chemical Co., Ltd), N, N′-Methylene bisacrylamide (NMBA, Tianjin Jinke Research Institute of Fine Chemical), NaOH and Sucrose were obtained from Beijing Chemical Factory. All of them were of analytical grade and used without further purification.

2.2. Characterization

The X-ray diffraction (XRD) patterns of the carbon materials were obtained with a Shimadzu 7000 diffractometer instrument using Cu K α radiation (λ= 0.15406 nm). The surface morphology of the carbon materials were observed by using a scanning electron microscope (SEM: HITACHI S-4800, Japan). The Fourier Transform Infrared Spectrophotometry (FTIR, IFS113V, Bruker) was used to identify the functional groups.

2.3. Sample preparation

In a typical experiment, barium titanate was added to the mixed solution of sodium hydroxide and hydrogen peroxide. After 10 min of reaction, barium peroxide was prepared by washing and drying. Then, the compound nanoparticles were prepared in the redox system, which used barium peroxide as an initiator and ferrous sulfate as a reducing agent.

3. Results and discussion

To study the surface changes of the $BaTiO_3$ nano particles, the functional groups were tested. Fig. 2 shows the Fourier-transform infrared (FTIR) spectra of pure barium titanate (BT) along with the products obtained by the surface treatment of barium titanate with the hydrogen peroxide solution (BT-OH) and the sodium peroxide solution (BTO), respectively. The graph shows that the broad peaks of the two surface-modified products at 3445 cm^{-1} are significantly enhanced, indicating that several hydroxyl groups[10] have been added to the surface of barium titanate. I

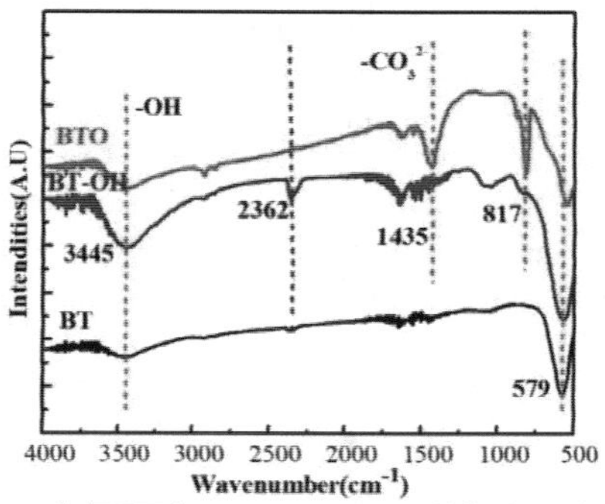

Figure 2. Fourier-transform infrared (FTIR) spectra of modified and unmodified barium titanate. The FTIR spectrum of BTO exhibits a new absorption band, indicating the appearance of a new chemical

To confirm the reaction process, the valence of the Ba, O and Ti are tested. Fig. 3 shows the XPS results of pure barium titanate and barium peroxide. The valence states of titanium and barium did not change while a new peak of 528.5 eV is generated in the O1s X- ray photoelectron spectrum of barium peroxide, which is not attributed to pure barium titanate. Because the energy of the peroxide group itself is higher than that of the more stable oxygen atom in the lattice, it can be presumed that the peak corresponds to the peroxide group.

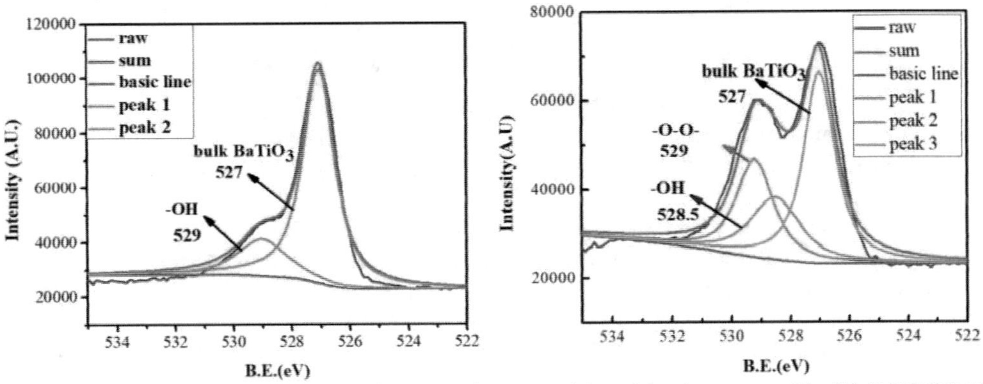

Figure 3. XPS comparison diagram for O1s elements (a) and barium peroxide (b) BOT@PAM

Fig. 4 shows the DTA and TG curves obtained by the differential thermogravimetric analysis of the polymerization products. According to the TG curve, the polymerization product has four obvious weightlessness stages. The first stage °Ccurs at 50–250 °C and the weight loss rate is about 13.18%.

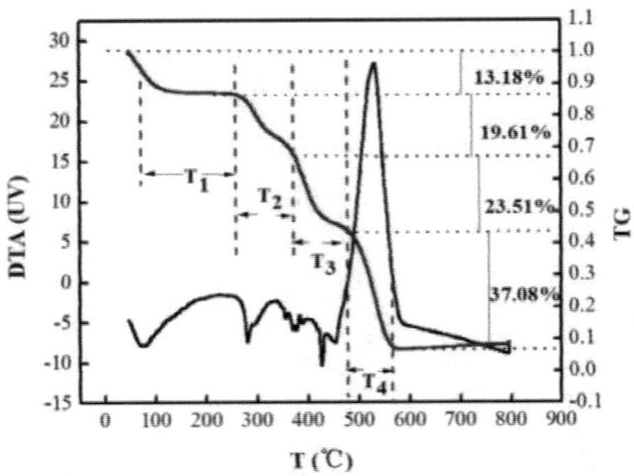

Figure 4. DTA and TG curves obtained by differential thermal - thermogravimetric scanning

Fig. 5 shows that the surface of pure barium titanate is smooth, while the surface of barium peroxide appears foggy because of the presence of some unevaporated solution after the oxidation treatment. The small black dots of barium titanate and several polymers growing around it can be clearly seen in the composite polymer.

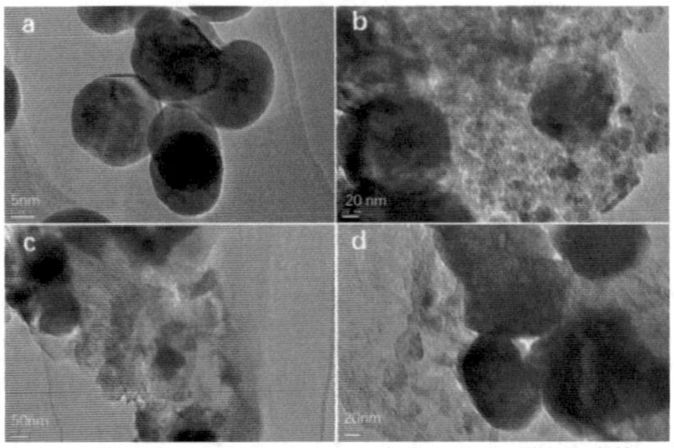

Figure 5. TEM images of (a)pure barium titanate particles BT, (b)oxidized barium titanate BTO, (c) BTO/polyacrylamide composite BTO@PAM, (d) BTO/polyacrylamide composite BTO@PAM.

Fig. 6 shows the influence of different amounts of barium peroxide on the composite nanoparticles. The figure illustrates that the maximum weight loss rate is reached when the dosage of barium peroxodioxide is 0.08 g, and the value of GP (representing the average polymer coverage of particles) is the highest, reaching 5.55. This proves that more reaction sites can be activated and more polymers can be grafted at this time, and that the initiated polymer polymerization is relatively uniform.

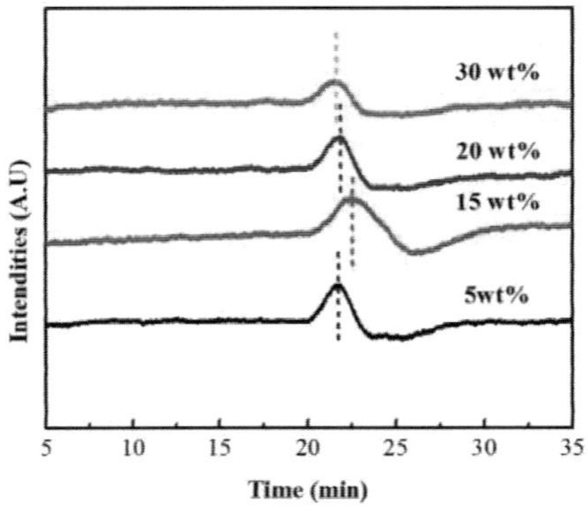

Figure 6. The GPC of the BTO@PAM composites

4. Conclusions

In summary, we have developed a easy process to prepare BTO@PAM nano hybrid material, which will enable a new polymerization mechanism. We believe that the phenomenon proposed in this study holds great promise for the study of surface properties of nano materials.

Acknowledgment.
This work was supported by Natural Science Foundation of China No. 21604007

References

[1] Kraus-Ophir, S., et al., Nanoparticle-imprinted polymers for size-selective recognition of nanoparticles. Angew Chem Int Ed Engl, 2014. 53(1): p. 294-8.
[2] Vinod, S., et al., Graphene Oxide Epoxy (GO-xy): GO as Epoxy Adhesive by Interfacial Reaction of Functionalities. Advanced Materials Interfaces, 2018. 5(2): p. 1700657.
[3] Kwon, O., et al., Direct Probing of Polarization Charge at Nanoscale Level. Adv Mater, 2018. 30(1).
[4] Zhou, X.F., W. Cheng, and R.G. Compton, Doping of single polymeric nanoparticles. Angew Chem Int Ed Engl, 2014. 53(46): p. 12587-9.

[5] Dencheva, N., et al., One-Step In Situ Synthesis of Polyamide Microcapsules With Inorganic Payload and Their Transformation into Responsive Thermoplastic Composite Materials. Macromolecular Materials and Engineering, 2016. 301(2): p. 119-124.
[6] Ghosh, A., N.K. Sheridon, and P. Fischer, Voltage-controllable magnetic composite based on multifunctional polyethylene microparticles. Small, 2008. 4(11): p. 1956-8.
[7] Nyström, G., et al., Amyloid Templated Organic-Inorganic Hybrid Aerogels. Advanced Functional Materials, 2018. 28(27): p. 1703609.
[8] Ibraeva, Z.E., et al., Preparation and Characterization of Organic-Inorganic Composite Materials Based on Poly(acrylamide) Hydrogels and Clay Minerals. Macromolecular Symposia, 2015. 351(1): p. 97-111.
[9] Kaczmarek, B. and A. Sionkowska, Chitosan/collagen blends with inorganic and organic additive-A review. Advances in Polymer Technology, 2018. 37(6): p. 2367-2376.
[10] Baek, C., et al., Formation and Accumulation of Intragranular Pores in the Hydrothermally Synthesized Barium Titanate Nanoparticles. Journal of the American Ceramic Society, 2016. 99(11): p. 3802-3808.

Journal of Physics: Conference Series **2194** (2022) 012046 doi:10.1088/1742-6596/2194/1/012046

Preparation of modified rare earth lanthanum oxide/polypropylene nonwoven fabric by meltblown method

Yunhan SUN[1], Xiaoning Wang[1], Qingxiu JIA[1*], Jian YANG[2]

[1] School of Materials Design and Engineering, Beijing Institute of Fashion Technology, Beijing, China 100029;

[2] Beijing Beihua New Rubber Special Material Technology Co., Ltd, Beijing, China 100029;

[*]Correspongding author's e-mail: jiaqingxiu@163.com,

Abstract—Lanthanum oxide was organically modified with dopamine and blended with polypropylene, and then lanthanum oxide (La_2O_3) /polypropylene (PP) nonwovens were prepared by melt blown method. The nonwovens were characterized by FTIR, SEM, TG, mechanical properties and X-ray shielding. The experimental results show that the addition of modified La_2O_3 makes the fiber arrangement of PP nonwovens more densely arranged and improves the thermal stability of PP nonwovens. The higher the amount of La_2O_3, the better the X-ray shielding performance of composite nonwovens.

1. Introduction

In recent years, with the development of nuclear science and nuclear industry, all kinds of radiation have an increasing impact on human and environmental production. The development and application of radiation protection materials have attracted more and more attention. At present, the common shielding material for X-ray is still lead products [1]. However, lead is toxic, and due to post use bending and incorrect suspension, the lead skirt is composed of layered thin layers, which has a common cracking problem in practice [2]. Besides, the absorption range of lead-containing composites is above 88 keV and below 40 keV [3, 4], but the energy range between 40 kev and 88 kev (especially the conventional energy region used for medical diagnosis) is not well covered. Most of the research on personal shielding materials for X-ray protection has focused on new non-lead and lightweight materials [5]. Kim et al [6] prepared a double-layer tungsten composite yarn containing $BaSO_4$ and polyethylene terephthalate (PET) fiber fabric. The shielding performance of the $BaSO_4$-containing tungsten composite and PET fiber fabric was 0.018 mmPb and 0.03 mmPb, respectively, suggesting that low-dose shielding may help protect aviation crew from scattered radiation. Kim et al [7] prepared a shielding fabric by wrapping polyethylene (PE) yarn around a 30 μm tungsten yarn woven with a pneumatic pressure of tungsten nanopowder applied to the fiber dispersion process. The fabric coated with tungsten nano-powder improved the shielding performance of ordinary tungsten fibers by about 15%, providing relatively effective low-dose radiation shielding at a distance of about 1.2 m from the X-ray imaging equipment.

The absorbed energy of k layer electrons of lanthanide rare earth elements ranges from 38.9 to 63.3 KeV, which effectively makes up for the weak absorption zone of lead and can be completely used as shielding filler for lead substituted materials in the field of radiation protection [8]. Liu et al [9] prepared $Gd(AA)_3$/natural rubber radiation protection materials. Sambhudevana et al [10] filled natural rubber with Gd_2O_3 modified by methacrylic acid to obtain X-ray protection material, and the modification of methacrylic acid improved the interface between Gd_2O_3 and rubber. Gu et al [11] found that the mass fraction of La_2O_3 was within 60%, and its nanoscale inhomogeneous dispersion in the PP matrix; with the increase of La_2O_3 addition and the corresponding increase of lead equivalent, the rare earth/PP composite protective fiber had better X-ray protection. Lou et al [12] found that the addition of rare earth oxides between the nonwoven layers can greatly improve the radiation protection properties of the composites. It was also found that used xylene method to remove PP matrix phase from melt-spun rare earth/polypropylene (PP)/ thermoplastic polyurethane (TPU) fibers, and prepared rare earth /TPU fibers with X-ray protection properties [13]. P. Vani et al [14] synthesized rare earth doped barium tellurate glasses and found that the increase of thulium doping in barium tellurate glasses can produce highly efficient high energy radiation shielding materials. At present, there are few studies on rare earth / polymer anti radiation fibers or fabrics.

In this paper, La_2O_3 was modified by dopamine, and then the modified La_2O_3/PP composite nonwovens were prepared by melt blown method. The effects of organic modification on the interfacial strength of the composite and the addition of La_2O_3 on the ray shielding properties of PP nonwovens were investigated.

2. EXPERIMENTAL

2.1. Materials

Polypropylene (PP), spinning grade, density 0.9g/cm3, melt index 25g/min(235℃,2.16kg), Sinopec Group; Lanthanum oxide (nano grade, ≥99.99%): China Beijing Nonferrous Metals Research Institute; Paraxylene (analytical pure, ≥99.7%), anhydrous ethanol (analytical pure, ≥99.7%), acetone (analytical pure, ≥ (99.5%), acetic acid (analytical purity, ≥ 99.5%): Beijing Tongguang Fine Chemical Company; trimethylolaminomethane (analytical purity, ≥ 99%), dopamine hydrochloride (analytical purity, ≥ 99.9%): Beijing Chemical Reagent Company.

2.2. Organic modification of La_2O_3 with dopamine

The proportion of dopamine modification solution was 50% absolute ethanol, 49% deionized water and 1% dopamine hydrochloride according to the mass percentage, and the pH of this solution was adjusted to 8 with appropriate amount ju trimethylaminomethane. Then, La_2O_3 particles equivalent to 25% of the mass ratio of the above solution were added to it, mixed with ultrasound for 30min, and then air dried at 40 ℃ to remove the solvent to obtain dopamine modified La_2O_3 particles.

2.3. Preparation of modified La_2O_3/PP nonwovens

In Hakee twin-screw torque rheometer, dopamine modified La_2O_3 was mixed with PP for 10 min at a mixing temperature of 210℃and a screw speed of 100 r/min, and then the spinning masterbatch was obtained by granulation. The masterbatch is spun by melt blown method to obtain nonwovens. Melt blown process conditions are as follows: spinning screw temperature 300℃, hot air 200℃, hot air pressure 1.5MPa, nozzle temperature 270℃, extrusion rate of metering pump 2kg/h and receiving distance 30cm. In the final nonwovens, the addition amount of modified Melt blown process conditions are as follows: spinning screw temperature 300℃, hot air 200℃, hot air pressure 1.5MPa, nozzle temperature 270℃, extrusion rate of metering pump 2kg/h and receiving distance 30cm. per 100 phr PP is 0 phr, 5 phr, 10 phr and 15 phr respectively, and the corresponding nonwovens are named 0#, 5#, 10# and 15#.

2.4. Material Characterization

Scanning electron microscopy (SEM) Surface observation of rare earth particles and composite fibers before and after modification was carried out using a Japan Electron JSM-7500F type field emission scanning electron microscope, and the surface gold spraying method was used to prepare the observation samples with a thickness of about 10 nm.

Electron transmission microscopy (TEM) Internal structure of the composite fibers was observed using a Japan Electron JEM-2100F field emission high-resolution transmission electron microscope, and the sections were prepared by the embedding method with a thickness of 50 nm.

Thermal decomposition test (TG) Thermal weight loss test of all samples was performed by Japan Seiko SII-6300 type thermogravimetric analyzer under nitrogen atmosphere, with a temperature rise rate of 20℃/min and temperature ranges of 30~600℃.

Differential scanning calorimetry (DSC) Differential scanning calorimetry analysis was performed with Seiko SII-6200 differential scanning calorimeter under nitrogen protection and gas flow rate of 50 ml/min. All samples were heated to 220℃ at room temperature, the scanning rate was 50℃/min, and maintained at 220℃ for 5 min to eliminate the thermal history. The reference material was α- Al_2O_3. The cooling curve was obtained by cooling from 220℃ to 30℃ at a scanning speed of 20℃/min; The heating curve was obtained by heating from 30℃ to 220℃ at a scanning speed of 10℃/min. The melting temperature and enthalpy of the sample were obtained in the second heating curve.

X-ray shielding performance test X-ray shielding performances of these nonwovens were tested with high stability MG324 X-ray machine of Philips, Germany and PTWDCI-8500 precision integral current meter of Austrian Scientific Research Center, radiation protection and nuclear safety medicine institute of Chinese Center for Disease Control and Prevention. The test sample size was 10cm × 10cm × 0.5cm, the voltage was 120kVp, and 2.5mm aluminum filter was adopted. X-ray shielding performance was expressed by lead equivalent.

3. Results and Disscussion

3.1. Morphology of dopamine modified La_2O_3/PP composite nonwovens

SEM photos of different modified La_2O_3/PP nonwovens are shown in Figure 1. It can be seen from the figure that the fiber diameter distribution of pure PP nonwovens is 2 ~ 5 μm, while the fiber diameter distribution of dopamine modified La_2O_3/PP nonwovens composite nonwovens is 2 ~ 10 μm. The addition of modified La_2O_3 widens the diameter distribution of nonwoven fibers, but the distance between fibers decreases and the number of fibers per unit area increases, indicating that the addition of La_2O_3 makes the structure of PP nonwovens more compact. Due to the higher viscosity and poor fluidity of La_2O_3/PP spinning system, the fiber stretching is not enough when it is drawn into fiber by high-speed hot air, resulting in the formation of coarse fibers and uneven diameter distribution in the obtained nonwovens. During the experiment, it can also be observed that the surface of PP melt blown nonwovens containing modified La_2O_3 is much rougher than that of pure PP melt blown nonwovens.

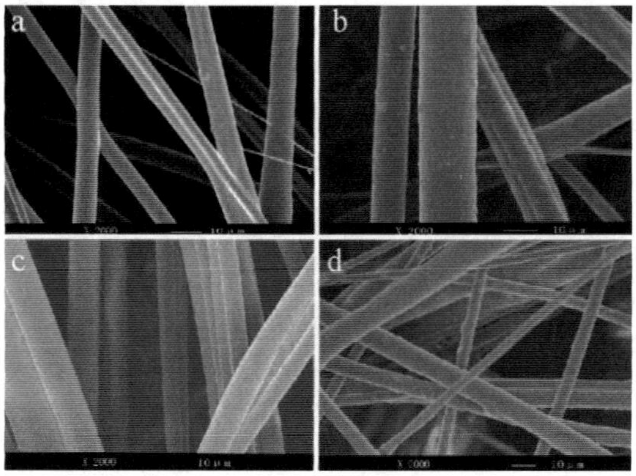

Fig.1 SEM images of dopamine modified La_2O_3/PP nonwovens. (a) 0#; (b) 5#; (c) 10#; (d) 15#

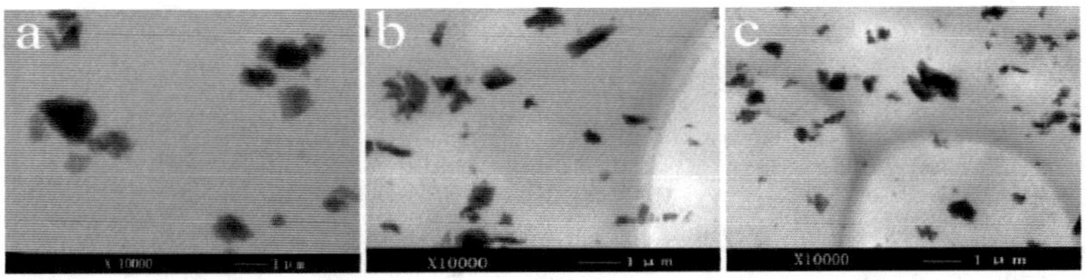

Fig.2 TEM images of dopamine modified La_2O_3/PP nonwovens. (a) 5#; (b) 10#; (c) 15#

Figure 2 shows the TEM image of dopamine modified La_2O_3/PP composite nonwovens. It can be seen from the figure that in these nonwovens, the average particle size of modified lanthanum oxide is between 200 ~ 700 nm. With the increase of the content of modified La_2O_3 in the composite fiber, the particle size distribution of La_2O_3 becomes larger. The particle size distribution in 15# samples ranges from tens of nanometers to 1 micron, which may be because the increase of filler fraction increases the viscosity of the system and leads to uneven stress during mixing in the screw. However, in general, dopamine modification improves the compatibility between La_2O_3 and PP matrix, making the overall distribution of particles more uniform. The smaller the particle size of La_2O_3, the more uniform the distribution, the higher the electron density of absorbing atoms in the composite fiber, the higher the photoelectron absorption efficiency and the better the anti radiation performance. The more uniform the distribution of La_2O_3, the better the uniformity of shielding performance of the composite fiber.

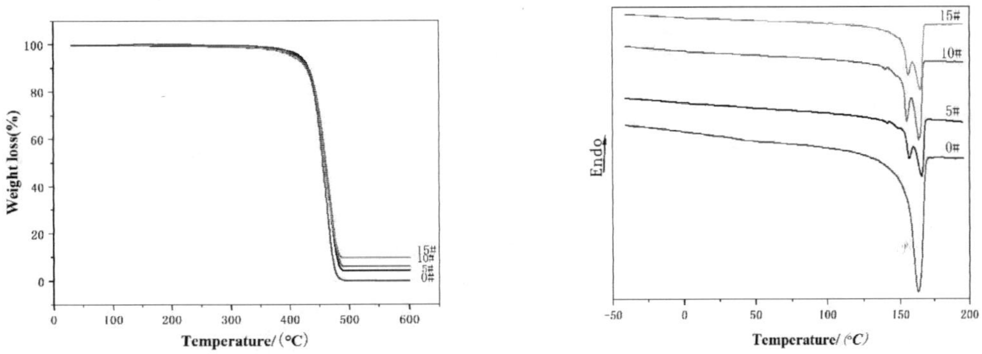

Fig.3 TG and DSC diagrams of polypropylene composite nonwoven fabric with 0#, 5#, 10# and 15# of dopamine modified lanthanum oxide addition, respectively

Figure 3 shows the TG and DSC of polypropylene composite nonwoven fabric with 0#, 5#, 10# and 15# of dopamine-modified lanthanum oxide addition, respectively. The thermal degradation temperature of polypropylene with different modified lanthanum oxide additions under nitrogen atmosphere was determined by thermogravimetric analysis (TG). The initial decomposition temperature of PP in meltblown nonwovens is around 410°C. When the temperature reaches 457°C, the decomposition rate of PP reaches the fastest. The heat loss residue of the nonwoven fabric is 0% at this point, indicating that no carbon compounds are present in the residue. The initial decomposition temperatures of the modified La_2O_3/PP composite nonwoven fabric were 412°C, 413°C and 412°C, respectively, and the heat loss residuals were 4.3%, 6.0% and 9.6%, indicating that the residuals contained modified La_2O_3. With the addition of dopamine-modified La_2O_3, the decomposition rate of the modified La_2O_3/PP composite nonwoven also decreased, and the residual amount of thermal weight loss also gradually increased, indicating that the addition of dopamine-modified La_2O_3 improved the thermal stability of the modified La_2O_3/PP composite nonwoven, which may be caused by the increase in molecular weight due to the formation of chemical bonds between dopamine-modified La_2O_3 and PP.

From the DSC plot, it can be seen that the addition of lanthanum oxide changed the melting peaks of polypropylene to two, and the melting points changed from 163°C to 157°C and 164°C, and two crystalline forms appeared. It may be because the La_2O_3 modified by dopamine surface combines with PP molecular chains to form a graft copolymer with increased intermolecular spacing. During the formation of PP crystalline nuclei, La_2O_3 modified by dopamine restricted the movement of polypropylene molecules, resulting in the formation of two crystalline forms. The crystallinity of the composite nonwoven is lower than that of the pure PP nonwoven, probably because the grafting of dopamine onto the PP substrate hinders the formation of PP crystallites.

3.2. Morphology of dopamine modified La_2O_3/PP composite nonwovens

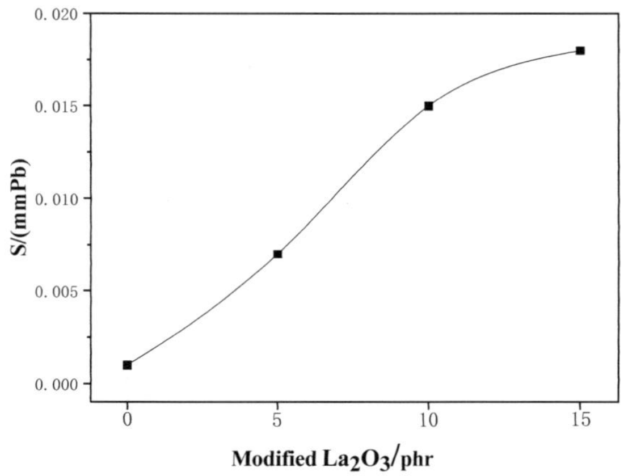

Fig.4 Modified lanthanum oxide/polypropylene composite nonwoven X-ray shielding performance test chart

The energy of most X-ray particles produced at medical X-ray tube voltages below 130 kVp is usually less than 88 keV [8]. To study the shielding effect of composite fibers, a tube voltage of 120 kVp was chosen for testing. x-rays have a strong penetrating ability and absorb energy only when they interact with the shielding material. the interaction between x-rays and matter depends mainly on the photoelectric effect. The photoelectric effect is the interaction of the incident photons with the internal electrons, which are fixed in number. Therefore, by increasing the density of composite fibers and increasing the number of electrons per volume, the probability of photoelectric effect of composite fiber materials can be effectively increased. Fig. 4 shows the test graph of X-ray shielding performance of dopamine-modified La_2O_3/PP composite nonwoven fabric. Fig. 4 Test graph of X-ray shielding performance of modified La_2O_3/PP composite nonwoven fabric. Combined with the TEM diagram, it can be seen that the protection effect of the fiber is positively correlated with the lanthanum oxide addition fraction, and the higher the La_2O_3 addition fraction, the corresponding increase in lanthanum oxide distribution density, and the better the X-ray shielding performance of the composite nonwoven fabric.

4. Conclusion

In this paper, lanthanum oxide was organically modified with dopamine and filled into polypropylene matrix, and then meltblown method was used to prepare rare earth/polypropylene composite nonwoven fabric. The addition of dopamine-modified lanthanum oxide makes the nonwoven fiber arrangement more dense and improves the thermal stability of the polypropylene nonwoven. Based on the random influence of fiber formation in the meltblown process, the amount of rare earth added to the nonwoven fabric was limited, and the limit of rare earth mass addition was 15 parts, but the higher the lanthanum oxide addition, the better the X-ray shielding performance of the composite nonwoven fabric. The interfacial compatibility of La_2O_3 modified with dopamine and PP matrix was significantly improved, and the X-ray shielding performance became significantly stronger with the addition of La_2O_3, which provides a reference for the subsequent research and industrial application of La_2O_3/polypropylene composite fibers.

Acknowledgments
The authors sincerely acknowledge the financial support from the Project of Top Young Talents of Beijing Excellent Talents Training Program, and the Innovation Teambuilding Program of Beijing Institute of Fashion Technology (BIFTTD201904). The authors declare that they have no conflict of interest.

References

[1] LYRA M, CHARALAMBATOU P, SOTIROPOULOS M, et al. Radiation protection of staff in 111In radionuclide therapy--is the lead apron shielding effective? [J]. Radiat Prot Dosimetry, 2011, 147(1-2): 272-6.
[2] OYAR O, KISLALIOGLU A. How protective are the lead aprons we use against ionizing radiation? [J]. Diagnostic and Interventional Radiology, 2012, 18(2): 147.
[3] BRANDT A M. Cement-based composites: materials, mechanical properties and performance [M]. CRC Press, 2005.
[4] GIL L. Cork composites: A review. Materials, 2 (3), 776–789 [M]. 2009.
[5] BOTELHO M, KüNZEL R, OKUNO E, et al. X-ray transmission through nanostructured and microstructured CuO materials [J]. Applied Radiation and Isotopes, 2011, 69(2): 527-30.
[6] KIM S-C, SON J S. Double-layered fiber for lightweight flexible clothing providing shielding from low-dose natural radiation [J]. Scientific Reports, 2021, 11(1): 1-9.
[7] KIM S-C. Development of a Lightweight Tungsten Shielding Fiber That Can Be Used for Improving the Performance of Medical Radiation Shields [J]. Applied Sciences, 2021, 11(14): 6475.
[8] POTTER K. Introduction to composite products: design, development and manufacture [M]. Springer Science & Business Media, 1996.
[9] LIU L, HE L, YANG C, et al. In situ reaction and radiation protection properties of Gd (AA) 3/NR composites [J]. Macromolecular rapid communications, 2004, 25(12): 1197-202.
[10] SAMBHUDEVANA S, SANKARA B, AA S, et al. SYNTHESIS AND CHARACTERIZATION OF ORGANICALLY MODIFIED GD2O3-NATURAL RUBBER COMPOSITES AS EFFECTIVE X-RAY SHIELDING MATERIALS [J]. SYNTHESIS, 2015, 1(1):
[11] Gu Chunyan,Sun Kuan,Zhang Yufang,Jia Qingxiu. Preparation and performance study of X-ray resistant rare earth/polypropylene composite fibers[J]. Rare Metal Materials and Engineering,2017,46(09):2633-2638.
[12] Lou Pengfei, Jia Qingxiu, Wang Fadong, Xiang Changlong. Preparation and performance study of radiation-proof laminated composites[J]. Journal of Beijing Institute of Fashion (Natural Science Edition),2019,39(02):10-15.
[13] LOU P, TENG X, JIA Q, et al. Preparation and structure of rare earth/thermoplastic polyurethane fiber for X - ray shielding [J]. Journal of Applied Polymer Science, 2019, 136(17): 47435.
[14] VANI P, VINITHA G, SAYYED M, et al. Effect of rare earth dopants on the radiation shielding properties of barium tellurite glasses [J]. Nuclear Engineering and Technology, 2021,

Poly-amino-based ionic liquids coupled with semi-coke for biogas upgrading

Yang Chen[1a*], Mei Liu[1a], Zhihui Qu[1a], Ziwei Su[1a]

[1]School of Materials and Environmental Engineering, Hunan University of Humanities, Science and Technology, Loudi, 417000, P. R. China

[a*]birdyang617@163.com

Abstract—Poly-amino-based ionic liquid (IL) $[TETAH]^+[BF_4]^-$ was mixed with sodium hydroxide solution for CO_2 separation from biogas in a spray scrubber packed with modified semi-coke. A CO_2 absorption capacity of 0.92mol was achieved when controlling the outlet CO_2 concentration below 3 vol%. The porous structure of semi-coke contributed to improving CO_2 absorption. Nitric acid oxidation and heat treatment could increase the specific surface area, enlarge the intrinsic pores and develop more micropores. Semi-coke firstly oxidized by nitric acid and then with heat treatment could not only exhibit higher specific surface area and more porous structure, but also show less degree of graphitization, which was suggested as possible to increase the reactivity to CO_2. It was assumed that the absorption system in this research showed promising application in CO_2 separation from biogas as well as waste treatment.

1. Introduction

To achieve the carbon peak and neutrality targets, increasing the share of non-fossil energy consumption is vital important. Biogas, one of the promising renewable energy has attracted worldwide interest as its potential applications in heating, electricity generation and as vehicle fuel [1-2]. However, the applications would be limited unless the carbon dioxide content is significantly reduced. Different technologies for CO_2 separation from biogas have developed, and each has its own advantages and disadvantages. For example, traditional approach such as aqueous amine solutions have the advantages of high reactivity, high surface tension, and low methane solubility, while show disadvantages such as equipment corrosion, high solvent loss, degradation and insufficient CO_2 capture capacity[3]. As a result, researchers are still looking for technology-effective and cost-effective methods for CO_2 separation.

Ionic liquids (ILs), the organic salts mainly composed of organic cations and inorganic/organic anions, have been considered as potential alternative solvents for CO_2 capture. Comparing to alkanolamines, the advantages of applying ILs in biogas upgrading should be appraised on the less corrosivity, high stability before and after absorption as well as the negligible vapor pressure, which would prevent the absorbent loss and the discharge of volatile chemicals [4]. A great many researches have been carried out so as to increase the CO_2 absorption capacity and absorption rate of ILs. Functionalized ILs grafted amino-group into cation or anion have found to be effective absorbents for CO_2 capture due to the greatly improved capacity and velocity of CO_2 absorption[5]. However, the competitiveness of applying ILs in large-scale biogas upgrading is limited by the high viscosities and high production costs[6]. The viscosities, originated from the strong interactions between their constituent ions, are even much higher for amine functionalized ILs. As a result, methods of blending amine functionalized ILs with commercially available solvents such as water, amine, ethanol and other non-ionic surfactants are widely adopted. Research results indicate that the mixtures can retain the desired capture effect of ILs, also, keep the viscosity and cost relatively low.

Besides solvents, many kinds of absorption materials were adopted for efficiently applying ILs in industrial manufacture of CO_2 separation. But most of were limited by the excessive price in commercial application.

Semi-coke, a product of coal pyrolysis at a relatively low temperature (873-973 K), has received a wide attention due to the low cost, well-developed porous structure and plentiful surface functional groups. The main ingredients of semi-cokes are carbon, ash and volatile compounds. The absorption activity can be improved by removing ash and volatile, changing composition of functional groups and altering the structure of carbon materials through modification. Although the surface area of semi-cokes is lower than activated carbon, the relatively higher mechanical strength of semi-coke facilitates the upgraded scale application.

In this paper, the poly-amino-based ILs selected could be easily synthesized by neutralization of polyamine and inorganic acid. The objective of this research is to determine the feasibility of using the ILs/NaOH solution as potential absorbent for biogas upgrading in a packed spray system. The absorption effects and the influencing factors of this dual system have rarely been reported. Semi-coke modified by nitric acid and heating treatment was used as structured materials to enhance the absorption effect. The effects of operational conditions such as mole ratio, absorbent concentration and semi-coke on CO_2 capture were investigated. The potential synergism for CO_2 absorption by poly-amino-based ILs/NaOH solution and semi-coke was analyzed.

2. Materials and methods

2.1 Materials

The $[TETAH]^+[BF_4]^-$ (purity>98%) was self-made following the literature [7]. Nitric acid (HNO_3, purity 68 %) and sodium hydroxide (NaOH, purity>99.5%) were supplied by Sinopharm Chemical Reagent Co., LTD. The simulated biogas consisted of 40 vol% CO_2 and 55 vol% CH_4 was obtained from Wuhan Specialty Gases Co., LTD.

The raw materials were milled and sieved to obtain fragments about 5-8 mm for the experiment. For the preparation of SN600, the sample was firstly washed with water to remove small particles and ash from the surface. And then sample was immersed in 45 wt% nitric acid and boiled for 2 hours at 363 K. Following that, the sample was treated with a tube furnace at 873 K for 2 hours under the atmosphere of N_2. For the preparation of S600N, the sample was firstly treated with a tube furnace at 873 K for 2 hours under the atmosphere of N_2, and then immersed in 45 wt% nitric acid and boiled for 2 hours at 363 K.

2.2 Experimental setup and operation

The absorption of CO_2 was carried out in a laboratory-scale spray scrubber packed with 350 mm semi-coke. As schematically shown in Fig.1, the scrubber consisted of a cylindrical perspex tower of 1000 mm in height and 80 mm in diameter. An atomizer with orifice diameter of 1 mm and spray angle of 0-100° was fixed at the upper part of the scrubber and the mean diameter of the spray was 20-30 μm when the pressure of the mechanical diaphragm metering pump was 0.5-1.0 MPa. The CH_4/CO_2 mixture was introduced from the bottom of the tower while the NaOH/ILs solution was sprayed from the top to form a counter-current flow. The mixed gas and the NaOH/ILs solution were heated to the designated temperatures before being introduced into the scrubber, and the temperatures were continuously monitored. The flow rates and accumulated volume of the gas were measured using a soap-film flow meter (Agilent Technologies Co. LTD.). The CO_2 concentrations were measured by a Biogas 5000 analyzer (Geotech, England).

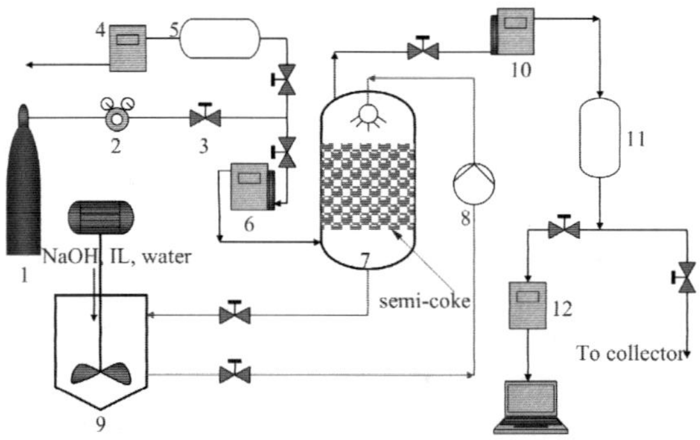

Fig.1 Schematic diagram of the CO_2 absorption facility by NaOH/ionic liquid. 1—N_2/CO_2 cylinder, 2—pressure reducing valve, 3—valves, 4,12—gas analyzer, 5,11—moisture filter, 6,10—foam-film flow meter, 7—spray scrubber, 8—mechanical diaphragm metering pump, 9—solution tank with heating system.

Raman spectra of semi-coke were obtained using the Raman spectroscopy (LabRAM HR800) with a 532 nm, 0.05 mW laser over two cycles of exposure (5 s for each cycle). Morphologies of semi-coke samples were studied with an SEM (JSM-5610LV, JEOL Co., Japan) at the accelerating voltage of 25 kV.

The CO_2 absorption capacity of the NaOH/ILs solution was calculated by the following equations:

$$n_{CO_2} = \left(\frac{C_{in} \times Q_{in} \times t}{V_{mol}} - \frac{\int_0^t Q_{out} \times C_{out} \times dt}{V_{mol}} - \frac{\frac{C_{out} - C_{ini}}{2} \times h \times \pi \times \frac{D^2}{4}}{V_{mol}} \right) \times \frac{1}{C_{ab} \times V} \qquad V_{mol} = \frac{P_0}{P} \times \frac{T}{T_0} \times V_0$$

where $n_{(CO_2)}$ was the CO_2 absorption capacity (mol CO_2/mol absorbent), C_{in} and C_{out} were the concentration of CO_2 in inlet and outlet gas (vol%), C_{ini} was the volumetric concentration of CO_2 in the spray scrubber before absorption, C_{ab} was the absorbent concentration in solution (mol/L), t was the absorption time (min), Q_{in} and Q_{out} were the flow rate of inlet and exit gas (L/min), D was the diameter of the spray scrubber, V was the volume of the solution, V_{mol} was the mole volume of gas at the temperature of T and pressure of P, T_0 was 273 K, P_0 was 1 bar, V_0 was 22.4 L/mol. We supposed that the concentration of CO_2 homogeneously varied along the direction of height when calculating the amount of CO_2 in the spray scrubber.

Table 1 The experimental conditions of the experiments.

Experimental parameters	Experimental conditions		
	Expt.3.1	Expt.3.2	Expt.3.3
Temperature(K)	288	288	288
Mole ratio of NaOH to ILs	4:1-1:4	4:1	4:1
Liquid-to-gas ratio	1:3	1:3	1:3
Volume of the NaOH/ILs solution(L)	3	3	3
liquid flow rate(L/min)	0.3	0.3	0.3
Concentration of the absorbent(mol/L)	0.4	0.2-0.5	0.4

3. Test Results and Discussions

3.1. The effect of IL mole ratio on CO_2 absorption

Fig. 2(a) shows the effect of mole ratio on CO_2 absorption. With the mole percentage of IL rising from 0 to 20 %, the CO_2 absorption mole ratio increased from 0.44 to 0.54 while controlling the outlet CO_2 concentration below 3 %, by which point the quality of the outlet gas could meet the standards of compressed natural gas as vehicle fuel (GB 18047-2000) in China. With the percentage of IL rose up to 40 %, the CO_2 capture capacity began to decrease, and reduced to 0.34 when the percentage of IL was 80%. Accordingly, the time to obtain qualified natural gas showed the same trend. It was revealed that the mixture of a small amount of IL and the bulk of sodium hydroxide solution could improve the CO_2 capture capacity because of the high absorption capacity of IL. However, this promotion may be limited by the high viscosity of IL and the reaction product when the percentage of IL continued to increase. Such high viscosity hindered the diffusion of CO_2 into the bulk liquid phase and seriously reduced the mass transfer rate. Further, the porous structure of semi-coke would be jammed by the reaction product that not timely diffused and formed foam layers. Therefore, the CO_2 capture ability experienced a reduction.

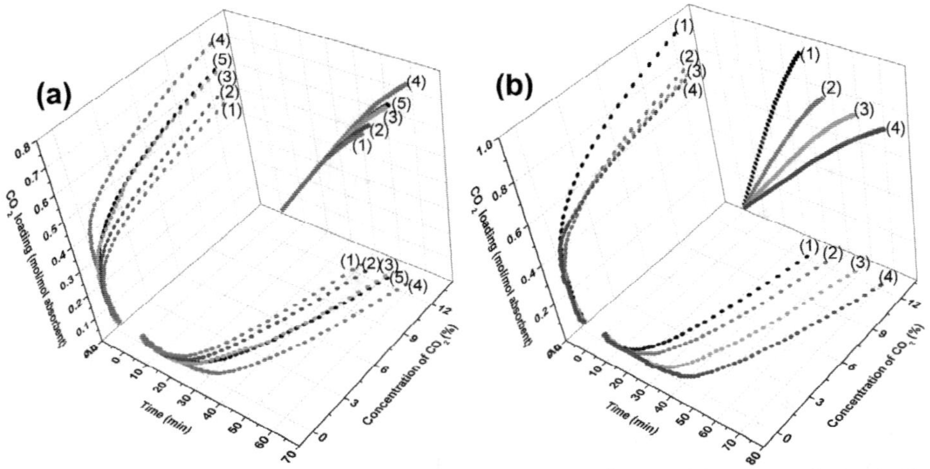

Fig.2 The effect of mole ratio and absorbent concentration on CO_2 absorption;**(a)**:(1) n(NaOH):n(IL)=1:4; (2) n(NaOH):n(IL)=2:3; (3) n(NaOH):n(IL)=3:2; (4) n(NaOH):n(IL)=4:1; (5) NaOH. **(b)**: (1) 0.2 mol/L; (2) 0.3 mol/L; (3) 0.4 mol/L; (4) 0.5 mol/L.

3.2 The effect of absorbent concentration on CO_2 absorption

Fig. 2(b) shows the effect of absorbent concentration on CO_2 absorption. When the absorbent concentration increased from 0.2 to 0.5mol/L, the time of obtain qualified natural gas with 3 L solution increased from 24 minutes to 50 minutes. However, the CO_2 absorption mole ratio was obviously decreased, which varied from 0.65 to 0.54 as the concentration increased to 0.3 mol/L when controlling the outlet CO_2 concentration below 3%, and not varied much as it continued to rise. The possible reasons could be explained as: on the one hand, sodium hydroxide exhibited higher reactivity to CO_2 than IL in solution of high pH values, though there were four amino groups in IL molecular. As a result, sodium hydroxide showed priority to IL when competed to react with CO_2. In the light of this observation, relatively low absorbent concentration was favorable for more IL participating in CO_2 capture, while high absorbent concentration may not protrude the high absorption capacity of IL. On the other hand, the foam layers on semi-coke were relatively thin in a low concentration solution. Thus, the porous structure of semi-coke was favorable for CO_2 capture.

3.3 The effect of semi-coke on CO_2 absorption

The effect of semi-coke on CO_2 absorption was shown in Fig. 3(a). Compared to the absorption without filler, the CO_2 absorption mole ratio as controlling the outlet CO_2 concentration below 3 vol% increased from 0.53 to 0.92 when SN600 was used. The different CO_2 absorption capacities when SWT, S600N and SN600 were used revealed that the structure and surface properties of semi-coke were vital important for CO_2 capture. It is possible that the porous structure and functional groups resulted from nitric acid and heating treatment supplied more active sites for CO_2 absorption.

3.4 Absorption mechanism of NaOH/IL solution

Similar to aliphatic amines, in the absence of solvent, CO_2 could react with both primary and secondary amines on IL to form unstable carbamic acid: $2AmH + CO_2 \rightleftharpoons AmCO_2^- + AmH_2^+$ (AmH is either primary or secondary amine)

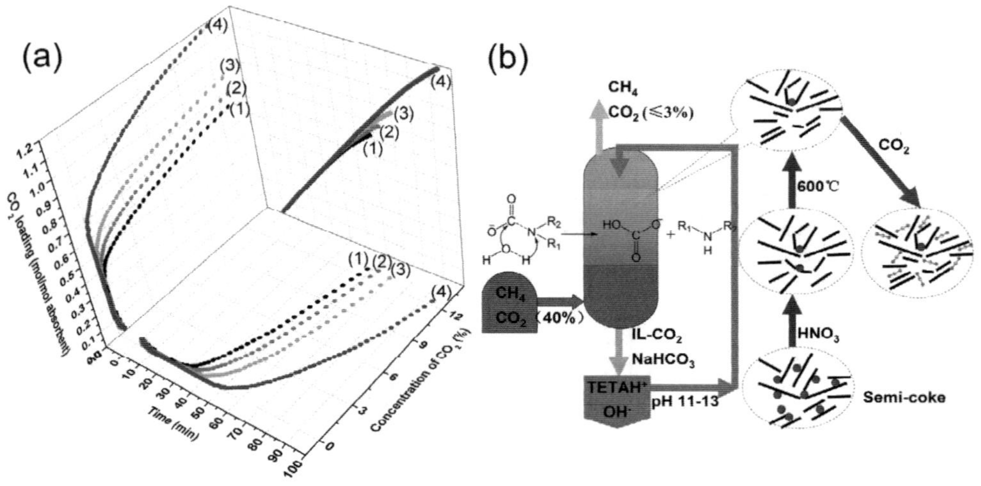

Fig.3 The effect of semi-coke on CO_2 absorption. **(a):**(1) WF—without filler; (2) SWT—semi-coke without treatment; (3) S600N—semi-coke was firstly with heat treatment at 873 K for 2 hours, and then treated with 45 wt% nitric acid; (4) SN600—semi-coke was firstly treated with 45 wt% nitric acid, and then with heat treatment at 873 K for 2 hours. **(b):** Schematic illustration of mechanism

It is indicated that two amino groups in the IL molecules interact with one CO_2 molecule.

With the presence of solvent, an alkyl carbonate is formed along with the protonated amine. As expressed: $AmH + CO_2 + ROH \rightleftharpoons ROCO_2^- + AmH_2^+$, the stoichiometric loading capacity is 1.

In this work, the CO_2 capture was taken place in NaOH/IL solution. Besides the reaction between NaOH and CO_2, it is noteworthy that water could participate in the absorption reaction. Additionally, water in solution could facilitate the secondary amines to be involved in CO_2 capture. The detail mechanism that accounting for the role of water in CO_2 capture by secondary amines was depicted in Fig.4(e).

IL-carbamate and protonated IL were formed by the nucleophilic attack of the amine lone pair to the CO_2 carbon. Subsequently, the nucleophilic reaction continued between the carbonyl atom of IL-carbamate and the oxygen of water, leading to the formation of bicarbonate. And then bicarbonate involved in the reaction with IL-amine to form IL carbamate and give back water. Additionally, when the concentration of NaOH was relatively high, there was a competing reaction for bicarbonate consuming: $OH^- + HCO_3^- \longrightarrow CO_3^{2-} + H_2O$

Therefore, much higher absorbent concentration would not exhibit higher CO_2 absorption capacity.

To confirmed the mechanism of the reactions and understand the characteristics of the reaction products, ^{13}C NMR analysis was performed in Fig. 4(d). The resonance in the range of 159.89-163.91 ppm confirmed the emergence of IL-carbamate (C_7, C_8). The peaks of 159.89 ppm and 168.15 ppm were attributed to the existence of sodium bicarbonate and sodium carbonate, respectively.

3.5 The role of semi-coke in CO_2 capture

In order to figure out the absorption mechanism of semi-coke, the surface physical properties were examined. In Table 2, the specific surface area of SN600 was 11.435 m^2/g, which increased by 10.488 m^2/g. And the total pore volume was 0.01 cm^3/g, which increased by 0.009 cm^3/g. Combined with the data of S600N, it is indicated that the modification method of combining nitric acid oxidation with heat treatment could effectively improve the specific surface area and porous structure. More micropores formed and the intrinsic pores were enlarged.

In modification process, nitric acid oxidation could dig pits and crevasses by penetrating into the surface. And the original pore structures of semi-coke would further be developed into larger pores. High temperature in the pyrolysis process could oxidize as well as release ash and organic volatiles jammed in the pores [8-9]. It seemed that semi-coke firstly treated with nitric acid was favorable for pore enlargement and surface area improvement, achieving a better absorption performance.

Table 2 Surface physical properties of semi-coke samples

Sample	S_{BET} (m^2/g)	Total pore volume (cm^3/g)	Average pore size (nm)	Microporous volume (cm^3/g)	Microporous average size (nm)
SWT	0.947	0.001	2.214	0.0004	1.6859
S600N	7.224	0.006	2.469	0.0013	1.6886
SN600	11.435	0.010	2.558	0.0010	1.6668

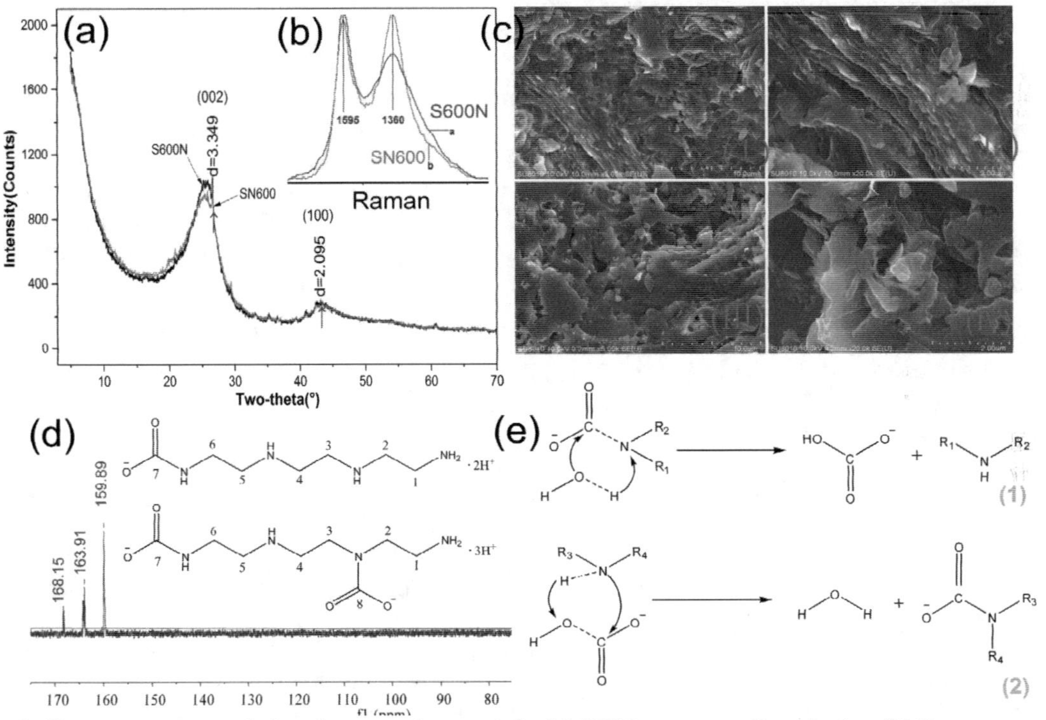

Fig. 4 *Characterization of absorbents and materials.* (a) XRD patterns of semicoke; (b) Raman spectra; (c) SEM images of semi-coke; (d) ^{13}CNMR spectra of solution; (e) Mechanism of CO_2 capture by IL.

Raman spectra were used to analyze the structural changes of semi-coke samples.As is illustrated in Fig. 4(b), two bands located at 1595 and 1360 cm^{-1} were detected in the spectra of S600N and SN600. The former one (1595 cm^{-1})corresponds to the E_{2g} species of single-crystal graphite, while the latter one (1360 cm^{-1}) is assigned to be the vibration mode of the graphite lattice of which the Raman activity is originated from the loss of translational symmetry at the borders of the crystalline areas [10]. The shape changes at bands 1595 and 1360 cm^{-1} revealed the structural differences resulted from the different degree of graphitization. This was consistent with XRD comparisons as showed in Fig. 4(a).

Fig. 4(c) shows the SEM images of the semi-coke samples S600N and SN600. It can be observed that the surfaces were dominated by pore-distributed structures. But the pore structures of the two types of semi-coke were quite different. For sample S600N, the folding on the surface was obvious, showing the layered structure in (I) and (II). Although concave and convex structures were observed, the cavities were relatively shallow. The interface between cavities was indistinct. It is indicated that the oxidation effects of nitric acid on graphitized carbon wall were not obvious. For sample SN600 in (III) and (IV), more obviously pore structure could be seen due to the formation of fragments when the sample was firstly treated with nitric acid for deashing. The fragments were mobile when heating, and were not tended to form graphitized structure [11]. The deep cavities created by the pieces of fragments were considered to increase the reactivity to CO_2. The detailed mechanism can be illustrated in Fig. 3(b).

4. Conclusion

The Poly-amino-based ionic liquid $[TETAH]^+[BF_4]^-$/NaOH solution was used for CO_2 separation from biogas in a spray scrubber packed with modified semi-coke.

(1) A CO_2 mole ratio of 0.92 was achieved as controlling the outlet CO_2 concentration below 3% with the incorporation of modified semi-coke. A mole percentage of 20 % was favorable for improving CO_2 absorption capacity and absorption rate, capturing CO_2 into carbamate and bicarbonate.

(2) The porous structure of semi-coke could contribute to improving CO_2 absorption. Combining nitric acid oxidation with heat treatment, more micropores formed and the intrinsic pores were enlarged. Semi-coke firstly oxidized by nitric acid and then with heat treatment not only exhibited higher specific surface area and more porous structure, but also showed less degree of graphitization, which was suggested as possible to increase the reactivity to CO_2.

(3) Considering the easy synthesis of the ILs, the excellent absorption performance of ILs/NaOH solution as well as the easy availability of semi-coke, it is assumed that the absorption system in this research showed promising application in CO_2 separation from biogas as well as waste treatment.

References

[1] Brunetti, A., Sun, Y., Caravella, A. (2015) Process intensification for greenhouse gas separation from biogas: More efficient process schemes based on membrane-integrated systems. Int. J. Greenh. Gas. Con., 35: 18–29.

[2] Al Mamun, M. R., Karim, M. R., Rahman, M. M. (2016) Methane enrichment of biogas by carbon dioxide fixation with calcium hydroxide and activated carbon. J. Taiwan. Inst. Chem. Eng., 58: 476-481.

[3] Krupiczka, R., Rotkegel, A., Ziobrowski, Z. (2015) Comparative study of CO_2 absorption in packed column using imidazolium based ionic liquids and MEA solution. Sep. Purif. Technol., 149: 228–236.

[4] Wang, C., Mahurin, S. M., Luo, H. (2010) Reversible and robust CO_2 capture by equimolar task-specific ionic liquid-superbase mixtures. Green. Chem., 12(5): 870-874.

[5] Sistla, Y. S., Khanna, A. (2014) Carbon dioxide absorption studies using amine-functionalized ionic liquids. J. Ind. Eng. Chem., 20: 2497–2509.

[6] Arellano, I. H., Huang, J., Pendleton, P. (2015) Synergistic enhancement of CO_2 uptake in highly ordered mesoporous silica-supported zinc-functionalized ionic liquid sorbents. Chem. Eng. J., 281: 119–125.

[7] Hu, H., Li, F., Xia, Q. (2014) Research on influencing factors and mechanism of CO_2 absorption by poly-amino-based ionic liquids. Int. J. Greenh. Gas. Con., 31: 33-40.

[8] Zhang, G., Qu, J., Du, Y. (2014) Hydrogen production from CO_2 reforming of methane over high pressure H_2O_2 modified different semi-cokes. J. Ind. Eng. Chem., 20: 2948-2957.

[9] Zhou, G., Wu, J., Miao, Z. (2013) Effects of process parameters on pore structure of semi-coke prepared by solid heat carrier with dry distillation. Int. J. Min. Sci.Technol., 23: 423–427.

[10] Wang, J., Yan, Z., Liu, L. (2014) In situ DRIFTS investigation on the SCR of NO with NH_3 over V_2O_5 catalyst supported by activated semi-coke. Appl. Surf. Sci., 313: 660–669.
[11] Kidena, K., Matsumoto, K., Katsuyama, M. (2004) Development of aromatic ring size in bituminous coals during heat treatment in the plastic temperature range. Fuel. Process. Technol., 85: 827–835.

Density Functional Theory Calculation of Electronic Structure of Fe-N-C, Fe-O-C, Fe-P-C, Fe-S-C Single-atom Catalysts Systems

JiangLu Ping[1*]

[1] Northfield Mount Hermon School, Gill 01354, USA

[*] jianglu4517@gmail.com

Abstract—Single atom catalysts (SACs) have emerged as a novel class of heterogeneous catalysts which exhibit superior catalytic performance in various reaction, CO oxidation, CO_2 conversion, oxygen reduction reaction (ORR), etc. It is challenging to explore electronic structure and catalytic mechanics of SACs when the single metal atom is bonding with different dopants via experimental methods, which influence the catalytic efficiency directly. In this paper, Density Functional Theory (DFT) calculations, an *ab initio* simulation methods, have been carried out to investigate the properties of Fe-N-C, Fe-O-C, Fe-P-C, and Fe-S-C with oxo or hydroxyl group, four common SACs models. Geometry structure of Fe-N-C, Fe-O-C, Fe-P-C, Fe-S-C, and after each has bonded with oxo or hydroxyl group are studied. Formation energy of the Single Atom Catalyst (SAC), oxygen formation energy, and hydrogen atom transfer energy of the four SACs models are also calculated. In order to investigate the electronic structure of the four SACs, we examined the HOMO LUMO gap values for each system. In addition, we plotted the corresponding HOMO and LUMO orbitals of each SACs models and with oxo or hydroxyl group. Through analyzing the data, I anticipated that Fe-N-C model is the easiest to synthesis and the most stable among the four models. Our work discloses the microscopic property of SACs and provides valuable suggestions to experimental scientists.

1. Introduction

Both academia and industry need high-performance heterogeneous catalysts with superior activity, selectivity, and stability. Conventional heterogeneous catalysts are subject to limited dispersion. Single atom catalyst (SAC) is an emerging class of catalyst that is able to incorporate the scalability and high dispersion [1]. Graphene is an ideal two-dimensional (2D) catalyst support because of its large surface area which allows it to have novel physical and chemical properties [2]. Metal species in SACs are all downsized into nanoparticles and are located dispersedly on a solid support, which increases the catalytic efficiency for the more exposed unsaturated reactive sites [3]. Since SACs is highly active, selective, and can be applied to various purposes, SACs has become a research hotspot in various fields such as energy and environmental protection [4]. Additionally, previous researches have shown that SACs are highly catalytically active in reactions such as CO oxidation, CO_2 reduction, CH4 activation, and oxygen reduction reaction (ORR).

Typically, the metal atom in the SACs is anchored to the N sites of C framework [5]. N is an ideal heteroatom dopant because its atomic radii is similar to that of for C atom and its notable electronegativity difference comparing with C atom. In general, there are three types of configurations for the N atoms doped in the graphene lattice: graphitic, pyridinic (two C-N bonds in a hexagon), and pyrrolic (two C-N bonds in a pentagon) [6]. Different from pyridinic N, pyrrolic N has an increased electron density on N-C sites because it has a pair of electrons in the unhybridized pz orbital [7]. Both experimental and DFT calculation have proved that pyridinic N is a more ideal doping type than graphitic or pyrrolic N because of its weaker binding energy [8]. Additionally, when trapping metal atoms, for example Fe, pyridinic N will generally have higher chemical reactivity due to its distinct charge distribution [9]. Also, pyridinic N is able to stabilize the metal center, and the activity of the catalyst will increase remarkably after adding the single metal atom.

Dodelet's group has designed a catalyst that has Fe doping with pyridinic N in the interstices of C sheet within the micropores [11]. This design has greatly increased the catalytic activity of Fe-N-C in ORR, and has narrowed the gap between commercial Pt/C and non-metal catalysts [10]. Although there have been a lot of improvements on Fe-N-C, such as optimizing the selection of precursors for Fe and modify the pyrolysis process, the gap between Fe-N-C and commercial Pt/c regarding of durability and catalytic activity still remains [11].

Changing the electronic structures of the active sites can also slightly tune the electrocatalytic activity of the SAC. Particularly, doping atoms such as O, S, P in the metal center can create different atomic structures for the SAC [12]. However, it is difficult to illuminate the catalytic performance of different doping atoms only by time-consuming experiments. Herein, we studied four different models of single atom catalysts, Fe-N-C, Fe-O-C, Fe-P-C, and Fe-S-C, having nitrogen, oxygen, phosphorous, and sulfur coordinate with iron (II) *via* density functional theory. OXO and hydroxyl are introduced as two chemical groups, and were bonded to the four SACs models. Our simulation can screen all the systems efficiently and gives guidance to experiment researchers.

2. Computational Method

In this work, SACs, SACs-oxo and SACs-hydroxyl with N, O, P and S dopants were considered. All the geometry optimizations were carried out with DFT method [13] using the ωPBEh/lacvps_ecp [14] as basis set in ORCA v.4.0 [15]. Constrained calculations (constrain the axis of the flake without Fe metal) were carried out in order to investigate the SACs models with planar structure.

$$\Delta E_{\text{formation energy}} = E_{\text{SACs}} - E_{\text{Flake}} - E_{\text{Fe}(+2)} \quad (1)$$

$$\Delta E_{\text{oxygen formation energy}} = E_{\text{SACs=O}} - E_{\text{SACs}} + E_{(N_2)} - E_{(N_2O)} \quad (2)$$

$$\Delta E_{\text{hydrogen atom transfer energy}} = E_{(\text{SACs-OH})} - E_{(\text{SACs=O})} + E_{(CH_3*)} - E_{(CH_4)} \quad (3)$$

To discuss formation energy of the SACs, E_{SACs}, E_{flake}, and $E_{Fe(+2)}$ are introduced, where ESACs is the total energy of single atom catalysts, EFlake is the total energy of the pristine flake, $E_{Fe\,(+2)}$ is the energy of Iron (II) in gas phase condition. The formation energy of oxygen is calculated using formula (2): E_{SACs}=O is the energy of SACs bonded with oxygen, E_{SACs} is the energy of SACs, $E_{(N2)}$ is the energy of nitrogen, and $E_{(N2O)}$ is the energy of nitrous oxide. Calculation process for the energy of hydrogen atom transfer is shown in formula (3): where $E_{(SACs\text{-}OH)}$ is the energy of SACs bonded with hydroxyl, $E_{(CH3*)}$ is the energy of methyl, $E_{(SACs=O)}$ is the energy of SACs bonded with oxygen, and $E_{(CH4)}$ is the energy of methane.

3. Test Results and Discussions

In Figure 1, (a), (b), (c), and (d) show the configuration for Fe-N-C, Fe-O-C, Fe-P-C, and Fe-S-C systems. All the four SACs keep relative planar structure after constrained the frame geometry optimization (Fe atom and the flake are nearly in the same plane). The second and third row shows the top and side configuration of the SACs models bonding with oxo and hydroxyl functional groups. Different from first row, Fe atoms in (e)-(i) obviously remain out of the plane of the flake. Notably, in (f), (j), and (i), Fe-OXO and Fe-hydroxyl are not bonding with the 4 coordinated atoms in flake, which will be discussed with detail later.

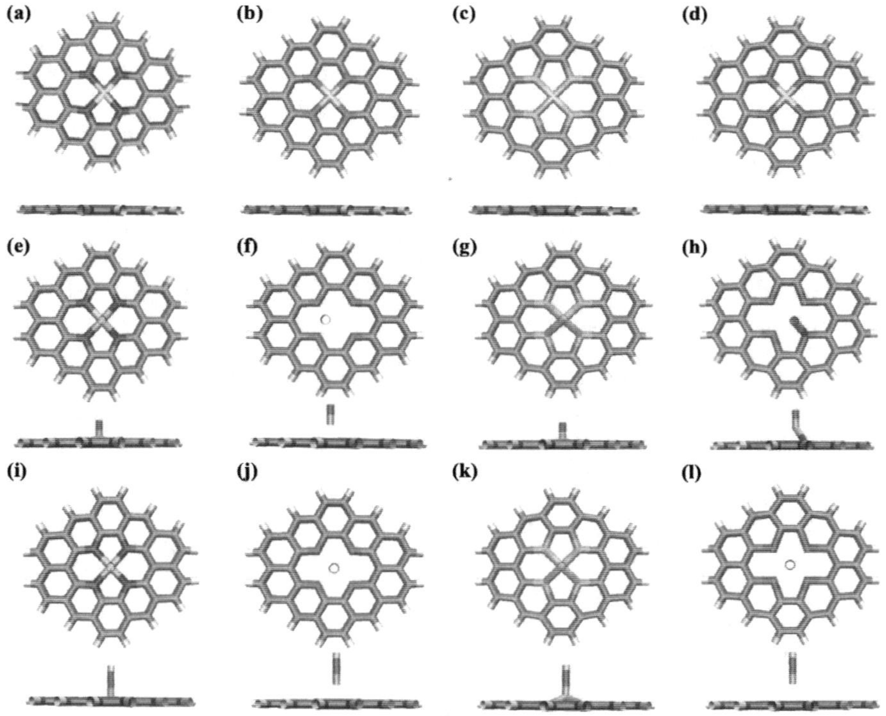

Fig.1 Optimized atomic structure of Fe-N-C, Fe-O-C, Fe-P-C, Fe-S-C with corresponding OXO and hydroxyl functional groups. White, grey, blue, orange, red, cyan, and purple stands for H, C, N, Fe, O, P, and S atoms, respectively.

Table 1 Bond length of Fe-N-C, Fe-O-C, Fe-P-C, and Fe-S-C optimized at ωPBEh/lacvps_ecp theory level. Bond length of Fe-OXO and Fe-hydroxyl, and bond angles θ_{12} (bond angle between dopant$_1$-Fe-dopant$_2$), θ_{23} (bond angle between dopant$_2$-Fe-dopant$_3$) are in unit: Å and θ: °.

System	Fe-dopant$_1$	Fe-dopant$_2$	Fe-dopant$_3$	Fe-dopant$_4$	Fe-OXO	Fe-hydroxyl	θ_{12}	θ_{23}
Fe-N-C	1.845	1.846	1.943	1.943	1.614	1.863	86.739	93.241
Fe-O-C	1.961	1.943	1.838	1.928	1.590	1.753	86.303	93.673
Fe-P-C	1.941	1.941	1.941	1.941	1.614	1.797	88.246	91.752
Fe-S-C	1.965	1.964	1.946	1.946	1.595	1.767	87.576	92.074

For Fe-N-C and Fe-S-C systems, the bond length of Fe-dopant$_1$ and Fe-dopant$_2$, Fe-dopant$_3$ and Fe-dopant$_4$ are similar to each other. In Fe-O-C system, bond length of Fe-dopant$_1$, Fe-dopant$_2$ and Fe-dopant$_4$ are all greater than bond length of Fe-dopant$_3$. The bond length of Fe with each phosphorus dopants are almost the same. Although not mentioned in the table, bond lengths of Fe with the four dopants of SAC-OXO and SAC-hydroxyl systems (SAC= Fe-N-C, Fe-O-C, Fe-P-C, and Fe-S-C) are also investigated. Bond length of Fe-dopant$_1$, Fe-dopant$_2$, Fe-dopant$_3$, and Fe-dopant$_4$ in Fe-O-C bonded with O ion are 2.132Å, 2.832Å, 2.834Å, and 2.134 Å, respectively. In the same order, bond lengths between Fe ion and the four dopants in Fe-O-C bonded with hydroxyl are 2.474 Å, 2.521 Å, 2.532 Å, and 2.499 Å. These bond lengths suggest that Fe ion was not able to form a bonding with O dopants, because their bond lengths exceed 2.5 Å. Same problem happened in Fe-S-C bonded with O ion and hydroxyl. Bond length of Fe-dopant$_1$, Fe-dopant$_2$, Fe-dopant$_3$, and Fe-dopant$_4$ in Fe-S-C bonded with O ion are 2.550 Å, 2.251Å, 2.359Å, and 2.842Å. The four bond lengths in Fe-S-C bonded with hydroxyl are 2.516 Å, 2.533Å, 2.532Å, and 2.513Å. Notably, most bond lengths in Fe-S-C bonded with oxo and hydroxyl are greater than 2.5Å. Therefore, the bonding between the metal ion and S dopants are considered fail. In later discussions, Fe-O-C with OXO or hydroxyl group and Fe-S-C with OXO or hydroxyl group will be excluded, since their models are not meaningful. The bond lengths of Fe with oxo (Fe-OXO) in Fe-N-C and Fe-P-C are greater than that of in Fe-O-C and Fe-P-C. Bond lengths of Fe-OXO in Fe-O-C is 1.590Å, and is slightly less than that of in Fe-P-C. The bond length of Fe-hydroxyl in Fe-N-C is 1.863Å, and is the greatest among all four systems. Fe-P-C system has the second greatest Fe-hydroxyl bond length, and Fe-O-C system has the least Fe-hydroxyl bond length. Bond angles between dopant$_1$-Fe-dopant$_2$(θ_{12}) and dopant$_2$-Fe-dopant$_3$(θ_{23}) in Fe-N-C and Fe-O-C systems are close. θ_{12} has the greatest value in Fe-P-C system, and second greatest value in Fe-S-C. θ_{23} is larger in Fe-N-C and Fe-O-C, and smaller in Fe-P-C and Fe-S-C.

Table 2 Formation Energy of SACs, Oxygen Formation Energy, Hydrogen Atom Transfer Energy, and HOMO-LUMO Gap of the four systems (unit: eV)

System	Formation Energy	Oxygen Formation Energy	Hydrogen Atom Transfer Energy	HOMO-LUMO gap
Fe-N-C	-34.775	-1.879	1.401	0.134
Fe-O-C	-29.094	-2.905	0.423	0.124
Fe-P-C	-32.926	-2.346	1.514	0.127
Fe-S-C	-29.867	-2.146	-0.522	0.125

The formation energy of Fe-N-C system is found to be -34.775eV, which is the lowest among all four systems, showing that Fe-N-C is the most stable and the easiest to synthesis among all four catalysts. In contrast, Fe-O-C system is found to have the highest formation energy of the SAC, which suggests that it is the least stable catalyst in the four. Formation energy of Fe-P-C and Fe-S-C are the second and third lowest. As shown in Table 2, oxygen formation energy is also calculated. This energy describes the amount of energy adding the oxo group on to the Fe atom would require. Ranking the four systems by having the lowest oxygen formation energy to the highest, it would be from Fe-O-C, Fe-P-C, Fe-S-C, to Fe-N-C. Although Fe-O-C possesses the lowest oxygen formation energy, its metal center did not successfully bond with the OXO group, as shown in Figure 1, (f) and (j). To investigate the stability of the four SACs after bonding with hydroxyl, we calculated the hydrogen atom transfer energy, which describes the amount of energy required to remove the H atom from its present bonding with the SACs. The larger the hydrogen atom transfer energy is, the more energy is needed to break the O-H bond. As shown in Table 2., Fe-P-C is found to have the largest hydrogen atom transfer energy as 1.514 eV, and Fe-N-C has the second greatest hydrogen atom transfer energy as 1.401 eV, indicating that Fe-N-C and Fe-P-C are more preferable catalysts. It is worth noting that the hydrogen atom transfer energy for Fe-S-C and Fe-O-S are remarkably less than that of for Fe-N-C and Fe-P-C, which suggest that Fe-S-C and Fe-O-C are a lot easier to lose the H atom. HOMO and LUMO are two types of molecular orbital: HOMO stands for highest occupied molecular orbital, and LUMO stands for lowest unoccupied molecular orbital. Since HOMO and LUMO are the two orbitals closest in energy among all orbitals, the HOMO-LUMO gap is where excitation is most likely to occur in a system. The smaller the HOMO-LUMO gap is, the better the conductivity the system has; the greater the HOMO-LUMO gap is, the better performance as semiconductors the system is. The HOMO LUMO gap of Fe-N-C, Fe-O-C, Fe-P-C, Fe-S-C catalysts are estimated to be 0.134 eV, 0.124 eV, 0.127 eV, 0.125 eV. The relatively small values of HOMO-LUMO gap suggest that all four catalysts have comparatively strong conductivity. Based on the date of HOMO-LUMO gap in Table 2, the conductivity of the four systems should follow this order: Fe-O-C $>$ Fe-S-C $>$ Fe-P-C $>$ Fe-N-C.

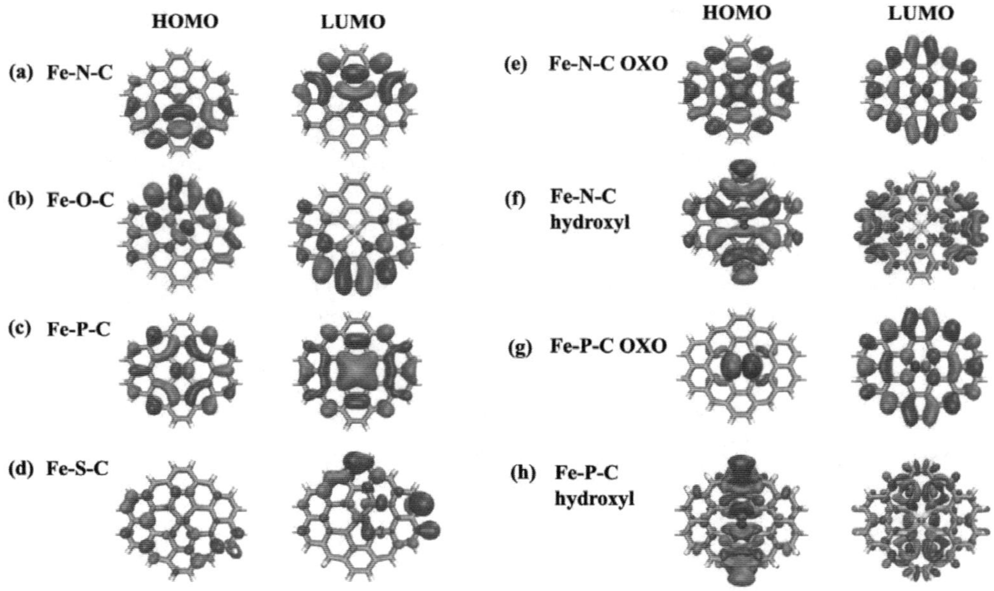

Figure.2 HOMO (red) and LUMO (blue) electron distribution of Fe-N-C, Fe-O-C, Fe-P-C, Fe-S-C with corresponding OXO and hydroxyl. The positive and negative density of electron are shown in red and blue in the figure, respectively. The isosurface value of 0.02 was used for all the systems.

The density of electrons of Fe-N-C, Fe-O-C, Fe-P-C, Fe-S-C, and after bonded with oxo or hydroxyl are show in Figure 2. Since the metal center did not successfully form a bond with the O or S dopants after OXO or hydroxyl was added in Fe-O-C and Fe-S-C, these four models are excluded from our discussion here, leaving eight models to investigate. The spatial distribution of HOMO and LUMO energy levels of those eight models are shown in Figure 2. We employed graphical representations because it is important to understand the properties and geometries of the HOMO and LUMO of each SAC model. From Fig.2 (a), it is clear that in Fe-N-C, the HOMO is localized in lower half of the system and the LUMO is localized in the upper half of the system. After bonding with the OXO group, the positively charged electrons in the HOMO of Fe-N-C OXO have increased notably, as shown in Fig2. (e). However, positively charged electrons in the HOMO of Fe-N-C hydroxyl have decreased significantly compared with that of in Fe-N-C OXO. It can also be observed in Fig2. (e) that the HOMO of Fe-N-C OXO is principally restricted on Fe and N. As shown in Fig.2 (b), the HOMO of Fe-O-C is mainly localized in the upper half of the system, and the LUMO of Fe-O-C is localized in the lower half of the system. HOMO and LUMO in Fe-S-C are mainly localized on the graphene nanoflake. Compare the HOMO of Fe-P-C and Fe-P-C OXO, it can be seen that the positive and negative density of electrons become centralized in the center after the OXO group is bonded to Fe-P-C, and HOMO is mainly localized in Fe in Fe-P-C OXO. In contrast, HOMO in Fe-P-C OXO is not as centralized as it is in Fe-P-C OXO. It can also be observed from Fig2. that both the HOMO and LUMO of Fe-N-C only have a vertical line of symmetry. However, after bonded with the OXO or hydroxyl group, the HOMO and LUMO of Fe-N-C start to have two lines of symmetry, one vertically and one horizontally, as shown in Fig2. (e) and (f). Besides Fe-N-C OXO and Fe-N-C hydroxyl, Fe-P-C, Fe-P-C OXO, and Fe-P-C hydroxyl models also are found to have two lines of symmetry, as shown in Fig2. (c), (g), and (h). In addition, we found out that there is a straight relation between the energy gap of HOMO-LUMO and the stability of the SAC. Based on the results from Table 2, we can rank the four system from having the lowest formation energy to highest formation energy and the order is: Fe-N-C < Fe-P-C < Fe-S-C < Fe-O-C. This shows a trend that the higher the energy gap of HOMO-LUMO is, the lower the formation energy is, and the more stable the system is.

4. Conclusion

In this work, we employed DFT calculations to investigate the properties of Fe-N-C, Fe-O-C, Fe-P-C, Fe-S-C, and each after bonded with oxo or hydroxyl group. Fe-N-C is found to have the lowest formation energy of the SAC, which suggests that it is the easiest to synthesis and most stable among the four models. Also, we found that Fe-N-C has the highest oxygen formation energy and Fe-O-C has the lowest oxygen formation energy. The collected data also shows that Fe-P-C has the highest hydrogen atom transfer energy, Fe-N-C has the second highest hydrogen atom transfer energy, and Fe-S-C has the lowest hydrogen atom transfer energy. Thus, we can conclude that Fe-N-C is a more preferable catalyst than Fe-O-C, Fe-P-C, or Fe-S-C from synthesis and energy perspectives. In addition, we found that Fe-N-C has the largest HOMO LUMO gap value among the four SAC models, which indicates that it might be a good candidate for semiconductors. Fe-S-C is found to have the lowest HOMO LUMO gap, which suggests that its conductivity is the best among the four SACs. Our simulation screens all the systems efficiently and give guidance to experiment researchers.

Acknowledgments

I would like to express my appreciation to Haojun Jia, from Massachusetts Institute of Technology, for his guidance and valuable support.

References

[1] F. Liu, T. Yang, J. Yang, E. Xu, A. Bajaj, H. J. Kulik, Front. Chem. 2019, 7, 219.
[2] B. Bayatsarmadi, Y. Zheng, A. Vasileff, S.Z. Qiao, Small. 2017, 13, 1700191.
[3] T.T. Sun, L.B. Xu, D.S.Wang, Y.D. Li, Nano Res. 2019, 12, 2067–2080.
[4] J. Lin, A. Wang, B. Qiao, X. Liu, X. Yang, X. Wang, J. Am. Chem. Soc. 2013, 135, 15314.
[5] L. Zhao, Y. Zhang, L. B. Huang, X. Z. Liu, Q. H. Zhang, C. He, Z. Y. Wu, L. J. Zhang, J. Wu, W. Yang, L. Gu, J. S. Hu and L. J. Wan, Nat. Commun. 2019, 10, 1278.
[6] L. Zou, W. Chu, W. Cen, Appl. Surf. Sci., 2020, 531, 147310.
[7] C. Xu, A. Vasileff, D. Wang, B. Jin, Y. Zheng and S. Qiao, Nanoscale Horiz., 2019,
[8] Y. Fujimoto, S. Saito, Phys. Rev. B. 2011, 84, 245446.
[9] Y.-C. Lin, P.-Y. Teng, C.-H. Yeh, M. Koshino, P.-W. Chiu, K. Suenaga, Nano Lett. 2015, 15, 7408–7413.
[10] M. Lefevre, E. Proietti, F. Jaouen, J.P. Dodelet, Science 2009, 324, 71–74.
[11] Y. Chen, C. Wang, Z. Wu, Y. Xiong, Q. Xu, S. Yu, H. Jiang, Adv. Mater. 2015, 27, 5010–5016.
[12] Y. Hou, M. Qiu, M. G. Kim, P. Liu, G. Nam, T. Zhang, X. Zhuang, B. Yang, J. Cho, M. Chen, C. Yuan, L. Lei, X. Feng, Nat. Commun. 2019, 10, 1392.
[13] J. Heyd, G. E. Scuseria, The Journal of Chemical Physics 2003, 118, 8207.
[14] J. Heyd, G. E. Scuseria, The Journal of Chemical Physics 2003, 118, 8207.
[15] F. Liu, T. Yang, J. Yang, E. Xu, A. Bajaj, H. J. Kulik, Front. Chem. 2019, 7, 219.

Journal of Physics: Conference Series **2194** (2022) 012049 doi:10.1088/1742-6596/2194/1/012049

The Effect of UV Degradation of Ketoprofen and Its Influencing Factors

Huiwen Zhen[1], Junliang Liu[1,*], Tiejian Zhang[1]

[1]College of Urban and Rural Construction, Hebei Agricultural University, Hebei, China, 071000

*Corresponding author e-mail: cjljl@hebau.edu.cn

Abstract. This paper mainly studies the removal effect of ketoprofen in UV-degraded water and explores the influence of light turbidity, light intensity and other factors on the degradation efficiency of ketoprofen. The results show that deep ultraviolet treatment has a good degradation effect on ketoprofen, which accords with the pseudo-first-order kinetic equations. After irradiation with 1700 μw/cm2 UV lamps for 40 minutes, the maximum removal rate can reach 99.59%. Besides, UV lamps with different irradiation strengths (1300μw/cm2/1700μw/cm2) have similar treatment effects on ketoprofen. Low turbidity (1-5NTU) solutions will reduce the degradation rate of ketoprofen, while high turbidity (5-9NTU) solutions will enhance the degradation rate of ketoprofen. Solution turbidity is increased after UV degradation of ketoprofen.

1. Introduction

Non-steroidal anti-inflammatory drugs (NSAIDs) have been widely used in medicine because of their good analgesic effects and small side effects, while some unabsorbed drugs enter the natural environment with human excretion. The existence of non-steroidal anti-inflammatory drugs in the natural environment is recognized as a thorny problem. At a small amount of concentration, it can also have harmful effects on public health and environment; but the biochemical treatment process of traditional sewage treatment plants has no obvious effect on such substances [1]. There will even be a "negative removal" phenomenon due to the accumulation and release of sludge [2]. Ketoprofen (KET), as a typical NSAIDs, is currently widely used in rheumatism and rheumatoid arthritis, ankylosing spondylitis, rheumatism, dysmenorrhea, etc. It is also effective for trauma and post-surgery pain and its annual dosage is up to 92t [3]. Owing to its wide application of ketoprofen, it is often detected in sewage [4].

The literature points out that KET is difficult to be biodegraded or mineralized and requires advanced treatment to remove KET from sewage. [5]. Compared with other sewage depth treatment methods, ultraviolet depth treatment has the characteristics of small land area, mild reaction conditions, energy saving, simple equipment, convenient maintenance and small secondary pollution, which is suitable for application in rural micro-sewage treatment stations.

This study is based on the deep treatment of ultraviolet light to investigate the effects of ultraviolet light intensity, light time, water turbidity, and hydroxyl radical trapping agent p-hydroxybenzoic acid (HBA) on the removal rate of ketoprofen. In addition, the kinetic analysis of the ketoprofen degradation process is carried out [6].

2. Materials and methods

2.1. Instruments and reagents
Main instruments: U3000 high performance liquid chromatograph, 10W 254nm ultraviolet lamp, 24W 254nm ultraviolet lamp, UV-C single-channel ultraviolet irradiation, turbidity meter.

Main medicines: Ketoprofen, HBA, Formazin turbidity standard solution, preparation water is ultrapure water.

2.2. Analysis and monitoring methods
Ketoprofen, HBA concentration: high performance liquid chromatography; solution turbidity: turbidity meter; ultraviolet intensity: ultraviolet irradiator.

2.3. Drawing the standard curve
Precisely configure ketoprofen solutions with concentrations of 2mg/L, 4mg/L, 8mg/L, and 16mg/L, use high performance liquid chromatography to detect and read the peak area. The ketoprofen standard curve is drawn to use peak area concentration and ordinate, with the curve equation y=1.21157x-1.35474 and R2=0.9997.

2.4. Experimental setup
The experimental device is divided into two parts: (1) The reactor part. It is composed of a three-mouth glass reaction vessel (diameter 6 cm, length 25 cm), quartz glass sleeve (diameter 3 cm, length 30 cm), and ultraviolet lamp; (2) Stirring device composed of PTFE magnetic rotor and magnetic stirrer.

When studying the effect of UV intensity on the reaction, different lamp adjustments are replaced. With 24W ultraviolet lamp, the average ultraviolet intensity in the device is 1700μw/cm2; with 10W ultraviolet lamp, the average intensity is 1300μw/cm2. The ultraviolet lamp is energized for half an hour in advance before reaction and the experiment is performed after the light stabilizes. When studying the effect of turbidity on the reaction, formadrazid turbidity standard fluid is used to adjust solution turbidity. Samples are sampled according to time and a 22μm filter membrane. All samples are repeated three times and take the average for graphing.

3. Results and discussion

3.1. The influence of the initial concentration of ketoprofen
Experiment was performed using 24W UV lamps with an average UV intensity of 1700w / cm2. The initial concentration of ketoprofen was set to five concentration gradients of 16mg/L, 8mg/L, 4mg/L, 2mg/L, and 1mg/L for the experiment. The results of experiment are shown in Figure 1 and Table 1.

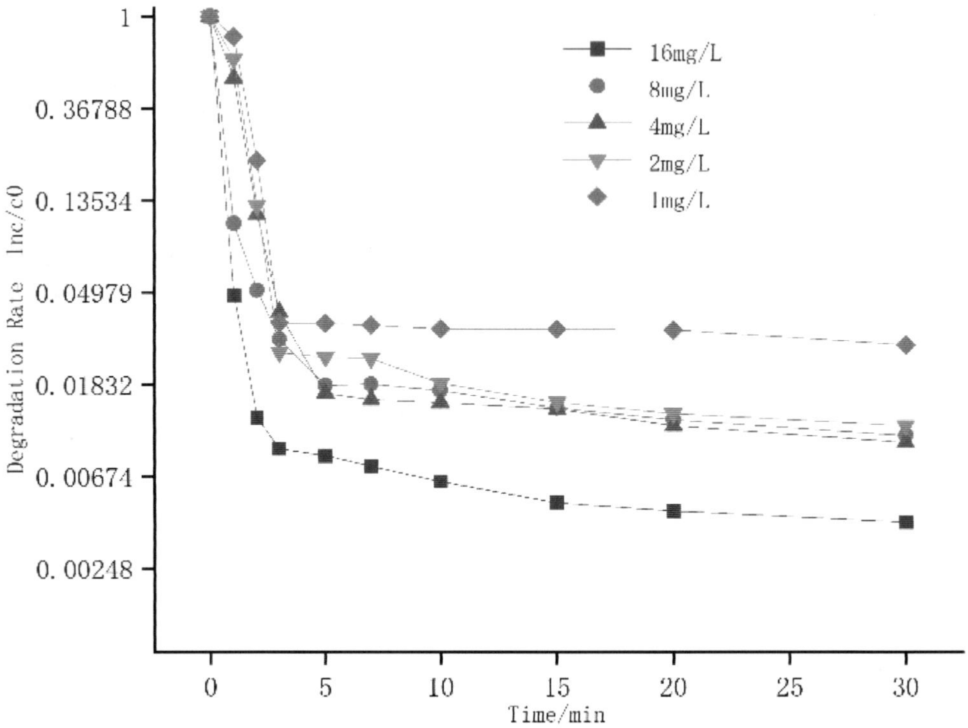

Figure 1. 24W UV degradation effect

Table 1. The kinetics of 24W UV degradation of different concentrations of ketoprofen

Mass concentration	**Quasi-first-order kinetic equation**	**R2**
16mg/L	ln c/c0=3.01009t	0.99953
8mg/L	ln c/c0=2.1631t	0.99562
4mg/L	ln c/c0=0.85458t	0.98623
2mg/L	ln c/c0=0.76399t+0.03651	0.96258
1mg/L	ln c/c0=0.61886t+0.6828	0.92553

Ultraviolet rays have a good degradation effect on ketoprofen. Ultraviolet rays can degrade ketoprofen at concentrations of 16mg/L, 8mg/L, 4mg/L, 2mg/L, and 1mg/L to 0.91%, 2.98%, 1.75%, 2.18% and 3.55% within 3 minutes. They degraded to 0.46%, 1.25%, 1.17%, 1.33%, 3.30% after 20 minutes and no longer decreased. The final degradation rate of ketoprofen decreases with the increasing ketoprofen concentration. Kinetic analysis shows that UV degradation of ketoprofen complies with the pseudo-primary reaction kinetics and that the ketoprofen degradation rate (Kobs) promotes with increasing concentration. Therefore, the degradation of ketoprofen by ultraviolet rays is a direct photolysis and the degradation rate varies greatly between each concentration. It shows that under the condition of 24W UV lamp irradiation, it is the concentration of ketoprofen that limits the degradation rate. The degradation rate of high-concentration ketoprofen solution is higher than that of low-concentration solution, which indicates that after increasing the concentration, the absorption effect of ketoprofen on UV light is enhanced and then increases the reaction rate.

3.2. The influence of UV light intensity on the removal rate of ketoprofen

24W and 10W UV lamps were used to irradiate 16mg/L, 8mg/L and 4mg/L ketoprofen solutions respectively to compare the degradation effects of the two light intensities. The results are shown in Figure 2 and Table 2.

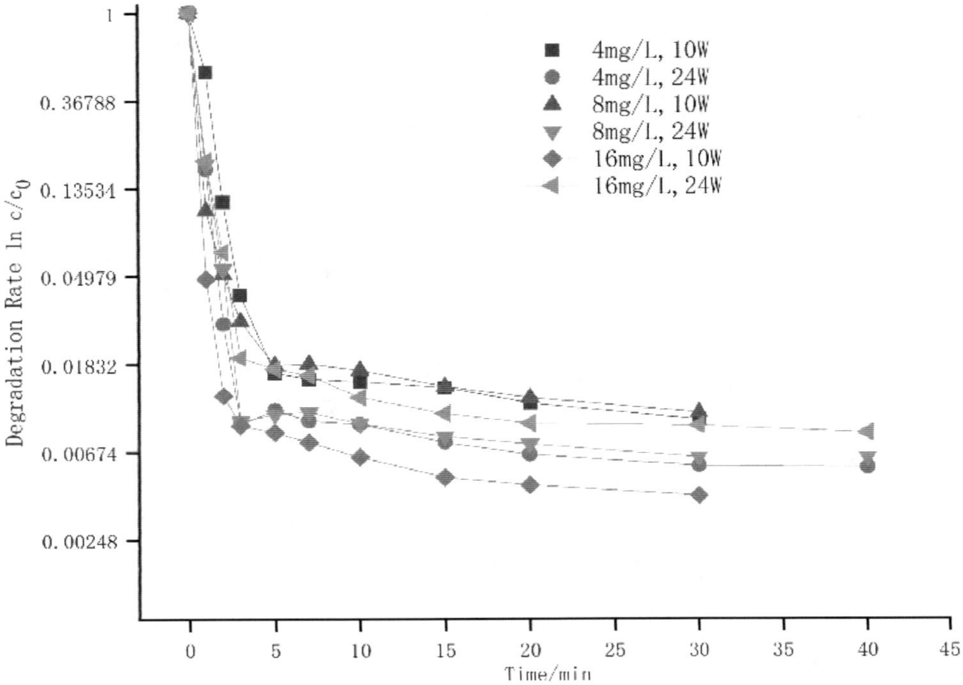

Figure 2. Degradation effect of different power UV lamps

Table 2. The kinetics of degradation of different concentrations of ketoprofen by 10W UV

Mass concentration	Quasi-first-order kinetic equation	R2
16mg/L	ln c/c0 = 1.6203t+0.00124	0.99761
8mg/L	ln c/c0=1.68064t	0.9989
4mg/L	ln c/c0=0.85458t	0.99943

The use of 10W ultraviolet lamp still has a good treatment effect on ketoprofen, which can respectively degrade ketoprofen of 16mg/L, 8mg/L and 4mg/L to 1.97%, 0.96%, 0.92% in 3min and it is finally degraded to 0.93%, 0.73% and 0.65% after 20min. Then it does not decrease any more. Kinetic analysis shows that the rate of ketoprofen degradation still complies with pseudo-first-order kinetic equation when irradiated with a 10W ultraviolet lamp. However, Kobs changes little at 16mg/L and 8mg/L, which indicates that at these two concentrations below, 10W ultraviolet lamp cannot have an effect on all the ketoprofen in the solution. At this time, the factor that affects the degradation rate is radiation intensity. Comparing the fitted kinetic equations in Figure 2 and Table 2, it is found that 4mg/L ketoprofen solution has the same Kobs under two different lighting conditions, which indicates that this degradation rate is the maximum degradation rate under the current conditions.

Journal of Physics: Conference Series **2194** (2022) 012049 doi:10.1088/1742-6596/2194/1/012049

3.3. The effect of solution turbidity on the degradation rate of ketoprofen

Use 24W UV lamp to degrade 16mg/L ketoprofen solution, adjust the turbidity of the solution with formazin turbidity standard solution. The turbidity is adjusted to 1NTU, 3 NTU, 5 NTU, 7 NTU, and 9 NTU respectively. The results are shown in Table 3.

Table 3. 24W UV degradation of 16mg/L ketoprofen with different turbidity effects and turbidity changes (Note: "-" means not detected or exceeded the detection limit.)

Turbidity / Time	Ultra-pure water （0.14NTU）	1NTU	3 NTU	5 NTU	7 NTU	9 NTU
1	4.82%	1.24%	7.18%	6.78%	2.06%	4.75%
2	1.27%	0.21%	1.73%	0.62%	0.18%	0.22%
3	0.91%	0.04%	0.08%	0.03%	-	-
5	0.84%	0.08%	0.02%	-	-	-
7	0.75%	0.05%	0.03%	-	-	-
10	0.63%	0.19%	0.01%	-	-	-
15	0.50%	0.08%	-	-	-	-
20	0.46%	0.10%	-	-	-	-
30	0.41%	0.01%	-	-	-	-
40	-	0.01%	-	-	-	-
Turbidity after 40 min	3.08NTU	6.56 NTU	10.61 NTU	13.76 NTU	17.86 NTU	19.74 NTU

After 15 minutes of irradiation for 3NTU solution, ketoprofen exceeded the detection limit and could not be detected; 5NTU solution could not be detected after 5 minutes of irradiation; 7NTU and 9NTU solutions could not be detected after 3 minutes of irradiation ketoprofen. Compared with degradation effect in aqueous solution and 1NTU solution, 1NTU solution was higher than in aqueous solution. After 40 minutes, 1NTU solution was degraded to 0.01%. It is better than 0.41% at the same time in aqueous solution. In 1-5NTU solution, ketoprofen degradation effect of phenanthrene decreases with the increase of turbidity. Under 5-9NTU, degradation effect increases with the increase of turbidity. The above experiment shows that higher turbidity can increase the utilization of ketoprofen to ultraviolet light. 1NTU ketoprofen solution still conforms to the pseudo first-order kinetic equation, the equation is $\ln c/c0 = 1.76768t+0.24412$ ($R2=0.95198$). However, Kobs decreases from 3.01009 to 1.76768, which indicates that the turbidity of solution increases and the UV degradation of ketoprofen have a negative impact.

3.4. The effect of free radical scavengers on the degradation rate of ketoprofen

20ml 1000 mg/L HBA was added to 16mg/L ketoprofen solution and with a fixed capacity to 1000ml. A 24W UV lamp was used for 24W UV degradation and loss of the hydroxytrap in the presence of hydroxytrap in solution. The results are shown in Figure 3 and Table 4.

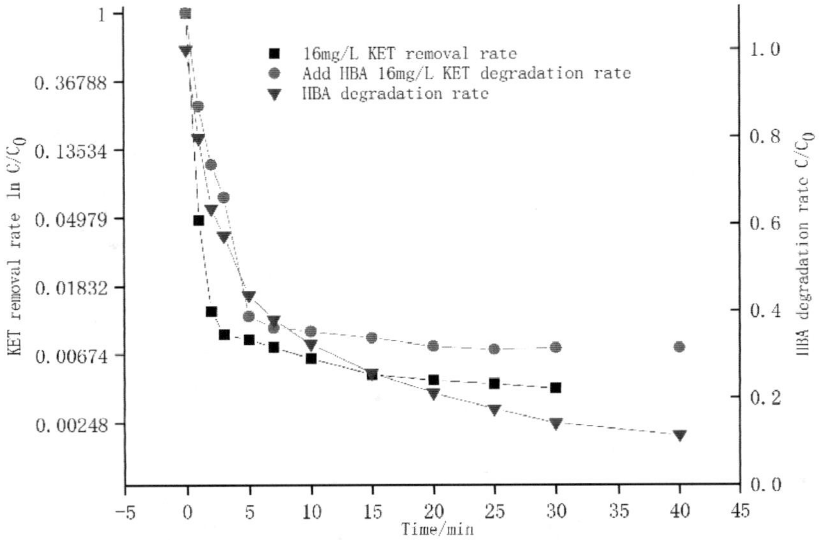

Figure 3. Degradation effect of 16mg/L ketoprofen solution after adding HBA

Table 4. Kinetics of 16mg/L Ketoprofen solution after adding HBA

Mass concentration	Quasi-first-order kinetic equation	R2
16mg/L KET	ln c/c0= 1.25871t+0.00438	0.99543
20mg/L HBA	ln c/c0=0.09074t+0.16886	0.90653

The degradation effect of ketoprofen changed greatly after adding HBA. Compared with no HBA solution, after adding HBA ketoprofen solution for 1 min, 3 min, and 20 min, ketoprofen surplus rate rose from 4.82% to 25.58%, from 0.91% to 6.74% and then from 0.42% to 0.76%. The remaining rate of HBA was 79.72%, 57.27%, and 21.17%. The Kobs of ketoprofen decreased from 3.01009 to 1.2587. The degradation of HBA also conformed to pseudo first-order kinetics and Kobs was 0.09074. The above results indicate that in addition to the direct photolysis of ultraviolet radiation, •OH produced by the degradation of ketoprofen will promote its own degradation during the degradation process of ultraviolet radiation. After adding HBA, •OH will preferentially combine with HBA to react. Advanced oxidation process that occurs under the action of ultraviolet is inhibited, so the degradation rate of ketoprofen is affected.

3.5. The effect of UV degradation of ketoprofen on water

Formaluid turbidity solution was added to the configured 16mg/L ketoprofen solution. Solution turbidity was adjusted to 1NTU, 3NTU, 5NTU, 7NTU and 9NTU respectively. Changes in turbidity were measured and recorded after 40 min of irradiation with a UV lamp at 24W. The results are shown in Table 4: the turbidity of 16mg/L, 8mg/L, 4mg/L, 2mg/L and 1mg/L ketoprofen solution during the degradation process was measured. Moreover, the changes were recorded. The results are shown in Figure 4.

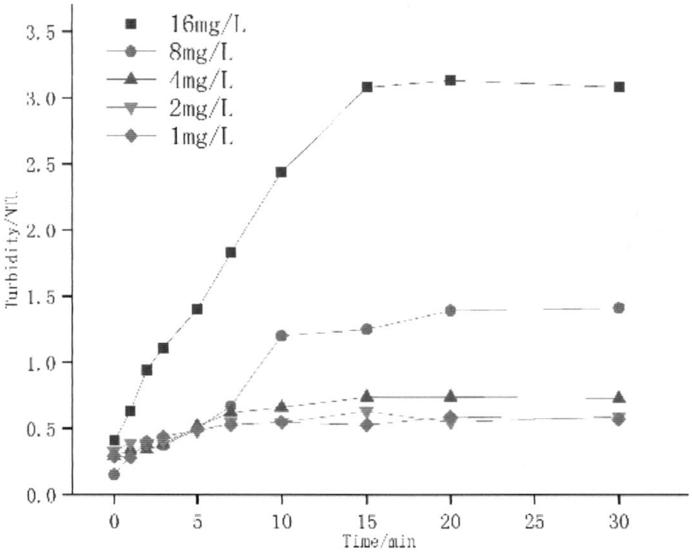

Figure 4. Turbidity change trend of ketoprofen solution of various concentrations

Above results show that the turbidity of ketoprofen at different concentrations changes after UV degradation. It can be seen from Table 3 that after 40 minutes of reaction, the turbidity of high-concentration solution changed greatly. The turbidity of 16mg/L and 8mg/L solutions increased from 0.41, 0.15 to 3.08, 1.41. The turbidity of low-concentration ketoprofen solution changed little. The turbidity of 4mg/L, 2mg/L, and 1mg/L solutions increased from 0.29, 0.33, 0.29 to 0.73, 0.59.0.57. Besides, the color of solution in the reactor gradually changed from colorless and transparent to brown and yellow within 0-15 minutes. The magnitude of high concentration ketoprofen change was significantly greater than that of low concentration ketoprofen solution, consistent with turbidometer measurements. In other ketoprofen solutions with adjusted turbidity, the greater the turbidity of solution before the reaction, the greater the range of change in the turbidity of solution.

3.6. Analysis of the mechanism of UV degradation of ketoprofen

Ketoprofen, as a multifunctional compound containing a benzene ring, its ultraviolet degradation process is mainly composed of two parts: direct photolysis and hydroxyl radical oxidation [7]. Among them, direct photolysis is that it is exposed to ultraviolet light after ketoprofen, hydroxyl group connected to the carbonyl group in the structure is prone to breakage after absorbing UV to form •OH and •OH will further promote its own advanced oxidation reaction [8]. Therefore, a higher concentration of ketoprofen will produce more •OH, so after 20 minutes of UV irradiation, the final removal rate of ketoprofen solution with a high initial concentration is also higher.

The carbonyl group used to connect benzene ring will form a conjugated double bond with carbon-carbon double bond on the benzene ring and deepen the chromaticity of water. The hydroxyl group as an auxochromic group will further increase the turbidity of aqueous solution, so after the ketoprofen solution is irradiated with ultraviolet light, the color and turbidity will change greatly [9]. In addition, under the action of stirring, H2O2 will diffuse to water layer far away from the lamp tube, which results in advanced oxidation of UV/H2O2. Finally ketoprofen in the region of lower UV intensity will produce advanced oxidation. The degradation rate of ketoprofen becomes lower after adding hydroxyl trapping agent HBA because hydroxyl radicals generated by the photolysis of ketoprofen are preferentially combined with HBA during the reaction. HBA is consumed and UV/H2O2 advanced oxidation reaction is also inhibited. Therefore, when HBA is still present in large quantities, the degradation rate of ketoprofen is greatly reduced. As HBA content decreases, UV/H2O2 advanced oxidation process begins to gradually recover, so the degradation rate begins to increase. However, because most of the ketoprofen has been photodegraded at this time, •OH has been consumed in large quantities and UV/H2O2 advanced oxidation occurs. The reaction is still small. At the same time, unreacted HBA has the effect of robbing •OH. Therefore, under the same conditions, the final degradation effect of the ketoprofen solution with HBA added is less than the final degradation effect of the ketoprofen solution without addition.

4. Conclusions and recommendations

1)UV light degradation rate of ketoprofen was over 96%, and each concentration of ketoprofen met the quasi-primary kinetic equation (R2> 0.9), and when the ketoprofen concentration increased from 1 to 16 mg/L, Kobs rose from 0.61886 to 3.01009. Univariate experiments showed that the maximum rate of 4 mg/L ketolprofen for UV degradation by UV254 was Kobs=0.85458.

2) When the ketoprofen concentration is gradually increased from 1 to 16 mg/L, the solution turbidity increases from 0.29 to 0.08 after 30 i n m of the reaction.Therefore, in addition to conventional indicators, close attention should be paid to the color, turbidity and other factors that affect the look and feel of effluent from the advanced treatment in practical applications.

3) The high-efficiency degradation of ketoprofen by advanced ultraviolet treatment is due to the structure of ketoprofen. The same process cannot achieve similar effects when processing other drugs. Therefore, more research is needed when performing advanced ultraviolet treatment to degrade other substances.

4) Different deep processing methods handle the same NSAIDs, but they have different effects and products. In future research, the decomposition process and products of specific drugs should be studied in depth.

Acknowledgments
This work was financially supported by Hebei Water Conservancy Science and Technology Plan Project (2020-55), Hebei Agricultural University Science and Technology Fund Project (LG201629) fund.

References

[1] Li Haiquan, Liu Sijia, Qiu Siwei, Sun Lei,Yuan Xiangjuan, Xia Dongsheng. Catalytic ozonation oxidation of ketoprofen by peanut shell-based biochar: effects of the pyrolysis temperatures [J]. Environmental technology, 2020.
[2] Ding Zirong, He Dechun, Wan Dajuan, Wu Genyi, Zhang Sukun, Fang Jiande. Distribution and removal of drugs and personal care products in medium-sized sewage treatment plants [J]. Environmental Engineering Journal, 2015, 9 (03): 1247-1252.
[3] Cai Le, Shu Zhankun, Ye Yahuai, Dong Jun, He Jianhui, Feng Rui, Zhao Keying, Li Hongdao. Analysis of the prevalence and influencing factors of main chronic diseases in Luoping County, Yunnan Province [J]. Chinese journal of Epidemiology, 2010, 31 (12): 1442-1443.
[4] Zhang Pan, Wen Xianghua, Wang Bo, Yu Kaichang, Xue Tao. Operation effect analysis of nanofiltration production reclaimed water demonstration project [J]. Environmental Engineering Journal, 2017, 11 (09): 4985-4992.
[5] Hu Xiang, Li Jin, Pi Yunzheng, Wang Jianlong. Study on the reaction mechanism of nonylphenol oxidation in water by ozone [J]. Environmental Science, 2007 (03): 584-587.
[6] Xue Juanqin, Jiang Meng, Yu Lihua, Luo Yao, Tang Changbin. The effect of trapping agent on the detection of hydroxyl radicals in electrochemical oxidation system [J]. Chinese Journal of Analytical Science, 2015, 31 (05): 606-610.
[7] Tu Xiang, Liao Haiqing, Chen Shaohua, Song Yonghui. Study on pseudo-first-order steady-state kinetic model of target pollutant degradation in UV/Fenton process [A]. Spring Nature Switzerland AG, 2019.
[8] Shao Wanting, Wang Wenlong, Du Ye, Wu Qianyuan, He Zhiming, Fu Zhimin, Hu Hongying. Dual-wavelength ultraviolet (VUV/UV) enhanced removal characteristics and principles of organic pollutants [J/OL]. Environmental Science Research, 1-14.

Journal of Physics: Conference Series **2194** (2022) 012049 doi:10.1088/1742-6596/2194/1/012049

[9] Lei Lirong, Li Youming, Ma Liming. Pollution characteristics and characteristic pollutants of southern hardwood chemical pulp bleaching waste water [J]. CIESC Journal, 2012, 63 (01): 244-250.

Analysis on Influencing Factors and Control Measures of Metal Materials Heat Treatment Deformation

Yanli Yu[1*]

[1]Jiangxi Vocational College of Mechanical & Electrical Technology, Jiangxi 330013, China

[*]1910012008@stu.jci.edu.cn

Abstract—The scientific application of heat treatment technology is aimed at improving the performance of metal materials, which is very important for improving the utilization rate of metal materials. In the process of h meat treatment, metals are easily affected by factors and deform, which leads to theetal materials can't be used normally. Therefore, it is necessary to deeply analyze the factors affecting heat treatment, and then formulate effective control measures to reduce the probability of metal material deformation and improve the service performance of metal materials. Based on this, this paper explores the influencing factors and control measures of heat treatment deformation of metal materials.

1. Introduction

With the deepening of China's economic system reform, all walks of life are showing a trend of vigorous development, and the level of industrialization is obviously improved. Especially since entering the new era, the application of metal materials is more and more extensive, involving industrial manufacturing, aerospace, fire safety, medical services and many other fields, so the quality requirements of metal materials are becoming more and more strict. In order to meet the use requirements of metal materials in different fields, it is necessary to change the properties of metal materials through heat treatment. However, the heat treatment process will lead to the deformation of metal materials, which will lead to the failure of metal materials to achieve the expected use effect. How to take effective measures to control the deformation of metal materials during heat treatment has become an urgent problem to be solved. In the process of heat treatment, metal materials have extremely high requirements on the treatment environment and technology. It is necessary to comprehensively analyze the causes that may lead to the deformation of metal materials, and optimize the operation process according to actual use requirements, so as to achieve high-quality heat treatment effect.

2. Overview of heat treatment of metal materials

Heat treatment is an important means of metal material processing, which can eliminate the non-uniformity of chemical composition, reduce the internal residual stress, avoid the defects of metal materials, and then promote the change of internal structure and improve the performance of materials. However, if the heat treatment is improper, it will not only make the metal materials unable to meet the use requirements, but also cause waste of manpower and material resources. According to the practice, it is found that the internal structure of metal materials will change during the heating and cooling process, and the high-temperature phase of the quenched metal materials will change into hard phase, which will enhance the hardness of materials [1]. At the same time, in the process of heat treatment, it can also protect the metal materials, so as to avoid reducing the carbon content of the materials by oxidation. In 1960s, plasma field effect was used for plasma nitrocarburizing and low temperature nitriding. Up to now, there are many kinds of heat treatment processes for metal materials. The commonly used heat treatment process is to heat metal materials to a certain temperature, and then slowly cool down after a certain period of time, or quickly cool down in water or oil after heating to a certain temperature to change the properties of materials [2]. There are three common methods of metal heat treatment: integral heat treatment, surface heat treatment and chemical heat treatment. Surface heat treatment mainly adopts flame heating quenching, induction heating quenching and laser heating treatment, while chemical heat treatment mainly adopts carburizing and nitriding. Heat treatment has a detailed technological process, which can be divided into the following steps:

Table 1 Heat treatment process

Technology name		Function	Applied range
Anneal	Homogenizing annealing	Component uniformity	Steel castings, forged and rolled parts with component segregation
	Full annealing	Tissue refinement, reduce hardness	Forging, welding and rolling parts of hypoeutectoid steel
	Isothermal annealing	Tissue refinement, reduce hardness, and prevent white spots	Carbon steel, alloy steel, high alloy steel, stamping parts, etc
	Under annealin	Tissue refinement, reduce hardness	Medium and high carbon steel, low alloy steel
	Relief annealing	Relieve internal stress	Castings, welded parts, forged and rolled parts
normalizing		Improve performance, increase hardness and reduce surface roughness	Low-carbon steel
		Uniform structure, grain refinement, and preparation for quenching	Medium carbon steel and alloy steel
		Elimination of reticular carbide	High carbon steel, high alloy steel
		Eliminate the network carbide in the infiltrated layer	Carburizing steel
		Eliminate abnormal structures such as coarse grains	Castings and forgings
		Final heat treatment	Low-requirement carbon steel
Quench	Single medium quenching	Quench in a single cooling medium such as oil, water and air to improve hardness and strength	The most common method
	Double medium quenching	Quench in two quenchant to form hardened layer and reduce deformation	Medium and high carbon steel parts Large alloy steel parts
	Step quenching	Reduce deformation and cracking	Parts with complex shapes and strict deformation requirements
	Bainite isothermal quenching	Reduce deformation and cracking, and obtain good comprehensive performance	High carbon die steel parts

Temper	Low temperature tempering	Tempering at 10-250 ℃ , reducing quenching internal stress and brittleness, and maintaining high hardness after quenching.	Ultra-high strength steel, measuring tools, bearings, carburized parts
	Medium temperature tempering	Tempering at 30 to 500℃, improve the elastic limit and yield strength, keep a certain toughness.	Spring, hot forging die
	High temperature tempering	Tempering at 50 to 650℃, meet the requirements of high strength and good toughness.	Nitrided parts, structural steel parts
	Multiple tempering	Eliminate internal stress, stabilize ruler	Ultra-high strength steel, tool and die steel, high speed steel

3. Analysis of Metal Material Deformation Mechanism

3.1. Internal Stress Shaping

In the process of heat treatment of metal materials, heating, heat preservation and cooling should be carried out. In many cases, the temperature of heating and cooling can't be effectively controlled, and the uniformity of metal materials in solid phase transformation is poor, which easily leads to internal stress and plastic deformation of metal materials [3]. There are many reasons for the formation of internal stress, among which the structural stress is mainly due to phase transformation, while the thermal stress is due to the thermal expansion of metal materials caused by heating and cooling. Different metal parts have different use requirements, different shapes and sizes, and different conditions for hot working plasticity, resulting in various deformations, which easily lead to defects in metal materials.

3.2. Specific Volume Deformation

Because the internal structures of different metal materials are obviously different, the slight changes in size and volume during heat treatment are called specific volume deformation. Although the degree of deformation is slight, it has great influence on the iron content, hardenability and alloy content of the material. Specific volume deformation is more uncertain than internal stress plasticity, and the number of thermal deformation directly affects the deformation form.

3.3. Creep Factor

Creep is mainly caused by the increase of plastic deformation of metal materials in hot state. If effective protection measures are not taken, the continuous influence of gravity will easily cause a large degree of bending deformation. In the process of heat treatment, the creep of metal materials is different from the plastic deformation. Plastic deformation is usually the deformation that occurs when the stress exceeds the elastic limit of metal materials, while creep, even if the stress reaches the material limit, is prone to creep as long as the stress acts for a long time [4]. Generally speaking, metal materials will show creep properties under certain conditions, especially in the process of heat treatment, in order to change the shape of materials, stress needs to be continuously applied, and creep is easy to occur.

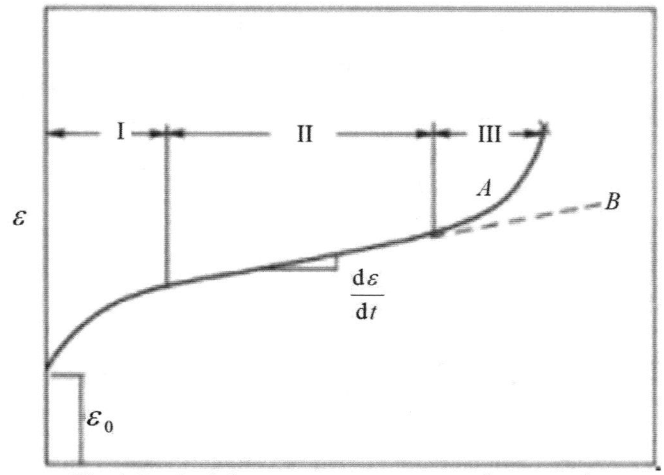

Fig.1 Typical three-stage creep curve (ε-t curve)

4. Influencing Factors of Metal Heat Treatment Deformation

4.1. Stress Factor

For the heat treatment of metal materials, professional technologies and methods are needed to standardize the operation. However, due to the corresponding particularity of each metal material and the influence of various external factors, the temperature distribution in the heat treatment process can not be fully guaranteed, and there is obvious internal imbalance, which will inevitably have some adverse effects on metal materials [5]. During the process of heating and heat preservation of metal, the ambient temperature will also change, which will affect the internal stress of the material and lead to the deformation of the material. In addition, if the stress distribution applied by heat treatment is uneven, the probability of material deformation will increase, which will easily reduce the quality of metal materials.

4.2. Quenchant

During the heat treatment of metal materials, quenchant is one of the important factors that affect the heat treatment effect, and it is also closely related to the deformation of metal materials. If the selected quenchant lacks rationality or has quality problems, it cannot be guaranteed that the quenching process meets the standard. In addition, the lack of effective control of the stirring speed and operation method of the medium will adversely affect the heat treatment of metal materials, resulting in material deformation.

4.3. Pretreatment

Pretreatment process can effectively avoid the deformation of metal materials and should be strictly implemented before heat treatment. However, although pretreatment can achieve the effect of stress relief, due to the influence of environmental factors, it has adverse effects on the reactor cooling effect, and it is difficult to reach the cooling standard, which will also affect the structure of materials [6]. In addition, due to the difference of pretreatment technology, normalizing pretreatment method will increase the probability of material deformation.

5. Deformation Control Principle of Metal Materials Heat Treatment

5.1. Scientificity

In the process of heat treatment of metal materials, in order to effectively avoid deformation, we should pay attention to the application of control methods to ensure that the whole process of heat treatment is scientific. During the heat treatment, we should strictly abide by the basic principles of science, standardize the treatment process, scientifically apply technology and equipment, scientifically control the deformation factors of heat treatment, make the metal materials reach the standard category, and effectively improve the use effect of metal materials.

5.2. Practicability

As a non-renewable resource, the storage of metal materials is very limited. When using metal materials, practical principles should be strictly followed to avoid wasting materials due to improper processing. In the process of heat treatment of metal materials, it is necessary to scientifically and reasonably control the deformation, establish the idea of saving resources and protecting the environment, and realize the coordinated development of industry and ecology [7]. At the same time, based on the basic principle of practicality, the heat treatment process used should have high efficiency, effectively improve the utilization rate of metal materials, and achieve the expected benefits.

5.3. Convenience

In the process of metal heat treatment, we should reasonably choose the treatment process and site conditions to meet the heat treatment standard. At the same time, we should improve the fault-tolerant rate of deformation control scheme and effectively reduce the adverse effects of environmental factors. In addition, we should scientifically optimize the heat treatment technology, reduce the difficulty of operation, and make the heat treatment orderly, so as to reduce the possibility of metal material deformation.

6. Control Measures for Heat Treatment Deformation of Metal Materials

6.1. Pay Attention to Pretreatment Link

In the process of metal heat treatment, normalizing and annealing will lead to deformation of metal materials. The higher the normalizing temperature, the greater the internal deformation of materials, so it is necessary to control the normalizing temperature scientifically. From the practical application point of view, isothermal quenching method should be selected after normalizing to effectively control material deformation factors. Before heat treatment, we should effectively improve the molecular structure of materials and enhance the internal uniformity [8]. At the same time, we should reasonably select and apply the process according to the specific use function to ensure that the temperature will not affect the heat treatment, effectively reduce the possibility of metal material deformation, and lay a good foundation for material heat treatment.

6.2. Reasonable Selection of Quenchant

Quenching plays an important role in the heat treatment of metal materials, which can control the internal stress of materials and avoid the deformation of metal materials. In order to ensure the standard of quenching operation for heat treatment of metal materials, it is necessary to strictly abide by the treatment principles and processes and improve the treatment effect. In the quenching and cooling process of metal materials, it is necessary to scientifically control the temperature to prevent deformation due to poor cooling uniformity. In the quenching process, water and oil are the main media, and the cooling speed of water can reach 600℃/s (550-650 ℃ cooling condition), and even at 200-300℃ cooling condition, the speed can still reach 270℃/s. When the cooling speed is high, material deformation is easy to occur [9]. In order to reasonably control the deformation factors, we can use saline water, alkaline water and other media. The cooling rate of oil at 200-300℃ is about 20℃ /s, which can control the quenching deformation within a certain range and avoid serious material cracking. However, at the temperature of 550-650℃, oil cooling not only takes a lot of time, but also can't form martensite, which leads to great limitations in application.

6.3. *Select the Applicable Cooling Method*

At present, single-liquid quenching, double-liquid quenching and graded quenching are the most commonly used cooling methods. Among them, single-liquid quenching is relatively simple, it uses a single medium such as oil, water and air, and it has a high degree of mechanization, which is the most commonly used cooling method. However, there are some defects in quenching speed, which often can't meet the use requirements of materials, and material deformation defects are easy to occur. At the same time, there are some problems in single oil quenching, such as poor uniformity and low hardness, which can not meet the standard requirements. Double-liquid quenching is mainly cooled by water and oil, water and air, oil and air. First, the material temperature is reduced to 300℃ in the medium with fast cooling speed, and then put into another medium, which is suitable for medium and high carbon steel parts and large alloy steel parts, and can effectively control quenching crack and deformation [10]. Graded quenching is to form martensite after salt bath or alkali bath treatment, so that the temperature inside and outside the material is the same, and finally, it is cooled in the air. However, the effect of graded quenching is poor, and its application scope is limited. It is usually only suitable for small-size and high-precision metal materials.

6.4. *Use Scientific Clamping Method*

In order to ensure the deformation of metal materials to be well controlled, we should carry out heating and cooling according to the standard to ensure the uniform stress of the material structure. In the specific operation process, we should adopt appropriate clamping methods, give full play to the function of the fixture, meet the verticality standard for the disc workpiece, and adopt the vertical mounting method reasonably. In addition, we can properly apply gasket components, optimize the operating conditions of heat treatment by overlapping gaskets and supporting gaskets, and effectively control the influence of deformation.

6.5. *Improve Machining Technology*

For metal materials, there are obvious fixing characteristics in the heat treatment process, but due to the differences in material properties, the corresponding heat treatment processes will be different. Heat treatment is the key process of metal material processing, but due to the particularity of some metal materials, it is often impossible to achieve the expected heat treatment effect. This requires us to reasonably improve the machining technology, reduce the influence of residual stress inside the material, and control the deformation within a reasonable range. If the material after heat treatment exceeds the deformation range, it is necessary to comprehensively analyze the specific factors causing the deformation, ensure the scientificity of mechanical processing and effectively solve the material deformation problem.

Fig. 2 Mechanical heat treatment

6.6. Eliminate Heat Treatment Residual Stress
During the heat treatment of metal materials, it is particularly necessary to control the residual stress to ensure the shape, size and properties of the materials. Therefore, we should pay attention to the method of eliminating residual stress in heat treatment. According to the characteristics of metal materials, we can improve the stress resistance of materials by adding alloy elements, which can not only effectively control the deformation problem, but also improve the service performance of metal materials, so that they can meet the actual use requirements. In addition, before heat treatment of metal materials, we must carefully check the surface cracks, scratches and roughness of metal materials, and deal with the defects of metal materials themselves, usually by welding, high-temperature plastic deformation and other methods to lay the foundation for the heat treatment process [11].

6.7. Scientifically Configure Part Structure
During the initial heating of metals, we should pay attention to the scientific configuration of parts and control the thickness of materials according to production standards. In the production process, we should ensure the smooth surface of the material and avoid stress concentration to avoid structural deformation. In addition, in order to control the influence of cooling rate on metal materials, we should formulate corresponding heat treatment scheme according to the test data to further optimize the properties of metal materials. In this process, we should formulate perfect process indicators and operation procedures, and assign special personnel to supervise and guide the operation process to effectively control the influence of human factors.

7. Conclusion
To sum up, in the process of heat treatment of metal materials, we should pay attention to the influence of deformation factors, choose appropriate control methods, and improve the performance of metal materials, so as to ensure that the heat treatment deformation does not exceed the specified requirements to meet the use requirements of metal materials. Therefore, it is necessary to strengthen the analysis of deformation mechanism of metal materials, scientifically control internal stress, reasonably select quenchant and improve the standardization of heat treatment to achieve the expected heat treatment effect.

References
[1] Rong Kang.(2021)Factors and Control Measures Affecting the Heat Treatment Deformation of Metal Materials[J].Metallurgy and Materials,41(4):23-24.
[2] Xu Feng.(2021)Study on Influencing Factors and Strategies of Heat Treatment Deformation of Metal Materials[J].Metallurgy and Materials,41(3):23-24.
[3] Tian Jing.(2021)Impact Factors and Control Measures of Heat Treatment Deformation of Metal Materials[J].Modern Chemical Research,11:19-20.
[4] Xu Xinlong.(2021)Analysis and Countermeasures of Heat Treatment Deformation of Metal Materials[J].China Metal Bulletin,(12):43-44.
[5] Li Ming.(2021)Analysis of Influencing Factors and Control Measures of Metal Heat Treatment Deformation[J].World Nonferrous Metals,(7):125-126.
[6] Pan Meiling.(2020)Analysis of Influencing Factors and Control Strategies of Metal Heat Treatment Deformation[J].Technology Wind,(10):168.
[7] Lv Tiezheng.(2020)Analysis on Influencing Factors and Countermeasures of Heat Treatment Deformation of Metal Materials[J].Mechanical and Electrical Information,(11):93,95.
[8] Yu Lei, Zhang Haoming, Wang Lu.(2020)Influencing Factors and Control Strategies of Metal Material Heat Treatment Deformation[J].Urban and Rural Studies,(1):69.
[9] Gao Yan, Li Shuangzhi.(2020)Influencing Factors and Control Strategies of Metal Material Heat Treatment Deformation[J].The Merchandise and Quality,(36):80.
[10] Zhao Hang, Xiao Jie.(2019)Influencing Factors and Control Strategies of Metal Material Heat Treatment Deformation[J].China Metal Bulletin,(11):123,125.

[11] Zhang Hongxia.(2019)Influencing Factors and Control Strategies of Metal Material Heat Treatment Deformation[J].World Nonferrous Metals,(2):150,154.

AUTHOR INDEX

Institute of Physics
Dirac House, Temple Back
Bristol BS1 6BE UK

ISSN: 1742-6588
ISBN 978-1-7138-5161-5

CURRAN ASSOCIATES INC.
proceedings
.com